Proterozoic Orogens of India

Proterozoic Orogens of India
A Critical Window to Gondwana

TRK Chetty

CSIR-National Geophysical Research Institute, Hyderabad, India

ELSEVIER

Elsevier
Radarweg 29, PO Box 211, 1000 AE Amsterdam, Netherlands
The Boulevard, Langford Lane, Kidlington, Oxford OX5 1GB, United Kingdom
50 Hampshire Street, 5th Floor, Cambridge, MA 02139, United States

Notices
Knowledge and best practice in this field are constantly changing. As new research and experience broaden our understanding,
changes in research methods, professional practices, or medical treatment may become necessary.

Practitioners and researchers must always rely on their own experience and knowledge in evaluating and using any information,
methods, compounds, or experiments described herein. In using such information or methods they should be mindful of their own
safety and the safety of others, including parties for whom they have a professional responsibility.

To the fullest extent of the law, neither the Publisher nor the authors, contributors, or editors, assume any liability for any injury and/
or damage to persons or property as a matter of products liability, negligence or otherwise, or from any use or operation of any
methods, products, instructions, or ideas contained in the material herein.

British Library Cataloguing-in-Publication Data
A catalogue record for this book is available from the British Library

Library of Congress Cataloging-in-Publication Data
A catalog record for this book is available from the Library of Congress

ISBN: 978-0-12-804441-4

For Information on all Elsevier publications
visit our website at https://www.elsevier.com/books-and-journals

  Working together
to grow libraries in
developing countries

www.elsevier.com • www.bookaid.org

Publisher: Candice G. Janco
Acquisition Editor: Marisa LaFleur
Editorial Project Manager: Tasha Frank
Production Project Manager: Mohanapriyan Rajendran
Cover Designer: Greg Harris

Typeset by MPS Limited, Chennai, India

This book is dedicated to my parents
Late Smt. Talari Subbamma and
Late Sri. Talari Chinnagangulaiah

Short Contents

Full Contents

Foreword

India has played a formative role in our realization that Earth's present-day continents were previously connected and not always scattered across the globe in their current positions. Of particular importance is the Permo-Triassic Gondwana Series of sandstone, shale, and coal with its distinctive plant fossils, which was named after the Gondi people of central India—it also closely matches similar sedimentary sequences in Africa, Antarctica, Australia, Madagascar, and South America. It was this correlation that led Eduard Suess, the leading Austrian geologist, to postulate the former existence of a single southern landmass that he called Gondwana Land. Now often referred to simply as Gondwana, this so-called supercontinent is thought to have been Earth's largest landmass during the Paleozoic Era, and was the precursor of the modern southern continents before it broke apart in Jurassic to Cretaceous times.

In recent years, the focus of much supercontinent research has shifted to the ancient crystalline rocks of the continental shields, and in particular to the Precambrian orogenic belts that are thought to mark ancient collisional sutures produced by the periodic assembly of ancient supercontinents. These studies are critically dependent on precise radiometric age data, which provide the only reliable method of correlating events within and between Precambrian orogenic belts. Such data have identified four main periods of supercontinent assembly. The first of these was in the Neoarchean, when Earth's continents are thought to have assembled into either a single landmass (often called Kenorland) or a small number of distinct supercratons, and this was followed by the sequential formation and destruction of three Proterozoic supercontinents: Nuna (or Columbia) in the Paleoproterozoic, Rodinia at the close of the Mesoproterozoic, and Gondwana in the latest Neoproterozoic. Interest in these ancient cycles of supercontinent assembly and break-up has grown exponentially over the past twenty-five years as it became apparent that these events had a first-order influence on Earth's surface environment, with added suggestions that they triggered profound changes in climate, sea water chemistry, and biological evolution, leading ultimately to the great diversification of life in the Cambrian.

The role of India in this perpetual ebb and flow of crustal fragments between aggregated and dispersed continental arrangements has been less clear than that of many other regions. While it had long been recognized that Precambrian basement rocks of India comprised a series of orogenic belts that wrap around granite-dominated cratons, a paucity of robust isotopic data meant that many of these rocks were still attributed an Archean age simply on account of their crystalline nature, making it difficult to integrate their histories with any certainty to those established for other continents. This situation has improved markedly in recent years as modern geochronological techniques have been applied to Indian outcrops at an ever-accelerating rate, and India's basement orogens are now known to preserve Proterozoic tectonic events that correlate closely with the assembly of Nuna, Rodinia and Gondwana supercontinents. Coupled with burgeoning structural, petrological, geochemical and geophysical data, these age data are leading to exciting new models for the evolution of Proterozoic India and its tectonic and paleogeographic relationships to the supercontinent cycle.

It is this explosion of data and ideas that makes this book such a timely addition to the literature. Following a discussion of fundamental concepts, the reader is guided through India's Proterozoic orogens, weaving across the subcontinent from south to north before disappearing beneath the Indo-Gangetic plain. We begin in the south with the Southern Granulite Terrane, a broad region of high-temperature metamorphism and deformation that correlates closely with rocks of Madagascar and Sri Lanka. Next we follow the Eastern Ghats Mobile Belt along the eastern margins of the Dharwar, Bastar and Singhbhum cratons, an orogen that most closely matches rocks of East Antarctica, before swinging back westwards along the Central Indian Tectonic Zone and then turning sharply to the north along the Aravalli-Delhi Belt at the western edge of the Bundelkhand craton. Although each of these four belts has a distinct history, there are also commonalities both in tectonic style and age with abundant evidence for reactivation and reworking of older features by younger events. Deciphering such complex histories requires the integration of multiple data sets, and this book presents field observations, structural geology, petrology, geochemistry, geochronology and geophysics for all four orogens and concludes each chapter with a tectonic synthesis before a final chapter places these results and interpretations within the broader context of Proterozoic supercontinents. It is worth noting that these tectonic models represent a snapshot of an understanding that is continually evolving as new data are published, but despite this rapid turnover of ideas the relationships and interpretations summarized in this book provide a valuable stocktake of recent progress that will be of interest to a wide audience, including those focused specifically on the Precambrian history of India and those whose interests lie in other parts of Gondwana or other supercontinents of which India was once a part.

I was a relative late-comer to Indian geology, having cut my Gondwana teeth on the Proterozoic granulites of East Antarctica, followed by work on similar rocks in Australia, Madagascar and southeastern Africa. However, it soon became apparent that any attempt to place the Proterozoic geology of these regions in a wider content would require careful correlation with India. I therefore leapt at the opportunity to join an international field workshop in the Southern Granulite Terrain in 2004 led by Dr Chetty. With the aim of encouraging international collaboration and improving Gondwana correlations, this 8-day odyssey included a 2000 km field transect across southern India and numerous technical seminars that introduced me to tectonic and petrological issues that continue to enthuse and frustrate me in equal measure. I owe Dr Chetty a large debt for such an inspiring introduction to the subcontinent that has led me to return on many occasions and fostered a long-term interest in India's Proterozoic orogens. I hope that through this book Dr Chetty will have a similar impact on others.

Ian Fitzsimons
Curtin University, Perth, WA, Australia

Preface

Plate tectonics provide a unifying conceptual framework for the understanding of Phanerozoic orogens, and the recent syntheses of ancient orogens apply these principles (often controversially) as far back as the Early Archean. Many ancient orogens are, however, poorly preserved, and the processes responsible for their evolution are not well understood. The effects of processes such as delamination, subduction of oceanic and aseismic ridges, overriding of plumes, and subduction erosion are rarely identified in ancient orogens, although they have a profound effect on modern orogens. However, deeply eroded and well-exposed ancient orogens provide insights into the hidden roots of modern orogens and serve as potential examples of what the Himalayas might look like after they have  been deeply eroded to lower- or mid-crustal levels. Recent advances in analytical techniques, as well as in fields such as geodynamics, have brought to light fresh insights into ancient orogenic belts, so that realistic modern analogies can now be applied.

The Proterozoic orogens of India (POI), central to supercontinental reconstruction models, have been under focus by many national and international groups of researchers for greater details in the last two decades to arrive at a better understanding of their evolution. These orogens expose a unique window of a wide range of structural levels with variations in the geometry of structural features and the vergence of thrusts in space, pointing to the larger scale geodynamics responsible for their development.

Recent contributions, especially the refinements of many geological, geochemical, geochronological, and geophysical studies, have uncovered many new facets of the geology of the POI along with some proposed new lines of thinking of traditional concepts, which were previously developed but were lacking a substantial amount of information that we now have today. The application of plate tectonic paradigms have gained increasing acceptance to realize the comprehensive deep crustal processes that were involved in the evolution models of POI. The new tectonic synthesis of the POI reveals many interesting features that are analogous to those commonly referred to as geological evidences of plate-margin activities in many Phanerozoic orogens of the world. Therefore, it is now realized that the time is ripe to examine and synthesize the evolution of POI in terms of the state-of-the-art concepts of global tectonic processes. A major problem in such an attempt, obviously as in many other ancient orogenic belts, are the controversial issues that arise from complexities caused by superposed polyphase deformations, metamorphism, and igneous activities.

The book aims at summarizing the current understanding of the lithological assemblages, tectonic subdivisions, and associated shear zone systems, followed by a regional synthesis about the geological history of the Proterozoic orogens of the Indian shield and provides state-of-the-art reviews. In the present synthesis, much stress is laid on the large volume of the recently obtained field-based multiscale data as well as from laboratories that will hopefully clarify and address some of the earlier concepts and attempt to resolve many of the well-known controversies. As a consequence of the new knowledge, some revised interpretations and pathways for new research are also suggested.

Finally, it must be mentioned here that the choice of topics in the book reflects the understanding and interest of the author, and he would not claim to have seen and worked out all the orogens. Based on the author's field-based experience of more than three decades and the wisdom gained in conjunction with the enormous wealth of literature, an attempt has been made here to present an unprejudiced view with an open mind without fear or compulsions. There have been conflicting ideas and results obtained by several groups of workers because of lack of detailed field mapping, inconclusive databases, advancements in methodologies, the emergence of new concepts and ideas, and finally the absence of exploring alternate possible explanations.

An attempt has been made here to provide the reader the fundamental structural framework of orogens with as many multiscale illustrations on which any other geoscientific aspect such as mineral resources, magmatism, deformation patterns, engineering geology, natural hazards, and finally the geodynamics, can be dealt or built. It is anticipated that a state-of-the-art exposition of the POI may help to identify the key issues, find gaps in the existing knowledge base, and guide the interested researchers to pursue future investigations. The development of Proterozoic orogens of India and their linkage with the assembly and breakup history of the Gondwana supercontinent is another key objective of the present book.

The book will be useful both as a text and as a reference book for students as well as for researchers of earth sciences, and it should appeal worldwide to all professional geologists with an interest in academics, mineral exploration, seismology, and geodynamics. The book features a comprehensive index, an extensive up-to-date reference list, numerous field-based illustrations (some of them in color), and major questions that focus on the discussion as well as suggested avenues for future research.

Acknowledgments

I owe my great sense of gratitude and appreciation to the CSIR-National Geophysical Research Institute (NGRI) for making the entire four decades of my scientific career fully satisfying, exciting, and meaningful by providing excellent infrastructural facilities and a congenial atmosphere for my scientific pursuits. I am indebted to the Science and Engineering Research Board, Department of Science and Technology (DST-SERB), Government of India, for their generous and gracious support for awarding this book-writing scheme.

The inspiration and impetus to write this book came from my beloved friend, academician Prof. M. Santosh during one of our field geological meetings. Prof. Santosh also suggested the possible titles for the book in its infancy. My association and participation in many activities of the International Association of Gondwana Research (IAGR) since its inception has stimulated and encouraged me to venture into the task of writing this book. I am grateful to Prof. M. Yoshida and Prof. M. Santosh, the founders of IAGR, for their enthusiastic encouragement and unstinted support in my scientific endeavors.

I am grateful to my "guru" Dr. D.N. Kanungo (PhD 1956, Imperial College, London), who guided me during my PhD in the early years of my career and taught me the fundamentals of structural geology and their applications. I am thankful to Prof. C. Leelanandam (PhD, Cambridge University) who has consistently encouraged me to keep my spirits high throughout the progress of writing this book. Both of these legendary persons have greatly influenced my scientific attitude, thinking, and presentation.

I appreciate and acknowledge the following experts for sparing their valuable time and scholarly reviews of different chapters: Prof. M. Santosh (Southern Granulite Terrane), Prof. Saibal Gupta (Eastern Ghats Mobile Belt), Prof. A. Chattopadhyay (Central Indian Tectonic Zone), and Dr. Vijaya Rao (Aravalli-Delhi Orogenic Belt). I am thankful to Dr. M.R.K. Prabhakar Rao for critically reading all the chapters and for useful suggestions; Mr. M. Venkatarayudu for making DTM-based geological maps, Mr. M. Vittal for making many of the Corel drawings, and Dr. T. Yellappa for his help in finalizing some of the drawings.

I am grateful to many of my colleagues, Drs. D.S.N. Nurthy, Y.J. Bhaskar Rao, B.L. Narayana, and international collaborators, Profs. Alan Collins (Australia), T. Tsunogae (Japan), and M. Santosh (China) for their valuable discussions both in the field and in the laboratory. I am stimulated by my PhD students Drs. Nagaraju, Guru Rajesh, Vijay, Yellappa, Venkatasivappa, Mohanty, and Nagesh for their lively and interesting discussions and assistance in the field as well as in the lab during their tenure.

I thank Prof. G. Parthasarathy for his consistent moral support and helpful suggestions during the project. I owe a sense of gratitude to all the directors of CSIR-NGRI for their encouragement and support throughout my career in NGRI.

I appreciate and wish to acknowledge the high-quality and prompt support rendered by the Elsevier team (Marisa LaFleur, Tasha Frank, Mohanapriyan Rajendran, Greg Harris, and Rajesh Manohar) in transforming the manuscript and figures to its final shape of the book.

Last but by no means least, I am very grateful to my wife T. Ramadevi for her unflagging support throughout my career, and for her patience and forbearance in taking care of our children and family during my long and frequent field visits. I thank my beloved daughters, T. SreeSudha, T. Sowmya, and T. Sujani and their families for making my life always cheerful, joyful, and beautiful.

OROGENS

1

CHAPTER OUTLINE

1.1 INTRODUCTION

Orogens are the hallmarks of the interaction among lithospheric plates. Orogens in space and time are the potential sources of information in understanding the mechanism of episodic global material circulation on a whole-mantle scale. Tectonic evolution of orogenic systems is a fundamental research problem in understanding the earth's evolution, which in turn helps in better comprehension of mineral resources, seismicity patterns, and various geological hazards. There are principally two types of orogenic systems in the Earth's history from the Archean through to modern Earth, viz., accretionary orogenic systems and collisional systems. Some of the best-studied orogenic systems in the world, such as Grenville orogeny, showed that the Greenville province resulted from a Mesoproterozoic continental collision and consists of tectonically stacked slices of Archean

Proterozoic Orogens of India. DOI: http://dx.doi.org/10.1016/B978-0-12-804441-4.00001-8

Paleozoic, Mesoproterozoic rocks that are exposed at various crustal levels. Features such as deformation, metamorphism, and magmatism may vary in intensity along and across the length and breadth of the orogens. The orogens are the sites of complex zones of transpressive deformation in which the degree of obliquity controls the styles of structures and metamorphism.

Interactions among lithosphere plates drive orogenesis within a context of which many interconnected geological processes may occur, including: variation in angle and rate of subduction, collisions, slab breakoff, lithosphere delamination, partitioned flow, uplift, and exhumation. Collisions that occur during the movement of the lithosphere over the earth's surface may involve plates of the same type, i.e., oceanic−oceanic or continental−continental, or of different type, i.e., oceanic−continental. Thus, the large-scale effect of continent break-up, drift, ocean closure, and continent collision generate a series of interactions such that rocks of diverse origin, with different evolutions, may end up juxtaposed, or interleaved in an orogen, or remnants of these may be preserved in an ancient suture zone. The subduction of India beneath the Eurasian plate and the development of Himalaya and the Tibetan Plateau is an example of currently active continental collision, a process, which is widely inferred for major Phanerozoic orogens such as the Caledonian/Appalachian and Variscan orogens (Hodges, 2000).

Deeply eroded ancient orogens provide insights into the hidden roots of modern orogens. Recent advances in analytical techniques and in fields like geodynamics, provided fresh insights into ancient orogenic belts, so that realistic modern analogies can now be applied. The orogens offer the realms of a natural laboratory to address the nature of large Earth's processes such as the behavior of lithosphere, crust-mantle interaction, supercontinent formation, different geodynamic processes, and ultimately the Earth's history. Also, the orogenesis has direct bearing on magmatism, metamorphism, mineral resources, and deformational styles.

1.1.1 DEFINITION

Traditionally, the term orogen has been applied to describe a mountain belt composed of different types of rocks or rock strata forming a complex of variable size, typically tens to hundreds of kilometers wide and several thousand kilometer long, later fragmented during a younger geological time due to various processes (Miyashiro, 1961). During the 19th century, it was believed that stratigraphic successions in mountain ranges were often much thicker or of much deeper water facies than equivalent successions on undeformed continental cratons. This assumption led to the concept that mountain belts originated in extended linear subsiding troughs, termed as "Geosynclines" (e.g., Dickinson, 1971). These models were decimated with the advent of plate tectonics in 1960s that account for the vast majority of orogenic belts and their features, which are considered to form along consuming plate boundaries.

In modern geology, an orogen or orogenic belt can be defined as a major linear deformed zone, sandwiched between cratons with prolonged deformational history, repeatedly reactivated and associated with different events of magmatic pulses and metamorphic episodes in space and time (Fig. 1.1). Orogens reveal the processes of continental growth and deformation including terrane accretion, ophiolite obduction, terrane amalgamation, terrane dispersal, and crustal reactivation.

Orogens are well preserved as long, linear, and arcuate belts of deformation of the earth's lithosphere forming mountain ranges involving a great range of geological processes collectively called as orogenesis. The word "orogeny" is derived from Greece (oros for "mountain," genesis for

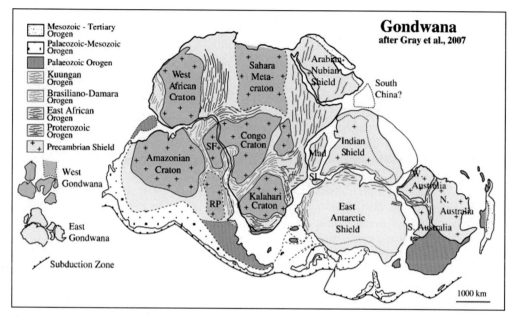

FIGURE 1.1

Distribution of orogens in the Gondwana supercontinent. West Gondwana is shaded in light blue and East Gondwana is shaded yellow. Neoproterozoic orogenic belts criss-cross the supercontinent. Those associated with the final amalgamation of the supercontinent are the East African orogen (red), the Brasiliano-Damara orogen (blue), and the Kuungan orogen (green).

After Meert, J.G., Lieberman, B.S. (2008). The neoproterozoic assembly of Gondwana and its relationship to the Ediacaran-Cambrian radiation. Gondwana Research, 14, 5–21 (Meert & Lieberman, 2008).

origin). The orogens are formed by the tectonic processes of subduction, where a continent rides forcefully over an oceanic plate (noncollisional orogens) or convergence of two or more continents (collisional orogens). They occur as mountain belts through crustal thickening, magmatism and metamorphism during more than one tectonothermal event (orogenies). An orogeny is an episode of orogenesis in a given mountain belt and the orogen may have had multiple episodes of orogeneses through time. They constitute a pronounced linear structural form resulting in terranes or blocks of deformed rocks, separated by suture/shear zones or dipping thrust faults. These thrust faults carry relatively thin slices of rock (which are called nappes or thrust sheets and differ from tectonic plateaus) from the core of the shortening orogen out toward the margins and are intimately associated with folds and the development of metamorphism.

The modern orogens such as Alpine-Himalayan chain and the ancient orogens Appalachians—Caledonian, Grenville, Trans-Hudson, Capricorn, Limpopo, etc.—are some of the classic examples of orogens. These orogens occur at plate margins in which deformation, metamorphism, and crustal growth took place in an environment of continuing subduction and accretion and such belts are known as accretionary orogens (also known as Pacific-type). Examples of these types of orogens include: Cordilleran, Pacific, Andean, Miyashiro, and Altaid. Accretionary orogens appear to have been active throughout much of the Earth's history and constitute major sites of continental growth (Cawood, Kroner, & Pesarevsky, 2006).

1.1.2 CLASSIFICATION

The orogens can broadly be grouped into collisional and accretional types (Windley, 1995). Accretionary orogenic systems (also described as "Pacific-type" or "Cordillera-type") are formed through on-going plate convergence during the period of supercontinent breakup and continental dispersal. Collisional orogenic systems (Himalayan-type) are generated when the ocean is closed during continental assembly and the formation of supercontinents. Collisional orogenic systems may be superimposed on accretionary systems, which can be described as subduction-to-collision orogenesis (e.g., Liou, Tsujimori, Zhang, Katayama, & Maruyama, 2004). In another type of classification, the orogens are divided into three types (Rogers & Santosh, 2004): (1) intercratonic, formed by the closure of ocean basins; (2) intracratonic, developed within continents where there was no preexisting oceanic crust; and (3) confined orogens formed by closure of small oceanic basins. Despite the different classifications, the evolution of orogens on a whole-earth scale suggests that they represent the surface manifestations of the motion of the earth's lithosphere and contributes to the generation of new continental crust through plate tectonics that is horizontally transported and eventually destroyed at subduction zones prior to orogenic "suturing." The subducted material accumulates at 660 km depth, being transformed from a curtain-like sheet to a large blob that drops vertically to the core−mantle boundary (Maruyama, Liou, & Zhang, 1994).

Four different types of orogens have been recently described on the basis of first-order structural and metamorphic characteristics: ultra-hot orogens (UHO); hot orogens (HO), mixed-hot orogens (MHO); and cold orogens (CO). The comprehensive details of these classes are described by many workers (Chardon, Gapais, & Cagnard, 2009). The widespread occurrence of ultrahigh-temperature (UHT) granulites in various terranes formed under extreme thermal conditions of 900−1100°C and pressures of 8−12 kbar were documented (Kelsey, Clark, & Hand, 2008). Based on temporal constraints, they can be further divided into modern and ancient orogens where in both cases of regional metamorphic belts occupy the orogenic core. In both the groups, the overlying and underlying units of the regional metamorphic belts are weakly metamorphosed or unmetamorphosed, forming either accretionary complex (Pacific-type) or continental basement and cover (collision-type) in origin. The structural features are similar and the major structures are subhorizontal with ocean ward convergence of deformation in both the types. Field observations of Proterozoic orogens show significant differences with modern orogens. The former are characterized by large areas of monotonous high-temperature-low-pressure metamorphic rocks with extensive magmatism at moderate depths often less than 25 km. Such orogens are formed atop of a hot mantle and remain extremely mechanically weak over protracted periods of deformation. The occurrence of blueschists and ultrahigh-pressure metamorphism (UHPM) of rocks is the hall mark of modern orogenic belts (Gerya, 2014). These belts are characterized by clockwise metamorphic pressure−temperature paths and display many UHPM terranes in the form of 1−5 km thick horizontal sheets worldwide. The UHPM rocks are preserved in the cores of antiformal nappe stacks that define structural domes of 5−50 km across.

In general, all orogens can be grouped into a spectrum of three interrelated end member types: collisional, accretionary, and intracratonic (Fig. 1.2). Collisional orogens form through collision of

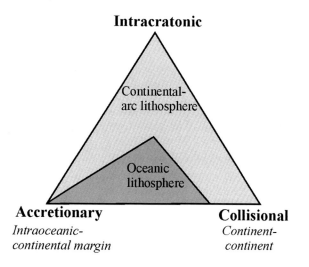

FIGURE 1.2

Diagram showing the classification of orogens: collisional, accretionary, and intracratonic.

Modified after Cawood, P.A., Kroner, A., Collins, W.J., Kusky, T.M., Mooney, W.D., & Windley, B.F. (2009). Accretionary orogens through Earth history. Geological Society, 319, 1—36. London, Special Publications.

continental lithospheric fragments from failed rifts (aulocogens). Accretionary orogens form at sites of continuing oceanic plate subduction (Fig. 1.3A). Intracratonic orogens lie within a continent away from an active plate margin. The intracratonic orogens are normally generated, possibly through the response from far-field stresses. The accretionary-type orogen (Fig. 1.3B) includes an accretionary complex comprising oceanic materials such as mid-oceanic ridge basalts, seamounts, ocean island basalt, and the carbonates that cap them, together with deep-sea sediments. All of these are finally capped by trench turbidites (Matsuda & Isozaki, 1991). These materials are subsequently incorporated to generate the accretionary complex. Some of them are tectonically transported to the mantle depths and regionally metamorphosed under low-T—high-P conditions (usually up to 700—800°C and at less than 60 km depth), and return to the surface to get incorporated within a shallow-level accretionary complex. Simultaneously, a huge batholith belt with felsic volcanic rock components forms on the continental side. Forearc basins are developed between these regions, displaying a normal stratigraphic sequence. Among the lithological components of the accretionary-type orogens, the volcano-plutonic sequences usually form with a width of 200—300 km, while the forearc basin deposits usually develop with a width of around 100 km. The regional metamorphic belts would be a few tens of kilometers wide and less than 2 km thick, with unmetamorphosed accretionary complex below and ocean ward. The total width of accretionary orogens is thus around 400 km, with the regional metamorphic belts at the center, and all the units are formed within a time span of 100—200 Ma (Fig. 1.3C), as in the case of Japan.

A recent synthesis by Cawood et al. (2009) provided a detailed evaluation of accretionary-type orogens. The formation of accretionary-type orogens is linked to intraoceanic and continental margin convergent plate boundaries and includes the suprasubduction-zone forearc, magmatic arc, and

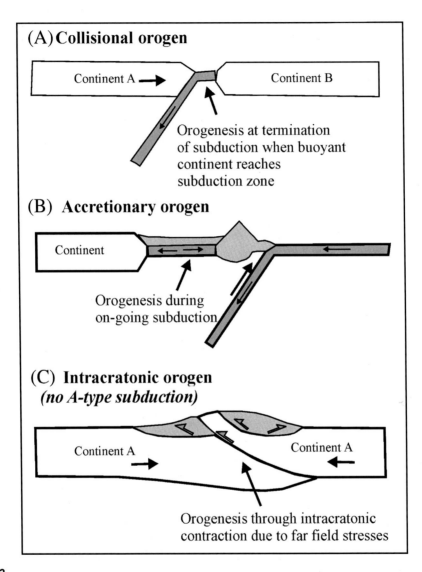

FIGURE 1.3

Schematic cross-sections through (A) collisional, (B) accretionary, and (C) intracratonic orogens.

Modified after Cawood, P.A., Kroner, A., Collins, W.J., Kusky, T.M., Mooney, W.D., & Windley, B.F. (2009). Accretionary orogens through Earth history. Geological Society, 319, 1–36. London, Special Publications.

back-arc components. The accretionary orogens are further classified into retreating and advancing types. While the modern western Pacific constitutes an example of a retreating orogen with a characteristic back-arc basin, the advancing orogens such as the Andes develop foreland fold-thrust belts with crustal thickening. The accretionary orogens are found to be active throughout the

Earth's history and were responsible for the major growth of the continental lithosphere through the addition of juvenile magmatic products. These orogens also mark major sites of consumption and reworking of continental crust through time by sediment subduction and subduction erosion. According to the model of Cawood et al. (2009), the net growth of the continental crust from the Archean is effectively zero because the rates of crustal growth and destruction are roughly equal.

Accretionary and continent—continent collisions occur at plate margins and form through the subduction of oceanic lithosphere with the former developing early at the sites of continuing subduction and the latter at the termination of subduction. However, there is a marked difference in size between the two major types of orogens. The thickness of regional metamorphic belts in the accretionary type is usually less than 2 km, and the width is around at least 100 km (much more in the case of Alaska; Kusky, Li, & Raharimahefa, 2004) and lateral extent of more than 1000 km. On the other hand, collision orogens contain relatively thicker regional metamorphic belts (up to 5 km), with widths of 200—300 km and lateral continuity of over a few thousand kilometers depending on the size of the collided continental mass.

1.1.3 GENERAL CHARACTERISTICS

Orogeny includes a collage of processes, such as: (1) magmatism, which generates continental crust; (2) rejuvenation and recrystallization by metamorphism where in the metamorphic belts occupy the orogenic core; (3) deformation to produce major structures of orogenic belts; and (4) sedimentation where the mountain-building activity takes place through the transportation of large volumes of sedimentary material. The proposal of Miyashiro (1961) on "paired metamorphic belts" in fact represents the parallel juxtaposition of two synchronous orogens of different thermal characters and represents a window to the processes that may be related to different levels and in different tectonic settings.

The important characteristics of accretionary orogens (Cawood et al., 2009) include: (1) presence of a range of mafic to silicic igneous rocks and their derivatives that develop on oceanic (e.g., West Pacific) or continental (e.g., Andes and Japan) lithospheric substrates during continuing plate convergence; (2) magmatic arc activity of typically calc-alkaline composition, but ranging from low-K tholeite to shoshonitic; (3) magmatic activity due to the interaction of the subducting and dehydrating slab of oceanic lithosphere with mantle in the wedge of the overriding plate that is generally initiated at a depth of around 100 km or more above the down going slab; (4) highly explosive arc magmas characterized by less percentage of water and other volatile phases and crystallization of the plutonic sections representing amphibole-bearing gabbros and ultramafic rocks; (5) existence of variably deformed and metamorphosed rocks due to tectonothermal events commonly in dual, parallel, high-T and high-P regimes up to granulite and eclogite facies; (6) presence of extensional or compressive deformational features that can be overprinted by short regional compressive orogenic events; (7) the still-evolving orogens such as the ones around the Pacific, have long and narrow aspect ratios, but completed orogens may be as broad as long (e.g., Central Asian orogenic belt or Arabian-Nubian shield; Superior and Yilgarn provinces); and (8) consisting of lithotectonic elements of an accretionary prism incorporating accreted and tectonically dismembered ocean plate strata, fore-arc basin and its substrate, magmatic arc and back-arc basin. Some of the orogens may also incorporate accreted arcs and oceanic plateaux and slices of convergent margin assemblages that have moved along the margin through strike-slip activity.

1.1.4 **ROLE OF PLATE TECTONICS**

It has been well established that the development of orogens involves plate tectonics through a variety of associated processes like subduction zones after the consumption of oceanic crust producing volcanoes and build island arcs magmatism. The other important associated processes include magmatism, metamorphism, crustal melting, and crustal thickening. However, these are dependent on the strength and rheology of the continental lithosphere and their change in their properties during orogenesis.

The arcuate nature of orogens are considered to be due to the rigidity of the descending plate and the island arc cusps are related to tears in the descending lithosphere. These island arcs may be accreted to continents during the orogenic building. The process of orogeny may take tens of millions of years to build mountains from plains or the ocean floor, and the topography is related to the principle of isostasy.

There has been a great debate on the beginning of plate tectonics in the earth's crust. It is now widely accepted that modern style plate tectonics can be tracked at least up to 3.0 Ga or even earlier and they were widespread by 2.7 Ga (Condie & Kroner, 2008). Based on petrotectonic assemblages and other plate tectonic indicators, it was established that plate tectonics were operational even during the Archean. Wilson (1966) was the first to propose the linkage between the plate movement and orogenesis. The cycle of ocean closing, orogenesis, rifting, and formation of a new ocean was described as the "Wilson cycle." The researchers later recognized the supercontinent cycles during Earth's history; such as Columbia (~ 1800 Ma), Rodinia (~ 1000 Ma), Gondwana (~ 550 Ma), and Pangea (~ 250 Ma).

Accretionary orogens develop along active continental margins where oceanic crust is subducted contributing strongly to continental growth owing to their high production rates of juvenile crust compared to Phanerozoic accretionary orogens. Many Precambrian accretionary orogens terminate by continent—continent collision during supercontinent assembly (Condie, 2007). Orogens involving oblique convergence are classified as an important part of collisional orogenic systems and are known to be either from middle crust or lower crust. They represent deeply eroded transpressional orogens representing the ductile lower crust. Such belts accommodate transpressional orogeny suggesting a common strain distribution across them incorporating strike-slip shearing in the internal parts of orogens, often with a mid-crustal scale median shear zone through progressively more oblique convergence to high angle overthrusting onto the foreland (Goscombe & Passchier, 2003). The classic examples of transpressional orogens include: Caledonian orogen in NE England, the Pan African Mozambique belt of East African Orogeny, and the Proterozoic orogens of southern India.

The role of plate tectonics with special reference to the nature of accretionary orogenesis are further described below. Accretionary orogens require convergent plate margins and hence their appearance in the geological record heralds the initiation of horizontal plate motions on the Earth. Recent studies of accretionary orogens reveal the following salient features (Cawood et al., 2006): (1) Convergent plate interaction, recycling and subduction of material from the earth's surface into the mantle has been active since at least 3.2—3.0 Ga and possibly even earlier; (2) well-constrained paleomagnetic data demonstrate differential horizontal movements of continents in both Paleoproterozoic and Archean times, consistent with the lateral motion of lithospheric plates; (3) well-preserved and unambiguous ophiolite association led to the inference that the plate subduction and the generation of ophiolites existed even as early as 3.8 Ga; (4) paired high-P and low-T; and high-T and low-P tectonothermal environments require plate subduction, which have been recognized as far back as Neoarchean; (5) preservation of greenstone sequences and several other lines of evidences indicate that plate tectonics were in operation since Paleoarchean; (6) identification of frozen subduction surfaces in the mantle in Neoarchean and Paleoproterozoic belts by Deep seismic reflection profiling; (7) geochemistry of the rocks of Eoarchean accretionary orogens of Isua, Greenland, show that they are similar to those in Phanerozoic convergent plate margins involving the subduction of a young, hot lithosphere; and (8) Jack Hill zircons from northern Yilgarn suggest that the subduction was established by 4.4 Ga ago.

The orogenic processes include continental rifting and ocean opening, oceanic and continental subduction, late-to-post orogenic extension, sedimentation, magmatism and metamorphism, exhumation of deep seated rocks, back-arc opening, and microcontinent rotation, etc. The interfering orogenic scale tectonics such as thrusting, folding, and shearing surface processes and deformation histories within orogenic belts can be treated as second-order processes. The details about these processes are beyond the scope of the book.

Plate tectonics has been an active component of the earth's processes possibly since the formation of the first continental crust at > 4.3 Ga (Ernst, 2007). Several distinct lines of evidence, in concert, established that the process of plate tectonics has been active since at least 3.1 Ga. The subduction zones, where the oceanic plates sink into the mantle, operated as "factories" since the beginning of plate tectonics on Earth. During this process, the raw materials such as oceanic sediments and oceanic crust get transformed into the formation of magmas and continental crust as products (Tatsumi, 2005).

At least four types of collisions are presently recognized: continent—continent (Alpine/Himalayan), continent—arc (Andean), arc—arc (Alaskan) collisions, and the fourth is a special category (Turkic-type), where there is a progressive accretion of small island arcs and migration of magmatic front that may produce sutures (Sengör & Natal'in, 1996). The Turkic-type orogens (typified by the Altaids) are characterized by continuous growth of subduction—accretion complexes across which arc magmatic axes prograde oceanic-wards in episodic jumps. Ophiolites occur in haphazard manner and carry no significance as structural markers or sutures to delineate former palaeotectonic entities (unlike the Himalayan- or Alpine-type orogens). Large strike-slip faults, which juxtapose assemblages formed in distant regions and metamorphosed at different structural levels, can also be erroneously reckoned as sutures. The common absence or rarity of blueschist- and eclogite-facies (high-P) metamorphic rocks in Precambrian subduction—accretion complexes may be attributed to elevated thermal gradients and shallow-angle subduction (Brown, 2009); hotter, less viscous, more buoyant, thicker and faster oceanic subducting Precambrian plates promote high-T metamorphism and slab melting (Polat & Kerrich, 2004).

The early slabs were too weak to provide a mechanism for UHPM and exhumation, and hence the absence of high pressure belts and blueschists in most of the Precambrian orogens (van Hunen & van der Berg, 2008). There is another view that all the Archean HP-UHP rocks were subducted to greater depths and the material never returned to the surface. The water acts as an effective lubricant and was available only near the surface of the earth in the high geothermal gradients prevailing in processes like plate tectonics as early as 3.8 Ga. On the other hand, availability of water in the whole upper mantle extruded much of the deeper rocks through the process of wedge extrusion in the younger Earth. Over the time, the controlling processes of asthenosphere circulation, lithosphere generation, and plate tectonics varied only in degree, not in kind (Ernst, 2005).

1.2 **PROTEROZOIC OROGENS OF INDIA**

The Precambrian Indian shield is a mosaic of Archean cratons and Proterozoic orogens. There are four major cratons, which are characterized by Archean low grade granite—greenstone sequences. The cratons are surrounded and separated by the Proterozoic orogens. The interface between cratons and orogens is marked by crustal-scale ductile shear zones. The Proterozoic period spans nearly 2 billion years, which can be divided into three eras: Palaeoproterozoic (2500—1600 Ma); Mesoproterozoic (1600—1000 Ma); and Neoproterozoic (1000—540 Ma). The Proterozoic is considered to be important because of great crustal stabilization marked by the development of global scale orogens. The Archean cratons of India include: Dharwar craton in the south, Bastar in the central, Singhbhum craton in the east, and Bundelkhand craton in the north, all welded together by a system of Proterozoic orogens/suture zones. The cratons remained tectonically fairly stable but for some epirogenic younger movements (Ramakrishnan & Vaidyanadhan, 2008).

The Proterozoic Orogens of India (POI) are defined by curvilinear, high-grade granulite-gneiss belts encompassing the entire Indian continent from Kanyakumari in the south to New Delhi in the north (Fig. 1.4). The POI comprises a sinuous chain of four major orogens from south to north in a counterclockwise direction that include: the Southern Granulite Terrane (SGT), the Eastern Ghats Mobile Belt (EGMB), the Central Indian Tectonic Zone (CITZ), and the Aravalli-Delhi Orogenic Belt (ADOB). The SGT occurs at the southern margin of the Dharwar craton and the EGMB is situated along the east coast wrapping around Dharwar and Bastar cratons. The CITZ in central India is sandwiched between Bundelkhand craton to the north and the Bastar and Singhbhum cratons to the south, while the ADOB occurs to the west of Bundelkhand craton and extends up to the Himalayan orogenic belt in the north. These orogens are also described as fold-thrust belts (Sharma, 2012) and their crustal evolution and mineral potential were described recently by Sarkar and Gupta (2012).

Earlier, the SGT was not included in the POI and the rest of the chain connecting EGMB-CITZ-ADOB was described as the Great Indian Proterozoic Fold Belt (GIPFOB) or "Mid-Proterozoic mobile belt" (Radhakrishna & Naqvi, 1986). Recent studies extended this belt and included the SGT also to make it a comprehensive picture of the POI (Chetty & Santosh, 2013). The POI may not represent one single entity of uniform characters despite several similarities. These orogens have been under focus by many national and international groups of researchers for greater details in the last two decades to arrive at a better understanding of the Proterozoic orogens of the Indian shield.

The southern Indian shield is skirted by two important Proterozoic orogenic belts, namely the Eastern Ghats Mobile Belt (EGMB) and the Southern Granulite Terrane (SGT) (Fig. 1.5). These orogens represent important collisional orogens in East Gondwana, exposing similar styles of tectonics, continued and extended large tracts of reworked areas ranging in age from Late Archean (2.5 Ga) to Neoproterozoic (0.5 Ga) with an intermediate orogenic event (1.0 Ga), popularly known as Grenvillian orogeny. These orogens also expose a unique window of a wide range of structural levels of orogenic belts.

A tectonic map of the south Indian shield (Fig. 1.5) has been compiled involving multiscale structural interpretation of Landsat TM data and reconnaissance field traverses in conjunction with the published geological maps. The major geological units in this part of the Indian shield are

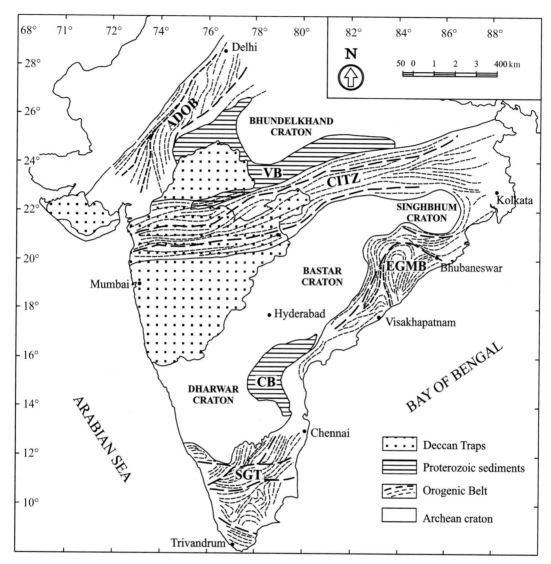

FIGURE 1.4

Map showing the Proterozoic orogens of India, wrapping around the Archean cratons and showing major shear zones and associated structural fabrics: *SGT*, Southern Granulite Terrane; *EGMB*, Eastern Ghats Mobile Belt; *CITZ*, Central Indian Tectonic Zone; *ADOB*, Aravalli-Delhi Orogenic Belt; *VB*, Vindhyan Basin; *CB*, Cuddapah Basin.

Archean granite—greenstone sequence of Dharwar and Bastar cratons, which are skirted by the EGMB along the east coast of India, and the SGT in the south. These cratons are separated by NW—SE trending Godavari rift filled with coal bearing Gondwana sediments. Narrow linear belts of high-grade gneisses shoulder the flanks of the Godavari rift on either side. The Mahanadi rift,

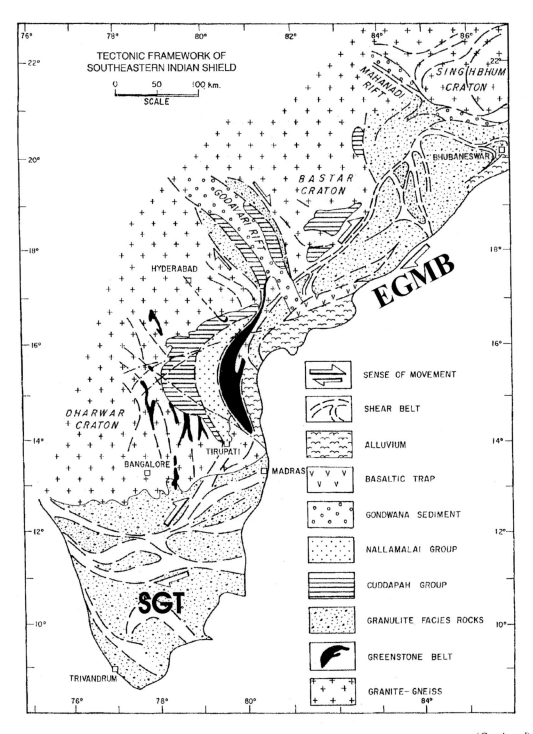

TECTONIC FRAMEWORK OF
SOUTHEASTERN INDIAN SHIELD

0   50   100 km.
SCALE

Legend:

- SENSE OF MOVEMENT
- SHEAR BELT
- ALLUVIUM
- BASALTIC TRAP
- GONDWANA SEDIMENT
- NALLAMALAI GROUP
- CUDDAPAH GROUP
- GRANULITE FACIES ROCKS
- GREENSTONE BELT
- GRANITE- GNEISS

SINGHBHUM CRATON

MAHANADI RIFT

BHUBANESWAR

BASTAR CRATON

GODAVARI RIFT

EGMB

HYDERABAD

DHARWAR CRATON

TIRUPATI

BANGALORE

MADRAS

SGT

TRIVANDRUM

(Continued)

comprising Gondwana sediments associated with coal seams, separates the Singbhum Craton to the north and Bastar Craton and the EGMB to the south. It can be observed that the NE−SW trending EGMB, gradually takes a southeasterly trend near Ongole and disappears at the Bay of Bengal. It reemerges again near Madras (currently known as Chennai) and forms a part of the SGT. The Proterozoic Cuddapah sedimentary basin, overlying the East Dharwar craton, occurs at the bend adjacent to these orogens suggesting its genetic link with the tectonometamorphic history of these collisional orogens (Chetty, 1999).

The most remarkable features in SGT and EGMB are the presence of a network of Proterozoic shear zones. The shear zones in SGT are broadly curvilinear but dominantly striking east west, while they show varied orientations in the EGMB despite their linear geometry. The shear zones trend NE−SW in the southern part, E−W trend in the northern part, and nearly N−S in the central part of the EGMB. The shear zones in both SGT and EGMB vary in their size, geometry, and their attitude in depth. The recognition of shear zones has led to major reinterpretation of structural history, tectonic style, resulting in new models providing insights of their evolution (Chetty & Murthy, 1998; Drury, Harris, Holt, Reeves-Smith, & Wightman, 1984). It has been well established that several processes like partial melting, mixing with fluids transmitted through shear zones, and strain variations combined with rheological properties affect the rocks in the shear zones.

Several studies show remarkable similarities in the tectono−magmatic, tectono−metamorphic, metamorphic P−T paths and common mafic magmatism during the Palaeoproterozoic period in the orogens of India such as ADOB, CITZ and EGMB. These results are also consistent with previous correlations, which predicted uniform Mesoproterozoic and Meso-Neoproterozoic crustal evolutionary histories of these orogens (Naqvi & Rogers, 1987).

It was pointed out that the association of felsic magmatism and coeval granulite facies metamorphism can act as a potential marker for inter-orogen correlations within the POI. Precise time constraints on the emplacement and subsequent modifications of these magmatic rocks can further complement recent tectono−metamorphic reconstructions (Bhandari, Pant, Bhowmik, & Goswami, 2011). These reconstructions suggest that Early Mesoproterozoic hot orogenesis was an integral component for the crustal growth and assembly in central, eastern, and north-eastern India. Knowledge of Palaeoproterozoic to Early Mesoproterozoic magmatic processes is thus central to modeling the growth of the Indian landmass between the Columbia and Rodinia supercontinent cycles.

The Proterozoic sedimentary basins that were closely associated with the POI include Cuddapah-Kurnool, Indravati, Khariar, Chattisgarh, Vindhyan, and Marwar basins consisting of

◄ **FIGURE 1.5**

Tectonic framework of southeast Indian shield showing the distribution of shear zone network and major geological units wrapping around the Archean Dharwar craton and linkage between the Proterozoic orogenic belts of the Eastern Ghats Mobile Belt (EGMB) and the Southern Granulite Terrane (SGT) and the other geological units. Structural interpretation is derived from multiscale satellite images and the lithologies are borrowed from the published geological maps of GSI.

After Chetty, T.R.K. (1999). Some observations on the tectonic framework of Southeastern Indian Shield. Gondwana Research, *2(4),*

651−653.

alternating sequences of sandstone, shale, and limestone with a conglomerate horizon at the base (Basu & Bickford, 2015). The sediments are mostly undeformed to mildly deformed while some of them are affected by magmatism in the form of basaltic flows, dolerite—picrite sills and dykes (Cuddapah), pyroclastic deposits and lamproites (Vindhyan). They appear to be erosional remnants of a single large basin of foredeep sediments frontal to the EGMB collisional belt (Radhakrishna & Naqvi, 1986). The evidence in favor of such a hypothesis are the boundaries of the basins facing the collisional front that are conformable in their structural trends and are thrust faulted and deformed, and the intensity of deformation decreases towards the foreland region.

The characteristics and similarities among the orogenic belts of POI suggest that the crustal architecture of India developed during the Proterozoic period possibly through stitching of several microcontinents, although the geographical location where such amalgamation occurred remains uncertain because of rather small paleomagnetic data base (summarized in Li et al., 2008). The presently exposed lower structural levels of some of the Precambrian orogens serve as potential examples of what the Himalayas might look like after they have been deeply eroded to lower- or mid-crustal levels.

During the assembly of Rodinia, a collisional orogeny of Grenvillian age welded Proto-India against Eastern Antarctica, and the resultant Eastern Ghats Belt-Rayner Complex terrane, which was not separated during the breakup of Rodinia, remained an entity until the fragmentation of Gondwana (Fitzsimons, 2003).

The application of plate tectonic paradigms have gained increasing acceptance to realize the comprehensive deep crustal processes that were involved in the evolution models of POI. Different segments of POI have dissimilar geological histories, uncorrelatable lithologic packages, unconformable structural fabrics, and disparate "time tables" of events. Tectonic evolution of the POI bears significance not only in understanding the geology of India but are also quite relevant to the formation and breakup of the three supercontinents: Columbia, Rodinia, and Gondwana (Rogers & Santosh, 2004).

All the orogens of POI show the following common characteristics: (1) occurrence of high-grade granulite facies rocks, (2) subjected to subduction—accretion—collisional processes, (3) fold-thrust tectonics and the presence of regional recumbent fold structures, (4) presence of dismembered ophiolite complexes, (5) association of manganese formations, (6) emplacement of anorthositic rocks, (7) geochronological ages ranging between 2000 and 500 Ma, and (8) the interface between orogens and the cratons is marked by a thrust-ductile shear system. The width of the POI varies from about 50 km to a few hundred kilometers. In general, the average dip of the orogens is away from the cratonic margins with the exception of the CITZ. The other striking common feature of POI is the association of Proterozoic sedimentary basins along their margins implying that the development of the basins may be genetically related to Proterozoic orogens. The variations in the geometry of structural features and the vergence of thrusts in different orogens in space and time are defined by the current disposition of orogens; pointing to the larger scale geodynamics responsible for the development of POI.

The proto-Indian continent figures in many reconstructions of the Palaeoproterozoic supercontinent Columbia. India's Palaeoproterozoic history begins with plume-related rifting of the Archean cratons, the formation of large basins, and the development of thick volcano-sedimentary sequences. Continued rifting in many domains resulted in continental fragmentation, the development of passive margins, and the birth of ocean basins. The development of POI involved the closure of the intervening oceans and the destruction of oceanic lithosphere in a prolonged subduction—accretion history culminating in final continent—continent collision. The evidence is well preserved in the geological rock record through ocean plate stratigraphy (OPS) imbricated within accretionary belts, extruded high-pressure and temperature-metamorphic orogens, as well as "frozen-in" subduction architecture imaged from geophysical techniques. The Palaeoproterozoic world in India traces a continuum through Meso- and Neoproterozoic, and has to be evaluated within the context of a prolonged "Wilson cycle" and supercontinent cycle. The Peninsular India thus preserves an unbroken record of a plate tectonic cycle from Pacific-type accretionary tectonics along the margins of the Columbia supercontinent to a Himalayan-style collisional assembly within the Neoproterozoic Rodinia supercontinent (Santosh, 2012). At the southernmost domain, the SGT a Neoproteorozoic accretionary orogenesis that culminated in a late Neoproterozoic-Cambrian collisional orogeny associated with the final assembly of the Gondwana supercontinent. The present book summarizes the main elements of all POI and highlights major advances in researches that took place in the last two decades. As a consequence of the new knowledge, some revised interpretations and pathways for new research are suggested.

The significant wealth of data in terms of geology, geophysics, and geochronology during the last two decades resulted in the revision and or refinement of many of the earlier concepts on its geological evolution. This necessitates a new tectonic synthesis of the POI that reveals many interesting features which are analogous to those commonly referred to as geological evidences of plate-margin activities in many Phanerozoic orogens of the world. Therefore, it is now realized that the time is ripe to examine and synthesize the evolution of POI in terms of the state-of-the-art concepts of global tectonic processes. A major problem in such an attempt, obviously as in many other ancient orogenic belts, are the controversial issues that arise from complexities caused by superposed polyphase deformations, metamorphism, and igneous activities. However, in the present synthesis, much stress is laid on the large volume of the recently obtained data from the field as well as from the laboratories that hopefully clarify and address some of the earlier concepts and resolve many of the well-known controversies.

1.3 TERMINOLOGY OF IMPORTANT GEOLOGIC UNITS

There have been remarkable advances in orogen related researches in the last two to three decades, and several new facets of terminology pertinent to orogens have come into the limelight because of the continued and persistent use of new as well as refined old terms used for some critical field geologic units. This shift in recognition and understanding the significance of such terms led to some rethinking of the traditional schemes, which resulted in revisits and reinterpretations of several orogenic belts. It is considered essential to comprehensively understand the terminology and the nature of some typical constituents or fundamental geological units associated with Precambrian orogens. The terms are briefly introduced and described below with an objective of better appreciation of the inclusive meaning of terminology and their significance in understanding the orogenic processes and particularly in the context of Proterozoic orogens of Indian shield. They include: suture zones, shear zones, metamorphism, sheath fold structures, UHT assemblages, melange structures, duplex structures, transpression and transtension, ophiolites, etc.

1.3.1 SUTURE ZONES

A suture zone is a linear belt of intense deformation, where distinct terranes, or tectonic units with different plate tectonic, metamorphic, and paleogeographic histories join together. The suture zones also provide the only record of deep oceanic crust and of ancient sea floor processes for roughly the first 90% of Earth's history. The study of suture zones provides a means to understand the end-product of plate tectonic processes in time and space. The suture zone is often represented on the surface by an orogen or mountain range comprising intensely deformed rocks similar to that of shear zones, but it is distinct from shear zones in representing the sites of former ocean basins within the orogenic belts. In plate tectonics, sutures are seen as the remains of subduction zones together with the terranes possibly representing fragments of different paleocontinents or tectonic plates. Outcrops of sutures may vary in width from a few hundred meters to a couple of kilometers. The suture zones can be networks of mylonitic shear zones or brittle fault zones, but are usually both. They are usually associated with igneous intrusions and tectonic lenses with varying lithologies from plutonic rocks to ophiolitic fragments. The Iapetus Suture from Great Britain, which is now concealed beneath younger rocks, and Indo-Tsangpo Suture well exposed in Himalayas are some of the best examples of suture zones. The following typical characteristics: (1) stratigraphically intact ophiolites, (2) chemically and mineralogically matured multicycle sediments, (3) high-pressure blueschist–eclogite belts, (4) ultrahigh-pressure metamorphic (UHPM) complexes, (5) deformed alkaline rocks and carbonatites, and (6) differences in the palaeolocation of crusts on both sides of orogen are considered as diagnostic tools to recognize a suture. Deep main faults, mega-lineaments, major thrusts, crustal-scale shear zones and terrane boundaries are occasionally described as palaeo, pristine, proto, or cryptic sutures, without compelling supporting evidence.

Argand (1924) was the first to come out with the idea that the orogens mark former sites of oceans, which received much attention with the advent of plate tectonics in 1960s by Wilson (1966). Later, Dewey (1969) described the orogenic belts (Caledonides and Urals) as the products of continental collision and developed the concept of suture zones. The term "Wilson cycle" was used for the cyclical or episodic processes of ocean opening and closing because Wilson was the first to appreciate the importance of plate tectonics. Suture zones, marking the sites of obliteration of oceanic lithosphere by subduction and the consequent intracontinental welding of continental masses, are rarely simple, single, easily recognizable lines. Although the obliteration of a major oceanic tract may be marked by an ophiolite-bearing suture or a cryptic suture, many mini-sutures with a wide variety of origins may be present within the convergent zone in addition to many kinds of intracontinental transform, graben, and fold/thrust zones in zones up to several thousand kilometers from the main sutures. This great variety of high-strain zones associated with wide zones of basement reactivation makes it difficult to recognize suture zones, particularly at deep structural levels in eroded orogenic systems.

The suture zones formed at terminal continental collision are likely to be extensively and progressively modified by the superposition of complex intense strains.

Sutures constitute complex zones along the area in which the continental and oceanic blocks collide. The great irregularity of colliding continental margins and intracontinental convergence are responsible for the production of deformation zones of extraordinary complexity and width, involving a great array of mini-sutures and transform-related high-strain zones. It is nearly impossible to recognize a line along which the continental masses may be snipped and dismembered, particularly at the deep structural levels of eroded old orogenic systems (Dewey, 1977). Recently, a new tool has been identified to map the old sutures. Burke, Ashwal, and Webb (2003) suggested that the deformed alkaline rocks and carbonatites (DARCs) sit on the old sutures and represent products of two well-defined parts (initial rift setting and subsequent collisional setting) of the Wilson cycle. DARCs thus "mark the places where vanished oceans have opened and then closed." Nepheline syenites (DARCs) indicate deep (lower) sections of sutures and hence are preserved in Precambrian terrains, whereas ophiolites are easily eroded, as they represent upper (shallower) levels of collisional sutures (Windley & Tarney, 1986).

The ophiolitic rocks, the characteristic features of suture zones, preserve a record of evidences for tectonic and magmatic processes from rift–drift through accretionary and collisional stages of continental margin evolution in various tectonic settings. The study of ophiolites is multidisciplinary involving structural, petrological, geochemical, and geochronological aspects, which would provide essential information on mantle flow field effects including plume activities, collisional induced asthenospheric extrusion, crustal growth, via magmatism and tectonic accretion. Several Phanerozoic to Precambrian parts of the globe have been revisited and reworked and are being appropriately reinterpreted.

The simplest kind of orogenic suture is a high-strain zone, containing mangled ophiolite remnants and, occasionally, blueschist melanges, that separates two continental terrains with dissimilar precollisional strain histories (Indus Suture, Zagros Crush Zone). The irregularity of colliding continental margins is shown in the Alpine System by sutures passing laterally into remnant oceanic tracts (Zagros Crush Zone-Eastern Mediterranean) and into subduction zones consuming large oceanic tracts (Indus Suture-Andaman Subduction Zone). Plate tectonic models of orogenic cycles involve a Wilson cycle of opening and closing of an ocean basin with deformation and metamorphism related to subduction-followed collision of continental blocks to generate mountain belts (Wilson, 1966).

1.3.2 SHEAR ZONES

Shear zones are the most significant structural features and represent deformation markers in orogenic belts. They are the places of preferred accommodation of deformation and the relative movement between the crustal blocks in both Phanerozoic and Proterozoic orogens. A majority of the published literature on shear zones deals with the methods and interpretations derived mainly from low-grade field examples. On the other hand, high-grade rocks have been traditionally studied extensively in terms of metamorphic petrology and geochemistry, but less commonly from a structural point of view. The main reason for this is the difficulty in the interpretation of complex fabric geometries of higher grade environments. Many researchers attempted their analysis by simple extrapolation of studies in low-grade to high-grade rocks, but this may result in erroneous interpretations.

Shear zones are, by simple definition, much more strongly deformed than the surrounding rocks. A shear zone is a planar zone of concentrated deformation which by itself, or in association with other zones, helps to accommodate, or wholly accommodates an imposed regional or local strain rate beyond the strength of the country rock. If the mode of deformation is predominantly by compression or extension, it is termed as pure shear (also coaxial deformation), and if the deformation occurs by tangential (wall parallel) displacements, it is called simple shear (also noncoaxial deformation). If the deformation of a shear zone consists of both, then it is general shear. Transpression and transtension deformations are the result of such a component of pure shear along with simple shear in a zone of deformation. Traditionally, the term shear zone was ascribed to denote only ductile shear zones to distinguish them from clean-cut faults. However, the term shear zone, as used by Ramsay (1980), encompasses both clean-cut faults and ductile shear zones.

The shear zones define the major boundaries of deeply eroded orogenic belts as well as zones of more intense deformation within them. The displacement along the shear zones may be of dip, oblique, or strike-slip type. The nature of these bounding shear zones is useful in constraining the kinematic evolution of very complex orogenic belts. This data would become the starting point in modeling the tectonic processes that fashioned the Panerozoic as well as Proterozoic orogenic belts. The nature, geometry, and other kinematic analysis of shear zones independently constrain the interrelationship between orogenic segments. This would greatly improve our understanding of the Proterozoic orogenic processes when combined with the available geochronological data. It is likely that large horizontal displacements dominated the evolution of Proterozoic orogenic belts. The displacements usually occur along major shear zones that link deformation at middle and lower crustal levels to that in high level foreland thrust belts (Daly, 1988). The level of erosion in most Proterozoic orogenic belts results in the extensive exposure of sheared gneisses and a common absence of foreland thrust belts.

Shear zones are very significant in several ways: (1) they are the prime targets for mineral exploration as mineralization is commonly associated with specific geometrical features such as bends and intersections; (2) they are the sites of very large strain and offer some of the strongest tools to unravel the complex deformation features of the earth's crust; (3) shear zones are also the sites for igneous intrusions like alkaline rocks, granite plutons, and anorthosites; (4) they are the only permeable path ways for the large continental crust and they act as effective fluid conduits during active deformation; (5) they often become the potential hazardous sites because of enhanced concentration of radon gas in soils, sometimes related to the uranium concentration. A possible correlation between the shear zone and U-Th content suggests a progressive increase in U enrichment with deformation, and the belts of extensive mylonitization, repeated reactivation, and chemical transfer.

According to Ramsay (1980), shear zones can be classified into three types: (1) brittle shear zones, in which tangential (wall-parallel) displacement takes place along brittle fractures and the wall rocks remain unstrained, (2) brittle-ductile shear zones in which tangential movement along the zone is associated with both ductile deformation and brittle fracture, and (3) ductile shear zones, where the tangential movement is associated with ductile deformation alone. Brittle shear zones or fault zones are a special variety of shear zones, where a clear discontinuity exists between the sides of the zone and the side walls are almost unstrained or at the most brecciated. Such fault zones are generally attributed to brittle failure controlled by the limiting elastic properties of the rock under orogenic stress. They are predominant in the upper and mid-crustal levels unlike the high-grade ductile shear zones of deep crustal levels. A narrow brittle shear zone made of discrete strike-slip faults may become wider in the deeper crustal levels and assume a form of a wide ductile shear zone in the lower crust and upper mantle depths. The study of fault pattern development in brittle shear zones aids in the correct kinematic analysis of multiply deformed shear zones. Brittle shear zones are characterized mainly by the occurrences of cataclasites and gouges. A cataclasite lacks a foliation and comprises angular clasts in a fine-grained matrix that consists of newly developed minerals, mainly white mica, chlorite, and/or calcite. A similar classification to that used in mylonites is applied to cataclasites. This can also grade into mylonite after initial cataclasis. Gouges are incohesive fault rocks that result from shallow level movements in a zone often with a weak foliation. They tend to be limited to narrow zones often within wider mylonitic or cataclastic zones.

The brittle-ductile shear zones are usually associated with some ductile deformation in the walls, which show permanent strain for a distance of up to 10 m on either side of the fault plane. There is a possibility that the ductile part of the deformation history formed at a different time from that of the fault discontinuity. Another type of brittle-ductile shear zone is the extension failure. The deformation zone shows an en-echelon array of extension openings, generally filled with fibrous crystalline material. The openings usually make an angle of 45 degrees or more with the shear zone and sometimes in a sigmoidal form.

The ductile deformation dominates and accommodates mainly in the form of ductile shear zones in the lower crust and upper mantle, which forms the base of the lithosphere with higher grade metamorphic conditions. Ductile shear zones are commonly described from vast areas of the high-grade terranes associated with Proterozoic orogens all over the globe. These zones are important in tectonic reconstructions as a source of information regarding the relative motion of large crustal blocks or plates in the geological past. High-grade terranes, formed at high pressure (8−10 kbar) and temperature conditions (700−1000°C) existing in the deeper crustal orogens. The ductile shear zones formed under high-grade conditions remain active continuously or intermittently during several episodes of tectonic activity. As a result, a superimposition of younger brittle-ductile and brittle deformations can also be identified in an earlier zone of ductile environment. Careful analysis is required to distinguish the polyphase deformations and their respective fabrics to deduce their corresponding deformational phases.

Ductile shear zones are typically characterized by the development of mylonitic fabrics. For instance, in granitic material, the fabrics are well defined in the form of closely spaced foliation by alternating layers of recrystallized quartz grains, milky ribbons of fine-grained, recrystallized feldspars and fine platy biotites. The foliation surfaces contain a very strong lineation (stretching lineation) defined by the elongation (and/or boudinage) of minerals like hornblende, micas, quartz, feldspar, etc., as well as mineral aggregates. S−C mylonites are very common, indicating noncoaxial deformation history. The amount of strain is highly variable resulting in the occurrence of mylonitic series (Proto- to ultramylonite). Retrogression, grain size reduction, development of new grain-growth, particularly biotite, kyanite, staurolite, and muscovite are typical.

1.3.3 SHEATH FOLDS

Sheath folds are noncylindrical folds with more than 90 degrees hinge line curvature, which are generated in a wide variety of tectonic settings involving noncoaxial deformation dominated by simple shear. These are typically considered to form by the gradual rotation of fold hinges towards the shear direction during progressive deformation (Cobbold & Quinquis, 1980). Folds are considered to initiate with gently curving hinges broadly normal to the shear direction, and may subsequently undergo opposing senses of rotation at either end of the hinge ultimately resulting in sheath fold geometries (Alsop, Holdsworth, & McCaffrey, 2007). Sheath folds are widely believed to develop by the passive geometric amplification of folds in which the layering has no mechanical influence during noncoaxial deformation. The existence of large-scale sheath folds has been recognized in the internal domains of major orogenic belts and from exhumed high pressure granulite facies rocks in subduction settings. Sheath folds are also reported to develop in several deformational regimes like pure shear (either plane strain or constrictional), diapirism, or three-dimensional differential (i.e., nonplanar) flow. However, recent models are mostly related to deformational regimes with a dominant component of subhorizontal simple shear as found in some nappe complexes (Skjernaa, 1989). Well-defined mega-sheath fold structures are rarely preserved in a deeply eroded Precambrian suture/shear zones.

1.3.4 DUPLEX STRUCTURES

Duplex structures are a system of imbricate thrust faults that branch off from a floor thrust below and curve upward to join a roof thrust at a branch line. The fault bound bodies of rock in the duplex are called horses. Duplexes have a variety of forms and sizes depending on the amount of displacement of the individual horses. Duplexes are formed through continued thrusting along a floor thrust with successive collapse of thrust ramps. Antiformal stacks are defined as systems of totally overlapping thrust horses that are characterized by a coincident trailing branch line. These stacks commonly occur in the cores of mountain belts, mainly in continent—continent or arc—continent collision zones where the subducted plate acts as an obstacle, forcing faulting upward. Duplex structures are also known as hallmarks of the existence of Ocean Plate Stratigraphy (OPS), whose presence may delineate suture zones and orogenic belts.

1.3.5 CONSTRICTIVE STRUCTURES

Constrictional structures are complex and localized in orogenic belts, which are of typical deformational pattern due to shortening from two directions. The structures arising from such constrictional deformation are distinctly different from those developed in other strain fields with systematic overprinting and geometrical relationships. In general, the interference pattern produced by general constriction can be recognized by: (1) the association of domes and basins with nonplane, noncylindrical folds, (2) by the occurrence of hair pin bends of hinge lines of open folds, (3) by the occurrence of amoeboid out-crop patterns, and (4) by the absence of a consistent overprinting relation among different sets of folds. Further, under special circumstances, constrictional strains also develop in transpression zones, giving rise to a spectrum of 3D deformation types (Fossen & Tikoff, 1998).

   In the field, constrictional strain is manifest as complex and systematic variations in orientation of both foliations and lineations and is dependent on: (1) the intensity of finite strain, (2) obliquity of simple shear component, and (3) the nature of kinematic partitioning within the deformation zone (Robin & Cruden, 1994). The interference patterns in constrictional deformation display diverse fold trends with less systematic interference geometry compared to a case of superposed folds (Ramsay & Huber, 1987). Such complex pattern of nonplanar or noncylindrical folds are often related to a single phase of constrictional deformation. In general, the study of constrictive structures has been the subject of extensive laboratory experiments (Ghosh, Khan, & Sengupta, 1995), and only a few field descriptions were reported. The notable example in India is an exceptionally well-exposed domal structure in a deeply eroded Precambrian crustal-scale shear system popularly known as the Cauvery Shear Zone system, the Southern Granulite Terrane (SGT) (see Chapter 2 for details). Similar tectonic scenarios have also been demonstrated along the Main Central Thrust in the Himalayan orogen as well as the Kaoko belt, Namibia (Goscombe, Gray, Armstrong, Foster, & Vogl, 2005). These features show that the lower crustal rocks can be extruded upward like toothpaste in a tube in convergent orogens, and the rate of extrusion is controlled by the geometry of the bounding shear zones and the rate of convergence of the lithospheric plates. Another recent example of constrictional deformation is in the subducted Sulu terrane that caused complex fold patterns in the overriding plate in the Liaodong peninsula, eastern China. This constrictive deformation is associated with subduction/collision where UHPM rocks are also generated (Yang, Peng, Leech, & Lin, 2011).

1.3.6 GNEISS DOMES

Gneiss domes are considered to be first-order structures in orogenic belts. Eskola (1949) was the first to systematically describe the setting and characteristics of gneiss domes in major orogenic belts around the world. According to him, a mantled gneiss dome consists of a metamorphic—plutonic complex in the core that is overlain by layered metasedimentary or metavolcanic strata. A mantled gneiss dome is typically represented by outward dipping foliations in the metamorphic core and layering of cover sequence. Since the cover sequence is normally eroded in Precambrian orogens, the mantled gneiss domes include all domal structures that are cored by metamorphic rocks and whose geometry is defined by concentric and outward dipping gneissic foliations, with or without cover sequence. Domal structures cored by metamorphic—plutonic rocks and surrounded by supracrustal rocks, often with lower metamorphic grade, and with a shear zone draping over the metamorphic core, are common in many orogenic belts. The type examples of Cenozoic gneiss domes are reported from North American Cordillera and the Himalayan orogens. The other deeply eroded mountain belts include Alps, Appalachians, Canadian Cordillera, Caledonides, etc. The nature of a gneiss dome can well be established after adequate documentation of field relationships. Crustal flow and gneiss dome formation are probably the most fundamental late orogenic processes because they accommodate variations in crustal thickness, topography, surface heat flow, and boundary conditions in thermally relaxed orogens. The gneiss domes can occur in extensional, convergent or strike-slip regimes. Crustal-scale flow is now considered as an important component of the evolution of orogenic crust through the exhumation of the lower/middle crust by channelized flow (e.g., Beaumont, Nguyen, Jamieson, & Ellis, 2006).

Many dynamic models were in vogue for the cause and kinematic development of gneiss domes. They include: (1) diapiric flow induced by density inversion, (2) buckling under horizontal constriction, (3) coeval orthogonal contraction and superposition of multiple phases of folding in different orientations, (4) instability induced by vertical variation of viscosity, (5) arching of corrugated detachment faults by extension induced isostatic rebound, and (6) formation of doubly plunging antiforms induced by thrust-duplex development. Diapiric flow model is the most popular mechanism for the formation of gneiss dome. This model will adequately explain the existence of outward dipping foliations and the development of stretching lineations, which are well preserved in some of the domal structures indicating a top outward sense of shear. However, a diagnostic link development between the observed geological setting of gneiss dome and the associated deformational processes remain poorly understood. Therefore, a new classification emphasizing the geometric and kinematic relationships between gneiss domes and ductile shear zones has been recently proposed to provide a transparent correlation between the individual classes of gneiss domes and geologic processes that were responsible for their formation (Yin, 2004). In the new scheme, gneiss domes can be divided into two fundamentally different classes: fault related and fault unrelated. However, it is well known that the gneiss domes in nature are often produced by superposition of several dome forming mechanisms, which makes it challenging to find out the real dynamic cause. In exhumed orogens, the signature of the rapid ascent of partially molten crust is a gneiss dome cored by migmatite $\pm$ granite. The large volume of material involved in the vertical transfer of partially molten crust indicates that the formation of gneiss domes is an efficient mechanism for heat advection during orogenesis (Teyssier & Whitney, 2002).

Gneiss domes commonly occur as a group in a large region of an orogenic system. They may be called gneiss dome systems and can be divided into evenly spaced and unevenly spaced systems. The occurrence of evenly spaced gneiss dome systems may be associated with instabilities induced by contrast in vertical density or viscosity and horizontal load causing the buckling. In contrast, unevenly spaced gneiss dome systems may have been associated with fault development or superposition of multiple deformational phases if the rock property is laterally homogeneous over the gneiss dome system. In Precambrian accretionary orogens, crustal flow may extend over long-lasting periods (c. 100 Ma), combining homogeneous thickening and three-dimensional mass redistribution in the viscous lower/middle crust leading to the formation of widespread dome and basin structures (Chardon et al., 2009).

1.3.7 METAMORPHISM

The interpretation of metamorphic rocks is critical for understanding their history along orogenic belts. The metamorphic changes on a regional scale help to identify the tectonic settings involved and the sequence of events that have produced a particular orogenic belt. It is also well recognized that there is a close link between the metamorphic record and global tectonic regimes (Brown, 2009).

Regional metamorphic belts occupy the core of orogenic belts and are hence one of the most important elements of orogens. The process to produce regional metamorphic belts is complex including the generation of metamorphic rocks, subsequent exhumation, emplacement at mid-crustal levels, and finally mountain building, which exposes the regional metamorphic rock units. Field relations, rock fabrics, mineral assemblages, reaction microstructures and mineral and whole-rock compositional variation provide a record of orogenic evolution.

The metamorphic petrology systematized through the generally accepted models (Thompson & England, 1984) faced challenges due to the advances in metamorphic petrology during the last three decades. The new discoveries warranted restructuring of the basic framework of metamorphic petrology in terms of geodynamics and tectonics of regional metamorphic belts. The discoveries of UHPM rocks from a few orogenic belts led to the reconsideration of the existing concepts of regional metamorphism from the narrow realm of metamorphic petrology to a much larger canvas, particularly with regard to incorporating processes at consuming plate boundaries (Maruyama et al., 2010). The traditional ideas and concepts of metamorphic petrology were synthesized and evaluated recently, and the obtained observations finally led to the proposal of some new concepts. The discovery of UHPM rocks from collision-type orogenic belts is one such concept that has revolutionized the classic interpretations of the following: (1) progressive regional metamorphism, (2) retrogressive and probably hydration metamorphism, (3) geochronology, (4) metamorphic textures and structures and the estimation of stress, (5) P-T-t paths, (6) metamorphic facies series, (7) exhumation tectonics, and (8) the behavior of metamorphic fluids. Many of these themes need to be addressed by future research in view of technological and conceptual advancements.

One of the important advancements is the discovery of UHPM rocks in Phanerozoic orogenic belts. The UHPM rocks are developed when the pressures exceed the stability field of quartz and are associated with the presence of either coesite or diamond and or by equivalent P−T conditions that can be determined using robust thermo−barometry. The UHPM rocks are generated from subducted continental crust during the initial collision between a subducting continental plate and an overriding plate. The material may melt, and if subducted past the point of "no return," it will be transported into the deep mantle and inhibit the generation of calk−alkaline magmas. It is not clear as to how these UHP rocks are preserved and exhumed back to normal crustal depths.

Another important advancement in metamorphic petrology is the discovery of UHT crustal metamorphism as a subdivision of granulite facies metamorphism, with temperatures in excess of $900°C$ at pressures of 7−13 kbar (Harley, 2004). UHT metamorphism is primarily recognized on the basis of mineral assemblages found in Mg-Al−rich pelitic rocks. Numerous origins have been proposed for Mg-Al−rich rocks including metamorphism of hydrothermally altered mafic to ultramafic rocks or high-Mg clays, sediments; metasomatic alteration of s-type granites, or formation of residual Mg-Al−rich domains in metasedimentary rocks via partial melting.

The concept of UHT was proposed in line with the new concepts of UHPM, implying that the crust can sustain thermally extreme conditions, which provides a new challenge to understand the tectonic and geodynamic drivers (Kelsey & Martin, 2015). UHT metamorphism (peak temperature exceeding $900°C$, with Opx + sillimanite + quartz; Saphirine + quartz or spinel + quartz; derived from metapelites) is common in ancient orogenic belts and is rare in post−Cambrian rock record. It is predominantly a Proterozoic phenomenon, probably associated with crustal aggregation into supercontinents. Four main periods of UHT metamorphism show a remarkable coincidence with the first formation of supercratons (Superior and Sclavia) and then the supercontinent cycle (Nuna, Rodinia, and Gondwana).

It is interesting to note the differences in metamorphic history between Phanerozoic and the Precambrian collisional orogens. The occurrence of blueschists and UHPM are common in Phanerozoic orogens, while they are rarely preserved in Precambrian orogens. The UHPM rocks are characterized by clockwise metamorphic P−T paths and exposed as 1−5 km thick subhorizontal sheets in Phanerozoic orogens. In contrast to that, counter clockwise P−T paths, followed by isobaric retrograde cooling are typical in Proterozoic orogens.

Eclogite, a rare and important rock along orogens, consists of a mass of light green pyroxene enclosing pink garnets. The typical eclogite mineral assemblage is garnet (pyrope to almandine) plus clinopyroxene (omphacite). They are generally formed from precursor mineral assemblages typical of blueschist-facies or amphibolite-facies metamorphism. Eclogite facies is determined by the temperatures and pressures required to metamorphose basaltic rocks to an eclogite assemblage. It develops from high-pressure metamorphism of mafic igneous rock (typically basalt or gabbro) plunging into the mantle in a subduction zone. Eclogites are helpful in elucidating patterns and processes of plate tectonics because many represent oceanic crust that has been subducted to depths in excess of 35 km and then returned to the surface. Eclogite under shallow conditions is unstable and is often subjected to retrograde metamorphism by producing amphibolite or granulite during exhumation.

Blueschist is a relatively rare metamorphic rock dominated by glaucophane with basaltic protolith. It is typified by high-pressure, low-temperature conditions and developed in association with subduction zones. The presence of blueschist was first reported in the Neoproterozoic era and may be associated with low-temperature eclogites (Maruyama, Liou, & Terabayashi, 1996).

1.3.8 MÉLANGE STRUCTURES

Mélange structures, the common constituents of orogenic belts, are defined as mappable chaotic bodies of mixed rocks with a block-in-matrix fabric whose internal structure and evolution are intimately linked to the structural, sedimentary, magmatic, and metamorphic processes attending its origin (Festa, Pinib, Dilekc, & Codegonea, 2010). They are the common features in collisional and accretionary orogenic belts and are distributed widely around the world (Wakayabashi & Dilek, 2011). In general, mélanges are formed along the convergent margins where ocean plate subducts beneath continental margin or island arc system.

The mélanges may comprise both exotic and native clasts of different shapes and sizes. The clasts would be, in general, angular, and range in size from decimeter to several tens of meters. There can be subordinate, smaller; nonangular clasts randomly distributed in a fine-grained matrix (could be pelitic). The clasts comprise blocks of different ages and origin, commonly embedded in an argillitic, sandy, or serpentinite matrix showing high stratal disruption and a chaotic internal structure. The mélange structures have attracted much attention in field-based structural studies since the nineteenth century because of their significance in understanding the mélange-forming processes in the geological record and the tectonic evolution of orogenic belts.

Several new advances have emerged recently on the concept of mélange, diverse mélange types, and mélange-forming processes. They are largely preserved in different tectonic settings of different ages and origins around the globe. The tectonic settings include rift–drift cycles, ocean subduction, continental collision and intracontinental deformation. They preserve the records of complex interplay of tectonic, sedimentary and diapiric processes during their formation. Depending on their internal setting and composition, fabric, and genetic processes they are classified into different groups by Festa et al. (2010), which can be related to: (1) extensional tectonics; (2) passive origin; (3) strike-slip tectonics; (4) subduction; (5) collision; and (6) intracontinental deformation.

Tectonic mélanges are the structures with fabric elements and structures resulting from contractional settings. They develop mainly at the base of accretionary wedges such as along the main decollement zone, within subduction channels, in the lower plate below the main decollement and in the zones of underplating in the accretionary wedges. These include zones of protracted off scraping and stacking of thrusts, zones of steeply dipping beds, and out-of-sequence thrusting. Such structures are mostly represented by metamorphic, high P–T mélanges inside the nappe stacking of collisional orogens and in strike-slip fault zones. Sedimentary mélanges are normally associated with extensional tectonic regimes and passive margin settings. They are commonly found at the base within the shallow nappes in intracontinental deformation zones of the ancient, submarine collisional orogens and are mostly studied in young orogenic belts such as Alpine and Himalayan ranges, representing classic collisional tectonic regimes.

The melange structures are the common constituents of orogenic belts indicating their tectono—stratigraphic significance because they form in different geological environments. Particularly, in collisional orogens, they display a record of various long-lived geological events accompanied by complex processes during which mélanges and olistostromes are developed. These include subduction of oceanic material and local emplacement of ophiolites during precollisional stages, thickening of the crust and lithospheric mantle during collisional stages and gravitational instability formed as a consequence of continued plate indentation during postcollisional stages (Dilek, 2006).

The significance of mélanges was well described in a paper published more than 25 years ago (Raymond, 1984), which continues to influence research on mélanges even today. Mélanges are characteristic features of modern and ancient convergent plate boundaries, and rank with ophiolites and high-pressure—low-temperature metamorphic rocks as critical recorders of convergent plate margin processes. Mélanges provide critical insights into sedimentary and structural evolution in the accretionary prism and forearc basin environments, including evidence for large-scale material movement (particularly in cross-sectional view) in accretionary wedges.

A close relationship has also been well established between the mélange types and the geodynamic settings of their formation. In general, the mélange structure is the end product of several processes involving cumulative complex interaction and superposition of different tectono—sedimentary processes in a single geodynamic setting. Therefore, understanding the evolutionary history of mélanges requires a complete knowledge and documentation of the spatial and temporal relationship between the mélange forming process and the dynamics of the tectonic setting of their origin. In addition to recording subduction- and collision-related sedimentary and tectonic processes, mélange formation may also include presubduction tectonics, including deformation along abyssal fracture zones (Shervais, 2001) and at oceanic core complexes as well as suprasubduction zone oceanic crust evolution.

The nature of exotic blocks and the protoliths in melanges led to many far reaching inferences. For instance, the protoliths of melanges in Jurassic accretionary complexes is OPS, which is composed of fragments of seamounts, sediments on ocean floor and trench turbidites. Further, the OPS indicate the travel history of an oceanic plate from the "birth" at an oceanic ridge to the "death" at a trench (Kozi walkita, 2012). Basalt is a component of seamount while lime stone was deposited as coral reef that covered seamount under tropical conditions. The chert is a radiolarian ooze in a pelagic environment. Siliceous shale is a mixture of radiolarian cherts and very fine terriginous clayey grains. The matrices of mélanges are normally pervasively sheared through multistage deformation. The pelitic matrix is deformed into scaly fabric dominated by mesoscopic penetrative shear fractures. Some of the clasts are rotated along the shear planes with mudstone injection. The quartz veins cut some of the shear planes and get displaced by the later stage shear planes. Mélanges also offer major insights into the most extreme vertical movements along convergent plate margins: the exhumation of high-pressure metamorphic rocks in ancient orogenic belts.

All geological processes (both crustal and deep earth) are involved in the formation, mechanisms, and interplay of mélanges. The well-preserved, on-land examples of ancient orogenic belts are therefore highly important to conduct three-dimensional studies of chaotic bodies of mixed rocks and mélanges at various scales and to better document different processes and their superposition during progressive evolution of accretionary complexes and orogenic belts. Field studies form the foundation for all the researches in the study of mélange structures in view of their three-dimensional complexity.

1.3.9 **OPHIOLITES**

Ophiolites are critical geological windows into the history of Earth to examine the mode and nature and the interplay between various igneous, metamorphic, sedimentological, hydrothermal, and tectonic processes during generations of oceanic lithosphere. They provide essential information on the mechanics and kinematics of orogenic processes. The connection between subduction zones and ophiolite emplacements is the strong linkage in the development of orogenic belts. Ophiolites make up the structural "roof" of paleosubduction zones, and ophiolite-marked subduction sutures have been considered the most important first-order structures in orogenic belts. They provide important clues as to how ocean basins formed and disappeared in the past and how the dynamic Earth's paleogeography (distribution of continental masses and oceans) appeared many millions of years ago (Moores, 2003).

Ophiolites are the remnants of ancient oceanic crust and upper mantle that were tectonically emplaced into continental margins during the closure of ocean basins. Originally, based on the field conference, an ophiolite was defined as a sequence of upper mantle peridotites, layered ultramafic–mafic rocks, cumulate to isotropic gabbros, sheeted dykes, extensive rocks and a sedimentary cover with laterally continuous contacts (Anonymous, 1972). Subsequently, it has been later recognized that ophiolites display compositional, structural and geochemical heterogeneities at different scales in contrast to the original definition. These heterogeneities are strongly influenced by the spreading rate and geometry in these settings and further dependent on the proximity to mantle plumes or trenches, the mantle composition, fertility and temperatures, availability of fluids, and recycled crustal material beneath their spreading centers, and finally the nature of mantle melting and magmatic differentiation patterns. Considering these developments, Dilek and Furnes (2011) offered a new definition for the ophiolites as "suites of temporally and spatially associated ultramafic to felsic rocks related to separate melting episodes and processes of magmatic differentiation in particular oceanic- tectonic environments." The ophiolites are further divided into subduction related and subduction unrelated categories depending on their variation. The former include suprasubduction type and volcanic arc type and the later include continental margin, mid-oceanic ridge, and plume-type ophiolites.

Ophiolites constitute a major component of the plate tectonics paradigm, and ophiolite-decorated sutures are the hallmarks for the operation of Wilson cycle events. A vestige of Earth's oldest ophiolite complex (albeit controversial) was recently discovered in the ∼3.8 Ga Isua supracrustal belt in southwest Greenland. The obvious implication of this discovery is that the sea floor spreading and subduction processes of Phanerozoic-like plate-tectonics were in existence at ∼3.8 Ga. Further, the operation of plate tectonic–like processes as early as 3.8 Ga in the Archean was supported by the presence of the suprasubduction zone ophiolites in the Archean greenstone belts. The ophiolites are interpreted to form in a very wide variety of plate-tectonic settings, like oceanic-spreading centers, back-arc basins, fore arcs, arcs, and other extensional magmatic settings, including those in association with plumes (Kusky et al., 2004). Depending on the tectonic environment, ophiolites, *sensu lato* represent a random record of the Wilson cycle of opening, widening, narrowing, and closing oceans. Present models suggest that many ophiolites originated above subduction zones (suprasubduction ophiolites) or in back-arc basins rather than in major oceans. "Transitional ophiolites" are those formed within the transition from rifted continental margins to ocean spreading centers during early stages of ocean spreading, which were later structurally detached and/or deformed and incorporated into convergent margins during ocean closure. The tectonic interpretation and tectonomagmatic evolution of such ophiolites is a difficult task, although their recognition is not so difficult. In collision zones between two Precambrian continental blocks, it is naive to expect unambiguous, characteristic evidence for the former's existence and later's extinction of the intervening ocean. Rock associations vary in different structural levels of collisional sutures. Consequently, the young (Phanerozoic) and old (Precambrian) suture zones representing higher and lower levels respectively, look differently. Of course, some late-Phanerozoic suture zones, such as the Western Alps, have been very extensively exhumed, and now reveal deep sections of the lower continental crust.

The absence of ophiolites, a common trait of the Precambrian terrains, need not be a hindrance or obstacle in our attempts to recognize suture zones. The rarity or lack of sheeted dykes can be explained by much higher spreading rates and higher mantle temperatures in the Archean than at present or in the Phanerozoic. The formation of sheeted dykes in an ophiolite complex depends on the relative rates of magma supply versus extension; if the magma supply is less than rate of extension, then dykes will be scarce or even absent (Moores, 2003). Ophiolite fragments are common; quasi, transitional, and partial ophiolites are less common; full ophiolites are rare. For these reasons, it was recognized many years ago that the absence or extreme scarcity of ophiolites in the deeply eroded orogens is not unusual, and should not be construed that the orogens are ensialic (Windley, 1992). In general, the ophiolites represent upper levels of collisional sutures and get easily eroded and the chances of their survival are practically absent in the cryptic terminal sutures of deeply eroded Precambrian orogens. In the absence of ophiolites, sutures can also be detected with other means like lithological, structural, geochronological, and metamorphic variations across them. In such situations, "additional" geochemical characterization of tectonic settings of the rock types is of great utility (Leelanandam, Burke, Ashwal, & Webb, 2006). Many Precambrian magmatic sequences have trace element signatures almost identical to those found in modern subduction environments, implying their formation in an analogous setting (Cawood et al., 2006). However, petrotectonic settings must be viewed cautiously in conjunction with the regional geologic setting because dependence on a single line of evidence may not give the true picture.

In modern orogenic belts like Alpine-Himalayas, the ophiolites dominantly comprise upper mantle peridotites and harzburgites of supradsubduction zone (SSZ) setting. The SSZ ophiolites display structural, petrological, and geochemical features in their crustal and upper mantle units that commonly indicate the time−progressive development of their ancient oceanic lithosphere in various stages of the Wilson cycle evolution of ocean basins (Pearce & Robinson, 2010).

Interestingly, in a very recent study, the ophiolites are found to contain diamonds, ultrahigh-pressure minerals, highly reduced phases and native elements and crustal minerals (e.g., zircon, corundum, kyanite, and rutile) in chromitites and peridotities of their mantle sequences (Dilek, Furnes, & De Wit, 2015). These features indicate their crystallization at depths of 150−300 km or even deeper, near the transition zone in the mantle. The presence of silicate minerals and diamonds suggest recycling of continental crustal material and surface carbon via subduction into the mantle transition zone.

It is well established that the ophiolites preserve the geological record and the evolutionary history of ocean basins from their rift−drift and sea floor spreading stages to the subduction initiation and final closure. The ophiolites are significant in many ways: (1) they are useful in understanding the nature of mid-oceanic ridge processes, (2) provide the chemistry of mantle rocks, (3) the paleo-geography of ancient ocean basins reveal tectonic clues about the evolution of orogenic belts, and finally (4) they boost the advancement of plate-tectonic theory. Ophiolites host a range of mineral deposits. Ultramafic and gabbroic rocks may contain deposits of chromium or platinum-group elements. Chrysotile asbestos occurs in serpentinites. Copper, zinc, cobalt, and nickel sulfides (marine exhalatives) may occur in economic amounts. Some ophiolites also host shear controlled epithermal or mesothermal gold mineralization.

1.3.10 TERRANES

Terranes are the building blocks of orogens. A terrane is a fault-bounded package of rock sequence of regional extent with distinctive geologic history (Coney, Jones, & Monger, 1980). It is well recognized that the present spatial juxtaposition of terranes in any orogen does not necessarily reflect their relative position prior to their assembly. Their modern analogs include microcontinents, magmatic arcs, intraplate volcanic chains, accretionary subduction complexes, and oceanic plateaus, which are brought together by plate motions. Their motion is seldom simple orthogonal convergence, as envisaged in the Wilson cycle. The motion involves complex patterns, not haphazard but dictated by the changing vectors associated with lithospheric motion oblique to the plate boundaries and migrating triple junctions (Jones, 1990). Identification of fundamental crustal blocks has increasingly gained importance in reconstructing the sequence of events in the evolution of an orogen. In recent years, several Precambrian orogens were described as the collages of juxtaposed terranes. Some of the important examples are the Archean southern west Greenland, Grenville Province of southeast Canadian shield, Arunta Inlier high-grade complex of Australia, Mesoproterozoic Sveconorwegian orogenic province of Scandinavia, Superior province of Canada, Archean Limpopo belt of South Africa, the Lewisian complex of northwest Scotland, etc. Recognition of terranes requires descriptive data on several aspects such as stratigraphy, petrological, geochemical, structural, geochronological, and geophysical characteristics that define a tectonostratigraphic terrane.

1.3.11 TRANSPRESSION AND TRANSTENSION

Transpressional regimes are widespread in orogenic belts and give rise to complex strain patterns. Transpression and transtension are strike-slip deformations that deviate from simple shear with a combination of shortening or extension orthogonal to the deformation zone. These deformations occur both at a local as well as a regional scale, located typically at plate boundaries principally in response to obliquely convergent or divergent relative plate motions on a wide variety of scales during deformation of the Earth's lithosphere. On the largest scale of planet Earth, this is an inevitable consequence of relative plate motion on a spherical surface because plate convergence and divergence slip vectors are not commonly precisely orthogonal to plate boundaries and other deformation zones. Crustal block margins are inherited features that act as zones of weakness, repeatedly reactivated during successive crustal strains often in preference to the formation of new zones of displacement. Any displacement zone margin that is significantly curvilinear or irregular is bound to exhibit oblique convergence and/or divergence unless it follows exactly a small circle of rotation. In addition to collisional orogens, transpression and transtension also occur widely in a large range of other tectonic settings: oblique subduction margins in the forearc (transpression), arc (transpression and transension) and back-arc (transtension) regions; restraining (transpression) and releasing (transtension) bends of transform and other strike-slip displacement zones; continental rift zones (transtension), especially during the early stages of continental breakup and formation of new oceanic lithosphere; during late orogenic extension (transtension); and in slate belts (transpression), where deformation may be accompanied by large-scale volume loss (Dewey, Holdworth, & Strechen, 1998).

Structural complexity along transpressional deformation zones makes coherent interpretations extremely challenging. The complexity greatly increases from the fact that the nature of deformation is heterogeneous and varies from place to place. Field observations have established that the structures that differ significantly in orientation may form simultaneously. But unfortunately, following traditional approaches, many of such structures are being interpreted as products of polyphase deformational events. The orientation of stretching lineation may show different geometrical relations with respect to the transport direction. Usually, the stretching lineations in many ductile shear zones are parallel to the tectonic transport. However, there have been several examples of stretching lineations reported even perpendicular to the tectonic transport direction from a number of subvertical transpressional shear zones.

The development of most geological structures in transpressional zones is controlled by the strain imposed by the boundary conditions, which implies that there may not be a simple or significant relationship between large-scale crustal deformation structures and the stress pattern. Transpression also involves extrusion tectonics that often redistributes rheologically weak material both laterally and vertically within the shear zones. It is important to note that syntectonically emplaced plutonic bodies along the active transpressive shear zones exhibit similar deformation patterns as those of the host rocks. Large-scale steeply dipping transpressive zones develop significant surface topographies both above and at the lateral terminations. These, in turn, would be sufficient to generate gravity driven deformation and associated exhumation of deep crustal rocks. Some of the well studied transpressional deformation zones are: the Alpine Fault Zone of northern South Island in New Zealand, the Pyrenees, the Mongolian Western Altai, the East Anatolian fault zone in Turkey, the Venezuela Andes, and central California. Examples of transtension include the Gulf of California, the Malawi Rift, and the North Aegean Sea in Greece.

Orogens involving oblique convergence are now believed to be an important part of collisional orogenic systems and are known to be from the middle crust. They represent deeply eroded transpressional orogens representing the ductile lower crust. Such belts accommodate transpressional orogeny suggesting a common strain distribution across them incorporating strike-slip shearing in the internal parts of orogens, often amid the crustal-scale median shear zone, through progressively more oblique convergence to high angle overthrusting onto the foreland (Goscombe et al., 2003). The classic examples include Caledonian orogen in NE Greenland, the Pan African Mozambique belt, forming a part of East African Orogeny (EAO); and the Proterozoic orogens of southern India, the prime subject of the present book. In view of the complexity of structures, one needs to be extremely cautious in interpreting structures in zones of transpressional tectonic regime, especially in the context of orogens.

Many of the characteristic orogenic geologic units described above have been recently recognized and extensively described in the following chapters dealing with the Proterozoic orogens of India.

LIST OF ABBREVIATIONS

ADOB	Aravalli Delhi Orogenic Belt
CB	Cuddapah Basin
CITZ	Central Indian Tectonic Zone
CO	cold orogens
DARCs	deformed alkaline rocks and carbonatites
EAO	East African orogeny
EGMB	Eastern Ghats Mobile Belt
GIPFOB	Great Indian Proterozoic Fold Belt
HO	hot orogens
MHO	mixed hot orogens
OPS	Ocean Plate Stratigraphy
POI	proterozoic orogens of India
SGT	Southern Granulite Terrane
SSZ	suprasubduction zone
UHO	ultra-hot orogens
VB	Vindhyan Basin

REFERENCES

Alsop, G. I., Holdsworth, R. E., & McCaffrey, K. J. W. (2007). Scale invariant sheath folds in salt, sediments and shear zones. *Journal of Structural Geology, 29*, 1585–1604.

Anonymous (1972). Penrose field conference on ophiolites. *Geotimes, 17*, 24–25.

Argand, E. (1924). La tectonique del'Asie, C.R. 13th international Geological Congress, 1, 171–372.

Basu, A., & Bickford, M. E. (2015). An alternate perspective on the opening and closing of the intracratonic purana basins in Peninsular India. *Journal Geological Society of India, 85*, 5–25.

Beaumont, C., Nguyen, M. H., Jamieson, R. A., & Ellis, S. (2006). *Crustal flow modes in large hotorogens* (268, pp. 91–145). London: Geological Society.

Bhandari, A., Pant, N. C., Bhowmik, S. K., & Goswami, S. (2011). ∼1.6Ga ultrahigh-temperature granulite metamorphism in the Central Indian Tectonic Zone: insights from metamorphic reaction history, geothermobarometry and monazitechemical ages. *Geological Journal, 46*, 198–216.

Brown, M. (2009). Metamorphic patterns in orogenic systems and the geological recordIn P. A. Cawood, & A. Krö Ner (Eds.), *Earth accretionary systems in space and time* (318, pp. 37–74). London: Geological Society, Special Publications.

Burke, K., Ashwal, L. D., & Webb, S. J. (2003). New way to map old sutures using deformed alkaline rocks and carbonatites. *Geology, 31*, 391–394.

Cawood, P. A., Kroner, A., Collins, W. J., Kusky, T. M., Mooney, W. D., & Windley, B. F. (2009). *Accretionary orogens through Earth history* (319, pp. 1–36). London: Geological Society.

Cawood, P. A., Kroner, A., & Pesarevsky, S. (2006). Precambrian plate tectonics: criteria and evidence. *GSA Today, 16*, 4–11.

Chardon, D., Gapais, D., & Cagnard, F. (2009). Flow of ultra-hot orogens: a view from the Precambrian, clues for the Phanerozoic. *Tectonophysics, 477*, 105–118.

Chetty, T. R. K. (1999). Some Observations on the Tectonic Framework of Southeastern Indian Shield. *Gondwana Research, 2*(4), 651–653.

Chetty, T. R. K., & Murthy, D. S. N. (1998). Regional tectonic framework of the Eastern Ghats Mobile Belt: a new interpretation. *Geol. Surv. India Special Pub*, *44*, 39−50.

Chetty, T. R. K., & Santosh, M. (2013). Proterozoic orogens in southern Peninsular India: Contiguities and complexities. *Journal of Asian Earth Sciences*, *78*, 39−53.

Cobbold, P. R., & Quinquis, H. (1980). Development of sheath folds in shear regimes. *Journal of Structural Geology*, *2*(1−2), 119−126.

Condie, K. C. (2007). Accretionary orogens in space and time. *Geological Society of America Memoirs*, *200*, 145−158.

Condie, K. C., & Kroner, A. (2008). *When did plate tectonics begin? Evidence from the geological record* (440, pp. 281−294). Geological Society of America.

Coney, P. J., Jones, D. L., & Monger, J. W. H. (1980). Cordilleran suspect terranes. *Nature*, *288*, 329−333.

Daly, M. C. (1988). Crustal shear zones in Central Asia: a kinematic approach to Proterozoic tectonics. *Episodes*, *11*, S-11.

Dewey, J. F. (1969). Evolution of the Appalachian−Caledonian orogen. *Nature*, *222*, 124−129.

Dewey, J. F. (1977). Suture zone complexities: a review. *Tectonophysics*, *40*, 53−67.

Dewey, J. F., Holdworth, R. E., & Strechen, R. A. (1998). Transpression and transtension zones. In: J. F. Dewey, R. E. Holdsworth, & R. A. Strachan (Eds.), *Continental transpressional and transtensional tectonics* (135, pp. 1−14). London: Geological Society of, Special Publication.

Dickinson, W. R. (1971). Plate tectonic models of Geosynclines. *Earth Planetary Science Letters*, *10*, 165−174.

Dilek, Y. (2006). Collision tectonics of the Eastern Mediterranean region: causes and consequences. In: Y. Dilek, & S. Pavlides (Eds.), *Postcollisional tectonics and magmatism in the Mediterranean region and Asia*. Geological Society of America. Special Paper 409, pp. 1−13. http://dx.doi.org.10.1130/2006.2409(1).

Dilek, Y., & Furnes, H. (2011). Ophiolite genesis and global tectonics: geochemical and tectonic fingerprinting of ancient oceanic lithosphere. *Geological Society of America Bulletin*, *123*(3−4), 387−411.

Dilek, Y., Furnes, H., & De Wit, M. (2015). Precambrian greenstone sequences represent different ophiolite types. *Gondwana Research*, *27*, 649−685.

Drury, S. A., Harris, N. B. W., Holt, R. W., Reeves-Smith, G. J., & Wightman, R. T. (1984). Precambrian tectonics and crustal evolution in south India. *Journal of Geology*, *92*, 3−20.

Ernst, R. E. (2007). Mafic-ultramafic large igneous provinces (LIPs): Importance of the pre-Mesozoic record. *Episodes*, *30*(2), 108−114.

Ernst, W. G. (2005). Alpine and Pacific styles of Phanerozoic mountain building: subduction-zone petrogenesis of continental crust. *Terra Nova*, *17*, 165−188.

Eskola, P. E. (1949). The problem of mantled gneiss domes. *Quarterly Journal of the Geological Society*, *104*, 461−476.

Festa, A., Pinib, G. A., Dilek, Y., & Codegonea, G. (2010). Mélanges and mélange-forming processes: a historical overview and new concepts International. *Geology Review*, *52*(10−12), 1040−1105.

Fitzsimons, I. C. W. (2003). Proterozoic basement provinces of southern and southwestern Australia, and their correlation with AntarcticaIn M. Yoshida, B. F. Windley, & S. Dasgupta (Eds.), *Proterozoic East Gondwana: supercontinent assembly and breakup* (206, pp. 93−130). Geological Society of London, Special Publication.

Fossen, H., & Tickoff, B. (1998). Extended models of transpression and transtension and application to tectonic settingsIn R. E. Holdsworth, R. A. Strachan, & J. F. Dewey (Eds.), *Continental transpressional and transtensional tectonics* (135, pp. 15−33). Geological Society of London, Special publications.

Gerya. (2014). Precambrian geodynamics: concepts and models. *Gondwana Research*, *25*, 442−463.

Ghosh, S. K., Khan, D., & Sengupta, S. (1995). Interfering folds in constrictional deformation. *Journal of Structural Geology*, *17*, 1361−1373.

Goscombe, B. D., Gray, D., Armstrong, R., Foster, D. A., & Vogl, J. (2005). Event geochronology of thespian-African Kaoko Belt, Namibia. *Precambrian Research*, *140*, 103−131.

Goscombe, B. D., & Passchier, C. W. (2003). Asymmetric boudins as shear sense indicators- an assessment from field data. *Journal of Structural Geology*, *25*, 575–589.

Harley, S. L. (2004). Extending our understanding of ultrahigh temperature crustal metamorphism. *Journal of Mineralogical and Petrological Sciences*, *99*, 140–158.

Hodges, K. V. (2000). Tectonics of the Himalayas and southern Tibet from two perspectives. *Geological Society of America Bulletin*, *112*, 324–350.

Jones, D. L. (1990). Synopsis of late Proterozoic and mesozoic terrane accretion within the Cordillera of western North America. *Philosophical Transactions Royal Society, London*, *331*, 479–486.

Kelsey, D. E., Clark, C., & Hand, M. (2008). Thermobarometric modelling of zircon and monazite growth in melt-bearing systems: examples using model metapelitic and metapsammitic granulites. *Journal Metamorphic Geology*, *26*, 199–212.

Kelsey, D. E., & Hand, M. (2015). On ultrahigh temperature crustal metamorphism: phase equilibria, trace element thermometry, bulk composition, heat sources, timescales and tectonic settings. *Geoscience Frontiers*, *6*, 311–356.

Kozi walkita., Mappable features of mélanges derived from Ocean Plate Stratigraphy in the Jurassic accretionary complexes of Mino and Chichibu terranes in Southwest Japan, Tectonophysics 568–569, 2012, 74–85.

Kusky, T. M., Li, J. H., & Raharimahefa, T. (2004). Origin and emplacement of Archean Ophiolites of the Central Orogenic Belt, North China CratonIn T. M. Kusky (Ed.), *Precambrian Ophiolites and related rocks* (13, pp. 223–282). Developments in Precambrian Geology.

Leelanandam, C., Burke, K., Ashwal, L. D., & Webb, S. J. (2006). Proterozoic mountain building in Peninsular India: an analysis based primarily on alkaline rock distribution. *Geological Magazine*, *143*, 195–212.

Li, Z. X., Bogdanova, S. V., Collins, A. S., Davidson, A., De Waele, B., Ernst, R. E., ... Vernikovsky, V. (2008). Assembly, configuration, and break-up history of Rodinia: a synthesis. *Precambrian Research*, *160*, 179–210.

Liou, J. G., Tsujimori, T., Zhang, R. Y., Katayama, I., & Maruyama, S. (2004). Global UHP metamorphism and continental subduction/collision: the Himalayan model. *International Geology Review*, *46*, 1–27.

Maruyama, S., Liou, J. G., & Terabayashi, M. (1996). Blueschists and eclogites of the world, and their exhumation. *International Geology Review*, *38*, 485–594.

Maruyama, S., Liou, J. G., & Zhang, R. Y. (1994). Tectonic evolution of the ultrahigh-pressure(UHP) and high-pressure (HP) metamorphic belts from central China. *The Island Arc*, *3*, 112–121.

Maruyama, S., Masago, H., Katayama, I., Iwase, Y., Toriumi, M., Omori, S., & Aoki, K. (2010). A new perspective on metamorphism and metamorphic belts. *Gondwana Research*, *18*, 106–137.

Matsuda, T., & Isozaki, Y. (1991). Well-documented travel history of Mesozoic pelagic chert in Japan: From remote ocean to subduction zone. *Tectonics*, *10*, 475–499.

Meert, J. G., & Lieberman, B. S. (2008). The Neoproterozoic assembly of Gondwana and its relationship to the Ediacaran-Cambrian radiation. *Gondwana Research*, *14*, 5–21.

Miyashiro, A. (1961). Evolution of metamorphic belts. *Journal of Petrology*, *2*, 277–311.

Moores, E. M. (2003). A personal history of the ophiolite conceptIn Y. Dilek, & S. Newcomb (Eds.), *Ophiolite concept and evolution of geologic thought* (373, pp. 17–29). Geological Society of America, Special Paper.

Naqvi, S. M., & Rogers, J. J. W. (1987). *Precambrian geology of India* (pp. 1–233). New York: Oxford University Press.

Pearce, J. A., & Robinson, P. T. (2010). The Troodos ophiolite complex probably formed in a subduction initiation, slab edge setting. *Gondwana Research*, *18*, 60–81.

Polat, A., & Kerrich, R. (2004). Precambrian arc associations: Boninites, adakites, magnesian andesites, and Nb-enriched basalts. In T. M. Kusky (Ed.), *Precambrian ophiolites and related rocks* (pp. 567–597). Amsterdam: Elsevier.

Radhakrishna, B. P., & Naqvi, S. M. (1986). Precambrian continental crust of India and its evolution. *Journal Geology*, *94*, 145−166.

Ramakrishnan, M., & Vaidyanadhan, R. (2008). *Geology of India* (vol. 1 Bangalore: Geological Society of India.

Ramsay, J. G. (1980). Shear zone geometry: a review. *Journal of Structural Geology*, *2*, 83−99.

Ramsay, J. G., & Huber, M. L. (1987). The techniques of modern structural geology, *Folds and fractures* (Volume 2, p. 700). London: Academic Press.

Raymond, L. A. (1984). Classification of melangesIn L. A. Raymond (Ed.), *Melanges: Their nature, origin and significance* (198, pp. 7−20). Boulder, Colorado: Geological Society of America, Special Paper.

Robin, P. Y. F., & Cruden, A. R. (1994). Strain and vorticity in ideally ductile transpression Zones. *Journal of structural Geology*, *16*, 447−466.

Rogers, J. J. W., & Santosh, M. (2004). *Continents and supercontinents* (p. 289). New York: Oxford University Press.

Santosh, M. (2012). *India's paleoproterozoic legacy*. Geological Society of London. *Special Publications*, 365, pp. 263−288, http://dx.doi.org/10.1144/SP365.14.

Sarkar, S. C., & Gupta, A. (2012). *Crustal evolution and metallogeny of India*. Cambridge: Cambridge Publishers.

Sengör, A. M. C., & Natal'in, B. A. (1996). Paleotectonicsof Asia: fragments of a synthesis. In A. Yin, & T. M. Harrison (Eds.), *The tectonic evolution of Asia* (pp. 486−640). Cambridge: Cambridge University Press.

Sharma, R. S. (2012). *Cratons and fold belts of India*. Berlin: Springer Publications.

Shervais, J. W. (2001). Birth, death and resurrection: the life cycle of suprasubduction zone ophiolites. *Geochemistry Geophysics Geosystems*, *2*, 2000GC000080.

Skjernaa, L. (1989). Tubular folds and sheath folds: definitions and conceptual models for their development, with examples from the Grapesvare area, northern Sweden. *Journal of Structural Geology*, *11*, 689−703.

Tatsumi, Y. (2005). The subduction factory: how it operates on Earth. *GSA Today*, *15*, 4−10.

Teyssier, C., & Whitney, D. (2002). Gneiss domes and orogeny. *Geology*, *30*, 1139−1142 . http://dx.doi.org/10.1130/0091-7613(2002)0302.0.CO;2.

Thompson, B., & England, P. C. (1984). Pressure-temperature-time paths of regional metamorphism. II. Their inference and interpretation using mineral assemblages in metamorphic rocks. *Journal of Petrology*, *28*, 929−955.

Van Hunen, J., & Van den Berg, A. P. (2008). Plate tectonics on the early Earth: limitations imposed by strength and buoyancy of subducted lithosphere. *Lithos*, *103*(1), 217−235.

Wakabayashi, J., & Dilek, Y. (2011). *Introduction: characteristics and tectonic settings of mélanges, and their significance for societal and engineering problems*. The Geological Society of America. Special Paper 480, p. v−x, http://dx.doi.org/10.1130/2011.2480(00).

Wilson, J. T. (1966). Did the Atlantic close and thenre-open? *Nature*, *211*, 676−681.

Windley, B. F. (1992). Proterozoic collisional and accretionary orogensIn K. C. Condie (Ed.), *Proterozoic crustal evolution. Developments in precambrian geology* (10, pp. 419−446). Amsterdam: Elsevier.

Windley, B. F. (1995). *The evolving continents* (3rd ed., 385 pp). Chichester: Wiley.

Windley, B. F., & Tarney, J. (1986). *The structural evolution of the lower crust of orogenic belts, present and past* (24, pp. 221−230). Geological Society of London.

Yang, T. N., Peng, Y., Leech, M., & Lin, H. Y. (2011). Fold patterns indicating Triassic constrictional deformation on the Liaodong peninsula, eastern China, and tectonic implications. *Journal of Asian Earth Sciences*, *40*, 72−83.

Yin, A. (2004). Gneiss domes and gneiss dome systemsIn D. L. Whitney, C. Teyssier, & C. S. Siddoway (Eds.), *Gneiss domes in orogeny* (380, pp. 1−14). Geological Society of America, Special Paper.

FURTHER READING

Bhowmik, S. K., & Dasgupta, S. (2012). Tectonothermal evolution of the Banded Gneissic Complex in central Rajasthan, NW India: Present status and correlation. *Journal of Asian Earth Sciences*, *49*(2012), 339–348.

Gray, D. R., Foster, D. A., Meert, J. G., Goscombe, B. D., Armstrong, R., Truow, R. A. J., & Passchier, C. W. (2008). *A Damaran Perspective on the Assembly of Southwestern Gondwana* (294, pp. 257–278). Geological Society of London, .

Whitney, D. L., Teyssier, C., & Vanderhaeghe, O. (2004). Gneiss domes and crustal flow. In: D. L. Whitney, C. Teyssier, & C. S. Siddoway (Eds.), *Gneiss Domes in Orogeny* (380, pp. 15–33). Geological Society of America, Special Papers.

THE SOUTHERN GRANULITE TERRANE

CHAPTER OUTLINE

Proterozoic Orogens of India. DOI: http://dx.doi.org/10.1016/B978-0-12-804441-4.00002-X

2.1 INTRODUCTION

The Southern Granulite Terrane (SGT), an important part of the Proterozoic orogens of India, occurs at the southern tip of the Indian shield. Being at the intersection of two global orogenies of East African orogen and the Kuunga orogen, the SGT is crucial not only in understanding the geodynamic history of the orogens, but also central to many reconstruction models of Rodinia and Gondwana supercontinents. In recent years, there has been much focus on the SGT and its relationship with the timing and the tectonics of amalgamation of Gondwana. The extensions of the SGT towards the northwest into Madagascar, East Africa, and further north to the Arabian shield in the form of East African orogen, can probably be correlated with Himalayan scale orogen (Collins, Clark, & Plavsa, 2014). The SGT also extends to southeast and northeast through Sri Lanka and Antarctica and witnessed 570−500 Ma old Kuunga orogeny (Meert & Lieberman, 2008).

The SGT has been one of the most intensely studied orogen in the last two decades by several national and international groups encompassing all aspects of geology and geophysics. Innumerable publications have brought out large volumes of data with several modern concepts and innovative ideas but with variable and contradicting interpretations (Chetty, Fitzsimons, Brown, Dimri, & Santosh, 2006). However, many of the controversial topics remain debatable even today, despite the accumulation of significant amount of geological, geochemical, geochronological, and geophysical data. The debatable points include: the definition of SGT and its extensions; transition zone where the low-grade tonalitic and granitic gneisses gradually transformed into granulite facies metamorphic charnockitic rocks; division and extensions of different tectonic blocks, suture zone/shear zones and their kinematics; existence of terrane boundaries; timing of subduction, accretion, and collisional processes, and so on. The plethora of these contrasting interpretations and the evolution and subdivision of crustal blocks within the SGT are a direct consequence of limited field observations, lack of field and structurally constrained geochronological data, and limitations of accessibility due to high elevation and dense vegetation.

Fermor (1936) was the first to propose the geological division of Peninsular India into "charnockite" and "noncharnockite" regions based on the presence of orthopyroxene isograd. The

Proterozoic orogens of the SGT and the Eastern Ghats Mobile Belt (EGMB) are currently regarded as the charnockite region, while the Archean granite–greenstone assemblage of the Dharwar craton as the noncharnockite region. The SGT is one of the largest exposed Precambrian deep continental crusts consisting of multiply deformed Archean and Neoproterozoic high-grade metamorphic and magmatic rocks (Fig. 2.1). However, some workers in recent years described the high-grade rocks

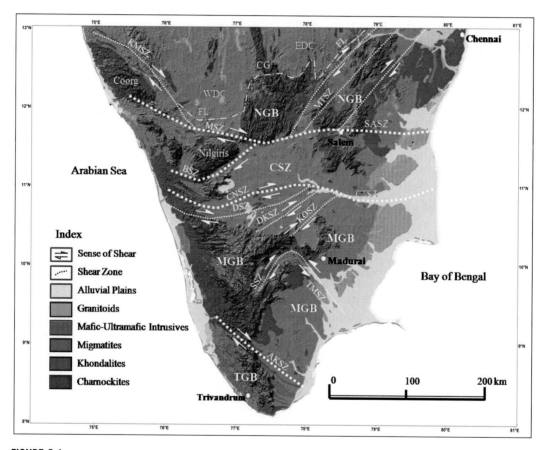

FIGURE 2.1

Geological and Tectonic map of the Southern Granulite Terrane showing major rock types and tectonic features along with the digital elevation model—*NGB*, Northern Granulite Block; *CSZ*, Cauvery shear zone; *MGB*, Madurai Granulite Block; *AKSZ*, Achankoil shear zone; *TGB*, Trivandrum Granulite Block; *EDC*, Eastern Dharwar craton; *WDC*, Western Dharwar craton; *CG*, Closepet Granite; *FL*, Fermor's line; *MTSZ*, Mettur shear zone; *KMSZ*, Kasargod-Mercara shear zone; *MSZ*, Moyar shear zone; *BSZ*, Bhavani shear zone; *CNSZ*, Chennimalai Noil shear zone; *SASZ*, Salem-Attur Shear Zone; *CTSZ*, Cauvery-Tiruchirappalli shear zone; *DSZ*, Dharapuram shear zone; *DKSZ*, Devattur-Kallimandayam shear zone; *KOSZ*, Kodaikanal Oddanchathram shear zone; *SSZ*, Suruli shear zone; *TMSZ*, Theni-Madurai shear zone.

that occur to the south of Cauvery shear zone as the Pandyan Mobile belt (Ramakrisnan & Vaidyanadhan, 2008).

The SGT includes high-grade metamorphic crustal blocks ranging in age from Mesoarchean to the Neoproterozoic and can be subdivided into discrete tectonic blocks separated by two important crustal-scale shear zone systems: the Cauvery suture/shear zone (CSZ) and the Achankovil shear zone (AKSZ). However, the significance of these shear zones and their relevance to the tectonic history of the SGT was brought to light only after the tectonic framework brought out by Drury and Holt (1980) and Drury, Harris, Holt, Reeves-Smith, and Wightman (1984). Chetty (1996) subsequently interpreted further finer details of large-scale features such as shear zones, regional fold forms, broad lithologies, lineaments and widely variable trends of structural fabrics from satellite data (Fig. 2.2). The high-grade metamorphic rocks of the SGT were earlier considered to have formed over a long span of time from early Archean to late Neoproterozoic through polymetamorphic cycle (Harris, Santosh, & Taylor, 1994). According to Ghosh, Maarten, de Wit, and Zartman (2004), the SGT witnessed at least seven thermo−tectonic events at 2.5 Ga, 2.0 Ga, 1.6 Ga, 1.0 Ga, 800 Ma, 600 Ma, and 550 Ma, and two distinct episodes of metasomatism/charnockitization (D1) between 2.50−2.53 and between 0.55−0.53 Ga. Deformation along major shear zones in the SGT occurred during Neoproterozoic to very early Paleozoic age, with an early phase (D2) concentrated between 700 and 800 Ma and a later phase (D3) between 550 and 600 Ma. Major Charnockitization (530−550 Ma) postdates D3, and is in turn, overprinted by granitization, retrogression, and uplift between 525 and 480 Ma.

The SGT is believed to be of lower crustal origin through a complex evolutionary history with multiple deformations, anatexis, intrusions and polyphase metamorphic events. The major rock types are high-grade granulite facies rocks that include essentially Neoarchean charnockites and their variably retrograded assemblages, pyroxene granulites, metasedimentary assemblages, which were subsequently intruded by Cryogenian anorthositic rocks, alkaline plutons, granitoids and mafic−ultramafic rocks including ophiolites. The metasedimentary assemblages include Banded Iron formations (BIF), calc silicates, and metapelites. Intense shearing and migmatization gave rise to a variety of amphibole-biotite−bearing migmatitic gneisses. But recent geologic, petrologic, and geochronologic studies confirmed that the Proterozoic block of the SGT was not accreted to the Archean Dharwar craton to the north until the latest Neoproterozoic coinciding with the phase of amalgamation of Gondwana supercontinent (Collins et al., 2007; Santosh, Maruyama, & Sato, 2009). They also proposed that the Cauvery Shear Zone (CSZ) represents the trace of suture developed by the closure of Mozambique Ocean during the late Neoproterozoic. An overview of geophysical results in the SGT shows that the average crustal thickness varies between 40 and 50 km and Poisson's ratio between 0.25 and 0.27 with the exception of the crust beneath the Nilgiri Hills where the crustal thickness is ∼60 km and Poisson's ratio of 0.28 (Gupta, Rai, Prakasam, & Srinagesh, 2003). These observations suggest a significant crustal shortening in Southern India during the Archean.

The analysis of the aeromagnetic data, collected at the reconnaissance scale to understand the tectonic evolution of the South Indian shield, displays E−W trending high-amplitude magnetic anomalies over the SGT due to magnetic sources at depth (Reddi, Mathew, & Naidu, 1988). A thin magnetic crust (∼22 km) was inferred suggesting that the deeper levels do not contribute much to the magnetic signatures. These magnetic trends are comparable to the broad structural trends of metamorphic gneissosity, lithological contacts, and the east−west trending tectonic features like shear zones and major fold forms in the region. Distinct magnetic highs are also associated with

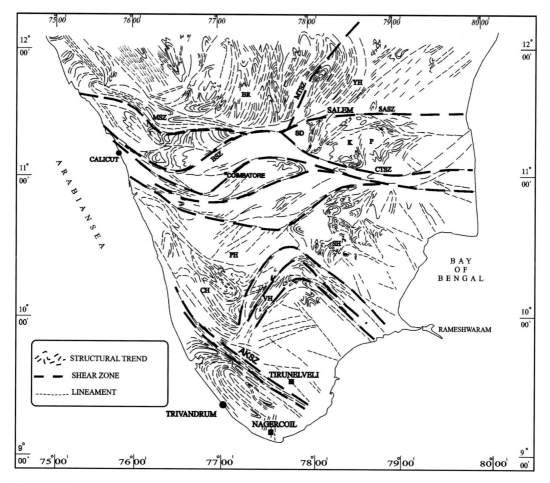

FIGURE 2.2

Structural interpretation of satellite data showing variable trends of structural fabrics, major shear zones, fold styles and other lineaments with in the SGT—*MSZ*, Moyar shear zone; *BSZ*, Bhavani shear zone; *MTSZ*, Mettur shear zone; *SASZ*, Salem-Attur shear zone; *CTSZ*, Cauvery-Tiruchirapalli shear zone; *AKSZ*, Achankoil shear zone; *YH*, Yarcaud Hills; *BR*, Biligirirangan Hills; *SD*, Sankaridurg; *P*, Pachamalai Hills; *K*, Kollimalai Hills; *PH*, Palani Hills; *CH*, Cardamom Hills; *SH*, Sirumalai Hills; *VH*, Varushanadu Hills.

Modified after Chetty, T.R.K. (1996). Proterozoic shear zones in southern granulite terrain, India. In: M. Santosh & M. Yoshida (Eds.), The Archean and Proterozoic terrain of Southern India with in Gondwana. Gondwana Research Group Memoirs 3, Field Science Publications 77–89.

the shear/suture zones, which are marked by high-pressure and ultrahigh temperature metamorphic assemblages suggesting their subduction related origin. Some of the highs may also be attributed to the metamorphosed clastic sediments, BIF, and mafic/ultramafic bodies resulting from the process of accretionary tectonics.

The Bouguer gravity anomalies are relatively strongly negative, reaching the minimum values of 100−120 mgal with similar east−west trends consistent with aeromagnetic anomalies. The local negative anomalies are attributed to the presence of low-density acid charnockites, while the regional negative field could be due to the effect of regional compensation of the elevated charnockitic massifs in the form of high land areas (Subramanyam & Verma, 1982). Large magnitude (steep) gravity gradient over the CSZ represents the boundary between the Dharwar Craton and the SGT. The following recent regional reviews are also suggested for further reading for a comprehensive understanding of the SGT (Braun & Kreigsman, 2003; Collins et al., 2014; Ghosh et al., 2004; Gopalakrishnan, 1996; Kröner et al., 2015; Plavsa, Collins, Foden, & Clark, 2015; Santosh et al., 2009).

Based on the recent developments and significant advances, the SGT can be divided into five distinct crustal/tectonic units based on lithological assemblages, structural styles, geochronological characteristics, and geophysical signatures. From north to south, they are: (1) Northern Granulite Block (NGB), (2) Cauvery suture/shear zone (CSZ), (3) Madurai Granulite Block (MGB), (4) Achankovil suture/shear zone (AKSZ), and (5) Trivandrum Granulite Block (TGB) (see Fig. 2.1). A comprehensive foliation trajectory map of the SGT is presented here (Fig. 2.3). The east−west trending tectonic features and associated structural fabrics broadly characterize the SGT. However, there are distinct variations in different segments as well as with in the shear zones. Apart from the crustal-scale shear zones, the map shows well defined foliation trajectories defining broad fold forms, variable trends and geometries at different places pointing to the existence of a mosaic of different tectonic blocks with in the SGT. The NGB, the northernmost block, is dominated by NE−SW trending structural fabrics with antiformal fold closures to the north. The CSZ, an east−west trending tectonic zone is characterized by structural heterogeneity. The MGB, the largest block bound by crustal scale shear zones, displays foliation trajectories in all directions. But, the predominance of particular direction is a striking feature in some of the segments. For instance, in the northern part of MGB, NE−SW trending fabrics dominate in the western segment, while intense folding is observed in the eastern segment reflecting the presence of metasediments. In the southern part of MGB, NW−SE trending fabrics are distinct in the western sector, while near E−W fabrics dominate the eastern sector reflecting the basement. Although the AKSZ as well as the TGB are characterized by NW−SE trending structural fabrics, the variations in their geometry and intensity of deformation provide clues for the recognition of the shear zone. The details of geological, geochronological, and geophysical characteristics of each tectonic unit are described below.

2.2 NORTHERN GRANULITE BLOCK

2.2.1 INTRODUCTION

The NGB, includes all the high-grade crustal blocks that occur to the north of Moyar-Bhavani shear zone (MBSZ) and Salem−Attur shear zone (SASZ), together defining the northern boundary of the CSZ. The NGB is mostly occupied by the high land areas of Biligirirangan (BR) hills, Shevaroy hills, Yarcaud hills, etc., extending in an east−west fashion with a width of about 80 km

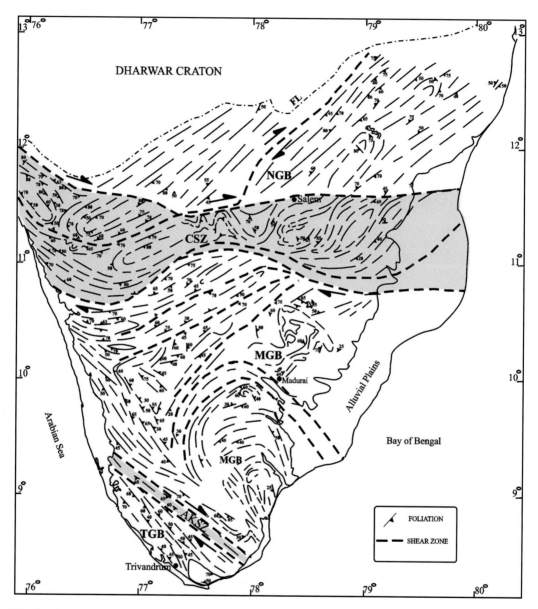

FIGURE 2.3

Comprehensive structural and tectonic map of the SGT showing crustal-scale shear zones, broad fold forms, foliation trajectories in different tectonic segments—*FL*, Fermor's Line; *NGB*, Northern Granulite Block; *CSZ*, Cauvery shear zone; *MGB*, Maduari Granulite Block; *AKSZ*, Achankoil shear zone; *TGB*, Trivandrum Granulite Block. The other details are same as in Fig. 2.1.

(see Fig. 2.1). Orthogneisses and charnockite massifs together with minor mafic granulites and high-grade metasedimentary rocks dominate the NGB. These hills are surrounded by low-lying areas of granitic orthogneisses and paragneisses, which experienced amphibolite to granulite grade metamorphism. The contact between the low-grade and high-grade terrane is intermixed by a zone of intense migmatization with predominant K-feldspar fenitization. This fenitized zone is well exposed north of Shevaroy hills and Yarcaud hills (together described as Salem block) in an east—west direction extending for about a kilometer. Some of the earliest recognizable foliations strike east—west and dip distinctly to south with gentle to moderate dips suggesting the presence of relict E—W trending isoclinal folds within the fenitized zone.

2.2.2 TRANSITION ZONE

A narrow zone of transition is described at the southern margin of the Archean Dharwar craton between the low-grade Peninsular gneiss in the north and the high-grade massif charnockites in the NGB. The important rock types of the transition zone include hornblende—biotite gray gneiss, pink granites and migmatites with a minor association of metasedimentary assemblage, which are traversed by a number of mafic dykes. The tonalitic and granitic gneisses, locally migmatized, predominate the northern amphibolite facies terrain with abundant incipient charnockitization while massive charnockites and their gneissic counter parts dominate in the NGB. This led to the notion that the lower amphibolites facies rocks were transformed into granulite facies conditions gradually from north to south (Janardhan, Newton, & Hansen, 1982; Pichamuthu, 1965). The progressive increase in metamorphic grade from tonalitic gneisses in the north to charnockitic gneisses in the south is marked by a systematic change in mineral compositions. Geothermometry and geobarometry also show a prograde metamorphism of 5.5 ± 1.5 kb and $730 \pm 40°C$ in the transition zone near the orthopyroxene isograd in the north to $\sim 8 \pm 1.5$ kb and $775 \pm 30°C$ in the south (Rameswara Rao, Narayana, Charan, & Natarajan, 1991). The pressure estimates indicate burial depths of 14—23 and 23—33 km in intermediate and lower parts of the crust respectively indicating the Archean crust (~ 2.44 Ga old) to be at least 50—68 km thick. All the features described above were accounted for by a hypothesis that the Indian shield tilted northward thereby exposing the mid-crustal granulite facies rocks in the south by deep erosion.

The process of transformation of peninsular gneissic rocks into charnockite is known as "charnockite in making." Progressive transformation was also reported at many other places all over the SGT (Ravindra Kumar, Srikantappa, & Hansen, 1985). However, some other workers have proposed an opposite view that the charnockite patches at several places exhibit gradation to gneisses indicating that they were produced by the breakdown or retrogression of the preexisting charnockites through the process known as "charnockite in breaking" (Devaraju & Sadasivaiah, 1969; Mahabaleswar & Naganna, 1981). Interestingly, Yoshida and Santosh (1987) emphasized that "charnockite in breaking," caused by retrogressed metamorphism, is as widespread as "charnockite in making" in the SGT. Such contrasting opinions and ambiguity are still being debated (Newton & Tsunogae, 2014; Peucat et al., 2013). Therefore, deciding the relative role of prograde and retrograde phenomenon regarding the genesis of charnockite is challenging and is a matter of debate, which may possibly be resolved only through the innovative integration of metamorphic petrology and field structural geology with multiscale perspectives.

2.2.3 STRUCTURE

The charnockitic gneisses in the NGB show well defined metamorphic gneissic foliation and the strike varies from N−S to NE−SW with easterly and south easterly moderate dips. Despite the dominant presence of near horizontal foliations in these rocks, some of the published geological maps grossly understate the common prevalence of gentle to subhorizontal gneissic foliation in the region. The early deformation D1 is evidenced by progressive folding of gneissic foliation S1 into tight and isoclinal fold forms often giving rise to interference and/or sheath fold geometries. These are preserved on a meter-to-kilometer scale. Detailed mapping of mesoscopic structures in a small outcrop (50 m^2) exhibit flat-lying, well-foliated charnockitic gneiss in the northern part of Shevaroy hills, east of Dharmapuri (Fig. 2.4). The orientation of hinge-line of F2 folds gradually change from east−west in the northern part to southwest in the southern part pointing to the geometry of a possible planar sheath fold with closure to south. There are also isolated outcrops of charnockites displaying gently dipping foliations to south with fold closures to north. Besides, complex fold styles with significant strain variation on meso- to macroscale are common in the region. This is evident from: (1) the consistency of fold axis orientation despite large variations in the amplitude of folds and the attitude of axial planes, (2) truncation of overstepping of gneissic foliation, (3) oblique and closed fold structures of sheath geometry, and (4) consistent northwesterly vergence indicated by regionally occurring gentle to moderately dipping foliation fabrics and shear planes and related fold forms observed in appropriate vertical sections in the field (Chetty, Bhaskar Rao, & Narayana, 2003).

Several NNE−SSW trending parallel shear zones occurring in the NGB deflect and merge with the northern boundary of the CSZ suggesting dextral kinematics. The shear zones also show dextral sense of movements and extend to the north−east for about 150 km. Prominent among them is the western most shear zone described as Mettur Shear Zone (MTSZ), which separates the BR hills and Javadi hills (Fig. 2.5). The MTSZ extends over a strike length of ∼200 km with a width of 20−30 km. The NNE−SSW trending MTSZ is considered as an extended branch of the MBSZ. The major lithologies of the MTSZ include epidote−hornblende gneisses, quartzo−feldspathic gneisses and variably retrograded charnockitic rocks. Alkaline dykes of different sizes, varying in composition from shonkinite, lamprophyre to perthite−syenite are associated with siderite−ankerite veins, arfvedsonite−riebeckite veins, white albitites and quartz−barite veins and traverse the entire MTSZ. Migmatization, shearing effects, vestiges of west-verging, thrust-related recumbent fold structures and possible sheath folds are also described. The MTSZ hosts a series of Neoproterozoic syenite, alkali granite, and carbonatite plutons (Kumar, Charan, Gopalan, & Macdougall, 1998). The plutons include Sevattur alkaline−carbonatite complex, Pakkanadu carbonatite complex, Yelagiri alkali syenite−pyroxenite complex, and alkali syenite of Angadimogar and ultrapotassic granite of Peralimala. Some of them were dated showing broadly Cryogenian ages (750−800 Ma). Undeformed igneous textures are well preserved in these younger plutons. The source for the alkaline rocks was suggested to be the melts derived at high P−T conditions from the c. 2.5 Ga old delaminated/eclogitized slab, which must have enriched the mantle (Santosh et al., 2014b). Detailed geochemical studies of Yelagiri complex suggest them to have "shoshonite" character and is considered to have originated from metasomatized K- and LILE-enriched lithospheric mantle and arc- related magma in a subduction zone tectonic setting (Renjith et al., 2015). Dismembered and agmatised bodies of gabbro, anorthosite, pyroxenite, norite are also recorded indicative of thrust-transported ophiolitic rocks within the MTSZ (Gopalakrishnan, 1996). All this evidence

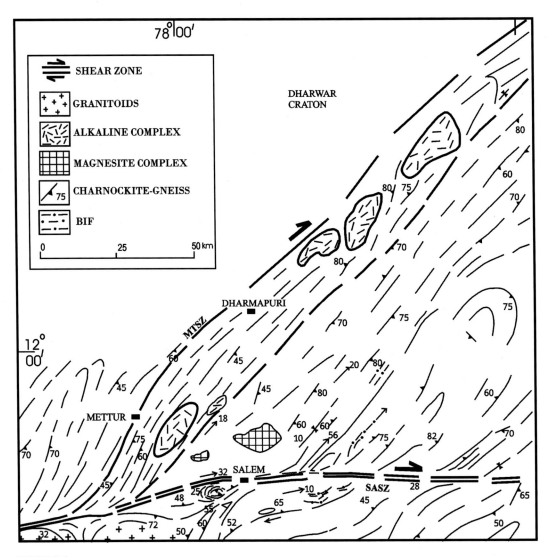

FIGURE 2.4

Geological and structural map of Mettur shear zone and the adjacent region within the NGB, MTSZ- Mettur shear zone, and the alkaline bodies.

suggests that the MTSZ is a suture zone, which has been reactivated probably during the extensional tectonics, and the available geochronological data suggests their emplacement during 770−560 Ma (Kovach et al., 1998). The MTSZ extends further north and coincides with the western boundary of the Proterozoic Nallamalai fold belt and the Sileru suture zone, the western margin of the EGMB. This suture zone reflects one of the largest suture zones (∼1500 km) of Indian shield and forms the interface between the Archean Dharwar, Bastar, and Singhbhum cratons and the Proterozoic granulite terranes of southern India.

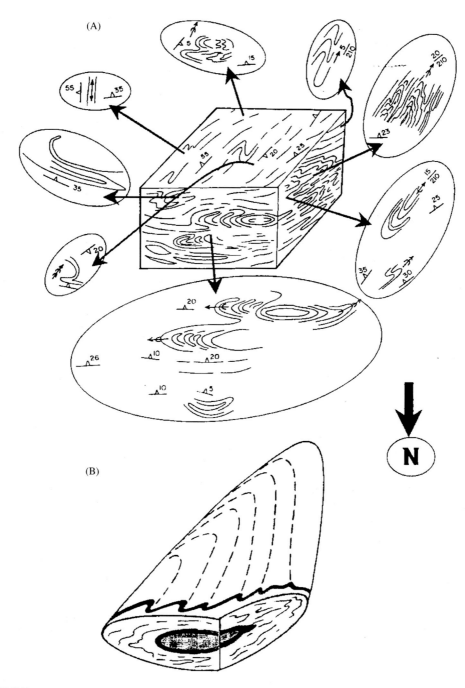

FIGURE 2.5

Structural analysis of mesoscopic structures (A) in an outcrop of possible sheath fold (B), east of Dharmapuri in the northern part of the NGB.

Adapted from Chetty, T.R.K., Bhaskar Rao, Y.J., & Narayana, B.L. (2003). A structural cross-section along Krishnagiri-Palani corridor, southern granulite terrain India. In: M. Ramakrishnan (Ed.), Tectonics of Southern granulite terrain, Kuppam-Palani Geotransect. Geological Society of India, Memoir 50, 255–277.

2.2.4 METAMORPHISM

The rocks in NGB were metamorphosed to granulite facies conditions during Neoarchean to Palaeoproterozoic, with relatively high-pressure (HP) ($\sim$14 kbar) metamorphism occurring during the early Neoproterozoic (Peucat, Mahabaleshwar, & Jayananda, 1993). Although igneous textures were not observed, the homogeneity, massive nature, and the presence of enclaves within the charnockite gneisses of the NGB indicate an igneous origin for the protoliths. Petrological studies of metagabbros from the southwest periphery of the Biligirirangan Block reveal clockwise prograde and retrograde metamorphism in a subduction zone setting at a HP of 18−19 kbar and temperature of $\sim$840°C (Ratheesh-Kumar et al., 2016). Sensitive High Resolution Ion Microprobe (SHRIMP) U−Pb isotope analyses of zircon from a charnockite and a charnockite-hosted leucosome reveal that the felsic magmatism occurred at 2.53 Ga, which was followed by high-grade metamorphism and anatexis at 2.48 Ga (Clark, Collins, Kinny, Timms, & Chetty, 2009). These ages and the chemistry of charnockites are consistent with the formation by accretionary processes at the southern margin of the Dharwar craton. The geochemical characteristics of granites from the NGB indicate a dominant calc−alkaline nature consistent with their formation in a convergent margin in an island arc setting at around 2.5 Ga. The rocks from the NGB appear to be free from a pervasive Neoproterozoic high-grade metamorphic event associated with the amalgamation of Gondwana.

2.2.5 GEOPHYSICAL SIGNATURES

The reflectivity pattern derived from the seismic reflection studies across the NGB suggest a deep penetrating, prominent south dipping reflection band extending up to Moho depth at the northern margin of the CSZ. The seismic reflection pattern is distinct on either side of the MTSZ with a characteristic divergent reflection fabric, which is typical of suture zones (Reddy et al., 2003). The crustal thickness across the NGB from the southern margin of Dharwar craton upto CSZ varies between 41 and 45 km with 3−4 km up warp across the MTSZ. High anomalous conductivity ($\sim$100 ohm) and a paired Bouguer gravity anomaly with a steep gravity gradient were recorded across the MTSZ. All these features suggest that the MTSZ is the site of collisional processes between the Dharwar craton and an unknown continent giving rise to the development of SGT.

2.2.6 TECTONIC SYNTHESIS

The complex fold styles including the multiscale sheath fold geometries, prominent horizontal foliation fabrics with gentle dips to south, and southward dipping fold closures indicate that the granulite facies rocks of the NGB represent north-verging frontal fold-thrust belt emerging from the south (Chetty et al., 2003). The root zone of the NGB was estimated to the south with a northward movement as a subhorizontal tectonic translation during exhumation. These structures from NGB can also be explained alternatively in a superstructure−infrastructure model of hot fold nappe (Williams, Jiang, & Lin, 2006) with a concept of the existence of different structural levels. The middle level is dominated by parallel folds; the upper level with fault zones, and the lower level with flow, nonparallel folds and widespread tectonic foliation. The model illustrates the contrasting styles of deformation in infrastructure and superstructure. According to the model, the lower crust is forcibly expelled outward over a lower crustal indentor to create fold nappes that are inserted into the mid-crust. Recumbent folds are restricted to the infrastructure. Upright folds are developed in the upper levels (superstructure) as well as in the upper part of infrastructure in association with sharply defined ductile thrusts. The upright folds have lower amplitude and are superimposed on recumbent fold structures. Tight recumbent folds, rootless folds, boudinage, and strongly foliated migmatite, ortho and paragneisses are common in infrastructure. The sheath fold structures and other associated complex fold structures described in NGB may occur at the base of fold-nappe system, which may be related to the basal thrust with significant displacements towards the foreland. It is also possible that the exposed surface geology shows the remnants of eroded nappe structures that occur as shear zone bound hill masses, such as Biligirirangan hills, Yarcaud hills, Shevaroy hills, etc. From the available age data, the granulite facies rocks of NGB are interpreted to have been derived possibly from the Neoarchean subduction related crustal thickening and palaeoproterozoic thrusting.

A spatial variation of metamorphic grade through a fold nappe is common in orogenic belts. During the process and emplacement of thrusts and nappes in NGB, the role of fluids derived from the underlying low-grade metamorphic rocks was not considered adequately when the high-grade metamorphic rocks were juxtaposed with the former. The circulation of fluids along the plate boundaries is suggested to be more important than the pressure–temperature changes (Maruyama et al., 2010). Accordingly, the so-called mineral isograds defined on the maps of regional metamorphic belts as transition zones such as in the NGB were the result of misunderstanding of the progressive dehydration reaction during subduction due to large-scale obliteration of the progressive minerals in pelitic-psammitic and metabasic rocks by extensive late stage hydration. Extensive hydration of high-grade metamorphic rocks occurs due to fluid infiltration underneath on the low-grade rock units at mid-crustal level. The presence of discontinuous boundary was believed to be a gradual and the continuous gradation from lower grade rocks into high-grade metamorphic rocks seems to be apparent. The structural boundary on a regional scale is not that distinct between the two distinct rock units of Dharwar craton and NGB. On outcrop scale, identifying a sharp boundary between the two is a challenging task due to the lack of marked difference in deformation styles and metamorphic grade. The presence of possible ophiolitic rocks, alkaline magmatism, thrust-related structures, uplift of Moho, bipolar gravity anomalies, southerly dipping refraction and reflection banding, suggest that the MTSZ is a well-defined suture zone and that the convergent tectonics must have occurred between the southern margin of the Dharwar craton and the SGT.

In the light of the above, the traditional concept of buoyancy-driven upward exhumation, deep erosion, and the exposition of high-grade metamorphic rocks in the NGB needs to be reviewed. Further, the concept of transformation of granite gneisses into charnockites needs to be studied with an integrated study of structural and petrological studies.

2.3 CAUVERY SUTURE ZONE

2.3.1 INTRODUCTION

Recent decades have witnessed the significance and wide acceptance of geological processes such as subduction–accretion–collision that manifested orogenic suture zones and the associated key lithologies such as ophiolites in many Precambrian terranes all over the globe (Kusky, 2004). Suture zones are the sites of oceanic lithosphere and are rarely simple, single, and easily recognizable. The simplest kind of orogenic suture is a high-strain zone containing deformed ophiolite remnants and tectonic mélanges (Burke, Dewey, & Kidd, 1977). The suture zones are extensively and progressively modified by the superposition of complex intense strains associated with terminal suturing. Sutures that result from the closure of oceans provide critical information, not only on paleogeography, but also on material that was transported to the trench, and the pressure—temperature conditions it underwent during subduction and exhumation. Thus, the presence of diagnostic HP rocks such as eclogites, blueschists, and whiteschists provides critical evidence for a suture to be closely associated with a former subduction zone. The deformation varies greatly in the proximity and away from the site of suturing based on the post suturing convergence, the duration and the shape of the colliding continental margins.

The Cauvery suture zone (CSZ) is the most significant and deeply eroded east—west tectonic belt (350×70 km). It separates the Archean Dharwar cratonic rocks to north and the Proterozoic granulites to the south within the SGT. The CSZ constitutes a pervasive network of shear zones and is collectively described as the Cauvery suture/shear zone (Bhaskar Rao, Chetty, Janardhan, & Gopalan, 1996; Chetty et al., 2003; Gopalakrishnan, 1996). The CSZ has been variedly interpreted by different authors as: (1) a dextral strike-slip shear zone based on the rotation of north—south fabrics of the rocks from Dharwar craton to near east—west disposition along the Moyar-Bhavani and Salem-Attur shear zones, (2) a collapsed marginal basin, (3) a collision zone and a cryptic suture zone evident from the occurrence of possible remnants of oceanic crust in the form of tectonic slivers and thrust slices and associated marginal sequences of microcontinents, (4) an analog of the central part of Limpopo mobile belt, (5) an Archean-Neoproterozoic terrane boundary, (6) a zone of Palaeoproterozoic and Neoproterozoic reworking of Archean crust, (7) a crustal-scale "flower structure," (8) a zone of pure shear-dominated transpression accompanied by down-dip stretching/shearing within the shear zones, and (9) as an extended join with Madagascar and east Antarctica.

A new structural architectural map of the CSZ on 1:250,000 scale has been recently compiled (Fig. 2.6) by making use of different spatial data sets like Landsat data, Google Earth map, published geological maps from Geological Survey of India, and the author's own field observations from the last two decades (Chetty, 2015). The granulite facies rocks within the CSZ are dominantly represented by charnockites (~ 2.5 Ga), migmatitic gneisses including biotite—hornblende gneisses, mafic, and ultramafic intrusives followed by Neoproterozoic younger syenitic—granitic plutons. The map shows major curvilinear shear zones, structural trends, fold patterns, major geological boundaries, circular features, regional duplex structures, etc. The map reveals dominant structural fabrics of metamorphic gneissic foliation (S1), which are, in general, parallel to lithological contacts. This is well reflected in satellite data on different spatial scales and is helpful in the delineation of major structural elements such as shear zones, large-scale fold forms and the disposition of intrusive bodies. The northern boundary of the CSZ is marked by MBSZ, while the southern margin is defined by Chennimalai-Noyil Shear Zone (CNSZ), which is also described by some as the Palghat-Cauvery shear zone. The fabrics along the MBSZ dip dominantly to south with steep values (70—80 degrees), while they dip to north with gentle to moderate dips along the CNSZ. The MBSZ is associated with the presence of north-verging thrusts dominated by the presence of ~ 2.5 Ga old charnockitic rocks, and the CNSZ is marked by south-verging back thrusts comprising mostly mafic—ultramafic complexes and granite plutons (0.8—0.5 Ga). For the sake of clarity and

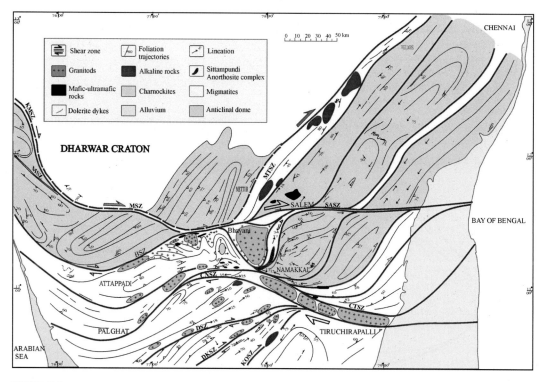

FIGURE 2.6

Map showing the structural architecture of the Cauvery suture zone.

Adapted from Chetty, T.R.K. (2015). The Cauvery suture zone: map of structural architecture and recent advances. Journal of
Geological Society of India, 85, *37—44.*

brevity, the CSZ is divided into two sectors for providing comprehensive details of structural synthesis as described below.

2.3.2 THE EASTERN SECTOR

In the eastern sector of the CSZ, the boundary shear zones are well defined in the form of east—
west trending SASZ in the north, the eastward continuation of MBSZ, the Cauvery-Tiruchirappalli
shear zone (CTSZ) in the south, and the extension of the CNSZ. A set of sigmoidal shear belts,
which are relatively narrow (~0.5 km wide) with subvertical planar fabrics, connect the boundary
shear zones (Fig. 2.7) (Chetty & Bhaskar Rao, 2006a). The important rock types in the region
include charnockitic massifs of Kollimalai and Pachamalai high land areas and a suite of late
Archean migmatitic gneisses and granulite grade Archean supracrustal rocks that include mainly
shallow marine arenaceous, calcareous, pelitic rocks and prominent banded iron formation units in
the low-lying areas. The 2.9 Ga old layered anorthosite bearing ultramafic—mafic complexes also
occur at the southern margin of the CSZ (Bhaskar Rao et al., 1996).

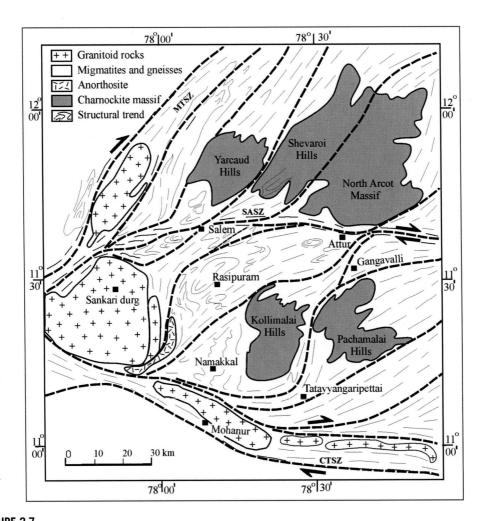

FIGURE 2.7

Regional structural map of the eastern part of the Cauvery suture zone.

The SASZ is an east–west trending 2–3 km wide shear zone along the Vasista river valley occurring between Salem and Attur and extends further east. It is characterized by steep southerly dipping mylonites of varying intensity. The SASZ is also characterized by ~1 km wide band of steeply dipping phyllonites over a strike length of 20 km between Salem and Attur. Regionally, the curvature of axial planar foliations in the north from NGB swerves into the SASZ indicating dextral strike-slip shearing. The SASZ shows extreme high aspect ratios of strain gradients and the effects of retrogression related to mantle-derived CO_2 and rich hydrothermal fluids (Wickham, Janardhan, & Stern, 1994). Detailed structural mapping of a small segment across the SASZ around Valayappadi (Fig. 2.8) shows intense transposition fabrics defining the two varieties of fabrics: the

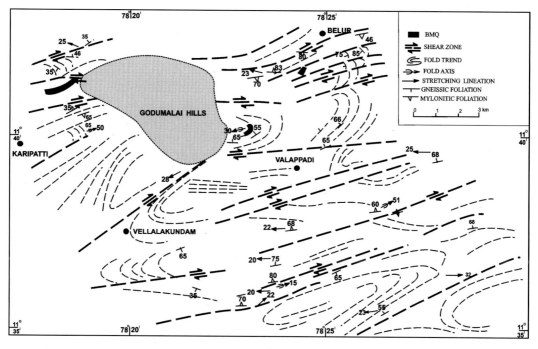

FIGURE 2.8

Map showing transposed fabrics and associated east—west trending mylonite zones in a small segment around Valayappadi across the Salem-Attur shear zone.

metamorphic gneissosity (S1), which was transposed giving rise to well-developed mylonitic fabrics (S2). A complex range of fold styles and elongated-closed structural forms are the resultant features of transposition. The intensity of transposition varies and in extreme cases, the S1 fabrics are completely transposed with new development of mylonitic fabrics often grading to ultramylonites (Fig. 2.9). The dip of gneissic foliation varies from gentle to steep, while the mylonitic fabrics strike ENE—WSW with steep dips defining shear zones. Another revealing feature is the abrupt ending of gneissic foliations against the margins implying that the Godumalai hills could possibly represent an allochthonous block. Although, gentle to moderate stretching lineations, plunging either east or west, are common along the SASZ, steep plunging lineations are also observed locally. The typical example is a well-preserved rare outcrop at Belur (Fig. 2.10) showing both shallow and steep lineations (Chetty & Bhaskar Rao, 1998) implying dextral transpressional shearing. The dextral kinematics are further substantiated by several other mesoscopic structures all over the SASZ in the region.

The CTSZ represents a near east—west trending 8—10 km wide zone where foliations dominantly dip moderately (40—60 degrees) to the north. The CTSZ consists of a number of subparallel shear zones characterized by mylonitic gneisses retrogressed from charnockites and pyroxene granulites. Profuse development of stretching lineations with consistent subhorizontal eastward plunges

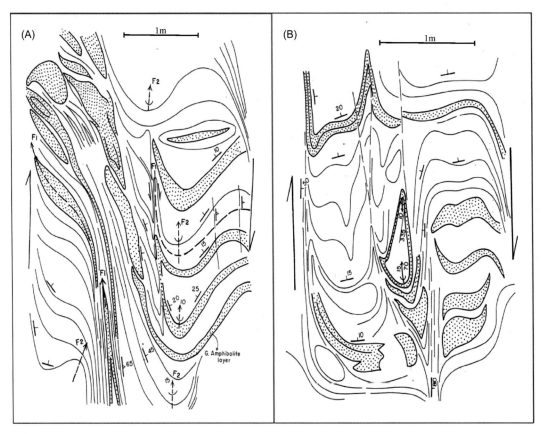

FIGURE 2.9

Large-scale sketches of transposed fabrics and the gradation of gneissic fabrics into mylonitic fabrics in retrogressed charnockites (A and B).

and other mesoscopic kinematic observations indicate dextral strike-slip movement. Deformed younger granitoids (700−500 Ma) occur all along the CTSZ.

2.3.2.1 Salem-Mohanur corridor

A detailed structural analysis of the N−S corridor between Salem and Mohanur in the eastern sector of the CSZ reveals insights into the structural history of the rocks preserving all stages of deformation during a protracted duration between Neoarchean and Neoproterozoic periods (Fig. 2.11). While Salem is located along the E−W trending northern boundary shear zone (SASZ), Mohanur is situated just south of the southern boundary shear zone (CTSZ). While the magnesite mineralization, 1 km away north of Salem, is associated with less-deformed mafic−ultramafic rocks along the SASZ, highly deformed mafic−ultramafic rocks with thin magnesite veins occur at the southern boundary shear zone in association with Neoproterozoic granitoids.

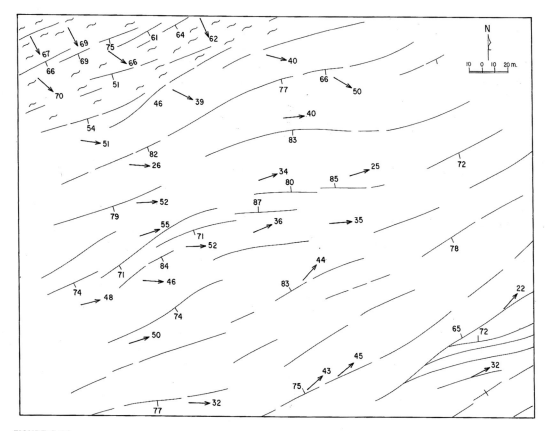

FIGURE 2.10

Belur outcrop map showing both steep and gentle stretching lineations and dextral sense of movement. Notice the presence of steep lineations in phyllonites in NW part of the map.

Modified after Chetty, T.R.K. & Bhaskar Rao, Y.J. (1998). Behaviour of stretching lineations in the Salem-Attur shear belt, southern Granulite Terrane, South India. Journal of Geological Society of India, 52, 443–448.

The corridor is transected by a series of near E−W trending shear zones that separate the fold dominated domains from the high-strain domains (see Fig. 2.11). These shear zones obtain sigmoidal geometry regionally linked to both the boundary shear zones. The earliest recognizable structure is E−W trending isoclinal folds, well preserved in low strain domains, which have been refolded with predominant eastward plunges (D1). These plunges have been rotated parallel to the development of sigmoidal shear zones during the D2 transpressional deformation. As a result, the fold plunges and the stretching lineations lie subparallel to each other. These E−W fold structures continue and merge with the N−S trending fold structures that occur to the east of Sankaridurg-Tiruchengodu granitoids. The N−S trending fold structures are in continuation with the geometry of the Sittampundi anorthosite body. The F2 folds, southwest of Namakkal, vary from tight isoclinal to broad open folds with gentle to moderate plunges to east with a few plunges to

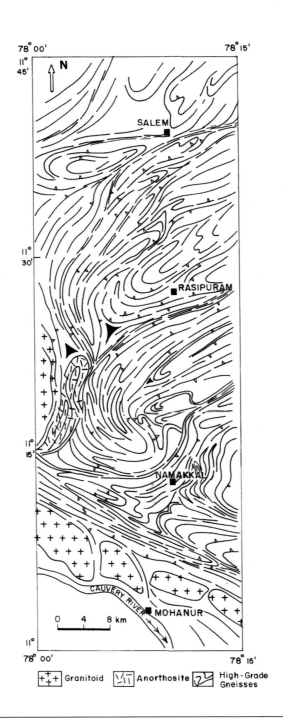

FIGURE 2.11

Geological and structural map of N−S trending Salem-Mohanur corridor across the CSZ.

the west. Several outcrops displaying meter to km scale sheath fold geometry are also well pre-served in this corridor. For instance, the Kanjamalai sheath fold structure (Fig. 2.12) occurs to the west of Salem (10 × 4 km) bound by shear zones on all sides (Mohanty & Chetty, 2014). The detachment zone at Kanjamalai Hills is characterized by a complex variety of fold styles with the predominance of tight isoclinal folds with varied plunge directions, limb rotations, and the hinge-line variations often leading to lift-off fold like geometries and deformed sheath folds (Fig. 2.13).

The central part of the corridor comprises dominantly BIFs, mafic dykes, and two-pyroxene granulites. These pyroxene granulites represent structural markers to define fold patterns. The BIFs also show the tight isoclinal fold structures in association with pyroxene granulites. Mafic dykes show a width of 100 - 200 m and extend for a strike length of a few km that occur mostly in the proximity of BIFs, pyroxene granulites, and particularly in the hinge regions of major fold

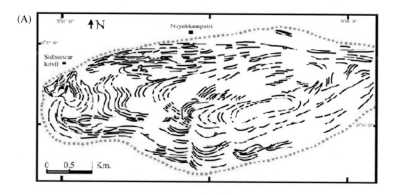

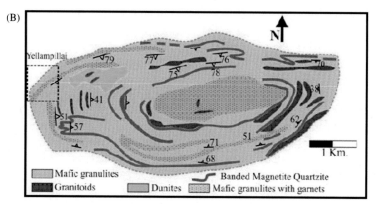

FIGURE 2.12

(A) Structural interpretation of the Google image of Kanjamalai Hills, (B) Broad lithological map of Kanjamalai Hills.

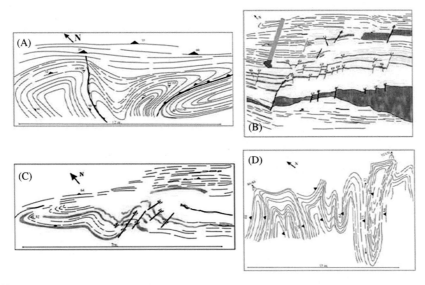

FIGURE 2.13

Field sketches in the detachment zone in the north western part of Kanjamalai hills showing a wide range of complex structures: (A) box-type folds, (B) small-scale faults with both sinistral and dextral movements within the lower limb of a fold, (C) tight isoclinal folds with curvilinear hinges, and (D) disharmonic folds.

After Mohanty, D.P. & Chetty, T.R.K. (2014). Possible detachment zone in Precambrian rocks of Kanjamalai Hills, Cauvery Suture Zone, Southern India: Implications to accretionary tectonics. Journal of Asian Earth Sciences, 88, 50–61.

structures. Many of the fold hinges are well exposed around Namakkal with varying fold geometries of noncoaxial deformation (see Fig. 2.11). Rotation of fold hinges progressively toward the transport direction resulted in the development of curvilinear sheath folds. These folds reflect a sequential development of centimeter-to-kilometer scale curvilinear folds often displaying three-dimensional exposures.

The interpreted structural cross section along the Salem-Mohanur corridor (Fig. 2.14) shows opposite senses of kinematic history: top to north in the northern part and top to south in the southern part. While the former is related to ∼2.5 Ga old north-verging frontal thrusts and associated fold structures, the later is described as a part of the back thrust system comprising major mafic−ultramafic complexes. Interestingly, many UHT assemblages are also reported along the southern margin of the CSZ mostly around Mohanur (Tsunogae & Santosh, 2007). The distribution of high pressure granulites within the CSZ are characterized by retrogression related to the transpressional tectonics of the CSZ. Sheath folds, common in all rock types, were developed at several places but predominantly in the vicinity of high-strain zones. They vary in scale size from cm to km, and the shapes range from a symmetrical to an asymmetrical nature. The presence of sheath folds is interpreted as the evidence of progressive shear during thrusting. When the size is larger, the symmetry is dominant as evidenced from the mega sheath fold structure preserved in Mahadevi hills (Fig. 2.15) (Chetty et al., 2012a). Such large-scale sheath fold structures are described from many HP metamorphic regions that are genetically connected to either subduction or exhumation

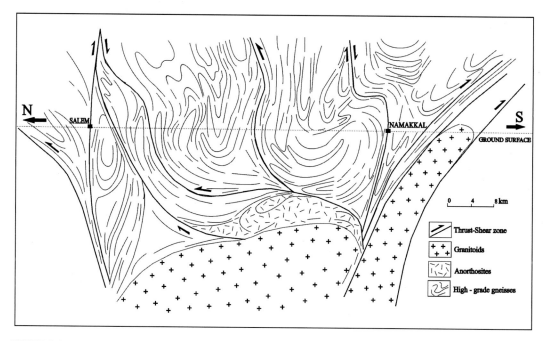

FIGURE 2.14

An interpreted structural cross section along the Salem—Mohanur corridor across the CSZ showing divergent kinematic senses.

(Stanek, Maresch, Grafe, Grevel, & Baumann, 2006). The sheath folds in the region might have developed during Neoarchean as well as Neoproterozoic periods. All these structures described above can be attributed to thrust tectonics possibly related to southward subduction.

2.3.2.2 Namakkal-Mohanur section

The Namakkal-Mohanur section forms the southern part of Salem-Mohanur corridor, where a ~6 km long-rail cutting section between Namakkal and Mohanur is exposed. Detailed geological and structural mapping along the section provides a natural geological cross section perpendicular to the trend of structural fabrics for closer examination (Venkatasivappa, 2014; unpublished thesis). The section can be divided into six zones based on distinct lithological assemblages separated by thrust/shear zones (Fig. 2.16). Each zone displays a characteristic lithological unit and is classified from north to south as follows: hornblende gneiss—pegmatite association, felsic—mafic—ultramafic rock association (dominance of peridotites, pyroxenites, metagabbros, and gneisses), charnockite—ultramafic rock association, mafic—ultramafic rock association (dominance of metagabbros, amphibolites), mafic—felsic rock association (amphibolites and gneisses) and finally Neoproterozoic granitoids in the south. The rocks trend WNW—ESE and are isoclinally folded with gentle to moderate dipping axial planes. The migmatitic rocks show intense folding associated with imbricate structures in some of the zones (Fig. 2.17), while tectonic mélanges are recorded,

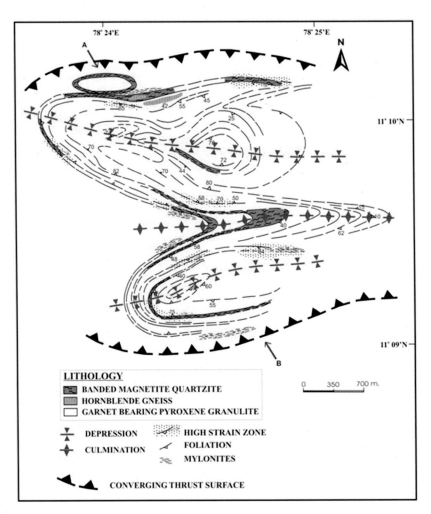

FIGURE 2.15

Mega-sheath fold structure displayed by Mahadevi hills.

After Chetty T.R.K., Yellappa, T., Mohanty, D.P., Nagesh, P., Sivappa, V.V., Santosh, M., & Tsunogae T. (2012a). Mega sheath fold of the mahadevi hills, Cauvery suture zone, Southern India: Implication for accretionary tectonics. Journal of Geological Society of India, 80, 747–758.

often in the matrix of serpentinite and in deformed mafic–ultramafic rocks in some others (Fig. 2.18). Tight isoclinal folds and recumbent fold structures are dominant in amphibolites and mafic–ultramafic associations with intrusive veins of plagiogranites and pegmatites (Fig. 2.19). The fold plunges vary from very gentle to moderate values (20–40 degrees) dominantly to the east. The foliations exhibit very gentle dips to north (10–20 degrees) at many places indicating

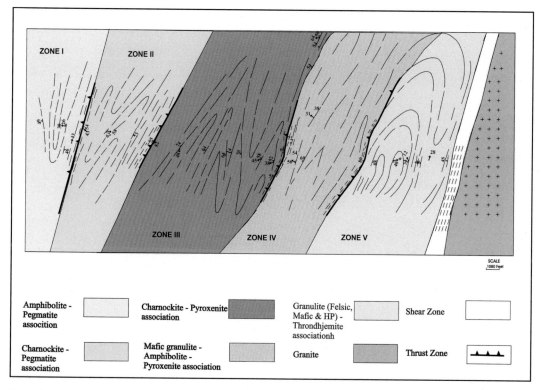

FIGURE 2.16

Geology of rail cutting section between Namakkal and Mohanur.

Modified after Yellappa, T., Venkatasivappa, V., Koizumi, T., Chetty, T.R.K., Santosh, M. & Tsunogae, T. (2014). The mafic-ultramafic complex of Aniyapuram, Cauvery Suture Zone, Southern India: Petrological and geochemical constraints for Neoarchean suprasubduction zone tectonics. Journal of Asian Earth Sciences, 95, 81–98.

that the rocks witnessed south-verging recumbent folding in the early stages of deformation. The metamorphic gneissic foliation (S1) is superimposed by mylonitic fabrics (S2) in the proximity of shear zones. The structural section shows a series of north dipping thrust/shear zones and distinct fold styles separating distinct lithological assemblages (Fig. 2.20).

Field relations, lithological assemblages, petrographic investigations, and geochemical characteristics of mafic—ultramafic complexes indicate that they may represent a possible Neoarchean ophiolite suite (Yellappa et al., 2014). The petrological and geochemical characteristics reveal that the magmas originated from island arc settings associated with suprasubduction affinity and their source magmas might have been derived from spinel—peridotite source with a relatively higher degree of mantle melting. The structural geometry of the complex also reveals south-verging thrust/ shear zones associated with small-scale thrust duplexes forming a major constituent of south-verging back thrust system of the crustal-scale "flower structure."

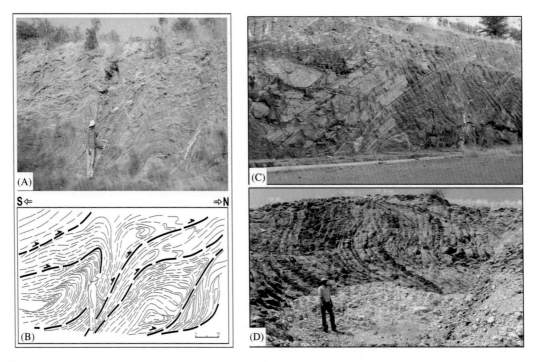

FIGURE 2.17

Field photographs showing: (A) folds and duplex structures in migmatitic gneisses (B) tracing of structural features from the field photograph displaying sheath folds and imbricate structures, (C) tight−isoclinal folds in amphibolites (width of the photograph is ∼8 m), and (D) duplex structures and pegmatite intrusions.

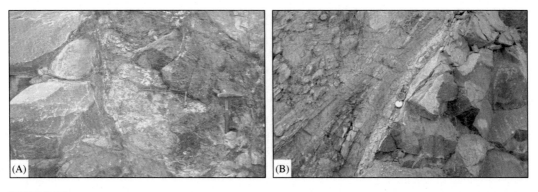

FIGURE 2.18

Field photographs showing: (A) tectonic mélange structures in mafic−ultramafic rocks of fragmented isotropic gabbros and deformed amphibolites, (B) matrix of highly sheared serpentinite with ultramafic clasts of different sizes and the adjacent isotropic gabbros.

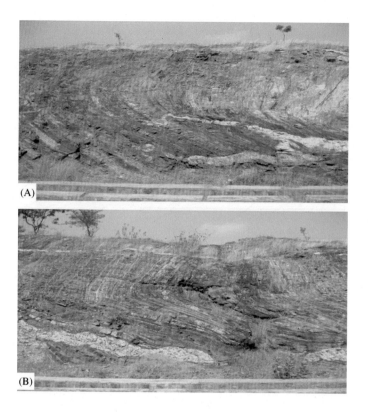

FIGURE 2.19

Field photographs (A) and (B) showing recumbent nature of folding in mafic−ultramafic rocks invaded by plagiogranites and pegmatites.

2.3.2.3 Mahadevi-Manamedu corridor

The Manamedu-Mahadevi corridor, located in the southeastern part of the CSZ, represents an important N−S sector across the major back thrust system and hosts ophiolitic complexes of Devanur and Manamedu. Detailed structural mapping reveals a series of strike parallel ridges with moderate heights and widths varying from a few meters to a few hundreds of meters (Fig. 2.21). The southern margin of the corridor is bound by the E−W trending CTSZ, which hosts elongated and discontinuous stretched Neoproterozic granite plutons surrounded by migmatites. The plutons exhibit high-strain fabrics at the margins and massive nature in the central parts.

The major rock units in the corridor include garnet-bearing pyroxene granulites, mafic−ultrama-fic associations, pyroxenites, chert−magnetite horizons (BIF), amphibolites, metacarbonates, and charnockites, all of which were folded and sheared. The primary foliation (So) is rarely preserved in hinge zones, while the metamorphic gneissosity (S1) is well developed and pervasive throughout. The lithological boundaries are mostly tectonised displaying mylonitic fabrics and associated dex-tral kinematics. The garnet bearing pyroxene granulites are intermixed with bands of meta-chert that often give rise to zebra-like structures in the rocks in some of the ridges. The pyroxenites,

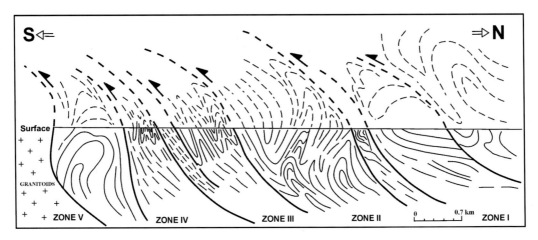

FIGURE 2.20

Structural cross-section between Namakkal and Mohanur showing intense folding and south-verging imbricate thrusts in mafic—ultramafic rocks.

amphibolites, and garnet-rich leucosomes (possible HP rock assemblages), occur as lenses/boudins within the matrix of garnet-bearing pyroxene granulites. These boudins are stretched extending up to 2—3 m with an aspect ratio of 1:20. However, it can be visualized that the fold hinges lie subparallel to the elongation of these boudin structures.

The Manamedu Ophiolite Complex (MOC) occurs at the southern margin, while the Devanur Ophiolite Complex (DOC) is situated in the central part of the Mahadevi-Manamedu corridor. The field observations and lithological assemblages led to the inference that both the complexes can be regarded as a dismembered complete ophiolite sequence and are originated from an island arc origin of suprasubduction zone setting. All the rock types exhibit gentle to moderately dipping foliations and structurally overlie older hornblende bearing migmatitic gneisses probably of the Archean basement. The details of these complexes would be described later.

A N—S structural cross section along the Mahadevi—Manamedu corridor (Fig. 2.22) reveals a series of E—W trending isoclinally folded sequence with gentle plunges on either side. The thrust zones are marked by well-developed down dipping stretching lineations suggesting that the rocks were tectonically transported with top to the south. The chert—magnetite horizons seem to represent tectonic markers and a series of south-verging thrust—imbricate structures could be identified displaying a stack of duplex structures reflected by the presence of outcrop scale south-verging duplex structures in the field. These observations indicate that the south-verging thrusts and associated duplex structures form a part of regional back thrust system in a crustal-scale "flower structure" described from the CSZ.

2.3.2.4 Regional synthesis

The complex deformational history in the eastern sector of the CSZ can be explained in terms of superposition essentially of two finite strain patterns referred to here as D1 and D2. The regional map pattern with reference to foliation trajectories is a reflection of predominantly D2-strain that shows considerable partitioning. The D2 shear zones are reflected in the form sigmoidal shear

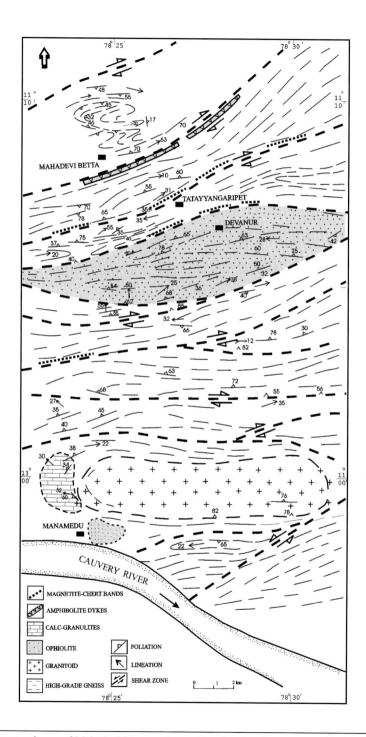

FIGURE 2.21

Geological and structural map of Mahadevi-Manamedu corridor showing the presence of Devanur and Manamedu ophiolite complexes, shear zones, and north dipping structural fabrics.

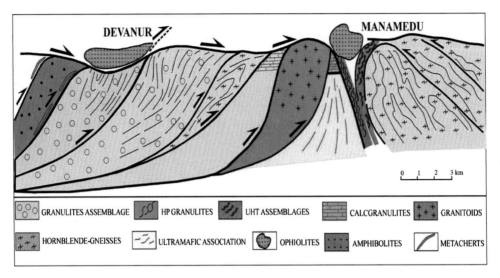

FIGURE 2.22

Structural cross-section along Mahadevi-Manamedu corridor showing ophiolite complexes and associated thrust−imbricate structures.

zones connecting the boundary shear zones of SASZ and CTSZ, which delineate the fold domi-nated domains characterized by low D2-strain. The boundary shear zones SASZ and CTSZ can be interpreted as complementary structures of the same shear system converging into a major crustal decollement at depth (Chetty et al., 2003). In a regional map view of the eastern part of the CSZ, the sigmoidal shear belts linking the two boundary shear zones presents duplex-like structures, while in a sectional view they occur as imbricate thrusts in the form of a positive "flower structure" (Fig. 2.23).

2.3.3 THE WESTERN SECTOR

The western sector of the CSZ predominantly consists of deformed and variably retrograded Neoarchean charnockitic gneisses associated with biotite- and hornblende-bearing migmatite gneisses (also known as the Bhavani gneisses). Supracrustal rocks (the Satyamangalam group) that include calc−silicate marbles, metapelites, quartzites, and metabasic rocks (amphibolites and mafic granulites) are common. These are associated with abundant sheets of granitoids. The striking fea-ture in the region is the presence of two layered complexes at Sittampundi comprising meta anorthosites and metagabbros. Sm−Nd whole rock isochron ages of c. 2.94 Ga for the layered anorthosites (Bhaskar Rao et al., 1996) and a similar zircon U−Pb age for a correlated unit (Ghosh et al., 2004) indicates that the predominant supracrustal gneiss association of the CSZ is at least c. 3.0 Ga old. However, zircon U−Pb geochronology and Hf isotopes of Sittampundi anorthosites suggest Neoarchean suprasubduction zone arc magmatism (Ram Mohan et al., 2012).

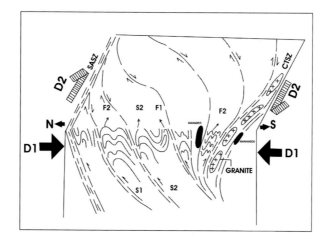

FIGURE 2.23

A cartoon from the eastern sector of the CSZ depicting the duplex structures in map view and positive "flower structure" in sectional view.

Adapted from Chetty, T.R.K. & Bhaskar Rao, Y.J. (2006a). Strain pattern and deformational history in the eastern part of the Cauvery shear zone, southern India. Journal of Asian Earth Sciences, 28, 46–54.

2.3.3.1 Kasaragod-Mercara shear zone

A new major shear zone was recognized in the northwestern part of the CSZ (Fig. 2.24). The shear zone extends from Kasaragod in the west coast, wrapping around the northern contact of the Coorg massif, and passing through Mercara, which finally converges with the Moyar shear zone in the east (Chetty et al., 2012b). This curvilinear shear zone is termed here as Kasaragod-Mercara shear zone (KMSZ) with a strike length of more than 100 km and a width of 20–30 km exhibiting dextral displacements. The foliations in the KMSZ have steep dips (70–80 degrees), and the plunges are mostly to the southeast, subparallel to the trend of the shear zone. The important rock types include high-grade gneissic rocks with variable amounts of biotite and hornblende. Charnockites and kyanite–mica schists occur as elongated and conformable enclaves showing dextral sense of movements. Syenite and granite plutons also occur as intrusions along the KMSZ. The KMSZ is also marked by steep gravity gradients, similar to those of other regions of craton–mobile belt interface in the southern Indian shield. The Coorg block, bound by the Moyar and Mercara shear zones in the north western part of the CSZ, has been speculated as the Mesoarchean exotic block that has escaped the other major thermal events witnessed by the other crustal blocks (Santosh et al., 2014a). The Nilgiri Block, occurring just south of Coorg block and separated by the Moyar shear zone, was built primarily through the late Archean arc magmatism in a convergent margin setting providing the evidence for continental growth through subduction–accretion processes during the Neoarchean (Samuel, Santosh, Shuwen Liu, Wei Wang, & Sajeev, 2014). It was also suggested that the arc magmatism was followed by early Paleoproterozoic high temperature and HP granulite facies metamorphism due to the crustal thickening and suturing of the Nilgiri Block onto the Dharwar Craton.

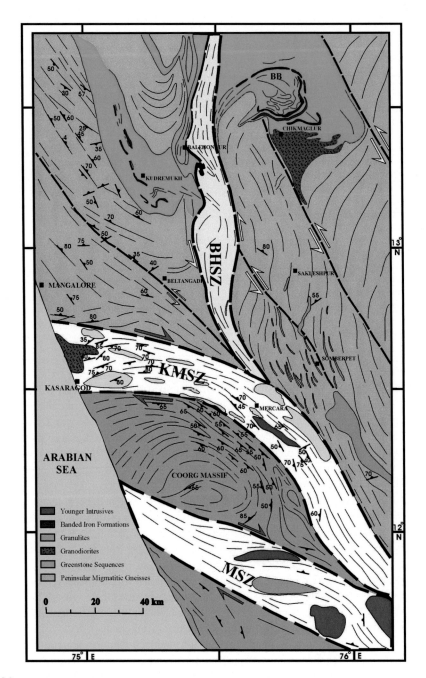

FIGURE 2.24

Map showing the structural architecture around Coorg granulite massif and the adjoining region displaying major shear zones. *BHSZ*, Balehonnur Shear Zone; *KMSZ*, Kasaragod-Mercara Shear Zone; *MSZ*, Moyar Shear Zone; *BB*, Bababudan Hills.

Adapted from Chetty et al. (2012b).

2.3.3.2 Crustal- scale "flower structure"

A north—south corridor between Bhavani and Palani in the western sector of the CSZ reveals that the D2-strain is partitioned into a set of major east-west—trending shear zones of varying width and geometry. Broadly, the shear zones present an anastomosing pattern with local convergence or divergence with near east—west trends with a broad tendency to extend towards northwest in the west and to the east extending to the eastern part of the CSZ. The map of foliation trajectories (Fig. 2.25) reveals the existence of shear zones representing parallel linear belts of high-strain delineating domains of variable but relatively low strain. Several east-westtrending subparallel shear zones were mapped in the western part of the CSZ. They include: (1) MBSZ, marking the northern boundary of the CSZ, continuing through SASZ; and (2) CNSZ, extending through CTSZ, broadly representing the southern boundary. The other important shear zones in the region include: (1) Dharapuram shear zone (DSZ), (2) Devattur-Kallimandayam shear zone (DKSZ), and (3) Karur-Oddanchatram shear zone (KOSZ). The strain gradient within these shear belts can broadly be related to the trend as well as width of shear zones. For instance, the width of the CNSZ varies from ~ 1 km to over 10 km in the central part. The MBSZ and CNSZ show near east—west trends and are subparallel over long stretches, while the DSZ, DKSZ, and KOSZ show NE—SW trends and merge with the CNSZ.

The consistent dextral kinematics and the convergence and divergence of branches indicate that all the shear zones are genetically and kinematically interconnected. The foliation fabrics along the MBSZ show steep dips to south, while the foliations in CNSZ, DSZ, and DKSZ exhibit moderate dips to the north (40—60 degrees). The regional disposition, structural geometry, consistent dextral kinematics, complex behavior of foliation trajectories and stretching lineations, heterogeneous strain patterns and the contemporaneity of mylonitic fabrics (750—500 Ma), and the crustal architecture of the CSZ are interpreted as a crustal-scale positive "flower structure" (Fig. 2.26), typical of transpressional tectonics in a convergent regime (Chetty & Bhaskar Rao, 2006b). The geological and structural characteristics of the corridor and associated shear zones and their geometry are summarized in a schematic block diagram representing a north—south cross section (Fig. 2.27) that divides the corridor into northern, central, and southern domains of contrasting structural and geological characteristics. MBSZ represents a north-verging frontal thrust belt with steep southerly dips, while the other shear zones dip northerly at gentle to moderate angles representing the complementary back thrust system.

2.3.3.3 Perundurai dome

A "gneiss dome" structure, described as Perundurai dome, was recognized between the MBSZ and the CNSZ (Chetty & Bhaskar Rao, 2006c). It is an elongate domal structure with the penetrative amphibolite facies foliation showing shallow dips in the center gradually steepening towards margins (Fig. 2.28). Quartzo—feldspathic migmatitic gneisses predominate in the core, while narrow bands of concordant high-grade supracrustal gneisses and amphibolites are increasingly abundant towards the periphery. Units of charnockite gneiss showing a variable degree of retrogression are also common along the peripheral zones. The foliated migmatite gneiss in the core is also associated with the shallow dipping sheets of stratoid granite—pegmatite intrusions. Commonly, interlayered and boudinaged amphibolite bands concordant with the gneisses are deformed and migmatized together. Variation in the intensity of strain is evident in contortions, mylonitic fabrics,

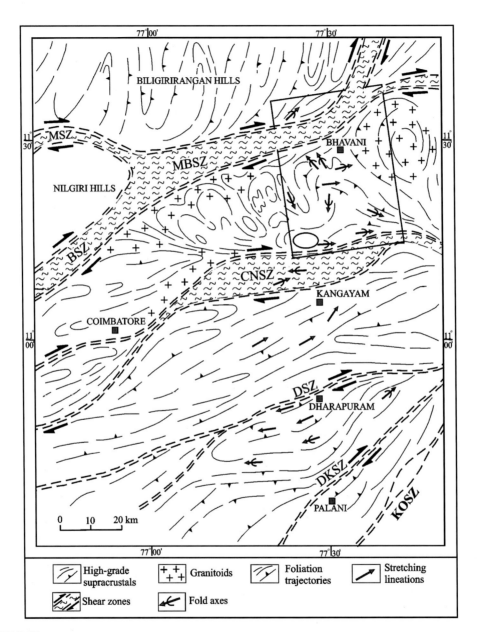

FIGURE 2.25

Geological and structural map of N−S trending Bhavani-Palani corridor across the CSZ.

Adapted from Chetty, T.R.K. & Bhaskar Rao, Y.J. (2006b). The Cauvery Shear Zone, Southern Granulite Terrain, India: A crustal-scale flower structure. Gondwana Research, 10, 77−85.

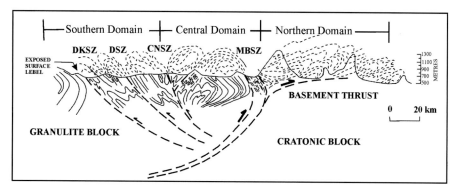

FIGURE 2.26

A N—S structural cross section between Bhavani and Palani in the central part of the CSZ.

Modified after Chetty, T.R.K., Bhaskar Rao, Y.J., & Narayana, B.L. (2003). A structural cross-section along Krishnagiri-Palani corridor, southern granulite terrain India. In: M. Ramakrishnan (Ed.), Tectonics of Southern granulite terrain, Kuppam-Palani Geotransect. Journal of Geological Society of India, Memoir, 50, *255–277.*

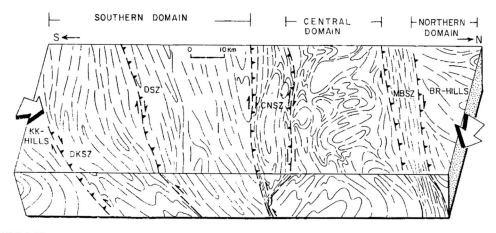

FIGURE 2.27

A schematic block diagram showing the general disposition and geometry of major shear zones along Bhavani and Palani corridor, depicting dextral transpressional tectonic regime.

Adapted from Chetty, T.R.K., Bhaskar Rao, Y.J., & Narayana, B.L. (2003). A structural cross-section along Krishnagiri-Palani corridor, southern granulite terrain India. In: M. Ramakrishnan (Ed.), Tectonics of Southern granulite terrain, Kuppam-Palani Geotransect. Journal of Geological Society of India, Memoir, 50, *255–277.*

and degree of melting. A striking feature is the presence of a regionally folded two-pyroxene granulite—amphibolite band at the western margin of the dome that serves as a marker horizon. A heterogeneous component of noncoaxial strain is indicated by widespread, though not pervasive, asymmetrical structural elements.

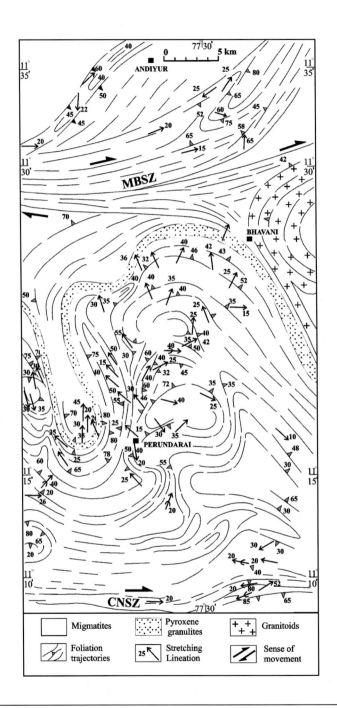

FIGURE 2.28

Map of Perundurai gneiss dome structure in the axial part of the CSZ displaying concentric foliation fabrics and radial stretching lineations.

Adapted from Chetty, T.R.K. & Bhaskar Rao, Y.J. (2006c). Constrictive deformation in transpressional regime: field evidence from the Cauvery Shear Zone, Southern Granulite Terrain, India. Journal of Structural Geology, 28, *713–720.*

The regional domal structure is well defined in the map of foliation trajectories (see Fig. 2.28) and constitutes a few closed forms of large-scale folds in the form of dome and basin structures. Well-developed mineral stretching lineations are defined by hornblende, biotite, quartz, and feldspar along the foliation planes. They display a down-dip orientation with moderate plunges (20−45 degrees) and show a remarkable radial pattern implying an upright conical geometry. Such a radial pattern in the field may be related to the erosional effect after the emplacement of migmatites preserved at the base of the nappe. The extruding−spreading nappes with similar strain patterns were described in Helvitic nappes (Ramsay, 1981, p. 302). The analog experimental studies of extrusion−spreading from a closing channel favor the radial displacement in the extruding−spreading zone; folds developed in such scenario would be noncylindrical and strongly asymmetric, if not sheath like (Gilbert & Merlie, 1987).

The dome is also characterized by a heterogeneous component of noncoaxial strain reflected in the form of widespread asymmetrical structural elements that include domes and basins, curvilinear, and hair-pin bends of hinge lines, ameboid forms, interfering folds, and associated planar and linear fabrics, which are typical of constrictive deformation (Chetty & Bhaskar Rao, 2006c). The bounding shear zones are marked by intense flattening strains together with the occurrence of stratoid Neoproterozoic granite intrusions and stromatic migmatites, while the interlying region exposes the Perundurai dome essentially characterized by inhomogeneous and extensive migmatites. Pods of relict, HP granulite facies assemblages are scattered and well preserved proximal to the bounding shear zones despite the widespread effects of retrogression, where pressures up to c. 11.8 kbar were obtained for garnet bearing granulites (Bhaskar Rao et al., 1996). The steep near isothermal decompressive P−T trajectories noted commonly for these assemblages indicate a rapid exhumation history related to transpressional deformation. This also leads to the inference that the spatial and temporal relationship between transpressive deformation, deep crustal melting, ascent of granite magmas and migmatization in the western part of the CSZ are akin to the process of channel flow demonstrated recently along the Main Central Thrust in the Himalayan orogen as well as the Kaoko belt, Namibia (Goscombe, Gray, Armstrong, Foster, & Vog, 2005).

2.3.4 GEOPHYSICAL SIGNATURES

Multiparametric geophysical measurements were made along a N−S transect across the western sector of the CSZ. Deep seismic studies (DSS) delineated four layered velocity structure with 6.1, 6.5, 6.0, and 7.1 km/s suggesting the presence of 10−15 km thick low velocity layer (LVL) at a depth of 20 km (Reddy et al., 2003). A prominent Moho uplift of about 4 km within the CSZ compared to the adjacent regions has been established. The Moho uplift is restricted to the boundary shear zones, while the presence of thick LVL is geologically correlatable with the extrusion of channel flow in the form of intense migmatites, granite plutons, and associated structures of gneiss dome. A significant gravity high is observed in both the residual and the regional components of the Bouguer anomaly across the CSZ indicating two source horizons, one at the shallow and the other at the deeper levels. The high gravity anomalies suggest the existence of a 10 km thick high density (2.89 g/cm^3) crustal body at a depth of 10−15 km and the upliftment of Moho by about

5 km (from 43 km to 38 km) indicating crust−mantle interaction (Singh, Mishra, Vijayakumar, & Vyaghreswara Rao, 2003). They also speculated Neoarchean continental collision and consequent delamination and asthenosphere upwelling, which were responsible for the current architecture of the CSZ.

The Magnetotelluric studies delineate the shear zone bound block structure on either side of the CSZ and the adjoining regions with contrasting electrical resistivity character (Harinarayana et al., 2003). The shear zones are marked by high conductivity zones extending to deeper levels. The high conductivity layer at mid-to-lower crustal levels can be correlated with LVL derived from DSS. The model shows a high resistive upper crust ($>20,000$ ohm.m) extending all along the profile to a depth of 15 km except for the region in the vicinity of shear/suture zones. The mid-crust is characterized by more than 10,000 ohm.m resistivity, while the lower crust together with upper mantle gave rise to a resistivity of 500−3000 ohm.m in general. The MT results in conjunction with the gravity model indicate a southward dipping low resistivity zone (1−100 ohm.m) and high density region at a depth range of 15−45 km (Naganjaneyulu & Santosh, 2010). The MT studies revealed a significant feature in the form of high conducting body in the deeper parts of the CSZ. These geophysical signatures can be attributed to the presence of the Perundurai gneiss domal structure associated with intense melting and migmatization, channel flow, and extrusion tectonics described earlier.

With a view to assess the regional synthesis of all geophysical signatures vis-à-vis the geological cross section across the CSZ, all the data sets are brought together onto the same section (Fig. 2.29). Magnetic and gravity signatures along with the results of DSS are stacked over the regional structural cross section to examine the correlations. Distinct magnetic anomalies, gravity high over the central part of the CSZ reflect the nature and spatial disposition with well-defined boundary shear zones. The DSS results show excellent spatial correlation with all the shear zones: the northern boundary shear zone of MBSZ comprising the north-verging frontal thrust and DSZ, DKSZ, and CNSZ the complimentary shear zones of the back thrust system. In general, the CSZ is characterized by divergent seismic reflection fabrics indicating heterogeneous seismic and resistive crustal structure; undulated LVL with low resistive mid-lower crust, the presence of high conducting features, broad gravity high (4 m/gal) and significant magnetic high anomaly; upwarp of Moho by about 5 km. The multiparametric geophysical signatures such as divergent seismic reflection fabrics and crustal variations in crustal velocity structure, bipolar Bouguer gravity, and magnetic anomalies, and high conductivity together with the presence of mantle−gneiss dome structures and the disposition of shear zones are related to southward subduction and collisional processes.

2.3.5 DUPLEX STRUCTURES

Duplex structures in orogens are known to be the hallmarks of the existence of ocean plate stratigraphy and the suture zones. A number of outcrop scale duplex structures are well preserved in different lithologies all over the CSZ (Chetty, Yellappa, & Santosh, 2016). They are well preserved in ortho-, and paragneisses, where the horses are defined by charnockitic rocks while the splays are marked by paragneisses. Intense structural duplication of quartz rich layers is well displayed with the splays constituting micaceous mineral assemblages. South-verging structures with small folds,

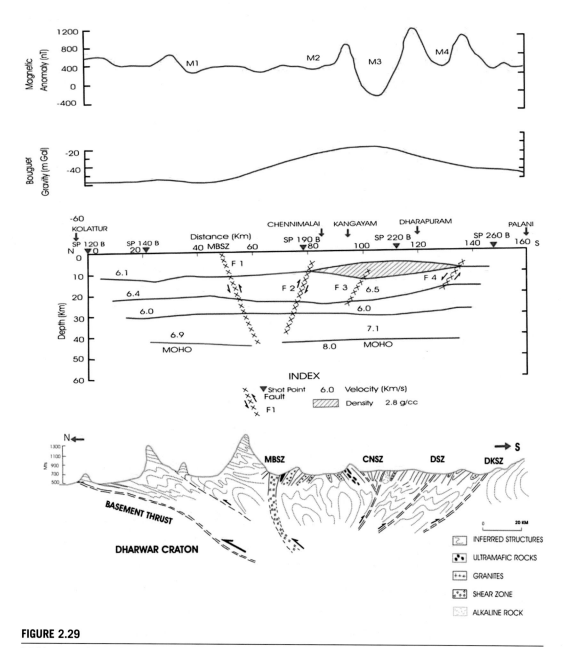

FIGURE 2.29

Multiparametric geophysical signatures along a profile across the CSZ showing close correlation with the "flower structure."

thrust faults, complex disharmonic folds and or/imbrications reflecting extreme structural duplication exhibit well defined duplex structure. Duplexes are also recorded in magnetite−chert bands with the iron mineral dominating the splay faults and the horses are composed of thickened cherty bands. Well-developed multistoried duplex structures are associated with intense structural duplication in calc−silicate rocks in the form of both ramps and flats in the vicinity of the MBSZ (Fig. 2.30). The abundant presence of duplex structures in mafic−ultramafic complexes with varying size and the geometry of horses suggest that the rocks of the CSZ represent typical "ocean plate stratigraphy" sequence.

A typical example of duplex structures is well documented from an east-west−trending geomorphic ridge near Chennimalai consisting of complex structures and mixed lithologies. The important rock types in the ridge include charnockites, amphibolites, BIFs, calc granulites, and thin layers of metasediments and metaigneous rocks in the form of lenses and bands. Thin layers of mafic and felsic material are intensely folded and transposed. Calc granulites occur in the form of bands and boulders on top of the ridge, while the other rock units apparently occur in the form of two horizons. The notable feature is the presence of well developed imbricate duplex structures related to south-verging thrust structures (Fig. 2.31). The duplex structures are well defined not only as interformational but also as intraformational. Small-scale duplex structures within a major duplex structure show multistoried duplex structures. Highly asymmetric nature of folding is a common feature within the duplex bound horses. All the duplex structures display hindward dipping geometry related to thrust zones. The lithologies and the structural fabrics resemble typical of ocean plate stratigraphic sequence.

FIGURE 2.30

Field photograph showing multistoried duplex structures in calc−silicate rocks.

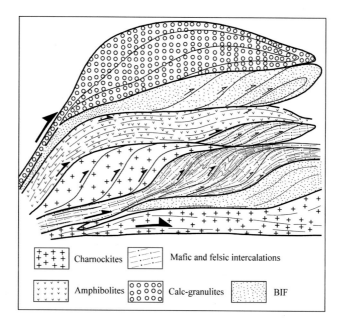

FIGURE 2.31

A schematic field sketch displaying typical duplex structures in a segment of the Chennimalai ridge.

2.3.6 OPHIOLITES

Many new reports of Precambrian ophiolitic rocks from the CSZ have come to light in recent years. They include Manamedu, Devanur, Agali Hill ophiolite complexes, and their characteristics are briefly described here. The Manamedu ophiolite complex (MOC), occurring ∼40 km east of Namakkal, in the south–eastern part of the CSZ (see Fig. 2.21), constitutes important lithologies such as highly altered ultramafic rocks/dunites, pyroxenites, gabbros, anorthosites, amphibolites, mafic dykes (dolerite), hornblendites, plagiogranites, calcareous rocks, and ferruginous cherts, extending over 2.5-km^2 area (Yellappa, Chetty, Tsunogae, & Santosh, 2010). The lithologies are tectonically intercalated resulting in typical tectonic melange structures comprising irregular, random and complexly scattered blocks of different rock types. The geometry of fold forms, foliation trajectories, and the field characteristics suggest that the rocks of the MOC are folded into tight isoclinal folds with north–south trending variable hinge lines pointing to possible sheath fold geometry (Fig. 2.32) (Chetty et al., 2011). Petrological and geochemical characteristics suggest that these rocks represent the remnants of oceanic crust, developed at shallow levels from mantle-derived arc magmas probably within a suprasubduction zone tectonic setting. The geochemistry of mafic dykes indicated the island arc signatures of suprasubduction zone settings. The $_{206}$Pb/^{238}U magmatic crystallization ages of zircons from gabbros and plagiogranite yielded 744 ± 11 to 786 ± 7.1 Ma and 737 ± 23 to 782 ± 24 Ma ages, respectively (Santosh, Xiao, Tsunogae, Chetty, & Yellappa, 2012), suggesting that the MOC could represent a part of Mozambique ocean floor evolved during the Neoproterozoic period. Structurally, the MOC overlies the south-verging back thrust system of the crustal-scale "flower structure" model of the CSZ. The MOC may represent the remnants of the

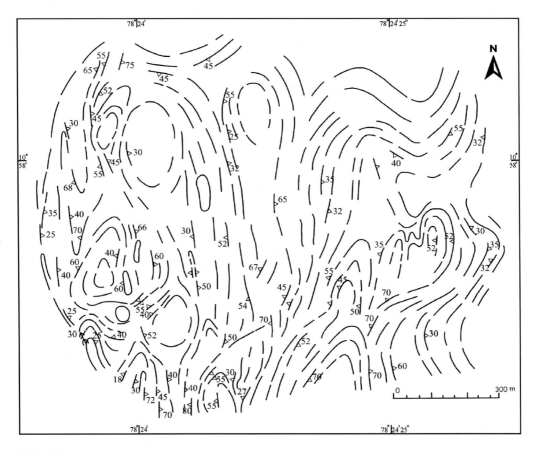

FIGURE 2.32

Structural map and foliation trajectories of Manamedu ophiolite complex.

Adapted from Chetty T.R.K., Yellappa, T., Nagesh, P., Mohanty, D.P., Sivappa, V.V., Santosh, M., & Tsunogae, T. (2011). Structural anatomy of a dismembered ophiolite suite from Gondwana: The Manamedu Complex, Cauvery suture zone, Southern India. Journal of Asian Earth Sciences, 42, 176–190.

Mozambique Ocean crust developed during the Rodinia breakup and which was later destroyed during the Cambrian period at the time of the Gondwana amalgamation.

The Devanur ophiolite complex (DOC), located about 20 km north of MOC, comprises dismembered outcrops along an east–west trending shear zone and represents typical oceanic sequences/ocean plate stratigraphy with mafic–ultramafic components, overlying felsic, and hornblende gneisses (Yellappa et al., 2012). The Devanur complex occurs in the form of a lensoid body and comprises rock types such as pyroxenites, gabbros, anorthosites, actinolite–hornblendites, amphibolite dykes, dolerites, pyroxene granulites, trondhjemites/quartz keratophyres and thin layers of ferruginous cherts with varied dimensions. The DOC extends for over a strike length of ~15 km with a maximum width of <1 km. The field and petrographic studies indicate that these lithologies are highly altered, sheared, and metamorphosed and obducted along shear/thrust planes with in the

CSZ (see Fig. 2.21) The geochemistry of mafic dykes shows basaltic-andesitic—type magmas with tholeiitic to calc—alkaline characteristics suggesting that these rocks were generated with island arc affinities along a suprasubduction zone tectonic setting. They are inferred to have been incorporated within a Neoarchean accretionary belt associated with continental collision. Trondhjemites from DOC yielded U—Pb zircon ages of 2528 ± 61 Ma and 2545 ± 56 Ma (Yellappa et al., 2012). Similar ages have been obtained from magmatic zircons in charnockites, anorthosites and orthogneisses in the adjacent regions in CSZ.

The Agali Hill ophiolite complex (AOC) in the north-western part of the CSZ near Attappadi consist of the metamorphosed equivalents of mafic—ultramafic group of rocks including dunites and pyroxenites with locally cumulate textures; gabbroic rock types including gabbro, gabbronorite, anorthosite; sheeted mafic dykes, amphibolite to metaandesite and plagiogranites/trondhjemites. A thin veneer of ferruginous chert is invariably associated with all the ophiolite complexes (Santosh et al., 2013). They have also reported 2547 ± 17 Ma and 2547 ± 7.4 Ma ages of zircons (U—Pb) of metagabbro and trondhjemites from AOC.

2.3.7 TECTONIC MÉLANGES

A range of mélange structures derived from different rock compositions have been recorded from different parts of the CSZ (Chetty et al., 2016). However, these are mostly recorded from mafic—ultramafic rock associations with some of them established as ophiolite complexes (see Fig. 2.18). A chaotic mixture of sedimentary layers with well developed tectonic fabrics, caught up in magmatic matrix consisting of amphibolites, gabbros, pyroxenites, and magnesite/calcareous horizons. Large rotated asymmetrical porphyroclasts of different origin and age in a fine-grained matrix are together deformed displaying cataclastic structures. The early metamorphic fabrics are completely superimposed by the later formation of cataclasites. Highly deformed and altered serpentinite occur as the matrix that envelopes the relict boulders of other mafic—ultramafic rocks. Gabbros and amphibolites often dominate while the serpentinite was sheared against the isotropic gabbros. Tectonic mélanges of mixed rock fragments of paragneisses, gabbros, pyroxenites, peridotites, and amphibolites are occasionally observed. Metasedimentary and metaigneous rocks are thinly interlayered and interfolded by folded pegmatite and quartz veins. The rock fragment clasts show irregular shapes and margins ranging in size from millimeter to decimeter scale. Well-foliated big clasts of orthogneiss with rectangular shape (2 × 3 m) are enveloped in a fine-grained mafic matrix.

2.3.8 TECTONIC SYNTHESIS

The most striking feature of the CSZ is the presence of a network of shear zones revealing the linear belts of high strain and domains of variable but relatively low strain. Broadly, the shear zones with near east—west trends with a tendency to extend towards northwest in the west and northeast in the east with a broad concavity to south indicating its north-verging, thrust-related tectonics. The NE-SW trending foliations north of the MBSZ were subjected to rotation and deflection parallel to the MBSZ, exhibiting a mega scale S—C fabrics, suggesting a dextral sense of movement. The CSZ represents a fundamental boundary between two geologically contrasting blocks and represents a major E-Woriented Gondwana suture zone with a top to the north direction of transport.

2.3.8.1 Sheath fold structures

A spectrum of the fold geometries varying from subcylindrical to intensely curvilinear sheaths on different scales recorded all over the CSZ depict continuum of fold shapes and geometric relationships suggesting continuous and progressive shearing. The field observations suggest that the folds must have been initiated by buckling, passively amplified and deformed to give rise to observed variety of sheath fold geometries. They exhibit complex three-dimensional shapes, which are probably reflections of original fold patterns, buckling instabilities and mechanical anisotropy. A mega sheath fold structure described as the Mahadevi Sheath Fold (MSF) is spectacularly well exposed with strong curvilinear hinges exposing typical parabolic fold geometry on their limbs rotated to near parallelism. The MSF is also characterized by strong extensional strain exhibiting well developed extension lineations and bound by flat-lying ductile shear zones (Chetty et al., 2012) and may be classified as recumbent sheath fold with peculiar geometry. Recent studies show that the type of curvature accentuation depends greatly on the nature of deformation, i e., simple shear vs. general shear vs. constrictive deformation, and each type of deformation produced characteristic 3D shapes and sheath structures (Alsop & Holdsworth, 2006). Although, these features are established in analog experiments, they are rarely documented in Precambrian orogenic belts. The existence of initial perturbation in multilayered strata explains the asymmetry and complexity of these fold structures. The kind of structural observations documented from the CSZ can be correlated with those seen in large-scale tectonic settings such as subduction zones in other regions.

2.3.8.2 Transpressional tectonics

Multiscale structural observations along the CSZ are suggestive of typical transpressional tectonics. The disposition, regional geometry, dramatic variations in foliation fabrics, behavior of stretching lineations, persistent dextral kinematic indicators on all scales of observation, the apparent contemporaneous development of mylonitic fabrics, and the presence of possible convex upward reverse or thrust faults suggest that the CSZ can be modeled as a crustal-scale "flower structure" typical of transpressional tectonics in a convergent regime akin to modern orogenic belts. These results are consistent with the geophysical interpretations from the CSZ. The expulsion or extrusion of the melts must have been triggered by vertical pressure gradient during tectonic contraction within the "flower structure" of the CSZ. The vertical extrusion, variations in the geometry, and the disposition of different shear zones with consistent dextral strike-slip displacements suggest that the YZ sections of the CSZ are vital in accommodating the shortening in obliquely convergent orogen of the SGT. The central part of the CSZ may constitute a deeply eroded part of orogen scale pop-up structure and that the highland areas within the CSZ could represent uplifted blocks during the transpression.

The CSZ shows several features typical of a deeply eroded transpressional orogen such as: high-grade metamorphism characterized by a clockwise P-T-t trajectory with a steep isothermal decompressive segment, ductile strike-slip shearing, convergence of crustal-scale shear systems at depth reaching the lithospheric mantle, evidence for significant Moho up-warp, heterogeneous strain variation, widespread melting, granite magmatism, and migmatization. Several recent studies have documented the presence of high pressure granulites within the CSZ with pressures in the range of 12−20 kb (Anderson et al., 2012; Saitoh, Tsunogae, Santosh, Chetty, & Horie, 2011; Sajeev, Windley, Connolly, & Kon, 2009) as well as ultrahigh-temperature mineral assemblages (Shimizu, Tsunogae, & Santosh, 2010; Tsunogae & Santosh, 2007).

Recent geophysical data across the CSZ involving seismic refraction and wide-angle reflection, gravity, and magnetotellurics have been interpreted in terms of late Archean collision tectonics and late Proterozoic/Phanerozoic extensional tectonics. The transpressive movements may cease to exist in the decollement zone at the base of the "flower structure," which is consistent with the 7−15 km-thick low velocity zone identified by the seismic data suggesting the presence of low-density ductile material. Further, the upward trend of the velocity-layered structure, south of the CNSZ, may indicate the south verging and flattening nature of the thrusts. The presence of domes and basin structures and many unusual structures in the region of Perundurai Dome are interpreted to be the resultant features of constrictive deformation in a larger transpressional zone. The dramatic variations in foliation trajectories and lineation behavior, vertical extrusion and heterogeneous strain patterns within the CSZ can be correlated with transpressive tectonic models of triclinic symmetry. The complex and systematic variations depend upon the intensity of finite strain, the obliquity of simple shear component, and the nature of kinematic partitioning within the deformation zone.

2.3.8.3 The CSZ: eastern and western sectors

Distinct differences can be observed between the eastern and western sectors of the CSZ. Both the sectors exhibit bounding shear zones that are structurally continuous (Fig. 2.33). While sigmoidal shear zones linking the boundary shear zones are prominent in the east, subparallel east−west trending shear zones are common in the west. Intense migmatization and extensive granitoids are striking features in the western sector, while the presence of mafic dykes and BIFs are common in the east. Gneiss domes and basin structures are dominant in the west; typical sheath fold structures of meter to km scale of varying sizes are documented in the east. The presence of high pressure granulites is a common feature in the west and possible eclogites are reported from the east. Widespread mafic−ultramafic assemblages and gold mineralization are reported from the west, isolated, and dismembered of both Neoarchean and Neoproterozoic ophiolite complexes that are described in the east. All the above features broadly show that the western sector of the CSZ is characterized by deeper structural levels, while relatively shallow structural levels are exposed in the eastern sector.

2.3.8.4 Subduction zone tectonics

The recently described Precambrian ophiolite complexes from within the CSZ provide a unique opportunity to unravel the different stages of a Wilson cycle of the Mozambique Ocean as well as the tectonics associated with the subduction−accretion−collision history during the amalgamation of the Gondwana supercontinent in the end of the Precambrian. Considering the geological characteristics and the geochronological constraints of ophiolitic complexes of the CSZ, at least two major episodes of oceanic crust generation and subduction followed by accretion and collision could be identified during the Neoarchean and Neoproterozoic, the two critical periods of Earth's history (Santosh et al., 2012). Whereas the suprasubduction complex in Devanur represents the remnants of a Neoarchean oceanic crust, the Manamedu-type occurrence belongs to the Neoproterozoic ocean closure. The geochemical characteristics of these ophiolites suggest a common suprasubduction zone setting related to the opening and closure of Mozambique Ocean. All the above studies confirm a Neoarchean-early Paleoproterozoic subduction system at the southern margin of the Dharwar Craton, the remnants of which were incorporated within a chaotic mélange of the Neoproterozoic CSZ.

The rock suite around the Attappadi region, in the western part of the CSZ, were interpreted to represent Neoarchean "Ocean Plate Stratigraphy," and the sequence is associated with arc and

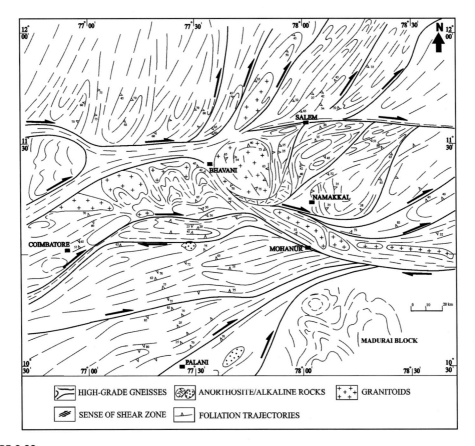

FIGURE 2.33

Regional tectonic framework of CSZ showing structural fabrics, geometry of shear zones and younger intrusions of granitoids.

exhumed subarc mantle material, followed by accreted remnants of suprasubduction-zone–derived ophiolites (Santosh et al., 2013). The presence of HP-granulites, UHT assemblages, alkaline, anorthositic, and granitic magmatism are the hallmarks of the CSZ confirming its nature of an ancient orogenic suture zone.

Banded Iron Formations and MORB-like compositions of mafic granulites of the Kanjamalai hills are related to the ~2.53 Ga subduction and island arc formation followed by accretion of arc and oceanic crust during ~2.48 Ga (Noack, Kleinschrodta, Kirchenbaura, Fonsecab, & Munker, 2013). Recently, Brandt et al. (2014) also reported the existence of Mesoarchean (c. 2850 Ma) to the late-Neoarchean (c. 2500 Ma) pulses of felsic magmatism, followed by HP-granulite facies metamorphism and partial melting of the crust in the earliest Paleoproterozoic (c. 2490–2450 Ma) within the Moyar-Bhavani-Cauvery suture zone. In another study, Praveen et al. (2013) obtained the U−Pb ages of 2567 ± 18 Ma; 2499 ± 19 Ma; 2555 ± 24 Ma; and 2576 ± 64 Ma from felsic volcanic tuffs around Attappadi region. Similar Neoarchean age equivalents were also described from several other locations, e.g., 2530–2540 Ma magmatic and 2470–2480 Ma metamorphic ages of charnockites in the Salem area, northern margin of the CSZ (Clark et al., 2009); 2528 ± 1.7 Ma age of orthogneiss from Kattur, south of Salem, (Ghosh et al., 2004); 2536.1 ± 1.4 Ma and 2532.4 ± 3.7 Ma ages of charnockites and quartzo–feldspathic garnet gneiss and 2443 ± 20 Ma emplacement ages of syntectonic granites around Salem (Sato, Santosh, Chetty, & Hirtata, 2011a), all revealing the dominant Neoarchean magmatism.

It is also pertinent to note the Neoproterozoic events in the CSZ described by many workers. The zircons from the plagiogranites of MOC yielded U–Pb ages of 800 ± 14 Ma (Sato, Santosh, Tsunogae, Chetty, & Hirata, 2011b). The magmatic cores of zircons from the plagiogranites and gabbros from Manamedu Ophiolite Complex yielded 737 ± 23 to 782 ± 24 Ma and 744 ± 11 to 786 ± 7.1 Ma ages (Santosh et al., 2012). The zircons from the Kadavur gabbro–anorthosite complex at the southern margin of CSZ also yielded the $^{207}Pb/_{206}Pb$ ages of 825 ± 17 Ma (Teale et al., 2011). Based on geochemical and isotopic systematics of alkaline rocks from the western part of the MBSZ, a petrogenetic model was invoked involving extensional setting, asthenospheric upwelling, melting of the enriched lithosphere, and interaction of magmas with lower crustal domains comprising subduction related components of various ages (Santosh et al., 2014b). This is further substantiated by the presence of ophiolites in several parts of the CSZ, recently described with a combination of field, petrological, and geochemical characteristics, developed within a suprasubduction zone tectonic setting. The U–Pb zircon geochronology has correlated the ophiolite formation with the Neoproterozoic subduction tectonics and the final collisional metamorphism during the Cambrian.

The understanding of subduction, accretion and collisional history along the CSZ within the SGT, together with a long lived transpressional tectonics, has been well reflected from several recent publications (e.g., Chetty & Santosh, 2013 and the reference therein). Santosh et al. (2009) proposed a plate tectonic model for the Neoproterozoic-Cambrian evolution of southern India involving southward subduction of the oceanic lithosphere of Mozambique Ocean and its final closure during the Cambrian assembly of the Gondwana supercontinent. The Cambrian ages from magmatic and metamorphic zircons from various rock units of the Cauvery Suture Zone reported by several workers are coeval to the amalgamation of the Gondwana supercontinent. However, the recent geochronological studies from different regions far and wide within the CSZ have yielded Neoarchean as well as Neoproterozoic ages, as described in the foregoing indicating multiple subduction events. The 3.0–2.8 Ga association of mafic seafloor relicts (Bhaskar Rao et al., 1996; Ghosh et al., 2004) and the later 2.9–2.5 Ga intrusive TTG rocks north of the CSZ are interpreted as products of Neoarchean–Paleoproterozoic subduction–accretion processes. Santosh et al. (2012) described a multistage evolution of the CSZ and suggested two subduction–accretion events: one at the Neoarchean-Paleoproterozoic boundary, and the other at the Precambrian-Cambrian boundary.

All the known ophiolitic rocks are characterized by a suprasubduction zone tectonic setting and thrust-nappe emplacement history. The subduction accompanied by the closure of a large ocean would develop into a sequence of accretionary complexes, and if a subduction of a mid-oceanic ridge were also involved, high temperatures and HP regional metamorphic belts would also be generated (Santosh & Kusky, 2010). It has been postulated that the Gondwana-forming orogens had consumed the intervening Mozambique Ocean during the Neoproterozoic period.

The seismic tomographic data from southern India might indicate asthenospheric upwelling at shallow regions above 200 km, particularly beneath the CSZ. One of the possible explanations would be a slab break-off and asthenospheric injection following the subduction—collision processes. Field relations, lithological assemblages, petrographic investigations, and geochemical characteristics reveal that the rocks of the CSZ are related to magmas that originated from island arc settings associated with suprasubduction affinity and their source magmas must have been derived from a spinel—peridotite source with relatively higher degree of mantle melting. Based on geochemical and isotopic systematics, a possible petrogenetic model would lead to an asthenospheric upwelling in an extensional setting, a melting of enriched lithosphere, and an intersection of magmas with lower crustal domains with subduction-related components of various ages (Santosh et al., 2014b). The nature of the southward-dipping low resistivity and high density anomaly may suggest a fragment of the subducted oceanic lithosphere, eclogitized and partly exhumed, doming up the sequences above as displayed by Perundurai dome, located in the axial part of CSZ. Considering all the above features, a regional geological and tectonic model of the CSZ (Fig. 2.34) is proposed here displaying the broad structure, regional geometry of major shear zone systems, and the distribution of arc complexes.

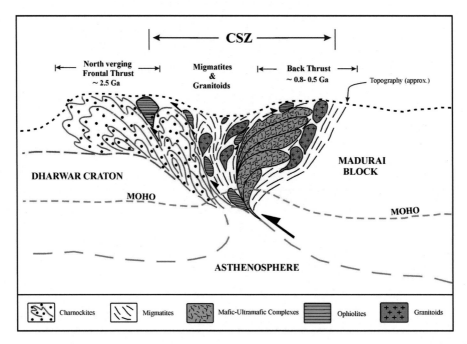

FIGURE 2.34

A schematic tectonic model showing the "flower structure" and the disposition of the subduction zone involving mantle—asthenosphere upwelling across the CSZ.

Adapted from Yellappa, T., Venkatasivappa, V., Koizumi, T., Chetty, T.R.K., Santosh, M., & Tsunogae, T. (2014). The mafic-ultramafic complex of Aniyapuram, Cauvery Suture Zone, Southern India: Petrological and geochemical constraints for Neoarchean suprasubduction zone tectonics. Journal of Asian Earth Sciences, 95, 81–98.

2.4 **MADURAI GRANULITE BLOCK**

2.4.1 **INTRODUCTION**

The Madurai Granulite Block (MGB) is the central and the largest crustal block in the SGT bound by the CSZ in the north and the AKSZ in the south (see Fig. 2.1). The MGB distinctly exposes elevated high land areas (>1000 m) of charnockite massifs and low-lying regions (<600 m) of migmatitic gneisses (both ortho- and para-gneisses) (Fig. 2.35), intruded by Neoproterozoic alkaline, anorthosite, and granitic plutons. Intimate association of charnockite massifs and sedimentary assemblages is well established with features like near conformable relationship, sharp contact, and an interbedded and interbanded nature before metamorphism. The migmatitic gneisses are of relatively low-grade amphibolite facies conditions often consisting of mafic and granitic sheets. On a large scale, the structural styles of deformation of these two high land and low land domains are strikingly different and hence they are described separately below.

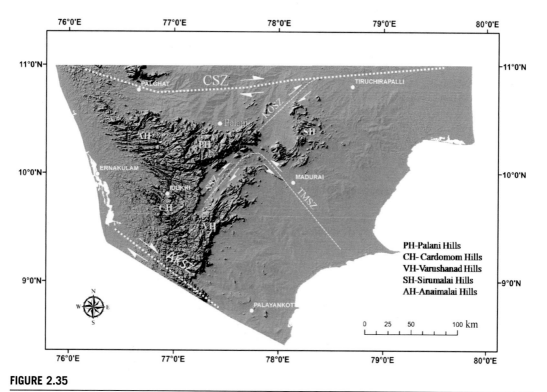

FIGURE 2.35

Digital elevation model of the Madurai granulite block. *Abbreviations*: *CSZ*, Cauvery shear zone; *KOSZ*, Karur-Oddanchatram shear zone; *SSZ*, Suruli shear zone; *TMSZ*, Theni-Madurai shear zone; AH, Anaimalai Hills; PH, Palani Hills; CH, Cardomom Hills; SH, Sirumalai Hills; VH, Varushanadu Hills

2.4.2 CHARNOCKITE MASSIFS

The important charnockite massifs consist of charno–enderbites, charnockites, and subordinate enderbites and occur in the form of Anaimalai-Palani Hills (Kodaikanal massif) in the northwestern part, Cardamom-Varushanad hills in the southwestern part, and Sirumalai hills in the northeastern part of MGB (Fig. 2.36). Enclaves of mafic granulites and swathes of metasedimentary associations with minor bands of quartzite, garnet–sillimanite gneiss, cordierite–sapphirine gneiss commonly occur within the massifs. The massifs are, in general, bound by shear zones all around displaying gradational sheared contact at a meter to decimeter scale marked by mylonitic fabrics and a range of kinematic indicators.

The Anaimalai hills in the west expose a sequence of complex granite gneisses which dip at low angles to the northwest. Biotite gneisses, migmatites, and K-rich granite sheets are dominant in the high-altitude regions. In the northern foothills, the composition of rocks varies from tonalitic to granitic, probably representing deeper crustal material. In the Palani hills, the regional structural fabrics of variably retrograded charnockites show a dominant NNE–SSW strike with steep dips to

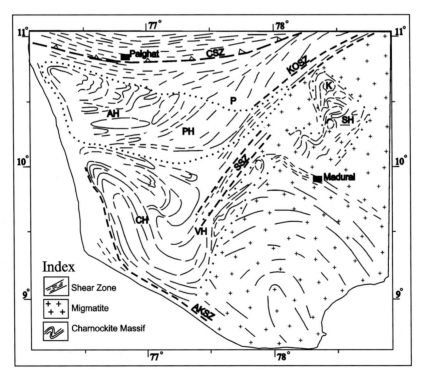

FIGURE 2.36

Structural interpretation of digital elevation model showing major shear zones, broad fold pattern, and important lithological assemblages in Madurai granulite block. *Abbreviations*: CSZ, Cauvery suture zone; KOSZ, Karur-Odhanchatram shear zone; P, Palani; K, Kadavur; AH, Anaimalai hills; PH, Palani Hills; SSZ, Suruli shear zone; SH, Sirumalai Hills; CH, Cardamom Hills; VH, Varushanadu Hills; AKSZ, Achankovil shear zone

WNW (often flat-lying), but the eastward swing is observed in the east. Large wavelength synformal fold structures occur between the Anaimalai hills in the west and the Palani-Kodaikanal hills in the east. Supracrustal rocks are dominant to the east and south of Kodaikanal and are interlayered with retrograded charnockitic rocks suggesting the possibility that these supracrustals lie tectonically below the Kodaikanal massif. The Varushanad and Andipatti hills are dominated by the presence of metasedimentary rocks in the form of broad isoclinal folds with wavelengths upto 15 km. The differences in the tectonic style between the charnockitic massif and their associated enclaves of interlayered supracrustal sequence must be, in large part, due to the contrast in competency and anisotropy between these two units. It is possible that the massive charnockites are of metaigneous origin, while the supracrustal sequence represents a pile of wet continental shelf sediments.

The Satellite image interpretation of MGB shows varying fold styles in different domains (see Fig. 2.36). The MGB, north of AKSZ, shows distinct foliation trends with dips ranging from shallow to steep values. The western part of MGB shows foliation trends and fold axial traces that are parallel to the AKSZ with dips to either side. The central part of the MGB around the Varushanad Hills is dominated by northwest trends in the southern part, while they are northeast in the northern part with steep dips towards east and southeast. Broadly, the MGB indicates a SE plunging regional fold closure.

The Cardamom hills display NNW−SSE trending refolded isoclinal folds in the western part. These fold structures seem to be extending and the eastern limb is represented in the form of Varushanad hills. The NE−SW trending shear zone between Cardamom and Varushanad hills marks the axial trace of this major fold. Sirumalai hills constitute predominantly complexly folded sedimentary associations of quartzites−carbonate−pelitic assemblages exhibiting a number of apparent structural basins and domes. Structural studies indicate the presence of remnants of an earlier regional recumbent folding in some parts of charnockite massifs. Structural details of Sirumalai hills will be described later under the title "Kadavur anorthositic complex."

Charnockitic massifs in the MGB dominantly occupy the highland areas and often structurally underlie the metasedimentary gneisses. All the charnockites (hypersthene-bearing granites) have been traditionally described as basic, intermediate, and acid variety based on the composition and textural characteristics with gradual and continuous variations. The basic variety shows medium to coarse grained (often with pegmatitic texture), dark, greasy, and homogeneous, and nongarnetiferous rocks without any tectonic fabrics superimposed on them. Such variety of rocks are well preserved in the central and upper parts of the massifs. The intermediate to the acid variety occurs at marginal and the boundary zones of massifs. These are characterized as being fine-to-medium grained, light-colored, with well-developed metamorphic gneissosity, and occasionally intense mylonitic fabrics showing signs of retrogression that are mostly devoid of Opx. Often, the charnockitic rocks also show variations into garnet−biotite gneisses, garnet−leuco gneiss to banded augen gneisses and irregular patches and veins of charnockites, which can be described as "charnockite in breaking." The intermediate variety occurs in between these two groups with the general development of metamorphic gneissosity. Deformed parts of massifs are garnetiferous, while less deformed/undeformed charnockites are, in general, nongarnetiferous. It is also pertinent to note the development of garnet-rich layers in the proximity of metasedimentary associations. All the above features may be related to variable retrogression of charnockites.

The earliest structures in charnockite massifs reveal the presence of flat-lying compositional banding and/subparallel gneissic foliation possibly representing relict recumbent fold structures.

These structures are occasionally overprinted thereby giving rise to complex structures, retrogression, and melting during subsequent deformation, which are well documented from the lower parts of massifs. They are frequently observed when they are in close proximity with metasedimentary associations; the best example could be the eastern part of Varushanad hills. The primary igneous textures are overprinted by mylonitic textures with granulite to hornblende−garnet amphibolite facies minerals indicating the possible influence of water content. Such evidences can lead to the interpretation that the charnockite massifs may represent large thrust-nappe structures. The upper most sections of the nappes are also often marked by the presence of deformational structures and high-strain fabrics suggesting that the massifs could have been subjected to multiple thrusting events. The boundaries of these massifs are, in general, broadly parallel to regional foliations. All the observations described above led to the inference that the charnockitic massifs in MGB must have been subjected to thrust-nappe tectonics, associated with top-to-the-north basal shearing giving rise to a series of north-verging multiple duplex structures involving multiple thrusting processes.

The chemical characteristics of charnockitic rocks of Cardamom hill massifs show continuous variations in major and trace element compositions suggesting a genetic link amongst all the varieties of charnockites (Rajesh & Santosh, 2004). They also inferred that the single source of basaltic magma gave rise to different varieties implying different conditions of temperature and water fugacity. Miller, Santosh, Pressley, Clements, and Rogers (1996) dated these charnockites using U−Pb in zircons that obtained an age of 588 ± 6 Ma.

Charnockites are generally magnesian calc−alkaline rocks with high Ba/Sr ratios, positive Sr anomalies and negativeNb, P, Ti anomalies and low K_2O- Na_2O ratios. The REE are strongly fractionated with low HREE concentrations. The alkali−silica relationship suggests that the intermediate charnockites have adakitic affinities. The Sm−Nd data indicate very little reworking and minimal crustal contamination (Tomson, Bhaskar Rao, Vijaya Kumar, & Mallikharjuna Rao, 2006). The charnockite massifs of MGB yielded Sm−Nd and Rb−Sr whole rock model ages in the range of 2200−3170 Ma (Bartlett, Dougherty-Page, Harris, Hawkesworth, & Santosh, 1998; Bhaskar Rao et al., 2003). Different models are proposed for the genesis of charnockites in the region. They include: (1) partial melting through basaltic underplating or charnockite as source of granite; (2) hydrous partial melting of subducted amphibolitic crust, and mixing of Archean crustal material with hydrous partial melt; and (3) juvenile magma. The charnockites of MGB represent a deeply eroded continental arc (Santosh et al., 2009). However, the source composition, timing, and mechanism of their genesis remain debated.

2.4.3 MIGMATITES

The low-lying areas of MGB are marked dominantly by migmatites of ortho- and paragneisses and granitic intrusives. They comprise a range of migmatitic gneisses that include hornblende−biotite gneiss, garnet−biotite gneiss, leptynite (garnetiferous leucogneiss), and cordierite gneiss. In some places, they are dominated by pink migmatitic biotitic gneisses. Hornblende−biotite gneisses are generally granodiorite to tonalite in composition, and may include both retrograded charnockites and orthogneisses. Garnet−biotite gneisses and leptynites may possibly represent migmatized paragneisses after khondalites. Often, they occur in the form of pelitic cordierite−garnet−spinel−sillimanite migmatites and graphite bearing garnet−biotite−sillimanite gneisses. The paragneisses occur together with aluminous, ferruginous, and orthoquartzites, calc silicates and calcite marbles.

Carbonate rich and calc–silicate layers vary in thickness with a maximum up to 5 m and act as dislocation or movement planes within the supracrustal sequence. They were completely recrystallized to carbonate–diopside–sphene–forsterite assemblages but often contain tectonic enclaves of adjacent lithologies in the form of aligned and detached layers, pods, and isolated fold hinges. The migmatitic gneisses are also often interlayered and deformed together with sparse mafic–ultramafic sheets and pods. Migmatites show extremely variable trends of both compositional banding and tectonic fabrics interspersed with imprints of several mesoscale shear zones. However, despite complex and intense deformation, the early developed stretching lineations show consistent shallow plunges but vary in orientation, possibly suggesting an original shallow plunges.

Detailed mapping of migmatic terrain in the north–eastern part of MGB reveals marked presence of well-preserved, flat-lying foliation fabrics and local imprints of high strain, particularly around Dindigul. Mafic bands/enclaves lie conformably with the flat-lying foliations. Sheets of charnockites with well-preserved, first-generation folds structurally overlie the garnet–sillimanite gneisses and show well-developed and closely spaced N–S trending shear planes with differential kinematics. At some places, local high-strain zones are associated with the development of melt producing metatexitic and stromatic migmatites. Leucosomes occur as pods, lenses, and thin layers often in association with 1–2 m wide pegmatite veins both along and across the foliation planes. Isoclinal rootless folds with horizontal axial planes are common. Closed structural forms are also recorded resembling sheath fold geometries. The structural features, emplacement of leucosomes and pegmatites, and the nature of foliation fabrics with in the migmatitic rocks from the MGB led to the interpretation that they have experienced intense shearing with an overall top-to-the-north sense of shear. Interplay between folding, shearing, and the melt generation played an important role during this process. The rocks, in general, experienced partial melting often with high temperature deformation resulting in coarser gneissic fabric and near complete annealing of the earlier structures.

An interpretative N–S structural section in the northeastern part of the MGB including the Kadavur anorthosite complex (KAC) shows a series of north-verging imbricate thrust structures related to large scale thrust-nappe structures (Fig. 2.37). While the competent layers of quartzites occur as detachment folds, the incompetent meta-carbonate rocks are highly disharmonically folded often acting as matrix. The calc–silicate rocks also occur as the detachment folds of various shapes from asymmetric kink like anticlines to nearly symmetric box-type folds. The fold structures in meta-sediments are further deformed due to intense synthrusting deformation (D2) accompanied by the emplacement of leucogranite–gneiss, and a variety of granitoids that occurred during 766_8 Ma. The structures developed in these rocks may be similar to the mylonitic fabrics, probably developed during the basal shearing at the base of the nappes. The U–Pb age spectra of detrital zircons from the metasedimentary gneisses of northern MGB indicate that their protoliths were deposited after ∼1700 Ma and for some rocks possibly as late as the mid-Neoproterozoic.

2.4.4 SHEAR ZONES

Four important shear zones have been recognized and described from MGB that include: Karur-Oddanchatram shear zone (KOSZ); Suruli shear zone (SSZ), and Theni-Madurai shear zone (TMSZ) (see Fig. 2.36). The KOSZ occurs to the east of Kodaikanal and extends to the northeast finally merging with the CNSZ. In general, the KOSZ trends NE–SW and dips moderately to

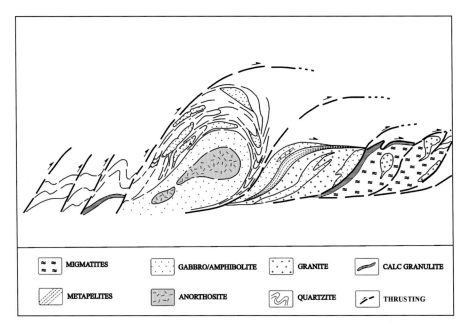

FIGURE 2.37

An interpretative N–S structural cross section in the north eastern part of Madurai granulite block across the Kadavur anorthosite complex depicting north-verging thrusts and associated imbricate structures.

northwest with dextral displacements. It also hosts Oddanchatram anorthosite body, which is syntectonically emplaced as a phacolith along the KOSZ. The Oddanchatram anorthosite (~560 Ma) is an elongate labradorite -type pluton (17×20 km) displaying intense mylonitic fabrics with no stratiform layering. The host rocks include charnockite, two-pyroxene granulites, and metasedimentary assemblages. Petrochemical studies reveal that it is a basic pluton with calc–alkaline basaltic character derived from parent magma of gabbroic anorthosite in composition.

The Suruli shear zone (SSZ) occurs in the form of a valley between Cardamom hills to the west and Varushanad hills to the east. The SSZ trends NNE–SSW starting from Kambam in the south and extending upto Ganguvarpatti near Theni. The SSZ comprises granulite grade pelitic and calc–silicate gneisses interbanded with pyroxene granulites with younger intrusives of granite–syenite–carbonatite plutons. In the eastern part of Kambam valley, khondalitic rocks occur as intricately folded bands interbanded with hornblende–biotite gneiss and charnockitic rocks. The SSZ is characterized by the presence of easterly dipping mylonitic fabrics and saphirine–cordierite–symplectite pods in Mg–rich, Opx–granulites. The orientation of SSZ varies from NE to SE with a curvature at the northern end near Theni and extends in SE direction through Madurai city upto Bay of Bengal. The later part of the shear zone is termed here as Theni-Madurai shear zone (TMSZ). Dextral displacements along the SSZ and sinistral kinematics along the TMSZ indicate that both may represent a single folded shear zone with closure to north, which is at variance with KKPTSZ.

A new "V"-shaped Neoproterozoic shear zone called the Karur-Kambam-Painavu-Trichur shear zone (KKPTSZ) has been described by extending the SSZ in a northwest direction starting from Kambam through Painavu-Trichur (Ghosh et al., 2004). The KKPTSZ was considered by some workers as a potential Neoproterozoic terrane boundary separating Neoarchean igneous rocks to the north and predominantly metasedimentary rocks to the south marking a noticeable isotopic boundary (Plavsa et al., 2012). Some others interpreted the KKPTSZ as decollement zone and the terrane boundary, while others as a tectonized basement/cover relationship with the isotopic shift being due to the relative inputs of more juvenile sources in the metasediments, as obtained from the detrital age spectra. Some of the syntectonic granite intrusives yielded ~567 Ma suggesting the possible timing of shearing along the KKPTSZ. However, the existence of the shear zone between Kambam and Trichur and the "V"-shaped geometry were questioned by many others citing the following features: (1) lack of field evidences, (2) absence of airborne magnetic reflections, (3) free from first order gravity gradient, and (4) the absence of post shearing effects in massive charnockites.

2.4.5 KADAVUR ANORTHOSITIC COMPLEX

Detailed mapping in the northeastern sector of Madurai block reveal that the terrain is dominated by medium-grade migmatitic, tonalitic, ortho- and paragneisses in low land areas surrounding the Sirumalai hills (Fig. 2.38). Migmatites vary widely in composition and consist of granitic gneisses, variably retrograded charnockites, biotite—hornblende gneisses, quartzites, and calc granulites. Often, they are interlayered, dominantly quartzo—feldspathic gneisses with subordinate tonalitic and metabasic gneisses (MORB composition). The general trend of foliations is NE—SW, but separated by some shear zones parallel to foliation fabrics finally merging with CNSZ. The migmatites are occasionally intruded by relatively younger monzonites, syenites, anorthosites, etc. Evidence of high strain in the field is well defined by several structural features like mylonitic fabrics, highly asymmetrical structures, and complex fold styles. The orthogneisses are complexly interfolded and intruded by undeformed to locally and weakly deformed granitic intrusions. Dominant southerly dips of foliations in the region suggest that they may be related to northerly verging thrusts. The migmatitic rocks show a range of deformational structures (variably banded and complex fold morphologies) indicating that they must have undergone prethrusting synmetamorphic and synplutonic folding, perhaps related to north-directed thrusting.

The Sirumalai hills are characterized by an intensely folded supracrustal sequence, including thick quartzites, calc silicates, and minor association of different types of metaigneous rocks (ultramafic rocks, differentiated gabbros, and anorthosites, etc.). Occasionally, charnockites and hornblende gneisses are also recorded. The sequence shows mappable complex fold structures with trends dominantly east—west and NE—SW, apparently showing structural domes and basins. The supracrustal sequence and associated charnockitic rocks of the Sirumalai hills seem to be structurally overlying the surrounding migmatitic gneisses. Considering the discordance of structural fabrics between charnockite massifs and the migmatites, it is inferred that the migmatitic rocks and the metasedimentary associations were possibly subjected to a prethrusting event of deformation along with the emplacement of gabbro—anorthosite and coarse-grained granite plutons. It has been observed that, at some places, the metasedimentary rocks underlie a thrust slab of intensely sheared and retrograded charnockitic rocks with mylonitic fabrics at the base. The close association,

FIGURE 2.38

Map showing the geology around the Kadavur anorthosite complex.

similarity in structure, and texture and the association of mineral assemblages established that both charnockites and the interlayered metasediments witnessed identical metamorphic conditions. This implies that both charnockites and the sediments must have had a conformable relationship, sharp contacts, and an interbedded and interbanded nature before metamorphism.

From the structural relationships and lithological assemblages, it can be inferred that the supra-crustal sequence of Sirumalai hills may represent an allochthonous block in the form of a klippen structure (eroded part of remnant large thrust sheet/nappe) and the migmatitic gneisses exposed in the central part of eroded hills may reflect the tectonic windows. This concept can also be extended to the basement migmatitic gneisses structurally underlying the charnockitic massifs of high land areas.

The KAC occurs in the central part and is believed to have been emplaced into the domal struc-ture defined by the quartzites (see Fig. 2.38). It covers an area of 70 sq. km and intrudes the Quartzite—carbonate—pelitic suite of Sirumalai hills. The KAC comprises gabbro, anorthosite, leu-cogabbro, and felsic gneisses, and the surrounding hills are marked by quartzites, calc—silicate rocks, and a few structurally conformable mafic bodies (ultramafic rocks, differentiated gabbros, amphibolites, and anorthosites). It is associated with cognate rapakivi granite and is devoid of metamorphism. Based on field, petrographic, and geochemical data, the KAC represents a differen-tiated massif type anorthosite pluton derived from a high A1 tholeiitic parental liquid and is compa-rable with that of Adirondack anorthosite suite.

These anorthositic rocks were crystallized at $\sim$829 Ma and originated in a suprasubduction zone setting (Teale et al., 2011). The interpretative structural section drawn across the KAC reveals a set of mega-duplex structures with thrust planes occurring at the lithological contacts, similar to Alpine structures, making KAC a possible nappe system, displaying tectonic windows, and typical klippen structures (Fig. 2.39). Deep erosion of antiformal fold structures can give rise to circular features often exposing the tectonic windows as typically represented by the KAC. Quartzites in the Kadavur region were derived from Palaeoproterozoic sources and are constrained as being deposited between 1926 ± 20 Ma and 829 ± 14 Ma. They locally show metamorphic zircon growth because of the intrusion of the gabbro—anorthosite complex. At the northern front of the KAC metasedimentary nappe system, UHT assemblages were documented within the migmatites. The Proterozoic metasedimentary rocks of KAC and the adjacent regions yielded Zircon ages of $\sim$2.6—1.8 Ga (Plavsa et al., 2014) pointing to the source from the southern continuation of East African orogen.

2.4.6 UHT METAMORPHISM

Ultrahigh-temperature (UHT) metamorphism is primarily recognized based on the presence of min-eral assemblages formed in Mg-Al—rich layers in pelitic and in more siliceous rocks. They are vol-umetrically insignificant in nature but well spread and widely reported in granulite facies terranes. They have been ascribed to metamorphism of hydrothermally altered mafic to ultramafic rocks or high Mg—clays, sediments; metasomatic alteration of S-type granites or in partial melting zones of residual Mg-Al—rich domains in metasedimentary rocks (Kelsey & Hand, 2015).

The metamorphic evolution of the MGB has been the focus of a number of recent studies spurred by the occurrence of UHT metamorphic assemblages from a number of localities (Brown & Raith, 1996; Mohan & Windley, 1993; Shimpo, Tsunogae, & Santosh, 2006). UHT metamorphism is mostly

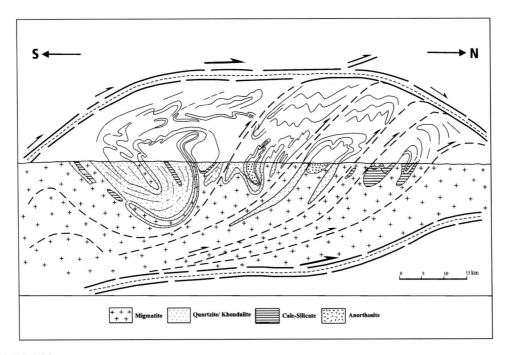

FIGURE 2.39

Interpretative N–S structural cross-section constructed across the Kadavur anorthosite complex showing the Alpine type imbricate and antiformal fold structures.

preserved in some pelitic gneisses and Mg–Al rocks intercalated with charnockites and hornblende–biotite gneisses. The host rocks in the northern part of MGB include quartz free Mg–Al rich rocks often containing aluminous minerals such as saphirine, spinel, and corundum. The southern part of MGB is characterized by bluish cordierite-bearing granulites occasionally intruded by pink alkali–feldspar granites. The central MGB comprises quartzo–feldspathic gneisses, charnockite, and intercalated metasedimantary rocks with a common mineral assemblage of garnet + sillimanite + biotite + cordierite + spinel (Tsunogae & Santosh, 2007). Some CO_2-rich fluids of primary origin also occur in UHT mineral assemblages as inclusions.

Peak conditions of pressure and temperature inferred for the UHT rocks in the MGB are in the range of 7–11 kbar and 950–1150°C. These rocks have yielded a range of ages between 600 and 480 Ma (Jayananda, Martin, Peucat, & Mahabaleswar, 1995) demonstrating that this metamorphism was synchronous with the final stages of Gondwana amalgamation. By contrast, Braun, Cenki-Tok, Paquette, and Tiepolo (2007) suggested that peak UHT metamorphism occurred earlier at c. 900 Ma based on thermal ion mass spectrometry (ID-TIMS) U–Pb dating of monazite inclusions in garnet.

Most of the Zircon grain cores from many of the UHT localities of MGB yielded relatively older ages between 2.3 and 2.6 Ga with a peak at 2.3–2.5 Ga. Monazites from the same locality show clear Late Neoproterozoic-Cambrian ages (540–480 Ma) from the rims of monazites (Santosh et al., 2006b). These results from Zircon rims were interpreted as juvenile additions during exhumation under high temperatures that isostructurally mantle the preexisting monazites. Based on the geochronological and geochemical data sets of zircon and monazite, the duration of UHT metamorphism from MGB was estimated to be c. 40 Ma with peak conditions achieved at c. 60 Ma at c. 610 Ma (Clark et al., 2015). This process occurred after the formation of an orogenic plateau related to the collision of microcontinent Azania with east Africa at c. 610 Ma.

2.4.7 GEOCHRONOLOGY AND SUBDOMAINS

There have been attempts to subdivide the MGB into distinct subdomains with the help of geochronological data in recent years. Based on contrasting primary igneous U−Pb ages of orthogneisses, two distinct terranes were identified: an Archean basement terrane to the north and west, and a Proterozoic terrane dominated by metasedimentary rocks to the south separated broadly by KKPTSZ (Plavsa et al., 2012). In the northern part, the charnockites recorded ∼2.7−2.5 Ga crystallization ages with metamorphic overprinting at ∼535 Ma indicating primarily mantle-derived origin in a suprasubduction zone setting with some Archean crustal components. The crystallization age in the southern domain is 1007_23 Ma and 784_18 Ma, suggesting significantly different isotopic evolution. The metasedimentary rocks from the southern domain of MGB are thrust over Neoarchean gneisses of the northern domain, similar to those in Anantanarivo Block of Madagascar (Plavsa et al., 2014). Tomson, Bhaskar Rao, Vijaya Kumar, and Choudhary (2013) delineated two distinct provinces within MGB with model Nd ages between 3.2 and 2.8 Ga to the north and those younger than 2.8 Ga to the east and south of the KKPTSZ based on Sm−Nd model age distribution. On the contrary to the division into northern and southern domains, Brandt et al. (2014), delineated the MGB into western and eastern domains based on the data of LA−ICPMS U−Pb zircon and U−Th−Pb monazite ages and geochemical data. The western domain comprises Late Neoarchean (2.53−2.46 Ga) subduction related charnockites, which were reworked during granulite facies metamorphism and partial melting (2.47−2.43 Ga). The eastern domain, a vast supracrustal sequence, which was deposited on a late Paleoproterozoic (1.74−1.62 Ga) basement of charnockites formed through reworking of underlying Archean rocks. Both the domains were intruded by mid-Neoproterozoic A-type charnockites and felsic orthogneisses (0.83−0.79 Ga) during the process of extensional rifting. According to them, these two west and east domains converged together during late Neoproterozoic (0.55 Ga) along the Suruli shear zone. The ambiguity and the complexity involved in the subdivision of MGB as described above may be resolved with an integrated study of field-based structural geology and geochronology so that alternate explanations can be explored.

2.4.8 GEOPHYSICAL SIGNATURES

A study of crustal reflectivity using multifold seismic reflections along a N−S trending profile across the MGB between the CSZ and the AKSZ shows conspicuous reflectivity pattern indicating the existence of several structurally distinct blocks. (Rajendra Prasad et al., 2007). Gentle and north dipping reflectors in the northern part and strong and steep southerly dipping reflections in the south were recorded. The reflectivity pattern was interpreted as a crustal dome in the central part of the MGB. Considering the fact that seismic profiles in the Precambrian crust indicate lower crustal reflectivity and tectonic processes, the reflections obtained from the profile together with the surface geology could be interpreted in terms of a series of north-verging thrust zones. The MGB is interpreted to lie directly over a south-dipping electrical feature in the form of a volcanic arc model (Naganjaneyulu & Santosh, 2010).

2.4.9 **TECTONIC SYNTHESIS**

The rocks from MGB, ranging in age from the Archean, Paleoproterozoic to the Neoproterozoic, are thought to have formed in an active convergent margin setting. The MGB comprises accretionary-stacked complexes with interleaved Mesoproterozoic and Neoproterozoic age slabs. The accretion of this complex involves thrusting associated with top to the north sense of shear and may broadly be related to the orogenesis forming part of E–W oriented East African Orogen. Erosion appears to be faster in the eastern part of MGB compared to the western block resulting in small klippen erosional remnants exposing tectonic windows. These features are well reflected in an E–W structural cross section drawn in the southern MGB (Fig. 2.40). The relative structural positions are well documented implying that the metasedimentary assemblages structurally underlie as well as overlie the thrust slabs of charnockites. This leads to the conclusion that the widespread charnockitic massifs in the MGB appear to be allochthonous in nature. During the translation of thrust-nappe, fluids are injected into the hot hanging wall rocks probably from dehydration

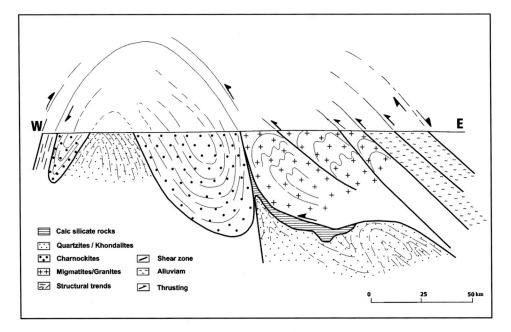

FIGURE 2.40

Interpretative E–W structural cross-section displaying the nature of metasedimentary assemblages both in the form of underlying and overlying structural units in relation to charnockitic rocks in the southern part of Madurai granulite block.

reactions in the floor as well as from sediments overridden by the hanging wall thrust sheets. This leads to a partial-to-complete hydration of the granulite grade assemblages in the hanging wall as well as contributing to the development of granite gneisses and migmatites. Granites are also generated by partial melting in the footwall terraces that occur typically 40−50 Ma after collision due to heat accumulation in the floor. However, many ambiguities regarding the age and the nature of the rock types within the MGB will remain unresolved until an integrated study of field geology, petrology, geochemistry, and geochronology is attempted.

The migmatitic structures must have undergone prethrusting synmetamorphic and synplutonic folding primarily in association with E−W trends during D1. These are further deformed during D2 due to intense synthrusting deformation accompanied by the emplacement of leuco−granitoids along the basal shear zone with top-to-the-north basal shearing. The mylonitic fabrics and other complex structures recorded in many parts of lowland migmatitic terrain may be related to the basal shear zone probably formed due to translation at the base of the nappes. Considering all the characteristics like relict recumbent fold structures, variation in the intensity of deformational fabrics, and associated melting, retrogression, and migmatization, the large charnockitic massifs in the MGB might have formed through large-scale thrust-nappe tectonics. The migmatitic gneisses in the lowlying areas are interfolded with thin sheets of pyroxenites and charnockites. The thrust-nappe movement resulted in the development of complex fold structures in metasediments and intense mylonitization in the upper sections of underlying migmatites as well as in the overlying lower sections of allochthonous charnockite massifs. Some of the larger antiformal fold hinges, when deeply eroded, apparently show circular features exposing the basement as tectonic windows, and the KAC represents a typical example of such features.

All the rocks from MGB are intensely deformed and mylonitized showing a variety of structures. These are distinctly different from those of the CSZ and the northern blocks with respect to metamorphic grade, structural style and lithological assemblages. It contains much lower volumes of mafic and ultramfic association. The MGB correlates well with those in central Madagascar and the Highland and Wanni Complexes of Sri Lanka. They are interpreted to have formed a part of the Neoproterozoic continent Azania (Collins & Pisarevsky, 2005) that lay above a south/west (present directions) dipping Cryogenian-Ediacaran subduction zone that facilitated closure of the Mozambique Ocean.

The wide range of ages in significant amounts obtained from U−Pb and Hf isotopic data for different lithologies of domains and the proposed tectonic divisions of MGB by different group of workers provide room for discussion and debate. It is felt that this kind of contrasting model might be due to lack of detailed field observations and comprehensive understanding of structural architecture, which would help in understanding the three-dimensional structure and geometry of the tectonic elements of MGB. The combination of structural architecture and the available geochemical and geochronological data sets would hopefully provide more insights about the tectonic models of MGB.

2.5 ACHANKOVIL SHEAR ZONE/SUTURE ZONE

2.5.1 INTRODUCTION

The Achankovil Shear Zone (AKSZ), located at the southern tip of SGT, separates the granulitefacies rocks of MGB to the north and TGB to the south. The AKSZ trends NW−SE and extends for a strike length of ~150 km with a width of ~20 km. It swerves to the north in the northwest and extends further subparallel to the west coast, while at the eastern end, it extends in an east—west fashion. It is well reflected in aeromagnetic maps as well as in satellite data displaying well-developed regional fabrics and fold patterns. The northern boundary of the shear zone is considered to be coinciding with the straight section of the Achankovil River while the southern boundary is confined to a long narrow valley passing through the Tenmalai village. Some workers consider these two boundaries as two independent shear zones, viz., the Achankovil and Tenmalai shear zones, respectively.

The important rock types within the AKSZ are garnet—biotite—sillimanite gneisses (khonda-lites) and garnet—biotite—quartzofeldspathic gneisses with associated migmatites and charnockites. Calc—granulites, quartzites, and ultramafic rocks occur as linear bodies and lenses parallel to the local structural trends (Fig. 2.41). Two varieties of charnockites could be observed from field studies, one being massive charnockite and the other being arrested or patchy charnockite. In some of the outcrops, the charnockite is retrogressed or bleached by the granitic and/or pegmatitic fluids. Charnockites are mostly garnet bearing, but a garnet-absent massif type is also observed. Patches and veins of incipient charnockite develop within gneisses in several localities. One of the characteristic lithological units is cordierite and orthopyroxene bearing charnockite showing various reaction textures indicating garnet breakdown and cordierite formation in a decompression setting. Calc—silicate rocks comprise both garnet-bearing and garnet-absent varieties.

2.5.2 STRUCTURE AND KINEMATICS

Regionally, the NW—SE trending AKSZ shows conformable metamorphic gneissic foliation and lithological banding, and the dips vary between 50 and 75 degrees to the southwest (Fig. 2.42). The rocks are folded into a series of upright isoclinal folds with their axial planes striking NW—SE. Rootless fold closures and transposition fabrics are common. Pronounced transposition is well mapped in a small segment of the AKSZ around Ambasamudram that reveals varying degrees of transposition of metamorphic gneissic fabrics grading into NW—SE trending mylonitic fabrics (Fig. 2.43). Well-preserved relict fold structures are common in the region. The morphology of water bodies and river courses follow the transposition fabrics. The foliations show steep dips to southwest. The stretching lineations show gentle to shallow plunges on either side. In the SE part of the AKSZ around Eruvadi, well-exposed quarries reveal the presence of granulite facies assemblages, which include predominantly quartzofeldspathic gneisses, garnet—sillimanite—graphite gneisses (khondalites), charnockites, and calc granulites with minor occurrences of mafic enclaves, granites, aplites, and pegmatites. The gneissic foliation is defined by alternating lighter, made of quartzofeldspathic layers with garnet, and darker bands constituting biotite, sillimanite, and graphite bearing layers. The layer thickness varies from centimeters to a few meters. Trains of garnet, biotite, graphite, elongated feldspars, sillimanite laths and quartz ribbons define the stretching lineations in the study area. The gneisses contain numerous kinematic indicators such as asymmetric boudins, porphyroclasts, flanking folds, "S"-shaped folds, shear bands, and stretching lineations.

The kinematics along the AKSZ are still under debate because of the dextral and sinistral structural features that have been observed. Different workers have interpreted different movement patterns such as: (1) sinistral strike-slip movement, (2) dextral strike-slip movements, and (3) zone of flattening with a component of thrusting. However, considering the multiple deformational history of the AKSZ, Rajesh and Chetty (2006) invoked an initial ductile deformation with dextral kinematics represented by asymmetric boudins, porphyroclasts, and S-C fabrics and flanking folds, followed by

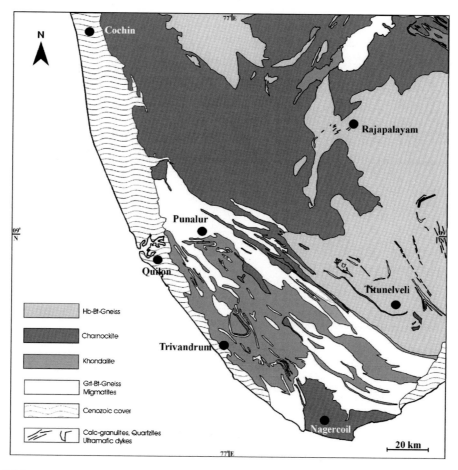

FIGURE 2.41

Regional geological map of AKSZ and the adjoining region.

Adapted from Rajesh, K. & Chetty, T.R.K. (2006). Structure and tectonics of the Achankovil Shear Zone, southern India, Gondwana Research, 10*(2006), 86–98.*

subsequent sinistral deformation of brittle–ductile nature as evidenced by "S"-shaped folds, associated shear bands and reworked Trapezoidal boudins. Further, the rheological differences characterized by ductile and brittle–ductile setting for the dextral and sinistral events and the overprinting relationships of the distinct structural features suggest temporal variation for the deformational events.

2.5.3 METAMORPHISM

The textural assemblage indicates that the rocks in the AKSZ were metamorphosed to granulite facies conditions of up to 900–950°C and 6–7 kbar (e.g., Cenki & Kriegsman, 2005; Santosh, 1987). The AKSZ is characterized by discontinuous strands of cordierite-bearing gneisses and

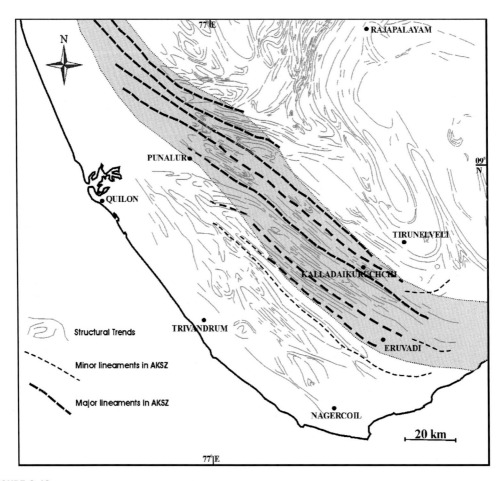

FIGURE 2.42

Regional structural map of AKSZ derived from satellite data interpretation.

Adapted from Rajesh, K. & Chetty, T.R.K. (2006). Structure and tectonics of the Achankovil Shear Zone, southern India, Gondwana
Research, 10*(2006), 86−98.*

preserves several peak and postpeak metamorphic assemblages. The UHT mineral assemblages were also documented from AKSZ and the peak metamorphic conditions were found to be 8.5−9.5 kbar and 940−1040°C (Ishii, Tsunogae, & Santosh, 2006). Following the regional geodynamic models, it is suggested that the UHT metamorphism is principally related to subduction, coupled with elevated crustal radiogenic heat generation rates.

2.5.4 INTRUSIVES

The AKSZ is marked by many syn- to posttectonic alkali granites that form an array of "suturing plutons" along the margin of the MGB. An undeformed ultramafic body (dunite) hosted by

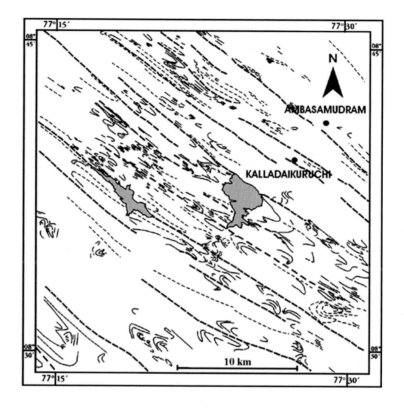

FIGURE 2.43

Detailed structural map showing the varying degree of transposition fabrics around Ambasamudram.

After Rajesh, 2008, unpublished thesis.

intensely deformed granulite facies metamorphic rocks has also been documented from AKSZ. The body consists of spinel—dunite, phlogopite—dunite, glimmerite, graphite—spinel—glimmerite, and phlogopite—graphite spinellite. (Rajesh, Arima, & Santosh, 2004). The geochemical data suggests that the body was formed by progressive crystallization of highly potassic CO_2-rich melts injected into lower crustal levels. K—Ar ages of phlogopite yielded 470 and 464 Ma, which are comparable to the phlogopite K—Ar ages reported from lithospheric shear zones in southern Madagascar indicating the nature and contiguity of AKSZ prior to the Gondwana breakup. This also implies widespread highly potassic CO_2-rich fluid/melt influx along shear zones in this part of East Gondwana.

2.5.5 GEOCHRONOLOGY

A range of U—Pb ages was reported from the rocks within the AKSZ using various techniques. Electron probe Th—U—Pb monazite data yielded intrusion ages for alkali granitoids (Pathanapuram granite) of 550 ± 25 Ma, with a younger age population of 515 ± 16 Ma interpreted to represent the timing of high-grade metamorphism (Santosh, Tanaka, Yokoyama, & Collins, 2005). Zircon ID—TIMS yielded ages varying between 548 ± 2 and 526 ± 3 Ma for charnockites (Ghosh et al.,

2004). Detrital zircon and monazite studies from metasedimentary protoliths yielded maximum depositional ages in the Neoproterozoic. U−Pb ages for zircon (Sato et al., 2010) from laser ablation inductively coupled plasma mass spectrometry (LA-ICPMS) yielded age peaks at 900−600 Ma and 560−480 Ma. These are interpreted as crystallization ages of the magmatic protoliths and the event of final suturing of the Gondwanan supercontinent.

Recent studies have identified similarities in the U−Pb and Hf isotopic composition of detrital zircon between the Achankovil shear zone and the southern MGB, and the region has been interpreted as a metamorphosed Neoproterozoic sedimentary sequence deposited in a rift setting (Collins et al., 2014). It is possible that the AKSZ represents an early to middle Neoproterozoic passive margin with sediments deposited prior to the collision of the MGB and TGB, or a basin formed during rifting in the late Proterozoic, similar to that proposed for Madagascar following its collision with East Africa.

2.5.6 GEOPHYSICAL SIGNATURES

A strong magnetic linear trending NW−SE coincides with the AKSZ that distinctly separates the khondalite belt of the TGB and the charnockite belt of the MGB. While the TGB is characterized by a magnetic high and a flat terrain, the MGB revealed a predominant magnetic low with a few E−W highs of linear nature that are oblique to the trend of AKSZ. The presence of a steep, southward-dipping basic intrusive at a depth of ∼2 km below the ground surface was also inferred that coincides with the known exposure of an undeformed ultramafic body. The results of tomographic study providing shallow (<8 km) anomalous high V_p/V_s ratio (>1.75), large variation of Poisson's ratio (0.25−0.29) of the upper crust of the AKSZ reveal the characteristic features of deeper levels (Rajendra Prasad, Behera, & Koteswara Rao, 2006). This is further substantiated by the studies of two parallel magnetotelluric (MT) traverses across the Achankovil shear zone (AKSZ). The MT results reveal the regional geo-electric average strike direction of N 40°W, consistent with the geological strike direction and moderately conductive (<500 ohm.m) crust below AKSZ (Dhanunjaya Naidu, Manoj, Patro, Sreejesh Sreedhar, & Harinarayana, 2011). Their model shows distinct high electrical resistivity (>1000 ohm.m) for the upper crust below the MGB with a gentle dip towards the south and a northerly dip below the TGB, suggesting that the AKSZ could represent a possible suture zone.

2.5.7 TECTONIC SYNTHESIS

A structural section across the AKSZ clearly displays the variations in the orientation of the lithounits and the geometry of structural fabrics on either side (Fig. 2.44). The TGB displays alternating antiforms, synforms, and isoclinal antiforms with axial traces subparallel to the AKSZ. The foliations from TGB trend NW−SE with dominant northeasterly dips (40−60 degrees) and the dips become steep to subvertical in the proximity of the AKSZ. The fabrics remarkably maintain consistent dip amount and direction throughout the width of AKSZ. The MGB is marked by a distinct structural setting with varying trends and gentle-to-steep dips. Shallow and north-verging thrust planes can well be visualized in the cross-section of MGB. Thus, the contrasting structural geometry is well reflected on either side of the AKSZ. Recent geological and geophysical studies have

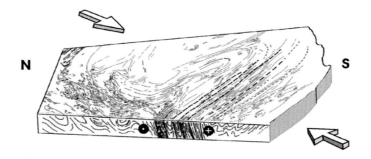

FIGURE 2.44

A 3D schematic structural section across the AKSZ.

Adapted from Rajesh, K. & Chetty, T.R.K. (2006). Structure and tectonics of the Achankovil Shear Zone, southern India, Gondwana Research, 10, *86–98.*

identified a dense, relatively conductive crust underlying the AKSZ, leading to the interpretation that the AKSZ represents a suture and that the MGB and TGB are discrete crustal terranes.

Integration of seismic, gravity, and heat flow data suggest a high density and moderately conductive mantle material brought up to the mid-lower level and a thermally eroded crust with a "flower structure" at depth (~ 10 km) suggest that the AKSZ represents the trace of a collisional suture. Further, the latest deformation along AKSZ is found to be sinistral transpression, which strengthens its correlation with the sinistral Ranotsara shear zone of Madagascar related to the East African Orogeny. In the light of this, the correlations with the shear zones of other Gondwana member terranes should be approached cautiously by dating different events of multiple deformations.

2.6 TRIVANDRUM GRANULITE BLOCK

2.6.1 INTRODUCTION

Trivandrum granulite block (TGB), also designated as the Kerala Khondalite Belt (KKB), is the southernmost crustal block in the SGT and occurs to the south of AKSZ (see Fig. 2.41). The TGB extends in NW—SE direction subparallel to the AKSZ and has a strike length of about 200 km and a width of ~ 100 km. The structural trends in the TGB are broadly similar in strike (NW—SE) to those of the AKSZ and dip moderately towards northeast.

The important rock types in TGB are garnet and sillimanite-bearing metapelitic granulites (khondalites) with or without cordierite, spinel, biotite and graphite. The metapelites are variably intercalated with garnet—biotite gneiss (leptynite), garnet- and orthopyroxene—bearing anhydrous granulites (charnockites) and subordinate bands, boudins, and lenses of pyroxene granulite and calc—silicate rocks (Santosh, Marimoto, & Tsutsumi, 2006a). Many localities in the TGB comprise the garnet—biotite gneisses showing progressive transformation into incipient or arrested charnockites in the form of patches and veins, and also on a massive (regional) scale. Small plutons with modal composition ranging from granite to alkali syenite and minor monzosyenite are common in the TGB. They comprise mafic minerals, alkali amphibole, biotite, titanite and apatite; Cpx and garnet occur occasionally. The TGB is bound by the Nagercoil charnockite massif to the south, which was described as a separate block by some workers.

2.6.2 STRUCTURE

In general, the rocks in TGB are strongly deformed and whose earlier fabrics were overprinted by intense ductile deformation, recrystallization, and migmatization. In general, the foliations trend NW−SE with moderate to steep dips to northeast (see Fig. 2.42). However, in some of the domains of low strain, they are near horizontal. Fold closures are rarely preserved in the region with the exception of the northwestern part. All these rocks are subjected to ductile deformation, migmatization, and anatexis superposing on to the earlier fabrics. The degree of intensity of deformation in TGB is relatively more than that of the fold dominated MGB as witnessed by the presence of transpositional fabrics.

The charnockite bands and interlayered khondalites and leucogneisses are all cofolded together. At several places, the rocks are intensely migmatized, sheared, and recrystallized giving rise to a new rock often described as leptynite (granitoid leucogneiss) or leucogneiss. These rocks are associated with minor bands of charnockites and mafic granulites. Leptynites include garnet-rich quartzofeldspathic gneiss with variable amounts of biotite and are rarely graphite. They are fine-grained granulite facies rocks with planar gneissic fabric predominantly consisting of alkalifeldspar, minor quartz, white mica, garnet, and tourmaline, and are reported to be of sedimentary origin. Many workers interpreted these quartzofeldspathic gneisses as a sequence of psammtic, psamopelitic, pelitic metasedimentary rocks and minor granitoid gneiss. These leucogneisses are occasionally interlayered with khondalites. Prograde incipient (patchy) charnockites are commonly reported in both khondalites and leucogneisses overprinting all the earlier fabrics.

The Nagercoil charnockite massif is medium-to-coarse grained with sporadic garnet. In contrast to TGB, the structural trends are predominantly NNE−SSW with easterly dips. Although, these rocks seem to be massive in nature, weak foliation is developed at several places. The foliation is more pronounced and conformable at the contact zone with khondalitic rocks or calc−silicate enclaves. The khondalites occur in the form of discontinuous patches and veins. Thin bands, lensoid bodies, and boudins of dark-colored, hornblende-rich mafic granulites also occur within the chanockites.

2.6.3 METAMORPHISM

Data from a migmatized metapelite raft enclosed within Nagercoil charnockite massif provides quantitative constraints on the pressure−temperature−time (P−T−t) evolution of the massif. An inferred peak metamorphic assemblage of garnet, K-feldspar, sillimanite, plagioclase, magnetite, ilmenite, spinel, and melt is consistent with peak metamorphic pressures of 6−8 kbar and temperatures in excess of 900°C. SHRIMP U−Pb dating of magmatic zircon cores suggests that the sedimentary protoliths were in part derived from felsic igneous rocks with Palaeoproterozoic crystallization ages. Three distinct stages of evolution (Ml−M3) during the Pan-African high-grade metamorphism were reported with possible temperature gradient from north to south of the TGB as detected from mineral phase equilibria thermobarometry. During the period between the metamorphic stages M1 and M2, the terrain experienced subisobaric cooling. Comparison of results from thermobarometry with data on absolute age determinations from geochronology of the metamorphic rocks in TGB allows the interpretation that the M1 metamorphic event took place during 540–600 Ma, M2 at about 530 Ma and M3 in the interval of 440–470 Ma (Fonarev, Konilovl, & Santosh, 2000).

The conventional geothermometry of massive charnockites have shown P–T conditions upto 930°C and 6.3 kbar. (Santosh, Yokoyama, Biju-Sekhar, & Rogers, 2003). However, the P–T conditions of khondalite–leptynite rock suite indicate high temperature metamorphism at 850–900°C and 6–8 kbar pressures (Marimoto, Santosh, Tsunogae, & Yoshimura, 2004). Detailed fluid inclusion and carbon isotopic studies reveal that the dehydration controlled by CO_2 advections along structural path is the principal mechanism of transformation of leucogneiss into incipient charnockite patches and veins. It still remains a debate whether CO_2 was released locally or derived from mantle fluids or graphite-bearing lithologies.

2.6.4 GEOCHRONOLOGY

Early geochronological results from the rocks of TGB established either metamorphic or depositional ages (Paleo-to-Mesoproterozoic), and the major geothermal event was during Neoproterozoic to Cambrian at c. 500–580 Ma (Bartlett et al., 1998). However, the magmatic protoliths of charnockites and granite gneisses have also yielded Paleo-to-Mesoproterozoic ages, which were attributed to the involvement of more ancient crust (Kröner, Santosh, & Wong, 2012). Recent studies of magmatic zircons from charnockites and granitoid gneisses from TGB yielded well-defined protolith emplacement ages between 1765 and 2100 Ma, with the metamorphic age to be around 540 Ma (Kröner et al., 2015). This implies that the enclaves of supracrustals within the charnockite massif must be older than c. 2100 Ma. The leucogranites yielded Pb–Pb whole rock isochron ages and U–Pb monazite ages at 510–530 Ma indicating the timing of melt generation and recrystallization. From the above, it can be argued that the rocks of TGB must have witnessed two thermal events during Paleoproterozoic and Neoproterozoic-Cambrian. However, it remains unclear whether or not the high-grade metamorphism occurred during Paleoproterozoic. The generation of small plutons of syenites and granites has been attributed to extensional tectonics between 770 and 560 Ma (Kovach et al., 1998).

The rock assemblage of TGB is comparable to the highland complex of Sri Lanka and the Paleoproterozoic terrane of southern Madagascar. Further, the similarities of the Paleoproterozoic khondalitic rocks in TGB and in the North China craton suggest that both the sequences might have been connected through supercontinent Columbia.

2.6.5 GEOPHYSICAL SIGNATURES

The high electrical resistivity of the upper crust in TGB displays northerly dip, in contrast to the southerly dip in MGB, which is consistent with the geometry of structural fabrics in both regions. The lower crust is resistive in both MGB and TGB, but it is moderately conductive along the AKSZ (Dhanunjaya Naidu, Manoj, Patro, Sreejesh Sreedhar, & Harinarayana, 2011). A moderate gravity low observed over the TGB might be due to the presence of low-density and less resistive khondalitic rocks. It needs to be studied further whether or not the TGB represents an extended part of the AKSZ.

2.6.6 **TECTONIC SYNTHESIS**

The khondalite—leptynite assemblage of TGB was interpreted as a sedimentary passive margin sequence with a maximum depositional age of 700 Ma (Santosh et al., 2009), but did not stipulate on which continental margin these presumed sediments were deposited. They have also considered that the TGB forms a part of Pan-African collisional orogen and the Nagercoil charnockite massif as the part of a continental arc in the realm of plate tectonic model. In another model, Rajesh, Arai, Satish-kumar, Santosh, and Tamura (2013) proposed that the khondalite—leptynite belt constitutes a Neoproterozoic-Cambrian (Pan-African) accretionary complex sandwiched between continental arcs of the MGB and Nagercoil block.

The available age data show that the high-grade metamorphism and granulite rock formation of TGB occurred during the late Neoproterozoic-Cambrian tectonothermal event associated with the final phase of the assembly of Gondwana supercontinent. The field observations and lithological characteristics suggest that the Nagercoil charnockite massif could be exotic in origin with distinct tectonic affinity compared to the other blocks that make up the SGT. Further, its clockwise P—T paths together with those of the other regions within the SGT have been interpreted to be the result of protracted Ediacaran-Cambrian orogenesis as the Neoproterozoic India collided with Azania-East Africa in a setting similar to modern day Tibet (Collins et al., 2014). The collision of the TGB with Azania is not conclusive, but the absence of Neoproterozoic magmatic protoliths in the Nagercoil massif implies that the Nagercoil massif may have formed the lower plate of this collision zone (Johnson, Clark, Taylor, Santosh, & Collins, 2015).

2.7 **TECTONIC EVOLUTION OF THE SGT**

2.7.1 **INTRODUCTION**

There has been growing evidence in recent years that the SGT was evolved through subduction—accretion and collisional processes. However, the polarity and the timing of subduction are open to discussion. A northward subduction model (Fig. 2.45) was first proposed based on satellite image

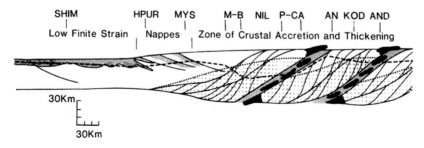

FIGURE 2.45

Northward subduction model for the tectonic evolution of the SGT.

Adapted from Drury, S.A., Harris, N.B.W., Holt, R.W., Reeves-Smith, G.J., & Wightman, R.T. (1984). Precambrian tectonics and crustal evolution in south India. Journal of Geology, 92, 3—20.

interpretation followed by large-scale geological and structural studies (Drury et al., 1984). According to this model of crustal thickening and accretion, the downward steepening and northward crustal underthrusting were considered to be the main mechanisms. In contrast, a southward subduction model was later invoked from the results of tomographic studies (Rai, Srinagesh, & Gaur, 1993). However, during the last two decades, southward subduction in the realm of plate tectonics has gained importance from different strings of geological and geophysical evidences from the SGT.

2.7.2 CAUVERY SUTURE ZONE

The CSZ has been described as the Gondwana suture zone and as a trace remnant of the Mozambique Ocean. Field relations, lithological assemblages including ophiolites, petrographic investigations, and geochemical characteristics reveal that the rocks of the CSZ are related to supra-subduction zone setting related to the opening and closure of Mozambique Ocean and their higher degree of mantle melting. The CSZ comprises north-verging frontal thrusts and complementary south-verging back thrusts making the CSZ as crustal-scale "flower structure" suggesting transpressional tectonic regimes and collisional processes akin to modern collisional belts (Fig. 2.46). Recently, remnants of oceanic crust within the CSZ were also reported suggesting two events of subduction—accretion events: one at the Neoarchean-Paleoproterozoic boundary, and the other at the Precambrian-Cambrian boundary.

Based on geochemical and isotopic systematics, a possible petrogenetic model suggests asthenospheric upwelling in an extensional setting, melting of enriched lithosphere and intersection of magmas with lower crustal domains with subduction related components of various ages (Santosh et al., 2014b). All the above studies confirm a Neoarchean-early Paleoproterozoic subduction

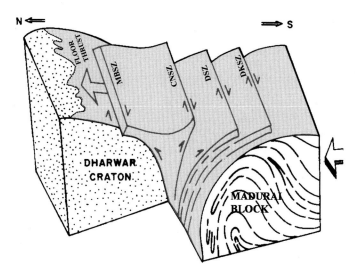

FIGURE 2.46

A 3D cartoon showing the "flower structure" across the CSZ.

Adapted from Chetty, T. R. K. Contrasting deformational systems and associated seismic patterns in Precambrian peninsular India, Current Science, 90, 7, 2006, 942-951.

system at the southern margin of the Dharwar craton, the remnants of which were incorporated within a chaotic mélange of the Neoproterozoic suture of CSZ (Chetty et al., 2016). Euhedral zircons with magmatic cores from the Banded Iron Formations from the CSZ yielded $^{206}Pb/^{238}U$ age of 760 ± 16 Ma probably marking the turning point from passive margin to active margin in the Wilson cycle and the construction of an arc-trench system with a southward subduction polarity (Sato et al., 2011b). The timing of the HP-UHT metamorphism in the CSZ and the MGB is constrained to be during 550−500 Ma. (Plavsa et al., 2015).

2.7.3 THRUST-NAPPE MODEL

A thrust-nappe tectonic model for the evolution of the SGT is proposed here based on multiscale field observations as well as the salient features derived from the structural cross sections of different tectonic blocks together with geophysical signatures and geochronological characteristics described in the foregoing sections. A regional ~ 500 km long N−S structural cross section has been constructed covering the entire spectrum of lithological formations and tectonic blocks of the SGT (Fig. 2.47). The NGB constitutes high-amplitude folds including large sheath folds and megaduplex structures are bound by floor thrust (Detachment zone) and the roof thrust. The floor thrust coincides with the basement of the Archean cratonic rocks. The presence of early recumbent folds in NGB suggests a zone of northward thrusting, thickening, and elevated crust. Subvertical shortening combined with continuing overthrusting may best explain the development of subhorizontal axial planar surfaces and the low magnitude of strain associated with the recumbent folds (Bastida

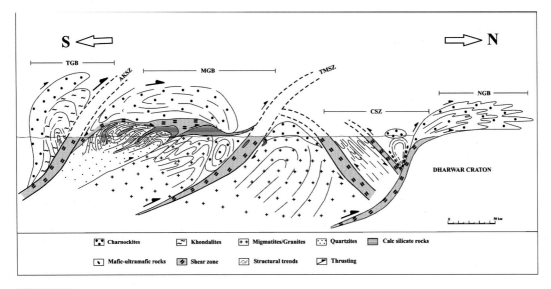

FIGURE 2.47

An interpretative regional N−S structural cross section depicting thrust-nappe tectonic model proposed for the evolution of the SGT.

et al., 2014). A well-defined crustal scale "flower structure" is well reflected in the section across the CSZ.

Sedimentary rock formations such as metapelites and metacarbonates are dominantly well exposed in the southern part of the MGB. These are highly deformed sheets of typically mixed lithologies and lie between thick slabs of granulites and appear to have been emplaced tectonically to resemble the intercalations of supracrustal rocks. It is likely that they must have been emplaced beneath crust-wide thrust zone during crustal shortening mixed with channel flow. The association of shelf sediments and stratiform basic igneous complexes suggest that they may be the sites of marginal basin closures when separate crustal masses collided and accreted. That the tectonized supraccrustal swathes in MGB may represent several such accreted crustal slabs through a northward sense of motion in the model is evident from southward dips of foliations within the MGB.

Widespread occurrence of extensive migmatites and granites along with isolated fragments of mafic—ultramafic igneous rocks are common in the northern part of MGB. These two domains are separated by TMSZ, which broadly coincides with the isotopic boundary. Regionally, the migmatites and the sedimentary sequences occur in close association and are overlain by charnockitic massif rocks.

The zone of migmatites may probably represent a basal detachment zone facilitating the translation of nappes through north-verging thrusts. The charnockitic masses in the form of highland areas are characterized by the presence of remnant recumbent fold structures representing large-scale thrust sheets and nappes. The sedimentary supracrustal swathes are highly deformed and lie between thick slabs of granulites and appear to have been emplaced tectonically beneath wide thrust zones marking lubrication and decoupling contributing to retrogression. All these processes gave rise to granite melt gneisses, granitoids and the development of migmatites.

The following features associated with charnockite massifs are critical in interpreting them as allochthonous in nature. They include: (1) existence of flat-lying foliations and recumbent fold structures; (2) generation of deformational fabrics, complex fold forms, and the presence of circumferential mylonitic fabrics all around the massifs; (3) significant continuous chemical-compositional variations, variation in the intensity of strain and deformation; (4) localized development of horizontal zones of melting and migmatization, grain size variation, presence or absence of garnet; and (5) the variation in the quantum of mafic content. All these features are spatially and genetically related to the thrust zones and the variability depends on their proximity to the thrust zones. The widespread presence of tectonic windows exposing the basement rocks surrounded by high land areas of MGB indicate the involvement of thrust-nappe tectonics in the evolution of the SGT. The generalized thrust-nappe model for the SGT also implies a complete allochthonous nature of all the supracrustals and granulitic units involving multiphases of thrusting. The evolution of the SGT involves accretion processes of island arc magmatic intrusives, thrust stacking, crustal coupling, granitic emplacement, and the obduction of ophiolite complexes indicating a complete range of geological processes encountered in modern orogenic belts. These features could thus be representative of multiple thrusting related to south-directed subduction processes followed by widescale collision of Gondwana fragments during the end phase of the Precambrian.

The recumbent structures depicted in the cross section represent the key structures, which are commonly developed in convergent orogenic belts (see Bastida et al., 2014). They are often large in SGT, and subsequent deformational structures make them difficult to establish their original geometries. The isoclinal character of these large folds and monotonous lithology, such as charnockite massifs, make them difficult in recognizing the recumbent structures. Further, recumbent sheath folds, if there are any, have a peculiar geometry and pose specific problems. The development of such large recumbent folds and thrusts involve ductile vertical shortening and horizontal stretching driven by lithostatic stress and a rock translation driven tectonic stress. The thrust-nappe tectonic model implies that there must have been two periods of transpressional deformation during the entire evolution history of SGT: the first, in the form of near-horizontal zones of transpressional displacements with top to the north during Neoarchean, and the second, a subsequent wrench-dominated transpressional tectonic regime along the E–W trending crustal-scale shear zones during the Neoproterozoic. This tectonic scenario may perhaps explain the geological complexity, and simple models may not fully explain the peculiarities of many specific issues involved in the tectonic evolution of the SGT.

In general, the SGT seems to expose synmetamorphic structural stacks of rootless, subhorizontal nappes that emerge from middle to lower crusts, tranported northwards against the Archean Dharwar craton during Neoproterozoic subduction–accretion–collisional processes. Considering the analogy from NE–Mozambique with similar thrust-nappe structures extending into the Dronning Maud land of East Antarctica and the available ages, the timing of nappe transport can be attributed to the final stage of Gondwana assembly between 480 and 600 Ma.

2.7.4 PLATE TECTONIC MODEL

The plate tectonic perspective was first attempted by Santosh et al. (2009) to explain the development of the tectonic evolution history of the SGT (Fig. 2.48) involving southward polarity of subduction along the CSZ with the counterparts of a foreland-fold-thrust belt developed to the north of

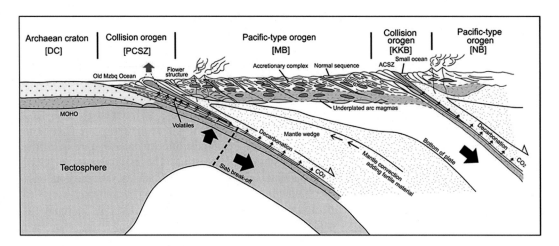

FIGURE 2.48

Plate tectonic model in understanding the evolution of the SGT.

Adapted from Santosh, M., Maruyama, S., & Sato, K. (2009). Anatomy of a Cambrian suture in Gondwana: Pacific-type orogeny in southern India? Gondwana Research, 16, 321–341.

the CSZ. They proposed a Pacific-type subduction in a rifted continental margin, representing the earlier orogeny that led to the consumption of an intervening Mozambique ocean during the later Cambrian collision akin to the Himalayan type, where in the continental margin of the Dharwar craton must have been subducted down to at least 60 km depth, which is evident from the presence of eclogite facies and HP granulite assemblages exposed in the region. The southward subduction is well defined by the existence of southward dipping and subparallel thrust surfaces all across the SGT. The structure of the lowermost units could be the foreland fold-thrust belts, the lower segments of which are represented by the currently exposed metasedimentary units with north-verging large-scale thrusts, which were subjected to intense deformation and migmatization. This scenario is broadly consistent with the thrust-nappe tectonic model and is evident from different structural cross sections presented in the foregoing.

The Zircon geochronology from plagiogranites, gabbros, and other associated rocks from the CSZ constrain the birth of the Mozambique Ocean floor prior to c. 800 Ma, probably marking the turning point from passive margin to active margin in the Wilson cycle and the construction of an arc-trench system with a southward subduction polarity. According to the model, initial Pacific-type orogeny culminated in the Himalayan-style collision along the CSZ, which was closely associated with the final closure of the Mozambique Ocean and the birth of the Gondwana supercontinent. It is well established that the Gondwana-forming orogens consumed the intervening Mozambique Ocean during the Neoproterozoic.

Based on the synthesis of available structural, petrological, geochemical, geochronological and geophysical data sets, the following composite schematic tectonic model proposed by Santosh et al. (2009) seems to be very appropriate in explaining the tectonic evolutionary history of the SGT. The model envisages an early rifting stage with the development of the Mozambique Ocean, followed by the southward subduction of the oceanic plate. The tectonic history of the SGT reveals a progressive sequence from Pacific-type to collision-type orogeny, which finally gave rise to a Himalayan-type Cambrian orogeny with characteristic magmatic, metasomatic, and metamorphic factories operating in subduction and collision setting.

The evolution of the SGT involves accretion processes of island arc magmatic suites, thrust stacking with duplex structures, and deformed sheath fold geometries, granitic emplacements, and obduction of ophiolite complexes: a complete range of processes like transpression associated with extrusion and exhumation, typical of modern orogenic belts. The SGT could be a classic representative of ancient subduction factory that witnessed large-scale collision related to Gondwana amalgamation.

The SGT is a distinct part of the East African Orogen (EAO), which comprises a collage of individual oceanic domains and continental fragments with different geodynamic settings of distinct orogenic styles between the Archean Sahara-Congo-Kalahari cratons to the west and the Antarctica in the east through Neoproterozoic India (Fritz et al., 2013). Taking the analogy from the EAO, the nappe assembly must have been completed around 620 Ma, and the Mozambique Ocean was closed in the part up to India. However, the oceans persisted between India and Antarctica beyond 620 Ma and were closed during 600−500 Ma, described as Kuunga orogeny. The entire EAO cycle spans during the period between break up of Rodinia (870−800 Ma) and the final amalgamation of Gondwana (500 Ma). The EAO is also considered as a "Transgondwanan supermountain range, whose formation and destruction had important implications for atmospheric oxygenation and for the evolution of higher organized organisms on the earth."

LIST OF ABBREVIATIONS

AKSZ	Achankovil shear zone
AOC	Agali Hill Ophiolite Complex
BIF	Banded Iron Formation
BR	Biligirirangan
CNSZ	Chennimalai-Noyil shear zone
CSZ	Cauvery shear/suture zone
CTSZ	Cauvery-Tiruchirapalle shear zone
DKSZ	Devattur-Kallimandayam shear zone
DOC	Devanur Ophiolite Complex
DSZ	Dharapuram shear zone
EAO	East African orogen
EGMB	Eastern Ghats Mobile Belt
KAC	Kadavur anorthosite complex
KKB	Kerala Khondalite Belt
KKPTSZ	Karur-Kambam-Painavu-Trichur shear zone
KMSZ	Kasaragod-Mercara shear zone
KOSZ	Karur-Oddanchatram shear zone
MBSASZ	Moyar-Bhavani-Salem-Attur shear zone
MBSZ	Moyar-Bhavani shear zone
MGB	Madurai Granulite Block
MOC	Manamedu Ophiolite Complex
MSF	Mahadevi sheath fold
MTSZ	Mettur shear zone
NGB	Northern Granulite Block
SGT	Southern Granulite Terrane
SASZ	Salem-Attur shear zone
SSZ	Suruli shear zone
TGB	Trivandrum Granulite Block
TMSZ	Theni-Madurai shear zone

REFERENCES

Alsop, G. I., & Holdsworth, R. E. (2006). Sheath folds as discriminators of bulk strain type. *Journal Structural Geology*, *28*, 1588−1606.

Anderson, J. R., Payne, J. L., Kelsey, D. E., Hand, M., Collins, A. S., & Santosh, M. (2012). High-pressure granulites at the dawn of the Proterozoic. *Geology*, *40*, 431−434.

Bartlett, J. M., Dougherty-Page, J. S., Harris, N. B. W., Hawkesworth, C. J., & Santosh, M. (1998). The application of single zircon evaporation and model Nd ages to the interpretation of polymetamorphic terrains: an example from the Proterozoic mobile belt of south India. *Contributions to Mineralogy and Petrology*, *131*(2−3), 181−195.

Bastida, F., Aller, J., Fernández, F. J., Lisle, R. J., Bobillo-Ares, N. C., & Menéndez, O. (2014). Recumbent folds: key structural elements in orogenic belt. *Earth Science Reviews*, *135*, 162−183.

Bhaskar Rao, Y. J., Chetty, T. R. K., Janardhan, A. S., & Gopalan, K. (1996). Sm-Nd and Rb-Sr ages and P-T history of the Archean Sittampundi and Bhavani layered meta-anorthositic complexes in Cauvery shear zone, south India: evidence for Neoproterozoic reworking of Archean crust. *Contributions to Mineralogy and Petrology, 125,* 237–250.

Bhaskar Rao, Y. J., Janardhan, A. S., Vijayakumar, T., Narayana, B. L., Dayal, A. M., Taylor, P. N., & Chetty, T. R. K. (2003). Sm-Nd model ages and Rb–Sr isotope systematics of charnockites and gneisses across the Cauvery Shear Zone, southern India: implications for the Archean-Neoproterozoic boundary in the Southern Granulite Terrain. *Geological Society of India Memoir, 50,* 297–317.

Brandt, S., Raith, M. M., Schenk, V., Sengupta, P., Srikantappa, C., & Gerdes, A. (2014). Crustal evolution of the Southern Granulite Terrane, South India: new geochronological and geochemical data for felsic orthogneisses and granites. *Precambrian Research, 246,* 91–122.

Braun, I., & Kriegsman, L. M. (2003). Proterozoic crustal evolution of southernmost India and Sri LankaIn M. Yoshida, B. F. Windley, & S. Dasgupta (Eds.), *Proterozoic East Gondwana: supercontinent assembly and breakup* (206, pp. 169–202). Geological Society of London, Special Publications.

Braun, I., Cenki-Tok, B., Paquette, J. L., & Tiepolo, M. (2007). Petrology and U–Th–Pb geochronology of the sapphirine-quartz-bearing metapelites from Rajapalayam, Madurai Block, Southern India: evidence for polyphase Neoproterozoic highgrade metamorphism. *Chemical Geology, 241,* 129–147.

Brown, M., & Raith, M. (1996). First evidence of ultrahightemperature decompression from the granulite province of southern India. *Journal of the Geological Society London, 153,* 819–822.

Burke, K. C., Dewey, J. F., & Kidd, W. S. F. (1977). World distribution of sutures: the sites of former oceans. *Tectonophysics, 40,* 69–99.

Cenki, B., & Kriegsman, L. M. (2005). Tectonics of the Neoproterozoic Southern Granulite Terrain, South India. *Precambrian Research, 138,* 37–56.

Chetty, T. R. K. (1996). Proterozoic shear zones in southern granulite terrain, India. In M. Santosh, & M. Yoshida (Eds.), *The Archean and Proterozoic terrain of Southern India with in Gondwana* (pp. 77–89). Gondwana Research Group Memoirs 3, Field Science Publications.

Chetty, T. R. K. (2006). Contrasting deformational systems and associated seismic patterns in Precambrian peninsular India. *Current Science, 90*(7), 942–951.

Chetty, T. R. K. (2015). The Cauvery Suture Zone: map of structural architecture and recent advances. *Journal of Geological Society of India, 85,* 37–44.

Chetty, T. R. K., & Bhaskar Rao, Y. J. (1998). Behaviour of stretching lineations in the Salem-Attur shear belt, southern Granulite Terrane, South India. *Journal of Geological Society of India, 52,* 443–448.

Chetty, T. R. K., & Bhaskar Rao, Y. J. (2006a). Strain pattern and deformational history in the eastern part of the Cauvery shear zone, southern India. *Journal of Asian Earth Sciences, 28,* 46–54.

Chetty, T. R. K., & Bhaskar Rao, Y. J. (2006b). The Cauvery Shear Zone, Southern Granulite Terrain, India: a crustal-scale flower structure. *Gondwana Research, 10,* 77–85.

Chetty, T. R. K., & Bhaskar Rao, Y. J. (2006c). Constrictive deformation in transpressional regime: field evidence from the Cauvery Shear Zone, Southern Granulite Terrain, India. *Journal of Structural Geology, 28,* 713–720.

Chetty, T. R. K., & Santosh, M. (2013). Proterozoic orogens in southern Peninsular India: contiguities and complexities. *Journal of Asian Earth Sciences, 78,* 39–53.

Chetty, T. R. K., Bhaskar Rao, Y.J., & Narayana, B.L. (2003). A structural cross-section along Krishnagiri-Palani corridor, southern granulite terrain India. In: M. Ramakrishnan (Ed.), *Tectonics of Southern granulite terrain, Kuppam-Palani Geotransect. Geological Society of India, Memoir, 50,* 255–277.

Chetty, T. R. K., Fitzsimons, I., Brown, L., Dimri, V. P., & Santosh, M. (2006). Crustal structure and tectonic evolution of southern granulite terrain, India. *Gondwana Research, 10,* 3–5.

Chetty, T. R. K., Mohanty, D. P., & Yellappa, T. (2012b). Mapping of shear zones in the Western Ghats, southwestern part of dharwar craton. *Journal Geological Society of India, 79,* 151–154.

Chetty, T. R. K., Yellappa, T., Nagesh, P., Mohanty, D. P., Sivappa, V. V., Santosh, M., & Tsunogae, T. (2011). Structural anatomy of a dismembered ophiolite suite from Gondwana: the manamedu complex, Cauvery suture zone, Southern India. *Journal of Asian Earth Sciences, 42,* 176–190.

Chetty, T. R. K., Yellappa, T., Mohanty, D. P., Nagesh, P., Sivappa, V. V., Santosh, M., & Tsunogae, T. (2012a). Mega sheath fold of the mahadevi hills, Cauvery suture zone, Southern India: implication for accretionary tectonics. *Journal of Geological Society of India, 80*, 747–758.

Chetty, T. R. K., Yellappa, T., & Santosh, M. (2016). Crustal architecture and Tectonic evolution of the Cauvery Suture Zone, Southern India. *Journal of Asian Earth Sciences, 130*, 166–191.

Clark, C., Collins, A. S., Kinny, P. D., Timms, N. E., & Chetty, T. R. K. (2009). SHRIMP U-Pb age constraints on the age of charnockite magmatism and metamorphism in the Salem Block, southern India. *Gondwana Research, 16*, 27–36.

Clark, C., Healy, D., Johnson, T., Collins, A. S., Taylor, R. J., Santosh, M., & Timms, N. E. (2015). Hot orogens and supercontinent amalgamation: a Gondwanan example from southern India. *Gondwana Research, 28*, 1310–1328.

Collins, A. S., & Pisarevsky, S. A. (2005). Amalgamating eastern Gondwana: the evolution of the Circum-Indian orogens. *Earth Science Reviews, 71*, 229–270.

Collins, A. S., Clark, C., & Plavsa, D. (2014). Peninsular India in Gondwana: tectonothermal evolution of the Southern Granulite Terrane and its Gondwanan counterparts. *Gondwana Research, 25*, 190–203.

Collins, A. S., Clark, C., Sajeev, K., Santosh, M., Kelsey David, E., & Matin, H. (2007). Passage through India: Mozambique Ocean suture, high pressure granulites and Palghat-Cauvery shear zone system. *Terra Nova, 19*, 41–147.

Devaraju, T. C., & Sadashivaiah, M. S. (1969). The charnockites of Satnur-Halaguru area, Mysore State. *Indian Minerals, 10*, 67–88.

Dhanunjaya Naidu, G., Manoj, C., Patro, P. K., Sreejesh Sreedhar, V., & Harinarayana, T. (2011). Deep electrical signatures across the Achankovil shear zone, Southern Granulite Terrain inferred from magnetotellurics. *Gondwana Research, 20*, 405–426.

Drury, S. A., & Holt, R. W. (1980). The tectonic framework of the south Indian craton: a reconnaissance involving LANDSAT Imagery. *Tectonophysics, 65*, T1–T5.

Drury, S. A., Harris, N. B. W., Holt, R. W., Reeves-Smith, G. J., & Wightman, R. T. (1984). Precambrian tectonics and crustal evolution in south India. *Journal of Geology, 92*, 3–20.

Fermor, L. L. (1936). An attempt at the correlation of the ancient schistose formations of Peninsular India. *Memoirs of the Geological Survey of India, 70*, 1–52.

Fonarev, V. I., Konilovl, A. N., & Santosh, M. (2000). Multistage metamorphic evolution of the trivandrum granulite block, Southern India. *Gondwana Research, 3*(3), 293–374.

Fritz, H., Abdulsalam, M., Ali, K. A., Bingen, B., Collins, A. S., Fowler, A. R., ... Viola, G. (2013). *Journal Asian Earth Sciences, 86*, 65–106.

Ghosh, J. G., Maarten, J., de Wit, & Zartman, R. E. (2004). Age and tectonic evolution of Neoproterozoic ductile shear zones in the Southern Granulite Terrain of India, with implications for Gondwana studies. *Tectonics, 23*, TC3006. http://dx.doi.org10.1029/2002TC001444.

Gilbert, E., & Merlie, O. (1987). Extrusion and radial spreading beyond a closing channel. *Journal of Structural Geology, 9*, 481–490.

Gopalakrishnan, K. (1996). An overview of Southern Granulite Terrain, India-constraints in reconstruction of Precambrian assembly of Gondwanaland. *Gondwana Nine 2* (pp. 1003–1026). Oxford and IBH Pub.

Goscombe, B., Gray, D., Armstrong, R., Foster, D. A., & Vog, J. (2005). Event geochronology of the Pan-African Kaoko Belt, Namibia. *Precambrian Research, 140*, 103–141.

Gupta, S., Rai, S. S., Prakasam, K. S., & Srinagesh, D. (2003). The nature of the Crust in Southern India: implication for Precambrian Crustal evolution. *Geo-Physical Research Letters, 30*, 1491. Available from http://dx.doi.org/10.1029/2002GL016770.

Harinarayana, T., Manoj, C., Naganjaneyulu, K., Patro, B. P. K., Karemunnisa Begam, S., Murthy, D. N., ... Virupakshi, G. (2003). Magnetotelluric investigation along Kuppam-Palani Geotransect, South India, 2D Modeling results. *Geological Society of India, Memoirs, 50*, 107–124.

Harris, N. B. W., Santosh, M., & Taylor, P. N. (1994). Crustal evolution in south India: constraints from Nd isotopes. *Journal of Geology*, *102*, 139–150.

Ishii, S., Tsunogae, T., & Santosh, M. (2006). Ultrahigh-temperature metamorphism in the Achankovil Zone: Implications for the correlation of crustal blocks in southern India. *Gondwana Research*, *10*, 99–114.

Janardhan, A. S., Newton, R. C., & Hansen, E. C. (1982). The transformation of amphibolite facies gneiss to charnockite in southern Karnataka and northern Tamil Nadu, India. *Contributions to Mineralogy and Petrology*, *79*, 130–149.

Jayananda, M., Martin, H., Peucat, J. J., & Mahabaleswar, M. (1995). Late Archean crust– mantle interactions: geochemistry of LREE-enriched mantle derived magmas. Example of the Closepet batholith, southern India. *Contributions to Mineralogy and Petrology*, *119*, 314–329.

Johnson, T. E., Clark, C., Taylor, R. J. M., Santosh, M., & Collins, A. S. (2015). Prograde and retrograde growth: an example from the Nagercoil Block, southern India. *Geoscience Frontiers*, *6*, 373–387.

Kelsey, D. E., & Hand, M. (2015). On ultrahigh temperature crustal metamorphism: phase equilibria, trace element thermometry, bulk composition, heat sources, timescales and tectonic settings. *Geoscience Frontiers*, *6*, e311–e356.

Kröner, A., Santosh, M., & Wong, J. (2012). Zircon ages and Hf isotopic systematics reveal vestiges of Mesoproterozoic to Archean crust within the late Neo protero zoic–Cambrian high-grade terrain of southernmost India. *Gondwana Research*, *21*, 876–886. Available from http://dx.doi.org/10.1016/j.gr.2011.05.008.

Kröner, A., Santosh, M., Hegner, E., Shaji, E., Geng, H., Wong, J., … Nanda-Kumar, V. (2015). Palaeoproterozoic ancestry of Pan-African high-grade granitoids in southernmost India: Implications for Gondwana reconstructions. *Gondwana Research*, *27*(1), 1–37.

Kovach, V. P., Santosh, M., Salnikova, E. B., Berezhnaya, N. G., Bindu, R. S., Yoshida, M., & Kotov, A. B. (1998). U-Pb zircon age of the Puttetti alkali syenite, southern India. *Gondwana Research*, *1*, 408–410.

Kumar, A., Charan, S. N., Gopalan, K., & Macdougall, J. (1998). A long-lived enriched mantle source for two Proterozoic carbonatite complexes from Tamil Nadu, southern India. *Geochimica Et Cosmochimica Acta*, *62*, 515–523.

Kusky, T. M. (2004). *Precambrian ophiolites and related rocks. Developments of Precambrian Geology* 13Amsterdam: Elsevier Publications.

Mahabaleswar, B., & Naganna, C. (1981). Geothermometry of Karnataka charnockites. *Bull Mineral*, *104*, 848–855.

Marimoto, T., Santosh, M., Tsunogae, T., & Yoshimura, Y. (2004). Spinel + quartz association from the Kerala khondalites, southern India: evidence for ultrahigh-temperature metamorphism. *Journal of Mineralogical and Petrological Sciences*, *99*, 257–278.

Maruyama, S., Masgo, H., Katayama, I., Iwase, Y., Toriumi, M., Osmori, S., & Aoki, K. (2010). A new perspective on metamorphism and metamorphic belts. *Gondwana Research*, *18*, 106–137.

Meert, J. G., & Lieberman, B. S. (2008). The Neoproterozoic assembly of Gondwana and its relationship to the Ediacaran–Cambrian radiation. *Gondwana Research*, *14*, 5–21.

Miller, J. S., Santosh, M., Pressley, R. A., Clements, A. S., & Rogers, J. J. W. (1996). A Pan-African thermal event in southern India. *Journal of Southwest Asian Earth Sciences*, *14*, 127–136.

Mohan, A., & Windley, B. F. (1993). Crustal trajectory of sapphirine-bearing granulites from Ganguvarpatti, South-India—evidence for an isothermal decompression path. *Journal of Metamorphic Geology*, *11*, 867–878.

Mohanty, D. P., & Chetty, T. R. K. (2014). Possible detachment zone in Precambrian rocks of Kanjamalai Hills, Cauvery Suture Zone, Southern India: Implications to accretionary tectonics. *Journal of Asian Earth Sciences*, *88*, 50–61.

Naganjaneyulu, K., & Santosh, M. (2010). The Cambrian collisional suture of Gondwana in southern India: a geophysical appraisal. *Journal of Geodynamics*, *50*, 256–267.

Newton, R. C., & Tsunogae, T. (2014). Incipient charnockite: characterization at the type localities. *Precambrian Research, 253*, 38−49.

Noack, N. M., Kleinschrodta, R., Kirchenbaura, M., Fonsecab, R. O. C., & Munker, C. (2013). Lu-Hf isotope evidence for Paleoproterozoic metamorphism and deformation of Archean oceanic crust along the Dharwar Craton margin, Southern India. *Precambrian Research, 233*, 206−222.

Peucat, J. J., Jayananda, M., Chardon, D., Capdevila, R., Fanning, C. M., & Paquette, J. L. (2013). The lower crust of the Dharwar Craton, Southern India: Patchwork of Archean granulitic domains. *Precambrian Research, 227*, 4−28.

Peucat, J. J., Mahabaleshwar, B., & Jayananda, M. (1993). Age of younger tonalitic magmatism and granulitic metamorphism in the South India transition zone (Krishnagiri area): comparison with older peninsular gneisses from the Gorur− Hassan area. *Journal of Metamorphic Geology, 11*, 879−888.

Pichamuthu, C. S. (1965). Transformation of Peninsular gneiss into charnockite in Mysore State. *Journal Geological Society of India, 2*, 46−49.

Plavsa, D., Collins, A. S., Foden, J. D., & Clark, C. (2015). The evolution of a Gondwanan collisional orogen: a structural and geochronological appraisal from the Southern Granulite Terrane, South India. *Tectonics, 34*(5), 820−857.

Plavsa, D., Collins, A. S., Payne, J. L., Foden, J. D., Clark, C., & Santosh, M. (2014). Detrital zircons in basement metasedimentary protoliths unveil the origins of southern India. *Geological Society of America Bulletin, 126*, 791−812, dio:10.1130/B30977.1.

Plavsa, D., Collins, A. S., Foden, J. F., Kropinski, L., Santosh, M., Chetty, T. R. K., & Clark, C. (2012). Delineating crustal domains in Peninsular India: age and chemistry of orthopyroxene-bearing felsic gneisses in the Madurai Block. *Precambrian Research, 198−199*, 77−93.

Praveen, M. N., Santosh, M., Yang, Q. Y., Zhang, Z. C., Huang, H., Singanenjam, S., & Sajinkumar, K. S. (2013). Zircon U-Pb geochronology and Hf isotope of felsic volcanics from Attappadi, southern India: implications for Neoarchean convergent margin tectonics. *Gondwana Research, 26*, 907−924. Available from http://dx.doi.org/10.1016/j.gr.2013.08.004.

Rai, S.S., Srinagesh, D., & Gaur, V.K. (1993). Granulite evolution in the South India - a seismic tomographic perspective continental crust of India. In: B.F. Radhakrishna (Ed.), Geological Society of *India, Memoir, 25*, 235−264.

Rajendra Prasad, B., Behera, L., & Koteswara Rao, P. (2006). A tomographic image of upper crustal structure using P and S wave seismic refraction data in the southern granulite terrain (SGT), India. *Geophysical Research Letters, 33*, L14301. http://dx.doi:10.1029/2006GL 026307, 1−5.

Ramakrishnan, M., & Vaidyanadhan, R. (2008). *Geology of India* (volume-1) Bangalore: Geological Society of India, 556 p.

Rajendra Prasad, B., Kesava Rao, G., Mall, D. M., Koteswara Rao, P., Raju, S., Reddy, M. S., ... Prasad, A. S. S. S. R. S. (2007). Tectonic implications of seismic reflectivity pattern observed over the Precambrian Southern Granulite Terrain, India. *Precambrian Research, 153*, 1−10.

Rajesh, H. M., & Santosh, M. (2004). Charnockitic magmatism in southern India. *Proceedings of the Indian Academy of Sciences: Earth and Planetary Sciences, 113*(4), 565−585.

Rajesh, V. J., Arima, M., & Santosh, M. (2004). Dunite, glimmerite and spinellite in Achankovil Shear Zone, south India: implications for highly potassic CO2-rich melt influx along an intra-continental shear zone. *Gondwana Research, 7*, 961−974.

Rajesh, V. J., Arai, S., Satish-kumar, M., Santosh, M., & Tamura, A. (2013). High-Mg low-Ni olivine cumulates from a Pan-African accretionary belt in southern India: implications for the genesis of volatile-rich high-Mg melts in suprasubduction setting. *Precambrian Research, 227*, 409−425.

Rajesh, K., & Chetty, T. R. K. (2006). Structure and tectonics of the Achankovil Shear Zone, southern India. *Gondwana Research, 10*(2006), 86−98.

Rameswara Rao, D., Narayana, B. L., Charan, S. N., & Natarajan, R. (1991). P-T conditions and geothermal gradient of gneiss-enderbitic rocks, Dharmapuri area. Tamil Nadu. *Journal of Petrology, 32,* 539−554.

Ram Mohan, M., Satyanarayan, M., Santosh, M., Sylvester, P. J., Tubrett, M., & Lam, R. (2012). Neoarchean suprasubduction Zone arc magmatism in southern india: geochemistry, zircon U-Pb geochronology and Hf isotopes of the sittampundi Anorthosite Complex. *Gondwana Research. Available from http://dx.doi.org/10.1016/j.gr.2012.04.004.*

Ramsay, J. G. (1981). Tectonics of the helvetic nappesIn K. R. Mc Clay, & N. J. Price (Eds.), *Thrust and nappe tectonics* (9, pp. 293−309). Special Publications of the Geological Society London.

Ratheesh-Kumar, R. T., Santosh, M., Qiong-Yan Yang, M., Ishwar-Kumar, C., Neng-Song Chend, & Sajeev, K. (2016). Archean tectonics and crustal evolution of the Biligiri Rangan Block, southern India. *Precambrian Research. Available from http://dx.doi.org/10.1016/j.precamres.2016.01.022.*

Renjith, M. L., Santosh, M., Li, T., Satyanarayana, M., Mahesh, K., Tsunogae, T., . . . Charan, S. N. (2015). Zircon U-Pb age, Lu-Hf isotope, mineral chemistry and geochemistry of Sundamalai peralkaline pluton from the Salem Block, southern India: implications for Cryogenian adakite-like magmatism in an aborted-rift. *Journal of Asian Earth Sciences. Available from http://dx.doi.org/10.1016/j.jseaes.2015.10.001.*

Ravindra Kumar, G. R., Srikantappa, C., & Hansen, E. C. (1985). Charnockite formation at Ponmudi in South India. *Nature, 313,* 207−209.

Reddi, A. G. B., Mathew, M. P., & Naidu, P. S. (1988). Aeromagnetic evidence of crustal structure in the Granulite terrain of Tamilnadu−Kerala. *Journal of Geological Society of India, 32,* 368−381.

Reddy, P.R., Rajendra Prasad, B., Vijaya Rao, V., Sain, K., Prasada Rao, P., Kare, P., & Reddy, M.S. (2003). Deep seismic reflection and refraction/ wide-angle reflection studies along Kuppam-Palani transect in southern granulite terrain of India. *Geological Society of India Memoir, 50,* 79−106.

Saitoh, Y., Tsunogae, T., Santosh, M., Chetty, T. R. K., & Horie, K. (2011). Neoarchean high-pressure metamorphism from the northern margin of the Palghat-Cauvery Suture Zone, southern India: petrology and zircon SHRIMP geochronology. *Journal of Asian Earth Sciences, 42*(3), 268−285.

Sajeev, K., Windley, B. F., Connolly, J. A. D., & Kon, Y. (2009). Retrogressed eclogite (20 Kbar, 1020°C) from the Neoproteorzoic Palghat-Cauvery suture zone, southern India. *Precambrian Research, 71,* 23−36.

Samuel, V. O., Santosh, M., Shuwen Liu, Wei Wang, & Sajeev, K. (2014). Neoarchean continental growth through arc magmatism in the NilgiriBlock, southern India. *Precambrian Research, 245,* 146−173.

Santosh, M. (1987). Cordierite gneisses of southern Kerala, India: petrology, fluid inclusions and implications for crustal uplift history. *Contributions to Mineralogy and Petrology, 96,* 343−356.

Santosh, M., & Kusky, T. (2010). Origin of paired high pressure-ultrahigh temperature orogens: a ridge subduction and slab window model. *Terra Nova, 22,* 35−42.

Santosh, M., Maruyama, S., & Sato, K. (2009). Anatomy of a Cambrian suture in Gondwana: Pacific-type orogeny in southern India? *Gondwana Research, 16,* 321−341.

Santosh, M., Marimoto, T., & Tsutsumi, Y. (2006a). Geochronology of the khondalite belt of Trivandrum Block, southern India: electron probe ages and implications for Gondwana tectonics. *Gondwana Research, 9,* 261−278.

Santosh, M., Collins, A. S., Tamashiro, I., Koshimoto, S., Tsutsumi, Y., & Yokoyama, K. (2006b). The timing of ultrahigh-temperature metamorphism in southern India: U−Th−Pb electron microprobe ages from zircon and monazite in sapphirine-bearing granulites. *Gondwana Research, 10,* 128−155.

Santosh, M., Tanaka, K., Yokoyama, K., & Collins, A. S. (2005). Late Neoproterozoic− Cambrian felsic magmatism along transcrustal shear zones in southern India: U−Pb Electron microprobe ages and implications for the amalgamation of the Gondwana supercontinent. *Gondwana Research, 8,* 31−42.

Santosh, M., Yokoyama, K., Biju-Sekhar, S., & Rogers, J. J. W. (2003). Multiple tectonothermal events in the granulite blocks of southern India revealed from EPMA dating, implications on history of supercontinents. *Gondwana Research*, 6, 29−63.

Santosh, M., Qiong-Yan Yang, Shaji, E., Tsunogae, T., Ram Mohan, M., & Satyanarayanan, M. (2014a). An exotic Mesoarchean microcontinent: the Coorg Block, southern India. *Gondwana Research. Available from http://dx.doi.org/10.1016/j.gr.2013.10.005.*

Santosh, M., Qiong-Yan Yang, Ram Mohan, M., Tsunogae, T., Shaji, E., & Satyanarayanan, M. (2014b). Cryogenian alkaline magmatism in the Southern Granulite Terrane, India: petrology, geochemistry, zircon U-Pb ages and Lu-Hf isotopes. *Lithos*, 208−209, 430−445.

Santosh, M., Shaji, E., Tsunogae, T., Ram Mohan, M., Satyanarayanan, M., & Horie, K. (2013). Neoarchean suprasubduction zone ophiolite from Agali hill, southern India: petrology, zircon SHRIMP U-Pb geochronology, geochemistry and tectonic implications. *Precambrian Research*, 231, 301−324.

Santosh, M., Xiao, W. J., Tsunogae, T., Chetty, T. R. K., & Yellappa, T. (2012). The Neoproterozoic subduction complex in southern India: SIMS zircon U-Pb ages and implications for Gondwana assembly. *Precambrian Research*, 192−195, 190−208.

Sato, K., Santosh, M., Tsunogae, T., Kon, Y., Yamamoto, S., & Hirata, T. (2010). Laser ablation ICP mass spectrometry for zircon U−Pb geochronology of ultrahigh-temperature gneisses and A-type granites from the Achankovil Suture Zone, southern India. *Journal of Geodynamics*, 50, 286−299.

Sato, K., Santosh, M., Chetty, T. R. K., & Hirtata, T. (2011a). U-Pb zircon geochronology of granites and charnockite from southern India: implications for magmatic pulses associated with plate tectonic cycles within a Precambrian suture zone. *Geological Journal*, 47, 237−252.

Sato, K., Santosh, M., Tsunogae, T., Chetty, T. R. K., & Hirata, T. (2011b). Subduction-accretion-collision history along the Gondwana suture in southern India; A laser ablation ICP-MS study on zircon chronology. *Journal of Asian Earth Sciences*, 40, 162−171.

Shimizu, H., Tsunogae, T., & Santosh, M. (2010). Petrology and geothermobarometry of Grt-Cpx and Mg-Al-rich rocks from the Gondwana suture in southern India: implications for high pressure and ultrahigh-temperature metamorphism. *Geoscience Canada, 2010, Abstract jd-650.*

Shimpo, M., Tsunogae, T., & Santosh, M. (2006). First report of garnet-corundum rocks from Southern India: implications for prograde high-pressure (eclogite-facies?) metamorphism. *Earth and Planetary Science Letters*, 242, 111−129.

Singh, A. P., Mishra, D. C., Vijayakumar, V., & Vyaghreswara Rao, M. B. S. (2003). Gravity-Magnetic signatures and crustal architecture along Kuppam−Palani Geotransect, South India. *Geological Society of India, Memoir*, 50, 139−164.

Stanek, K. P., Maresch, W. V., Grafe, F., Grevel, C. H., & Baumann, A. (2006). Structure, tectonics and metamorphic development of the Sancti Spiritus Dome (Eastern Escambray massif, Central Cuba). *Geologica Acta*, 4(1−2), 151−170.

Subramanyam, C., & Verma, R. K. (1982). Gravity interpretation of the Dharwar greenstone-gneiss-granite terrain in the southern Indian shield and its geological implications. *Tectonophysics*, 84, 225−245.

Teale, W., Collins, A., Foden, J., Payne, J., Plavsa, D., Chetty, T. R. K., ... Fanning, M. (2011). Cryogenian (∼830Ma) mafic magmatism and metamorphism in the northern Madurai Block, Southern India: a magmatic link between Sri Lanka and Madagascar. *Journal of Asian Earth Sciences*, 42, 223−233.

Tomson, J. K., Bhaskar Rao, Y. ,J., Vijaya Kumar, T., & Choudhary, A. K. (2013). Geochemistry and neodymium model ages of Precambrian charnockites, Southern Granulite Terrain, India: constraints on terrain assembly. *Precambrian Research*, 227, 295−315.

Tomson, J. K., Bhaskar Rao, Y. J., Vijaya Kumar, T., & Mallikharjuna Rao, J. (2006). Charnockite genesis across the Archean−Proterozoic terrane boundary in the South Indian Granulite Terrain: constrains from major-trace element geochemistry and Sr−Nd isotopic systematics. *Gondwana Research*, 10, 115−127.

Tsunogae, T., & Santosh, M. (2007). Ultrahigh-temperature metamorphism in Southern Indian Granulite Terrane. *IAGR Memoir10, 55–75.*

Venkatasivappa, V. (2014). Structural History and Geochemical Characteristics of Granulite Facies Rocks Around Namakkal, Southern Granulitie Terrane, India. (Unpublished PhD thesis).

Wickham, S. M., Janardhan, A. S., & Stern, R. J. (1994). Regional carbonate alteration of the crust by mantle derived magmatic fluids, Tamil Nadu, S. India. *Journal of Geology, 102,* 379–398.

Williams, P. F., Jiang, D., & Lin, S. (2006). *Interpretation of deformation fabrics of infrastructure zone rocks in the context of channel flow and other tectonic models* (268, pp. 221–235). London: Geological Society, Special Publications.

Yellappa, T., Chetty, T. R. K., Tsunogae, T., & Santosh, M. (2010). Manamedu complex: Geochemical constraints on Neoproterozoic suprasubduction zone ophiolite formation within Gondwana suture in southern India. *Journal of Geodynamics, 50,* 268–285.

Yellappa, T., Santosh, M., Chetty, T. R. K., Sanghoon Kwon, Chansoo Park, Nagesh, P., . . . Venkatasivappa, V. (2012). A Neoarchean dismembered ophiolite complex from southern India: geochemical and geochronological constraints on its suprasubduction origin. *Gondwana Research, 21,* 245–265.

Yellappa, T., Venkatasivappa, V., Koizumi, T., Chetty, T. R. K., Santosh, M., & Tsunogae, T. (2014). The mafic-ultramafic complex of Aniyapuram, Cauvery Suture Zone, Southern India: Petrological and geochemical constraints for Neoarchean suprasubduction zone tectonics. *Journal of Asian Earth Sciences, 95,* 81–98.

Yoshida, M., & Santosh, M. (1987). Charnockite in the breaking and making in Kerala, South India: tectonic and microstructural evidences. *Journal of Geosciences, Osaka City University, 30,* 23–49.

FURTHER READING

Dilek, Y., & Furnes, H. (2011). Ophiolite genesis and global tectonics: tectonic fingerprinting of ancient oceanic lithosphere. *Geological Society of America Bulletin, 123,* 387–411.

Lorraine, T., Aránzazu Piñán-Llamasb, & Charlotte Möller (2015). High-temperature deformation in the basal shear zone of an eclogite-bearing fold nappe, Sveconorwegian orogen, Sweden. *Precambrian Research, 265,* 104–120.

Prakash, D., Prakash, S., & Sachan, H. K. (2010). Petrological evolution of the high pressure and ultrahigh-temperature mafic granulites from Karur, southern India: evidence for decompressive and cooling retrograde trajectories. *Mineralogy and Petrology, 100*(1–2), 35–53.

Raith, M., Karmakar, S., & Brown, M. (1997). Ultra-hightemperature metamorphism and multistage decompressional evolution of sapphirine granulites from the Palni hill ranges, Southern India. *Journal of Metamorphic Geology, 15,* 379–399.

Rajesh, H. M. (2012). A geochemical perspective on charnockite magmatism in Peninsular India. *Geoscience Frontiers, 3,* 773–788.

Rajesh, H. M., & Santosh, M. (2013). Charnockites and *charnockites. Geoscience Frontiers, 3,* 737–744.

Sandiford, M. (1989). Horizontal structures in granulite terrains: a record of mountain building or mountain collapse. *Geology, 17,* 449–452.

Santosh, M., & Drury, S. A. (1988). Alkali granites with Pan-African affinities from Kerala, south India. *Journal of Geology, 96,* 616–624.

THE EASTERN GHATS MOBILE BELT

3

CHAPTER OUTLINE

3.1 INTRODUCTION

The Eastern Ghats Mobile Belt (EGMB), a Mesoproterozoic collisional orogen, occurs along the east coast of India with a strike length of over 900 km and a width varying from 50 km in the south to a maximum of 300 km in the north (Fig. 3.1). The margins of the EGMB are characterized by lithospheric shear zones at the contact with the Archean cratons of Dharwar and Bastar in the west and the Singhbhum craton in the north. The Eastern Ghats trend was described as a product of the Eastern Ghats orogeny (1570 Ma) and the strike was believed to be NE−SW over a large part of the EGMB with minor variations at southern and northern parts (Krishnan, 1961). The EGMB trends ESE−WNW to E−W in the north along the Mahanadi valley changing to near N−S in the

Proterozoic Orogens of India. DOI: http://dx.doi.org/10.1016/B978-0-12-804441-4.00003-1

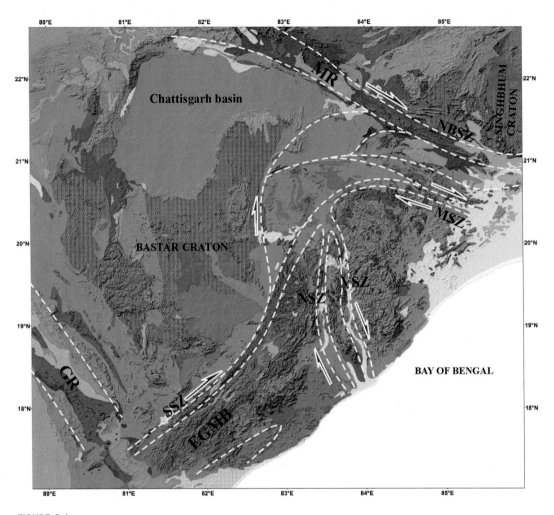

FIGURE 3.1

The geological map of the Eastern Ghats Mobile Belt (EGMB) superposed on the digital elevation model. *SSZ*, Sileru shear zone; *NSZ*, Nagavali shear zone; *VSZ*, Vamsadhara shear zone; *MSZ*, Mahanadi shear zone; *NBSZ*, Northern boundary shear zone; *GR*, Godavari Rift; *MR*, Mahanadi Rift. Lithologies of the EGMB: Orange— Khondalitic rocks; dark pink—Charnockitic rocks; light pink—migmatitic gneisses.

south. The EGMB is transected by two prominent NW−SE trending Precambrian-Cambrian rift structures: Godavari rift (GR) in the south and the Mahanadi rift (MR) in the north, both hosting Permo-Carboniferous coal-bearing sediments. Some workers proposed that GR and MR structures were formed as a consequence of large scale strike-slip motion along the preexisting Precambrian shear zones (Chetty, 1995a; Nash et al., 1996).

The EGMB principally consists of high-grade metamorphic gneisses, viz., khondalites (garnet−sillimanite gneisses), charnockites (hypersthene-bearing granites) and porphyritic gneisses.

The khondalitic group of rocks includes khondalites, quartzites, and calc granulites, while the char-nockitic group of rocks includes mafic granulites, enderbites, and variably retrograded charnockites (Fig. 3.2). All the rocks were metamorphosed to granulite facies conditions exhibiting well-developed gneissosity. Based on the distribution of dominant lithogroups/suites, the EGMB was longitudinally subdivided into four zones from west to east: (1) western charnockite zone (WCZ), (2) western khondalite zone (WKZ), (3) central charnockite—migmatite zone (CMZ), and (4) east-ern khondalite zone (EKZ) (Nanda & Pati, 1989). WCZ is broadly similar to the "basal charnockite zone" of Narayanaswami (1975), which is separated from the adjoining cratons by a "Transition Zone" containing diverse cratonic components with metamorphism grading to granulite facies.

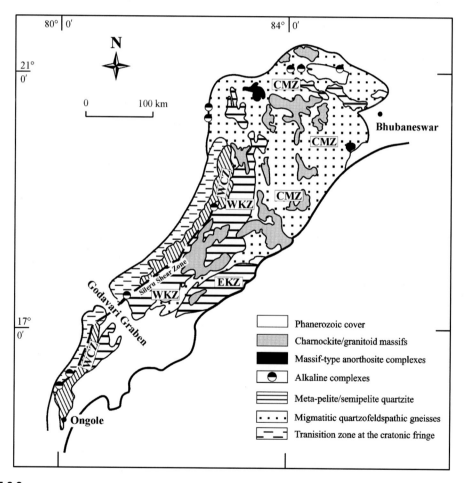

FIGURE 3.2

Geological map of the Eastern Ghats Mobile Belt (EGMB) showing different lithologies: *WCZ*, western charnockitic zone; *WKZ*, western khondalitic zone; *EKZ*, eastern khondalitic zone; *CMZ*, central migmatitic zone.

Modified after Gupta, S., & Bose, S. (2004). Deformation history of the Kunavaram complex, Eastern Ghats belt, India: Implication for alkaline magmatism along the Indo-Antarctica suture. Gondwana Research, 7, 1228–1335.

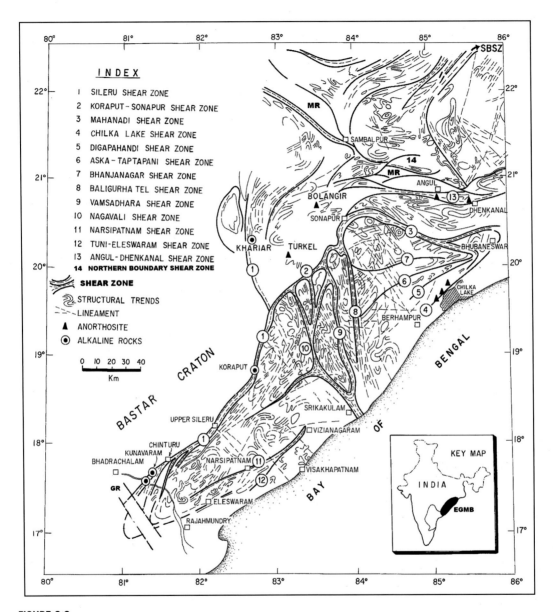

FIGURE 3.3

Map of the Eastern Ghats Mobile Belt (EGMB) showing the disposition and orientation of network of major shear zones and associated structural fabrics: 1. Sileru shear zone, 2. Koraput-Sonepur shear zone, 3. Mahanadi shear zone, 4. Chilkalake shear zone, 5. Digapahandi Shear zone, 6. Aska Taptapani Shear zone, 7. Banjanagar Shear zone, 8. Baligurha Tel shear zone, 9. Vamsadhara shear zone, 10. Nagavali shear zone, 11. Narsipatnam shear zone, 12. Tuni-Eleshwaram shear zone, 13. Angul-Dhenkanal shear zone, and 14. Northern Boundary shear zone. *MR*, Mahanadi Rift; *GR*, Godavari Rift. Note the locations of anorthosites and alkaline rocks.

However, recent studies concluded that the contact zone between the craton and the EGMB is marked by major ductile shear zones (Chetty & Murthy, 1993). The EGMB constitutes a network of major shear zones delineating different crustal blocks/tectonic domains within the EGMB. The internal structural homogeneity and continuity of structure with independent structural style and distinct lithological association characterize each domain indicating that the EGMB could represent a collage of juxtaposed terranes (Chetty, 2001), and a belt of distinct crustal provinces (Rickers, Mezger, & Raith, 2001) in variation with the longitudinal classification and division made by Ramakrishnan, Nanda, and Augustine (1998). Dobmeier and Raith (2003) proposed a classification of EGMB and the adjoining regions of lower metamorphic grade into four "provinces" based on integrated structural, metamorphic, and geochronologic information. These are: (1) Jeypore, (2) Rengali, (3) Krishna, and (4) Eastern Ghats Provinces. A "province" was defined as a crustal segment with a distinct geological history. Provinces are further subdivided into "domains" based on lithology, structure, and metamorphic grade, consistent with the observations of Nash et al. (1996). Simmat and Raith (2008) could further elucidate the complex tectono-metamorphic history involving multiple phases of deformation, metamorphism, and melting in different provinces based on U–Th–Pb monazite geochronometry.

The domain boundaries are genetically interrelated and are manifested with distinct magmatic activity. The domains may represent different thrust nappes as allochthonous tectonic sheets representing tectonostratigraphic terranes (Chetty, 2001). Several recent reviews have addressed various aspects such as the general geology (Gupta, 2004; Mukhopadhyay & Basak, 2009; Ramakrishnan et al., 1998; Vijaya Kumar & Leelanandam, 2008), shear zone network (Chetty & Murthy, 1998a; Chetty, 2001), geochronological history (Das, Bose, Karmakar, Dunkley, & Dasgupta, 2011; Upadhyay, 2008), thermochronological history, and the constituent domainal architecture of the EGMB (Dasgupta, Bose, & Das, 2013; Dobmeier & Raith, 2003; Gupta, 2012).

The network of major ductile shear zones (Fig. 3.3) was first recognized and described from the EGMB (Chetty & Murthy, 1993, 1998a). The shear zones vary in their extension and orientation, often deviating from the regional NE–SW trend of the EGMB. They include: near N–S trending shear zones south of GR, conformable NE–SW striking shear zones north of GR, orthogonal N–S shear zones in the central part, and strike parallel E–W shear zones in the northern part. The lithological contacts and the foliation trajectories are, in general, conformable to the shear zones in their proximity. For the sake of brevity and convenience, the EGMB has been described here by dividing the EGMB into three parts to consider the lithological associations, structural styles, metamorphism, and geochronology (Fig. 3.2). The parts from south to north include: EGMB-south, occurring to the south of GR; the EGMB-central, between the GR and the MR, and the EGMB-north, that occurs to the north of Mahanadi shear zone (MSZ).

3.2 EGMB-SOUTH

3.2.1 GENERAL GEOLOGY

The EGMB-south occurs to the south of GR structure extending up to Ongole before plunging into the Bay of Bengal. It is a 15–60 km wide linear zone of granulite facies rocks in association with migmatites situated to the east of the Cuddapah basin, which is described here as the Terrane

Boundary Shear Zone (TBSZ) (Fig. 3.4). The TBSZ hosts several elliptical to linear gabbro—anorthosites, nepheline syenite, and granitic plutons, which are aligned in a NNE—SSW trend with high-strain margins. The regional foliation trajectory map shows that the gneissosity in this zone is complex and swerves around the plutonic complexes suggesting that some of these plutons must have been emplaced passively into the tectonic voids provided by transpressional tectonic regime.

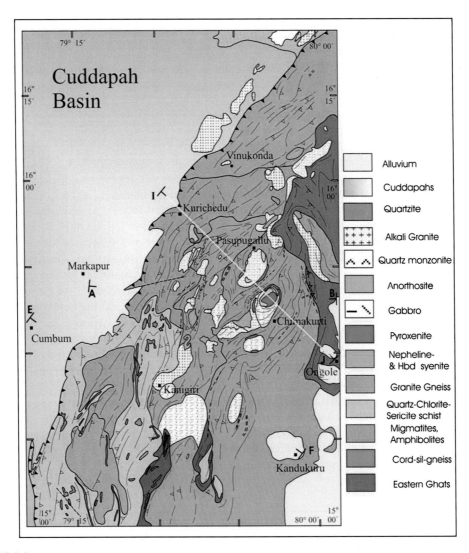

FIGURE 3.4

Structure and tectonic map of the EGMB-south (also described as TBSZ) displaying foliation trajectories, complex thrust/shear zones, and a chain of plutons along with Nallamalai thrust at the eastern margin of the Cuddapah basin (unpublished thesis of Nagaraju, 2007).

Important rock formations in the region from west to east are: Cuddapah sediments, Nellore schist belt (NSB), migmatized gneisses, and granulite facies rocks of the EGMB. A major curvilinear thrust trending NE−SW in the north and NW−SE in the south, marks the eastern margin of Cuddapah basin. Several shear zones exist at the contacts of all major tectonic units mimicking the geometry of Cuddapah marginal thrust. The region is intruded by alkaline, gabbro−anorthosite−ultramafic, and granitic plutons. Considering the lithological variations and the associated structural features, the entire region of the EGMB-south is also described as the TBSZ. The general foliation trend is N−S in the south and NE−SW in the north, which wraps around the plutons. The foliations in general show medium to steep easterly dips (50−75 degrees) and the exceptions are found near the plutonic margins, where the country rocks are also folded. The plutons in the southern part are circular/discoidal in shape, while the plutons in the middle part are ellipsoidal/lensoidal and those in the northern part are relatively more elongated. These geometrical variations show the heterogeneous distribution of strain along the TBSZ. The plutons represent apparently filled-up tectonic voids provided by the transpressional deformation along the TBSZ located between the Mobile Belt and cratonic units.

This part of the TBSZ has also been described as Krishna province (Dobmeier & Raith, 2003) with subdivisions of Ongole domain (granulite facies rocks) and Udayagiri domain (NSB). The presence of pervasive and en echelon shear zones and associated lithologies indicate that the TBSZ could be considered as the interface between the eastern Dharwar craton and the EGMB. The foliations in the region strike NE−SW with steep to moderate easterly dips and the stretching lineations are moderately plunging toward the southeast. The west-verging nature of regional fold structures and associated structural fabrics is consistent with the regional compression directed toward the northwest.

The TBSZ consists of intensely reworked migmatitic gneisses, high Mg−Al granulites (cordierite−sillimanite gneisses), leptynites, granitic gneisses, amphibolites, sporadic charnockites, and shares the litho−tectonic characteristics of both the adjacent terrains. These lithologies were overprinted by the subsequent deformational events resulting in the development of a complex zone consisting of mylonites, phyllonites, larger quartz reefs, pegmatite dykes, and swarms of anastomizing quartz veins. These are intruded by gabbro−anorthosites bodies, alkaline, granitic plutons, and dolerite dykes; thus, the zone has been aptly named as the Prakasam Alkaline Province (Leelanandam, 1989). These magmatic complexes show diverse chemical compositions and preserve various magmatic to deformational structures. The plutons, ranging in composition from mafic−ultramafic to granitic, quartz syenite to nepheline syenite, reveal not only their mantle-crust interactions but also help in understanding the complex tectonothermal events related to their emplacement and evolution. Interestingly, all plutonic bodies are confined to the shear thrust zones within the TBSZ. UHT metamorphism was also documented around some of these plutonic complexes. Available geochronological data for some of the plutons suggest that they range in age from 1400 to 1000 Ma (Sarkar & Paul, 1998). The lithounits of the EGMB in this domain (Ongole granulites) yielded metamorphic ages of ∼ 1670 Ma (Dobmeier & Raith, 2003). However, so far there are no reports of evidence for a Grenvillian event of granulite facies metamorphism, which is widespread in the other parts of the EGMB.

It appears that the TBSZ witnessed a short-lived, mantle-derived magmatism between 1.72 and 1.70 Ga followed by a prolonged polyphase granulite facies metamorphism between 1.65 and 1.55 Ga (Mezger & Cosca, 1999), and the event was correlatable with the collision between eastern

Dharwar craton and the Napier complex of East Antarctica. A second major igneous event, associated with extensive tholeiitic and alkaline magmatism, occurred between 1.4 and 1.2 Ga possibly in response to continental rifting and/or postcollisional relaxation of the thickened lithosphere (Leelanandam, 1989).

3.2.2 MAGMATISM

3.2.2.1 Gabbro plutons (Pasupugallu-Chimakurti)

The Pasupugallu and Chimakurti plutons, both elliptical in plan view and aligned in a NE−SW direction, are located in the central part of the TBSZ (see Fig. 3.4). The country rocks include cordierite−sillimanite quartzofeldspathic gneisses and migmatized amphibolites and granulites, with well-developed mylonitic foliations, stretching lineations, and sigmoidal porphyroclasts (Nagaraju, 2007). The foliation trajectory map shows that the plutonic fabrics got deflected in a dextral manner along with the mylonitic fabrics in the host rocks. A NE-SWtrending shear zone with moderate stretching lineations occurs in between the two plutons along which the deflection of plutonic as well as the host rock fabrics are well exposed (Fig. 3.5).

The Pasupugallu Gabbro Pluton (PGP) consists of gabbro, leucogabbro, and anorthosite units in the order of abundance, while the Chimakurti mafic−ultramafic pluton (CMP) is a concentrically zoned layered complex comprising gabbro, leuco-gabbronorite, anorthosite, and olivine−clinopyroxenite from margin to the core. The lithounits display rhythmic magmatic layering (steep dips in the margins and moderate/gentle in the central parts) and the intensity of layering increases towards the plutonic margins with the grain size reduction. Several dolerite and pegmatite dykes traverse the pluton with variable orientations. The structural mapping of PGP shows steep inward dipping magmatic foliations at the margins and moderate to gentle dips in the central parts (Fig. 3.6). The NE−SW trending shear zone separates the two plutons of Pasupugallu and Chimakurti and is marked by mylonitized cordierite−sillimanite−gneisses and quartzofeldspathic gneisses. The regional foliation trajectory map shows that the gneissosity wraps around the plutonic complexes, and displays local fold patterns due to the recurring magmatic episodes.

The magnetic lineation trajectories of PGP also define circular/spiral pattern in the form of lobes in both northern and southern domains. Lineations are mostly margin parallel, but radial distribution of moderately to sub horizontally plunging lineations at the northern and eastern margins imply a lateral flow of magma toward the margins from the central part in the northern domain. In the southern domain, the central lobes are characterized by steep plunges, while the margins are marked by moderate to gentle plunging lineations, indicating that the ascent of magma was restricted to the central parts. The upward migration and lateral spread of magma must have been generated from the same stress pattern. Broadly, the PGP shows helicoidal foliation geometry. The observed solid-state fabrics at the margins are likely to be coeval with the later stages of emplacement, but not related to the post-emplacement deformational event. These features define the characteristics for the syntectonic emplacement of plutons with regional deformation. The differences in the behavior of lineations, at some places, could be due to strain variation during the magmatic flow in the form of a helicoidal/spiral pattern, which could be related to transpressional deformation along the boundary shear zones (Nagaraju, Chetty, Vara Prasad, & Patil, 2008). All the features, like helicoidal flow trajectories, the transition from magmatic to mylonitic fabrics toward the margins, the presence of marginal dextral

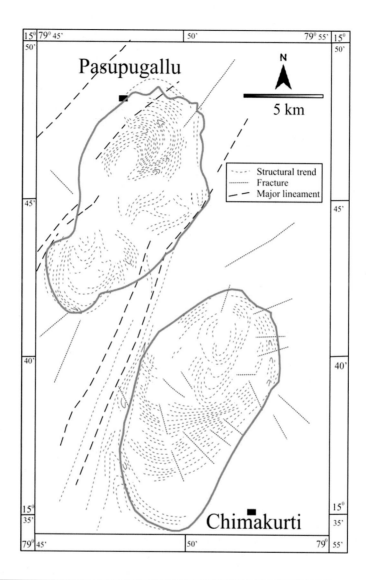

FIGURE 3.5

Image interpretation of digitally processed IRS-1D LISS3 + PAN merged data of area comprising Pasupugallu gabbro and Chimakurti mafic plutonic complexes on 1:50,000 scale.

Modified after Nagaraju,J and Chetty, T.R.K. Imprints of tectonics and magmatism in the south eastern part of the Indian shield: satellite image interpretation, Journal of Indian Geophysical Union 18,2, 2014, 165-182

strike-slip shear zone, the deflection of magnetic foliations and lineations along the Riedel shear zones within the pluton, internal flow folds, the shape and long axis of the pluton, remarkable changes in the lineation patterns at the margins and the dominant constrictional magnetic strain ellipsoids, favor the hypothesis of syntectonic emplacement of the PGP (Nagaraju & Chetty, 2005).

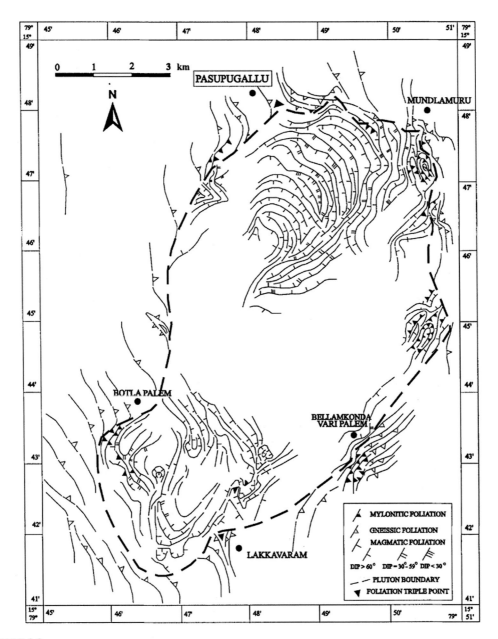

FIGURE 3.6

Map of foliation trajectories of the Pasupugallu gabbro pluton with dip variations, exhibiting the typical helicoidal structure within the pluton.

After Nagaraju, J. & Chetty, T.R.K. (2005). Emplacement History of Pasupugallu Gabbro Pluton, Eastern Ghats Belt, India: A structural study. Gondwana Research, 8, *87–100.*

The CMP exhibits concentric, circular, and inward-dipping foliation pattern (Fig. 3.7). Anorthosites and olivine−clinopyroxenites in the central part are very coarse-grained and generally massive. The gabbros at the marginal parts of the pluton are relatively fine-grained and more deformed, showing peripheral granulation, bent and erased twin lamellae giving rise to the solid-state fabric in the plutonic rocks. Petrographically, the rocks in the central parts consist of magmatic cumulate textures essentially of plagioclase and clinopyroxene (augite) with olivine, biotite, spinel and opaques as accessories. Petrography, mineral chemistry, geochemistry, and isotopic compositions suggest that the rocks of the CMP were derived from a high-Al tholeiitic parental magma. The ferrosyenite exhibits A-type characteristics and are interpreted as a differentiate or partial melt of a tholeiitic source (Vijaya Kumar, Carol, Ronald Frost, & Kevin, 2007). In view of their spatial and geochemical affinities, the Chimakurti gabbro−anorthosite pluton was considered to be equivalent of Errakonda ferrosyenite pluton which was dated as 1352 Ma by using the U−Pb Zircon method. Pasupugallu and Chimakurti lithologies show similar geochemical characteristics of the Island arc low-K tholeiitic nature indicating the partial melting process in the upper mantle generated by subduction-related regime for their genesis (Vijaya Kumar & Ratnakar, 2001). The regional foliation trajectories along the TBSZ show that the gneissosity wraps around the plutonic complexes, indicating that some of these plutons are likely to have been filled passively into the tectonic voids formed due to a combination of compression and strike-slip regime.

3.2.2.2 Granite plutons (Kanigiri, Podili, Vinukonda)

A number of granite plutons of Proterozoic age occur all along the TBSZ intermittently and the important plutons are located around Kanigiri, Podili, and Vinukonda. While Kanigiri and Podili plutons are spatially connected in the central part, Vinukonda is situated at the western margin of the TBSZ. The details of each pluton are described below. Kanigiri pluton (KGP), exposed in the form of hills, is leucocratic and massive biotite granite surrounded by the host rocks of quartz mica-schist and chlorite schist of the NSB (Fig. 3.8). The KGP trends NNE−SSW to NE−SW and the southern part exposes a ductile shear zone marked by intense deformation. Enclaves of fine-grained metabasic rocks of NSB in the central part of KGP show sharp contacts. Fluorite bearing quartzo−feldspathic veins also occur along minor shear zones within the pluton. Petrological and geochemical studies of KGP indicate the presence of rare metals and molybdenite. Fluorite is a conspicuous accessory mineral commonly noticed as discrete crystals, within the biotite granite, aplitic, and quartzo−feldspathic veins indicating crystallization of the host granite as later quartzo-feldspathic phases from a fluorine saturated magma suggesting a sedimentary source (Sesha Sai, 2004).

The Kanigiri biotite granite and Podili alkali granite plutons are emplaced along the contact zone between Udayagiri Group of upper NSB toward the west (chlorite schist, agglomerate tuffs and intercalated quartzite) and the sheared granite towards the east. Enclaves of rock units of older NSB are widely noticed across Kanigiri-Podili granite plutons (PdGPs), while deformed basement biotite granite gneiss are exposed intermittently as low-lying outcrops along the western margin of TBSZ.

PdGP, located to the south of Podili town, represents a deformed leucocratic alkaligranite pluton. The PdGP occurs in the form of hills in an apparent continuation with KGP and its host rocks of quartz−mica schist, chlorite schist and intercalated quartzites of NSB. Enclaves of chlorite schist and meta-acid volcanics of NSB are also preserved within the pluton. The presence of an

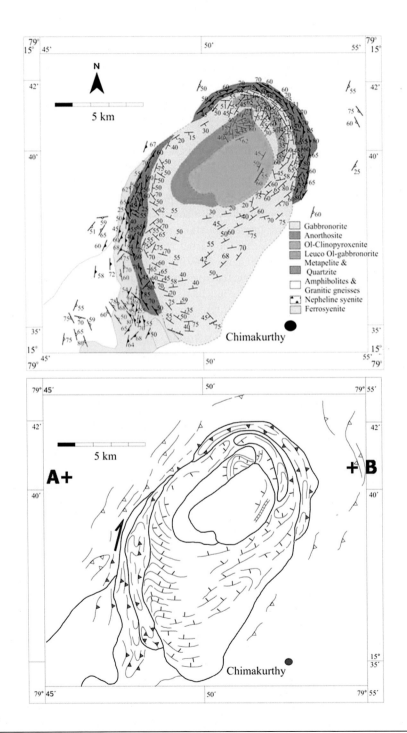

FIGURE 3.7

Map of Chimakurti mafic pluton showing major lithologies (A) and foliation trajectories (B) (unpublished thesis of Nagaraju, 2007).

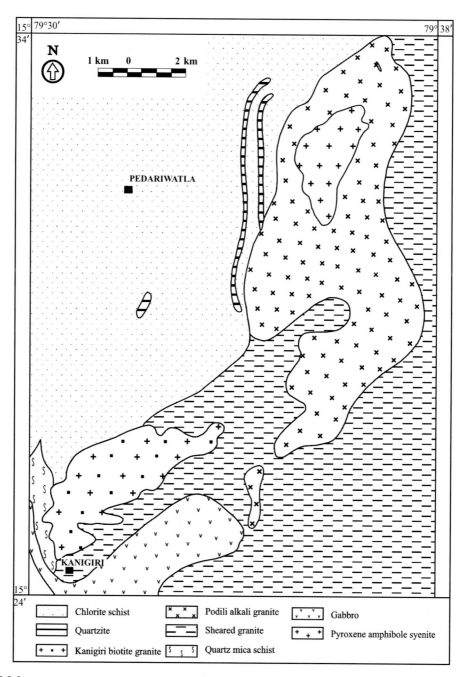

FIGURE 3.8

Geological map of granite plutons and the host rocks around Kanigiri and Podili (personal communication from Sesha sai, GSI).

undeformed pyroxene—amphibole syenite body in the northern part of PdGP is a striking feature. Tourmaline-bearing quartz veins traverse the pluton in the western part while blue quartz is conspicuous in the southern part. The deformational fabric trending N—S to NNW—SSE is well developed throughout the pluton and is relatively more deformed along the margins. The interlying region between the Podili and KGPs is occupied by the dominant presence of quartz—chlorite schist and quartzites of NSB. Intensely deformed hornblende—biotite gneiss with NNE—SSW deformational fabrics is well exposed to the east of Kanigiri—Podili plutons. A volcanic plug composed of rhyodacite is reported to the east of the PdGP. Enclaves of the rhyodacite are noticed in the central part of the PdGP, indicating that the volcanic plug is possibly a part of the preexisting suite of lithounits belonging to the older Archean NSB.

A small, semielliptical body of pyroxene—amphibole syenite occurs in the central part of northern part of PdGP. It is relatively undeformed, coarse-grained, massive, mesocratic and essentially composed of microperthite, plagioclase, and amphibole with subordinate quartz, biotite, and clinopyroxene while sphene, chlorite, monazite, apatite, carbonates, and ilmenite are observed as accessory minerals. The field observations, distinct mineralogy, and chemical characteristics suggest that both KGP and PdGP are deformed along the margins and were emplaced along the TBSZ in a late-orogenic phase close to the vicinity of a possible collision boundary zone. The chemistry of these granites shows that they are crystallized from a fluorine saturated magma derived from the partial melting of enriched continental crust along the TBSZ.

All the granite plutons including the KGP and PdGP are intensely deformed particularly along the margins, while the development of crude foliation is observed in the central parts. Petrographically, a majority of these granites vary from alkali feldspar granite to granite. The field observations, mineralogical association, and chemical characteristics suggest that the emplacement of these granite plutons was restricted to TBSZ possibly during the late-orogenic to anorogenic tectonic setting close to the vicinity of a collision boundary zone (Sesha Sai, 2013). Although both KGP and PdGP are spatially coexisting, coeval (Mesoproterozoic), and ferroan in nature, they are distinct in their mineralogical characteristics. The PdGP is riebeckite—arfvedsonite—biotite bearing hypersolvus granite with higher Na_2O/K_2O ratio, while the KGP is essentially a subsolvus two-feldspar biotite granite with lower Na_2O/K_2O ratio. Fluorite is a conspicuous accessory mineral in both the plutons. The chemistry of both the plutons shows typical characteristic features of anorogenic A-type granites. Rb—Sr dating yielded an isochron age of 1120 ± 25 Ma for Kanigiri granite (Gupta, Pandey, Chabria, Banerjee, & Jayaram, 1984), while Mesoproterozoic age of 1.33 Ga was attributed to the plagiogranite of Kanigiri ophiolite mélange (KOM) (Dharma Rao, Santosh, & Yuan, 2011) that occurs in the vicinity.

Vinukonda granite pluton (VGP) is located in the close vicinity of the Eastern Cuddapah thrust and immediately to the southwest of Vinukonda town. The VGP is leucocratic, medium to coarse—grained alkali feldspar granite in the form of a hill range (6 kmx 2 km) trending NW—SE with intense deformational fabrics along the margins. The VGP intruded the metamorphosed and strongly deformed granitic epidote—biotite gneisses, which were recrystallized during epidote—amphibolite facies metamorphism. The VGP consists of a medium to coarse—grained and weakly porphyritic leucocratic meta-granite with multigrain biotite blotches of up to two cm length that impart a spotted appearance. Widespread titanite—epidote amphibolite layers within the VGP are interpreted as metamorphosed basaltic dykes that intruded the plutonic precursors of the gneisses (Dobmeier, Lütke, Hammerschmidt, & Mezger, 2006). A supracrustal rock unit of magnetite—garnet—biotite schist (25 × 3 m) also occurs within the granitic gneisses. Fluorite is a common accessory phase in the white mica-bearing biotite—plagioclase—quartz—K—feldspar meta-granite. Accessory phases include apatite, magnetite, titanite, and zircon. Broadly, the foliations in the pluton show WNW—SSE trends. A shear zone with mylonitic fabrics occurs along the eastern margin of the pluton separating the spotted meta-granite and a medium-grained grayish meta-granite. The two-mica character of VGP indicates its subsolvus nature. Fluorite is noticed as conspicuous accessory mineral in the host alkali feldspar granite (Sesha Sai, 2013). The zircons from VGP yielded an age of 1590 Ma, which is interpreted as the emplacement age of the VGP (Dobmeier et al., 2006). The time of emplacement of the granitic precursor could be coeval with the emplacement of calc—alkaline plutons in the TBSZ (Ongole domain) between 1720 and 1704 Ma.

3.2.2.3 Alkaline plutons (Elchuru-Kunavaram)

Alkaline magmatism is restricted to the western part of the TBSZ, marked by a crustal-scale shear zone that lies at the junction between the regions of cratonic and granulite facies rocks. The magmatism is represented in the form of several alkaline plutons of different sizes and shapes located at Kunavaram, Elchuru, Purimetla, and Uppalapadu. A detailed review of the geology, petrology, genesis and nature of these plutons has been well presented by Leelanandam (1989). In the present context, Elchuru and Kunavaram alkaline plutons (KAPs) are chosen for more detailed explanation, described below.

The Elchuru alkaline pluton (EAP), an oval-shaped body with its long axis trending NE—SW, shows sharp contacts with the surrounding gneisses (Fig. 3.9). Nepheline syenite forms the major rock type in EAP with subordinate mafic varieties that include ijolite, malignite, shonkinite, melanocratic nepheline diorite and mesocratic nepheline syenite (Ratnakar & Vijaya Kumar, 1995). The rocks are dominated by melanocratic members and are massive in the central part while they are deformed and intensely foliated at the margins. At places, the mafic rocks are intruded by veins and apophyses of nepheline syenite on a centimeter scale. The mineral assemblages suggest that the metamorphism reached an amphibolite—facies condition. Steeply dipping lamprophyre dyke swarms, with small fragments of nepheline syenite, cut across all lithologies of the EAP. The melanocratic alkaline rocks contain abundant clinopyroxene and biotite with minor amounts of amphibole. The abundance of biotite suggests that the original magma was hydrous and rich in potassium. The mesocratic rocks contain predominantly amphibole, followed by clinopyroxene and biotite. The leucocratic rocks comprise all the three ferromagnesian minerals in subordinate amounts and in variable proportions, though the subsolvus varieties are generally devoid of clinopyroxene (Madhavan & Leelanandam, 1977). Presence of lamprophyric dykes and the absence of syenite, quartz syenite, and gabbro makes EAP different from the other alkaline plutons of the EGMB. The rocks of EAP yielded a whole rock Rb—Sr isochron age of 1242 ± 33 Ma and was interpreted as the age of magma emplacement (Subba Rao, Bhaskar Rao, Siva Raman, & Gopalan, 1989). The zircon and biotite ages from the EAP also indicate that they were deformed and metamorphosed during Pan-African tectonism, which was possibly related to the westward thrusting of the granulite facies rocks of the EGMB (Upadhyay, 2008).

The KAP occurs in the form of a chain of hills and ridges trending NE—SW and is confined to the Sileru shear zone (SSZ) at the western margin of the EGMB. South of GR, the KAP is flanked by the granite gneisses in the east and amphibolites in the west (Fig. 3.10). The dominant rock

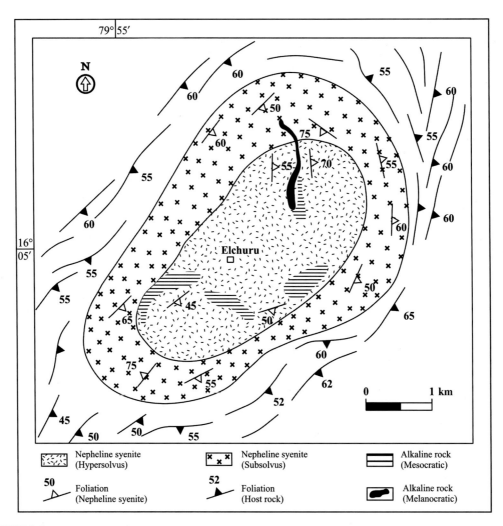

FIGURE 3.9

Simplified geological map of Elchuru alkaline pluton.

Modified after Leelanandam, C. (1989). The Prakasam Alkaline Province in Andhra Pradesh, India. Journal of Geological Society of India, 34, *25–45.*

types of the pluton are nepheline syenites composed primarily of microcline (perthite), plagioclase, and nepheline with subordinate biotite, hornblende, clinopyroxene, calcite, apatite, and sphene. Depending on the proportion of hornblende, biotite, and clinopyroxene, the nepheline syenites in Kunavaram have been classified into hornblende bearing, biotite-bearing and clinopyroxene-bearing varieties (Subba Rao, 1971). In addition to nepheline syenites, syenites (perthitic alkali

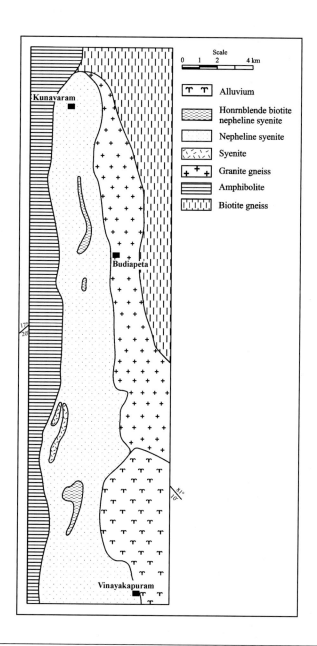

FIGURE 3.10

Simplified geological map of Kunavaram alkaline pluton.

Compiled and modified from Subba Rao, K.V. (1971). The Kunavaram Series-a group of alkaline rocks, Khammam District, Andhra Pradesh, India. Journal of Petrology, 12, 621–641 *and Gupta S. and Bose S., Deformation history of the Kunavaram complex, Eastern Ghats belt, India: Implication for alkaline magmatism along the Indo-Antarctica suture,* Gondwana Research **7**, 2004, 1228–1335.

feldspar, hornblende and/or biotite), and diorites (plagioclase and hornblende) are the other rock types in KAP. The complex is also traversed by coarse pegmatitic nepheline-bearing and granitic dykes. Nepheline-bearing pegmatites, with individual nepheline crystals having dimensions of 20−30 cm occur throughout the KAP, while granitic veins preserving graphic intergrowth textures are restricted to the west. Hornblende syenite is the most abundant variety with well-developed gneissic banding. Some of the paragneisses belonging to the EGMB occur as distinct bands and patches within syenites and amphibolites. The rocks in KAP strike NE−SW with varying dips (60−70 degrees) to the southeast. The mineral lineations show gentle plunges to the southwest.

While amphibole-bearing nepheline syenite characterizes the western part, the biotite nepheline syenite dominates the eastern part of the KAP. Outcrops of graphite-bearing syenite and corundum-bearing nepheline syenite often occur within the KAP. The rocks exhibit extreme mineralogical variability. Upadhyay (2008) and Vijaya Kumar et al. (2007) suggested that the parental magma for the alkaline complexes were of basanitic composition produced by partial melting of enriched mantle sources in the subcontinental lithospheric mantle; fractionation, with or without crustal assimilation, produced the different rock suites. The foliations and lineations within and outside the KAP suggest that the alkaline plutons are syntectonically emplaced and are confined to SSZ. However, there is also another opinion that the deformation in the rocks occurred after the emplacement of the plutons, based on evidence for superposed deformation within the complex (Gupta & Bose, 2004). The alkaline rocks of Kunavaram yielded a Rb−Sr whole rock age of 1244 Ma (Clark & Subba Rao, 1971). The combined SHRIMP and TIMS analyses of interiors of zircon grains from the KAP yielded an upper intercept age of 1384 ± 63 Ma (intrusion age) and a lower intercept age of 632 ± 62 Ma (metamorphism age) (Upadhyay & Raith, 2006a).

3.2.3 KONDAPALLI LAYERED COMPLEX

Kondapalli Layered Complex (KLC), located in the northern extremity of the EGMB-south is a discontinuous and dismembered mafic−ultramafic layered intrusion, occurring amidst charnockitic and other granulite-facies rocks (Fig. 3.11). The KLC consists of abundant mantle-derived mafic granulites, felsic plutonic rocks, and is considered as a unique ultramafic−mafic layered complex (Leelanandam, 1997). According to Leelanandam (1972), the chromite-bearing ultramafic rocks merely represent a component of the complex, and that there is neither consanguinity nor contemporaneity between the Kondapalli chromitites and enclosing charnockites. Chromite occurs as the dominant (cumulus) mineral in chromitites, and as a minor (intercumulus) mineral in orthopyroxenites, websterites, clinopyroxenites, dunites and harzburgites; thin bands of chromite are occasionally seen in orthopyroxenites. The mineral association suggests that the KLC directly witnessed granulite−facies conditions, similar to that of the country rocks, without being metamorphosed in the prograde part of the metamorphic cycle, and is unique in the entire EGMB (Leelanandam, 2015). An ultrahigh temperature (UHT) granulite facies metamorphism with a peak temperature >1000°C and pressure >10 kb, and a deep crustal heating and isobaric cooling trajectory are documented from the rocks at the contact zone of the KLC (Bhui, Sengupta, & Sengupta, 2007).

The regional geological setting, rock and mineral associations and their geochemistry indicate an Andean-type "continental magmatic arc" environment for the emplacement of the KLC and the chromite-bearing ultramafic cumulates form a part of the Mesoproterozoic arc-root complex (Leelanandam, 1997). These rocks yielded an age of c. 1630 Ma (Dharma Rao, Santosh, & Dong,

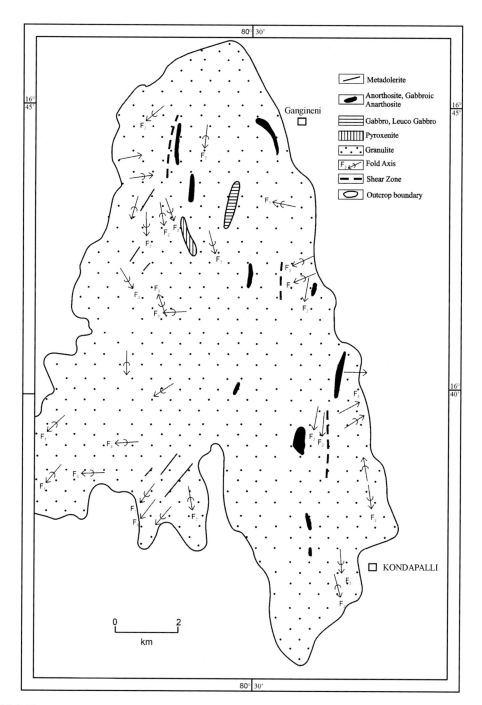

FIGURE 3.11

Geological map of the Kodapalli hill ranges.

Modified after Leelanandam, C. (1997). The Kondapalli layered Complex, Andhra Pradesh, India: A synoptic overview. Gondwana Research, 1, *95–114.*

2012). The KLC seems to have witnessed Paleoproterozoic rifting to Grenvillian/Pan-African collision tectonics, typical of subduction-related magmatic arcs and is construed as an arc-root complex in the deeply eroded EGMB.

3.2.4 KANIGIRI OPHIOLITE MELANGE

A NE−SW trending tectonic mélange around Kanigiri is well exposed with ophiolitic fragments, which is well described as KOM (Fig. 3.12). The KOM (120 × 30 m) shows a concordant relationship with the NE−SW trending NSB of the East Dharwar craton, and a thrust/ sheared zone at the contact with granites in the southeast. The KOM constitutes a complex ensemble of diverse metamorphic, sedimentary, and igneous rock units including ultramafic rocks, cumulate gabbros, mafic sills, pillow basalts, plagiogranites, pelagic sediments, and meta-chert with a near-complete Penrose-type ophiolite complex (Dharma Rao et al., 2011). These rocks are juxtaposed and variably deformed in a fine-grained and scaly argillaceous matrix, possibly indicating significant shearing either on the seafloor or during emplacement. Sedimentary structures such as graded beds, ripple lamination and sole markings have not been preserved within the meta-sedimentary members of the mélange. The rock sequences are dismembered, fractured, and sheared typically exhibiting a mélange fabric. Extensively foliated metabasalts have a typical mineral assemblage of epidote−chlorite−albite−ilmenite−clinozoisite−stilpnomelane−actinolite, calcite, and quartz. Well-preserved and deformed pillows locally exhibit vesicular structures and zoned pillow cores (Dharma Rao & Reddy, 2009). The pillow structures provide a strong evidence for ocean floor magmatism. The meta-gabbro fragments are rounded and occur as inclusions with sharp boundaries within the enclosing brecciated serpentinite. Plagioclase-rich leucocratic rocks occur as scattered bands/lenses/veins within the mosaic of the mélange. Such rocks are variously termed as plagiogranite, oceanic plagiogranite, trondhjemite, continental trondhjemite, and keratophyre. Meta-ultramafic rocks are represented by serpentinites and actinolite-tremolite−rich amphibolites. Serpentinites are composed of hydrated metamorphic assemblages giving rise to variable mineralogical compositions and include serpentine, talc, magnetite, tremolite, and rarely cummingtonite. These assemblages together may represent dismembered fragments of oceanic crust imbricated within an accretionary wedge. The constituent units of KOM show the characteristics of rocks often documented in subduction-related island arc suites and are correlated with many Proterozoic suprasubduction zone ophiolites of the world. The magmatic zircons in cogenetic gabbros and plagiogranites of KOM yielded U−Pb c. 1.33 Ga ages (Dharma Rao et al., 2011). These features of KOM ignite important clues for arc−continent collision during the final stages of amalgamation of the Columbia-derived fragments within the Neoproterozoic supercontinent assembly.

In summary, the rocks in the TBSZ must have obtained granulite facies metamorphism at around 1.60−1.65 Ga and was considered responsible for suturing of the TBSZ with proto-India and possibly with east Antarctica along the Napier Complex/Rayner complex (Vijaya Kumar, Leelanandam, & Ernst, 2011). A subsequent continental rifting with the formation of a new ocean basin was inferred by the emplacement of alkaline magma, carbonatite, and possibly the KOM at 1.5−1.3 Ga. Interestingly, the TBSZ has not witnessed any major event of tectonothermal reworking during the Grenvillian orogeny except for minor disturbances in the Ar−Ar radio clock at 1.1 Ga (Mezger and Cosca, 1999).

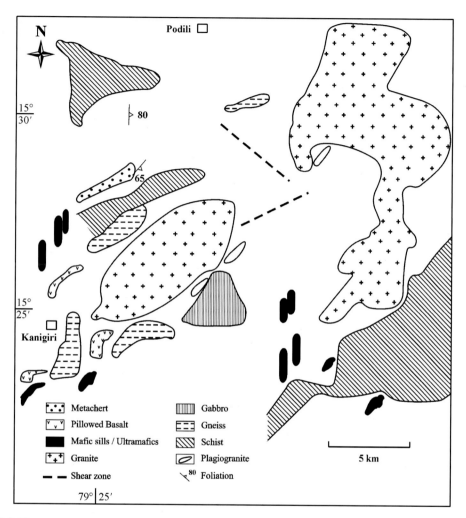

FIGURE 3.12

Geological map of Kanigiri ophiolite mélange.

Modified after Dharma Rao, C., Santosh, M., & Yuan, B.W. (2011). Mesoproterozoic ophiolitic Mélange from the SE Periphery of the Indian Plate: U−Pb Zircon Ages and Tectonic Implications. Gondwana Research, 19, 384−401.

3.2.5 STRUCTURAL CROSS SECTION

A NW−SE structural cross section between Kurichedu and Ongole, covering all the lithologies of the TBSZ including the Pasupugallu and Chimakurti mafic plutons, reveals many interesting structural features (Fig. 3.13). Among them, the Nallamalai fold-thrust belt occurs at the west, followed by schist belt and mafic plutons surrounded by migmatitic gneisses on both sides, and finally the granulite facies rocks of the EGMB near Ongole are exposed. The region comprises a series of

FIGURE 3.13

NW—SE structural cross section across the Terrane Boundary Shear Zone (TBSZ, section I-J in Figure 3.4) showing west-verging thrusts and associated gabbro plutons (PGP and CMP) (unpublished thesis of Nagaraju, 2007).

subparallel shear zones/thrust faults that often coincide with the major lithological boundaries. The major lithologies strike NE—SW with steep easterly dips. Both the plutons exhibit inward-dipping foliations and are enveloped by ductile strike-slip shear zones. The lithounits between the shear zones are isoclinally folded suggesting the NW-verging overthrust sense. Field and petrographic observations such as S—C fabrics and asymmetrical porphyroclasts suggest that Pasupugallu and Chimakurti plutons are enveloped by dextral shear zones.

The complex mineralogy of spinel—cordierite—orthopyroxene hornfelses and the prograde mineral reactions within metapelites near the plutonic rocks suggest that these host rocks have been subjected to UHT metamorphism (Dasgupta, Ehl, Raith, Sengupta, & Sengupta, 1997) during the emplacement of mafic magma. Considering all the lithologies, structural fabrics and the subparallel thrust zones, a dextral transpressional tectonic model was invoked for the emplacement of these plutonic complexes (Nagaraju et al., 2008).

3.2.6 TECTONIC SYNTHESIS

The TBSZ is marked by a series of subparallel thrusts/shear zones (see Fig. 3.4) with west-vergent tectonic transport of the granulite facies rocks of the EGMB over the Dharwar craton (Chetty & Murthy, 1994). The TBSZ has also been considered as the southern extension of the SSZ and a cryptic suture with an eastward-dipping major thrust zone.

The alkaline magmatism is strikingly confined to the western margin of the TBSZ and is represented by several alkaline plutons of different sizes and shapes with ages ranging between 1250 and 650 Ma for over a protracted period (Upadhyay & Raith, 2006a). An overview of geology, petrology, genesis, and nature of these alkaline plutons together with the low $^{87}Sr/^{86}Sr$ ratios of the rocks from alkaline and carbonatite complexes of the EGMB suggests a mantle origin with repeated ascent of mantle-derived material along major tectonic zones separating contrasting geologic terranes. Vijaya Kumar et al. (2007) suggested that the parental magma for the alkaline complexes

were of basanitic composition produced by partial melting of enriched mantle sources in subcontinental lithospheric mantle; fractionation, with or without crustal assimilation, produced different
rock suites. Geodynamically, it is generally believed that the alkaline magmatism has been generated as a result of an extension resulting from the cessation of subduction. However, the structural
features of gabbro—anorthosite plutons of PGP and CMP reveal concentric, helicoidal, and inward-
dipping magmatic/tectonic foliations suggesting syntectonic nature of emplacement involving a
dextral transpressional tectonic regime (Nagaraju & Chetty, 2005). The geochemical characteristics
of the plutons suggest that a low potassium island arc tholeitic in nature and support the
subduction-related magmatism. The field structural observations and AMS measurements indicate
that these plutons were emplaced into dilational jogs induced by a crustal-scale dextral transpressional tectonic regime (Nagaraju et al., 2008). The presence of marginal shear zones and their control over the distribution of plutons all over the TBSZ indicate the major role of transpressional
tectonics in controlling the emplacement of ascending magmas that have formed at different levels
of lower crust between Mesoproterozoic and Neoproterozoic periods. This hypothesis is consistent
with the magma emplacement models of Cruden (1998), where complex intrusive relationships,
pluton infilling by one or more vertical conduits associated with inward-dipping structures have
been predicted.

UHT metamorphic conditions were also reported from places such as the contact zones of KLC
and CMP. In both cases, the UHT conditions were ascribed to the heat supplied by mafic/ultramafic
magmatism, and are not considered to represent the conditions associated with the regional
granulite-facies metamorphism in the TBSZ. The granulite facies metamorphism in the TBSZ was
reported to be generally between c. 1700 and 1512 Ma (Kovach et al., 2001). The broad range of
ages is attributed to isotopic disturbance of monazite and zircon grains during prolonged periods of
granulite-facies metamorphism. Mesoproterozoic ($\sim$1352 Ma) tholeiitic and alkalic magmatism
was well recorded in three plutons of Chimakurti, Errakonda, and the Uppalapadu nepheline syenites pointing to the spatial and temporal association of both tholeiitic and alkaline endmembers of
continental rift magmatism (Vijaya Kumar et al., 2007). The alkaline plutons of the TBSZ were
also described as DARC's (Deformed Alkaline Rocks and Carbonatites) and considered the TBSZ
as a Proterozoic fossil plate boundary (Leelanandam, Burke, Ashwal, & Webb, 2006).

Examination of gravity signatures around the TBSZ shows a sharp rise in the gravity anomalies
of 50—60 mgal, which was interpreted as a cryptic suture and an eastward-dipping thrust
(Subramanyam, 1978). The eastward-dipping thrust was also delineated from a seismic profile
along Alampur-Koniki across the TBSZ and the Moho is estimated at a depth of 35 km in the
region and that it increases to 39—40 km in the west underneath the Cuddapah basin (Kaila et al.,
1987). The TBSZ is also characterized by bipolar gravity anomaly with a relative gravity high over
the EGMB and a low over the eastern part of the Cuddapah basin delineating the distinct geologic
units separated by a westward verging discontinuous thrust fault (Singh, Mishra, Gupta, & Rao,
2004).

Pan-African metamorphic cooling ages were also documented from the Mesoproterozoic nepheline syenite complexes (Elchuru, Kunavaram, and Khariar) having a similar geodynamic setting
along the suture (Aftalion, Bowes, Dash, & Fallick, 2000). Recently, a two-stage granulite formation with two generations of mineral growth in the evolutionary history of the rocks of TBSZ was
proposed (Sarkar, Schenk, & Berndt, 2015). The first event was believed to be associated with the
UHT metamorphism at low pressures caused by magma emplacement, while the second

metamorphic event was due to crustal thickening during collision of continental blocks. Major and trace elements together with Sr, Nd, and Pb isotopic data indicate a multistage magmatic history, generated from a part of the subcontinental lithospheric mantle with the opening of a NE−SW trending Mesoproterozoic rift along the present east coast- cratonic margin of India.

The eastward-dipping upthrusted block of EGMB-south along with the presence of ophiolite mélange and deep-seated intrusives and disturbed nature of sediments of Cuddapah basin represented by the Nallamalai fold belt suggest a west-directed compressive regime related to collisional processes along the south eastern margin of India. Further, the eastward-dipping crustal column under the eastern part of the Cuddapah basin suggests that the polarity of the associated subduction is towards the east and the Cuddapah basin was developed as peripheral foreland basin abutting the EGMB. However, westward subduction is also invoked by several workers in recent years based on many geological and geophysical characteristics. Therefore, in the light of contrasting views, comprehensive multidisciplinary studies are warranted for better understanding of the timing and polarity of the subduction in this part of the region.

Two contrasting evolutionary models were proposed for the evolution of the EGMB-south (TBSZ). The first model envisages subduction-related continental arc magmatism at Mesoproterozoic active continental margin along the southeastern part of India (Dharma Rao, Santosh, Zhang, & Tsunogae, 2013) and the accretion of a juvenile island arc; prior to terminal continent−continent collision in the Neoproterozoic amalgamation of the supercontinent Rodinia. The continental rifting and subsequent subduction−accretion−collision processes are correlated to the time span from the breakup of Columbia to the assembly of Rodinia (Dharma Rao et al., 2012). The other model involves Mesoproterozoic active margin tectonics leading to the accretion of an island arc, followed by the generation of a continental magmatic arc prior to terminal continent−continent collision at c. 1.6 Ga (Vijaya Kumar et al., 2011). Further, the younger Mesoproterozoic events (1.5−1.3 Ga) were attributed to rifting parallel to the accreted margins. The collisional regime was superposed subsequently during Pan-African tectonism, which was common in many parts of Gondwanaland. This is also consistent with the postulated final breakup of Columbia at about 1.3−1.2 Ga.

A comprehensive tectonic model was proposed by Henderson, Collins, Payne, Forbes, and Saha (2014) based on combined U−Pb geochronology and Hf isotope provenance study for the metapelitic rocks of the TBSZ with the following sequence: (1) a major episode of passive rifting and intracontinental mafic magmatism along the SE- margin of the Indian proto-continent at c. 1.89−1.87 Ga; (2) oceanic decoupling initiated subduction in the newly formed oceanic crust, leading to the formation of an oceanic island arc; (3) westward migration of subduction resulting in the accretion of the oceanic island arc to the rifted margin; (4) further rifting in the region led to the production of oceanic crust, represented by a Kandra Ophiolite Complex, in a suprasubduction-zone setting, typical of a continental back arc basin (highly speculative); and (5) closure of the proposed back-arc basin and the subsequent accretion of the crust of TBSZ at c. 1.6 Ga. This model accounts for the magmatism and metamorphism recorded in the EGMB- south between 1.72 and 1.6 Ga leading up to continent−continent collision. However, two collisional events were invoked that occurred during 1.85 and 1.33 Ga based on the U−Pb age correlations of KOM and mafic plutons along the TBSZ (Subramanyam et al., 2016).

The interpretation of the Ongole domain as a magmatic arc near the Indian plate during the growth of Columbia Supercontinent implies its allochthonous origin and argues against the derivation of protoliths from the adjacent Dharwar craton (Sarkar et al., 2015). In the light of the above discussion, it can be summarized that a suprasubduction zone tectonic setting was responsible for the oceanic plate subduction, multiple oceanic crust-arc accretions along the southeastern margin of Indian shield during the Paleoproterozoic that culminated in the Mesoproterozoic period. The volcanic and plutonic lithological associations along the TBSZ represent the juxtaposed geologic units over the eastern margin of proto-India during the accretionary tectonic regime in Cordilleran-style during the assemblies of Columbia and Rodinia.

3.3 EGMB-CENTRAL

The EGMB-central is the largest and the widest part of the EGMB. It is bound by the SSZ in the west, the MSZ in the north, the GR zone in the south, and the Bay of Bengal in the east (see Fig. 3.3). A network of major ductile shear zones was first recognized and described from the EGMB by Chetty and Murthy (1998a) that stimulated new insights, discussions, and researches. The shear zones vary in their extension and orientation, which often deviate from the regional NE−SW trend of the EGMB. They broadly include: conformable NNE−SSW striking shear zones in the EGMB-south, N−S shear zones in the EGMB-central, and strike parallel E−W shear zones in the EGMB-north. The lithological contacts and the foliation trajectories are, in general, conformable to the shear zones in the proximity.

3.3.1 SHEAR ZONES

The network of shear zones is strikingly the most prominent deformational feature in the EGMB-central (Fig. 3.14), which occur both at the margins as well as in the central region. They represent wide and linear geomorphic expressions mostly coinciding with the major river courses and hence are named after them (Chetty & Murthy, 1998a). The shear zones are well reflected in satellite images with closely spaced lineaments and characteristic tonal variations. Field observations, such as the presence of mylonites, flattened folds, well-developed stretching lineations, and many other kinematic features, confirm the existence of shear zones. The margins of the EGMB are marked by major ductile shear zones hosting alkaline plutons, anorthosites, and extensive migmatization and retrogression. These zones vary in width from 2 to 20 km and extend along the strike length for a few hundred kilometers.

Amongst all the shear zones, the most significant is the SSZ that occurs at the western margin of the EGMB-central in contact with Bastar craton. Besides the SSZ, the other important shear zones include Nagavali-Vamsadhara shear zone (NVSZ) and the MSZ at the southern margin of the EGMB-north with varied nature, scale, and geometry. A few other important but relatively narrow shear zones are also delineated in the region between the NVSZ and the MSZ: (1) the Baligurha-Tel shear zone, (2) the Chilka lake shear zone, (3) the Digapahandi shear zone, (4) the Aska-Taptapani shear zone, and (5) the Bhanjanagar shear zone (see Fig. 3.3). The details of each shear zone are described below.

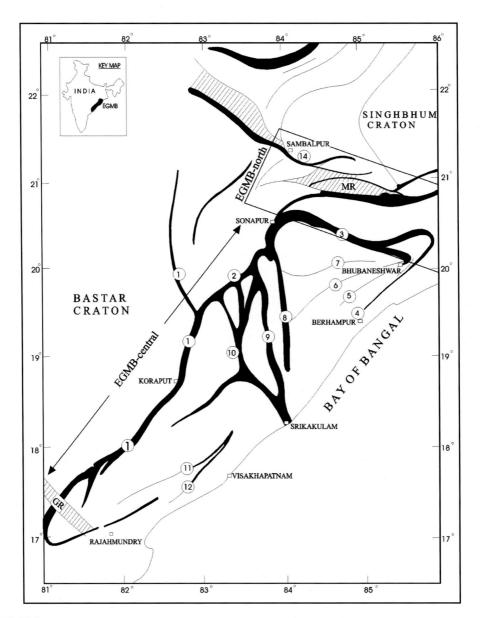

FIGURE 3.14

Map showing the network of major shear zones in the EGMB-central. 1. Sileru shear zone, 2. Koraput-Sonepur shear zone, 3. Mahanadi shear zone, 4. Chilka lake shear zone, 5. Digapahandi shear zone, 6. Aska Taptapani shear zone, 7. Banjanagar shear zone, 8. Baligurha Tel shear zone, 9. Vamsadhara shear zone, 10. Nagavali shear zone, 11. Narsipatnam shear zone, 12. Tuni-Eleshwaram shear zone, 13. Angul-Dhenkanal shear zone, and 14. Northern Boundary shear zone. MR—Mahanadi Rift, GR—Godavari Rift.

After Chetty, T.R.K. (2010). Structural architecture of the northern composite terrane, the Eastern Ghats Mobile Belt, India: Implications for Gondwana tectonics. Gondwana Research, 18, 565–582.

3.3.1.1 Sileru shear zone

The Sileru Shear Zone (SSZ), the most prominent among all the shear zones, occurs at the contact zone between the Archean cratonic rocks of Dharwar and Bastar cratons in the west and the granulite facies rocks of the EGMB in the east. It was earlier believed that the contact between the craton and the EGMB was not sharp, but transitional, with an intervening sandwich zone containing the rock types of both the units (Narayanaswami, 1975; Ramakrishnan et al., 1998). The SSZ varies in width from 3 to 10 km and extends over a strike length of ∼500 km with an average dip of about 50 degrees toward the southeast (Chetty & Murthy, 1993). The rocks in the SSZ preserve intensely developed foliations giving rise to a complete spectrum of mylonites varying from proto- to ultramylonite. Stretching lineations are well developed with gentle to moderate plunges (20−30 degrees) to southwest (Fig. 3.15). The mesoscopic structures such as high asymmetrical fold structures, S−C fabrics, and deformed augen gneisses indicate dextral sense of strike-slip movements. However, the behavior of stretching lineations and the heterogeneity of foliation fabrics indicate

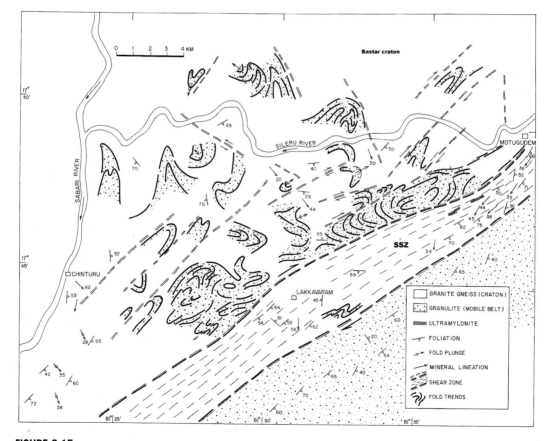

FIGURE 3.15

Map showing upthrusted granulite facies rocks of the EGMB over the Bastar craton as fragmented parts of a thrust sheet, west of the Sileru shear zone (SSZ) at the confluence of Sileru and Sabari rivers.

that the rocks in the shear zone witnessed a transpressive tectonic regime. Several segments of folded granulite facies rocks around Chinturu, lying over the Bastar craton occur to the west of SSZ, implying that they have been thrusted on to the craton from the east (Chetty & Murthy, 1998b). This hypothesis was substantiated and well documented in the northern part of the SSZ around Lakhna area where the rocks of both craton and the EGMB are well exposed in direct contact with each other. The contact zone exhibiting pronounced mylonite fabrics and the kinematic features suggests a westward- thrusting of granulite facies rocks of the EGMB over the craton (Biswal, Jena, Datta, Das, & Khan, 2000). The deformation pattern across the contact was correlatable with a modern style fold thrust belt, and the existence of large nappe structures bound by lateral ramps was established. This leads to the inference that the deformation across the SSZ was similar to the "thin-skinned" tectonics commonly observed in the frontal zones of modern orogenic thrust sheets. However, a slight different interpretation was proposed from an area around Deobagh located in the central part of the SSZ, suggesting that the thrusting of granulite facies rocks was synchronous with lower crustal metamorphism juxtaposing with upper crustal cratonic rocks (Gupta, Bhattacharya, Raith, & Nanda, 2000). The shear zone comprises hornblende gneisses and leptynites containing enclaves of Mg−Al granulites, mafic granulites, BIF, khondalites, and calc granulites, which are mostly mylonitized. Kinematic studies indicate west verging thrust implying the transportation of hot fluids over the cold craton displaying inverted metamorphism (Bhadra & Gupta, 2016).

Considering the rock assemblages and structural styles within the shear zone, contrasting geological features on either side and its association with alkaline magmatism, the SSZ has been described as a major Precambrian suture zone (Chetty & Murthy, 1998a). Gravity anomalies over SSZ are strongly negative, rising sharply up to the axial zone and then becoming increasingly positive toward the eastern boundary, indicating that the SSZ is a suture zone (Subramanyam & Verma, 1986). The SSZ was also interpreted as an east-dipping thrust based on Deep Seismic Sounding (DSS) studies by Kaila and Bhatia (1982). Alkaline magmatism ranging in age from 1250 to 650 Ma is restricted to the SSZ and is marked by the presence of a chain of nepheline syenite plutons (DARCs), indicating deep sections of a suture zone (Leelanandam et al., 2006). The SSZ extends further south of the GR and is described as the TBSZ, which has already been documented in the earlier section of this chapter. The SSZ also extends northward all along the western margin of the EGMB in the form of Koraput-Sonepur-Rairakhol shear zone (KSRSZ) and MSZ, and southward to join the Cauvery suture zone (CSZ of the Sothern Granulite Terrane). In summary, geological and geophysical characteristics undisputably suggest that the SSZ represents a typical Precambrian suture extending to lithospheric−mantle depths.

The SSZ also hosts Koraput alkaline complex (KAC), located five km northeast of Koraput town, in its northern extensions (Fig. 3.16). The alkalic suite comprises alkali−gabbro, in the core of the complex and the syenite bands of calc−alkali syenite, perthite-syenite (perthosite), and nepheline syenite on either side of the body (Bose, 1970). The central gabbroic unit preserves prominent igneous textures and is characterized by the random orientation of amphibole needles and plagioclase laths. The gabbro is successively intruded by dykes of syenite, mafics (now amphibolites), and pegmatoidal granites. The KAC preserves the magmatic/deformational fabrics, which were developed during the emplacement. The strain distribution was heterogeneous, and the strain partitioning was clearly widespread during the different deformation events. The nature and the development of fabrics in alkaline complexes has been differently interterpreted in the context of the time relationship between emplacement of the pluton and regional tectonics. Two distinct fabrics

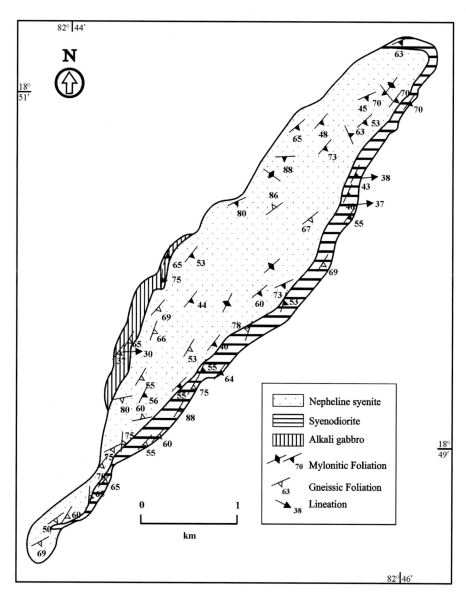

FIGURE 3.16

Geology and structure of Koraput alkaline complex.

Gupta, S., Nanda, J., Mukherjee, S.K., & Santra, M. (2005). Alkaline magmatism versus Collision tectonics in the Eastern Ghats Belt, India: Constraints from Structural studies in the Koraput Complex. Gondwana Research, 8, 403–419.

were recognized: the first being magmatic foliation preserved in the gabbroic core, and the second solid-state foliation fabrics at the margins of the complex. These observations led to the interpretation that the KAC witnessed solid-state deformation and metamorphism following emplacement

(Gupta, Nanda, Mukherjee, & Santra, 2005). The rocks from KAC yielded Rb−Sr whole-rock age of 856 ± 18 Ma, which was considered as the emplacement age (Sarkar, Nanda, Paul, Bishui, & Gupta, 1989). However, considering the ages obtained from the other alkaline complexes along the SSZ, it can be interpreted that the rocks of KAC were also subjected to the Pan-African tectonics (Nanda & Gupta, 2012; Nanda, Gupta, & Dobmeier, 2008).

It is well established that the occurrence of all the alkaline plutons is restricted to the SSZ, but their emplacement history with reference to shearing is still widely debated. One school of thought believes that the alkaline plutons are syntectonically emplaced (e.g., Biswal, Waele, & Ahuja, 2007; Chetty & Murthy, 1994, 1998b), while others describe post emplacement deformation and shearing produced solid-state metamorphic fabrics (e.g., Gupta, 2004; Gupta et al., 2005; Upadhyay & Raith, 2006a). However, it is widely accepted that the parental magma of these rocks must have been derived from partial melting of the enriched mantle sources. It also implies from the available ages of different alkaline plutons that there must have been multiple events of reactivation or accretion, which is a common feature of many ancient accretionary orogens.

3.3.1.2 Nagavali-Vamsadhara shear zones

The NVSZ occur in the widest central part of the EGMB (Fig. 3.17) representing two important crustal-scale shear zones, viz., the Nagavali shear zone (NSZ) and Vamsadhara shear zone (VSZ). While the VSZ is a near straight feature, the NSZ is sinuous, swerving to southeast in the south to merge with the VSZ and both finally disappear in the Bay of Bengal. Both the shear zones vary in their width from 2 to 10 km and extend for about 150 km strike length and coalesce with the SSZ in the north. The NVSZ comprises voluminous megacrystic gneisses and quartzo-feldspathic gneisses with mappable enclaves of charnockites, khondalites, and mafic granulites. The enclaves are mostly deformed and aligned subparallel to the regional gneissic foliation. The leucogneisses occur as thin parallel bands, small lensoid bodies, and often as small mounds and ridges with in megacrystic gneisses (Fig. 3.18). Interestingly, the leucogneisses are predominant in the axial zone of the NSZ and the foliations wrap around these bodies.

Complex fold forms and curvilinear structural trends mark the presence of major noncylindrical folds in the southwestern part across the NSZ around Parvatipuram (Fig. 3.19) (Chetty, Vijaya, Narayana, & Giridhar, 2003). The structural trends are, in general, conformable to the lithological contacts between the folded quartzite horizon and the migmatitic/ megacrystic gneisses (both hypersthene and biotite bearing) in the widened part of the NSZ around Parvathipuram. K-feldspar−bearing megacrystic granites, also containing the enclaves of metapelites, are dominant along the NSZ. An important granite body known as Sankarda granite occurs in peripheral parts of the NSZ, west of Rayagada, and was dated at 1000 Ma (Fig. 3.20). Four clusters of ages were obtained to understand the sequence of metamorphic, migmatitic, and magmatic events along the NSZ that include: (1) an intrusion of protoliths of mafic granulite and leptynite at 1450 Ma; (2) a major event of regional granulite facies metamorphism at 1000 Ma; (3) emplacement of megacrystic granite at 800 Ma; and (4) Pan-African shearing and fluid influx event at 550 Ma (Shaw et al., 1997). The granitoids at other localities of the NVSZ yielded ~900 Ma age (Kovach et al., 1998).

Thermobarometric calculations of the rocks from the NSZ near Rayagada revealed maximum P−T conditions of 950°C and 9 kbar, and the P−T path is characterized by near-isobaric cooling starting from a relatively higher temperature regime, followed by decompression. Corundum−quartz

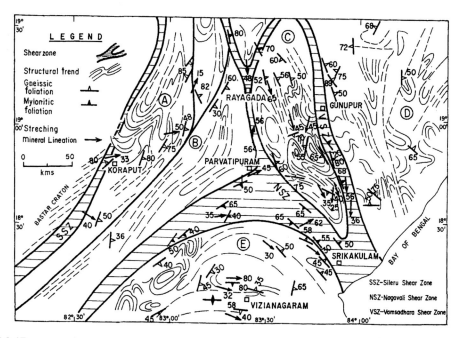

FIGURE 3.17

Map showing the structural framework of Nagavali-Vamsadhara shear zones (NVSZ) and the associated fold styles and other structural trends. Notice the presence of megastructural domes and basins and megacrystic granitoids restricted to NVSZ. A—E represent distinct structural domains.

After Chetty, T.R.K., Vijaya, P., Narayana, N.L., & Giridhar, G.V. (2003). Structure of the Nagavali shear zone, Eastern Ghats Mobile Belt, India: Correlation in the East Gondwana reconstruction. Gondwana Research, 6, 215—229.

assemblages were also reported from the region, which might put the peak metamorphic condition at higher P and T levels (Shaw & Arima, 1996).

The VSZ is characterized by the presence of quartz—feldspathic gneisses in the form of linear alignment of ridges with typical metamorphic/mylonitic foliation. The early formed gneissic folia-tion is well preserved with the alignment of megacrysts (dominantly K—feldspar), generally lying parallel to the boundaries of shear zones. The gneissic foliation shows moderate to steep dips and the mylonitic foliation is invariably steep suggesting its superimposition on the earlier fabrics. However, both of them often occur subparallel to each other indicating the role of transpositional fabrics.

The interlying terrane between the NSZ and the VSZ (Fig. 3.21) shows a lensoid form and a longer axis in a north—south direction. A variety of highly asymmetrical fold forms and elongated closed structural forms are evident. Mesoscopic shear zones are common within megacrystic gneisses. The feldspar porphyroclasts exhibit a rotated and winged shape of kinematic indicators suggesting noncoaxial dextral kinematic sense of movements. Both the shear zones show dextral sense of movements despite the variation in their geometry and is consistent with regional trans-pressive deformational history of the EGMB (Chetty & Murthy, 1998a). The shear zones separate

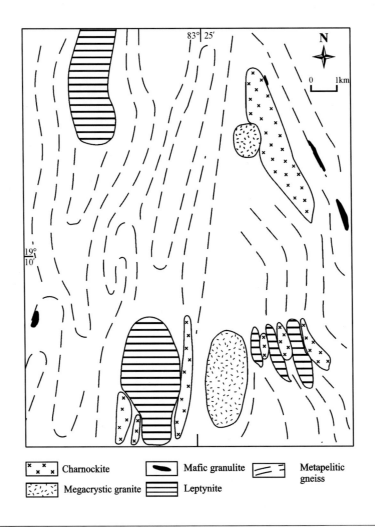

FIGURE 3.18

Map showing different lithologies of high-grade rocks within the Nagavali shear zone (NSZ) in the vicinity of Rayagada.

Modified after Shaw, R.K., Arima, M., Kagami, H., Fanning, C.M., Shairashi, K., & Motoyashi, Y. (1997). Proterozoic events in the Eastern Ghats granulite belt, India: Evidence from Rb-Sr, Sm-Nd systematics and SHRIMP dating. Journal of Geology, 105, 645–658.

the region around the NVSZ into distinct structural domains. While the domains are characterized by distinct structural pattern, fold styles, and differently oriented structural fabrics, the shear zones are marked by tight isoclinal folds, steeply dipping mylonitic foliations, and gently plunging stretching lineations. The structural domains signify the low-strain regions bounded by shear zones. Considering the structural observations around the NVSZ, it is suggested that both the shear zones NSZ and VSZ merge together at depth and dip eastwards probably in the form of a listric thrust

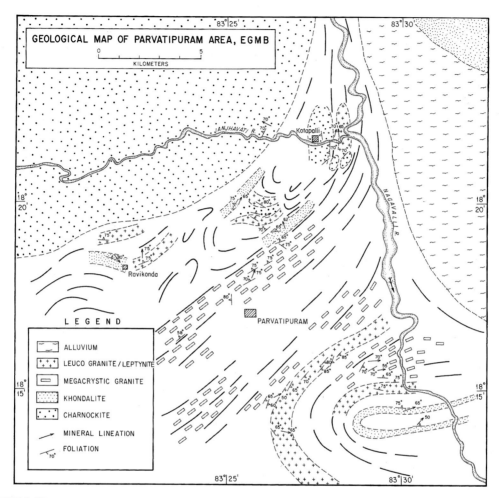

FIGURE 3.19

Lithological and structural variations along the Nagavali shear zone (NSZ) around Parvatipuram.

After Chetty, T.R.K., Vijaya, P., Narayana, N.L., & Giridhar, G.V. (2003). Structure of the Nagavali shear zone, Eastern Ghats Mobile Belt, India: Correlation in the East Gondwana reconstruction. Gondwana Research, 6, 215–229.

system. The NVSZ is also interpreted to have been extended into the Enderby Land of east Antarctica (Chetty, 1995a).

3.3.1.3 Other important shear zones

There are five other important shear zones (4, 5, 6, 7, and 8 in Fig. 3.3) in the triangular region bound by the NVSZ in the west, Bay of Bengal in the east, and the MSZ in the north. They include the Baligurha-Tel shear zone (no. 8), the Chilka Lake shear zone (no. 4), the Digapahandi shear zone (no. 5), the Aska-Taptapani shear zone (no. 6), and the Bhanjanagar shear zone (no. 7). In

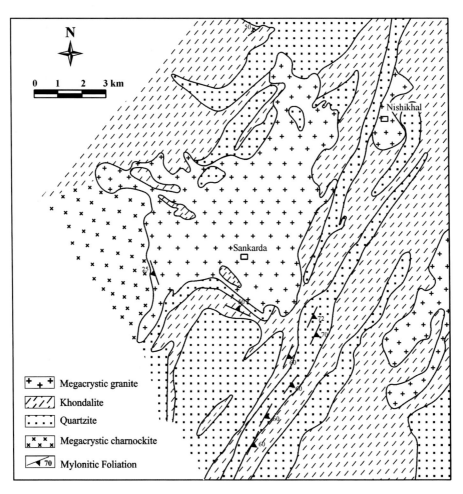

FIGURE 3.20

Map showing disposition of Sankarda granite (1000 Ma) and Tikhiri charnockite within the Nagavali shear zone (NSZ) with profuse development of megacrysts of k-feldspars and manganiferous horizons around Rayagada.

Modified after Nanda, J.K. (2008). Tectonic framework of Eastern Ghats Mobile Belt: An overview. Memoir, Geological society of India, *74, 63–87.*

general, the region is dominated by migmatitic gneisses, retrogressed charnockitic rocks, megacrystic gneisses, and isolated and intensely deformed khondalitic group of rocks in order of abundance. The early formed structures seem to be near east–west or NE–SW. Among these shear zones, the Baligurha-Tel shear zone (BTSZ) is unique in its characteristics. It occurs to the east of the NVSZ and trends N–S lying subparallel to the NVSZ. The BTSZ shows a width of ~8 km and swings westwards from Baligurha to join the course of Tel River. Unlike the NVSZ, the BTSZ is dominated by N–S trending brittle fractures overprinting the preexisting east–west metamorphic

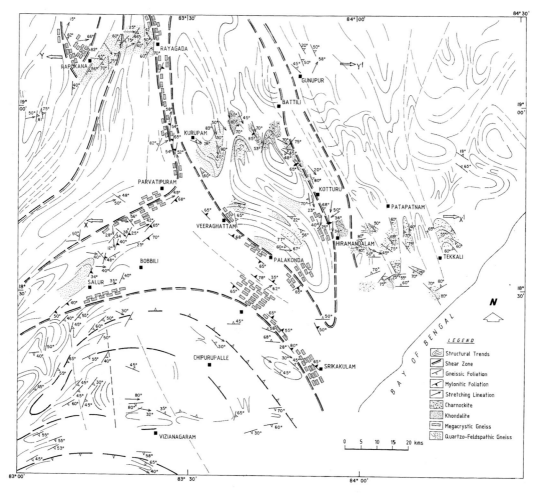

FIGURE 3.21

Geological and structural map of complex fold structures, closed structural forms, and different structural trends in different structural domains in and around the Nagavali-Vamsadhara shear zone.

After Chetty, T.R.K., Vijaya, P., Narayana, N.L., & Giridhar, G.V. (2003). Structure of the Nagavali shear zone, Eastern Ghats Mobile Belt, India: Correlation in the East Gondwana reconstruction. Gondwana Research, 6, 215–229.

gneissic foliations. The nature of deformation seems to be brittle–ductile in nature. Apart from the BTSZ, the other four shear zones form a distinct set trending near NE–SW and all of them emerge from the eastern end of the MSZ around Bhubaneswar and terminate at the BTSZ.

The Chilka Lake shear zone strikes NE–SW with a length of about 100 km passing through Chilka lake igneous complex and extends up to Bhubaneswar to join with the MSZ. All the anorthosite bodies around Chilka lake are confined to the shear zone. The presence of mesoscopic shear zones and boudin structures are common with in these igneous rocks. The Digapahandi shear zone, parallel to the Chilka lake shear zone, also follows a similar geometry but is characterized by the presence of megacrystic gneisses and joins the MSZ. The Aska-Taptapani shear zone, the site of hot springs, also trends ENE−WSW similar to other shear zones and extends for a distance of 150 km and finally merges with MSZ in the north east. The Bhanjanagar shear zone, which is relatively narrow compared to the other shear zones, trends near E−W and envelops the Phulbani charnockite block together with the MSZ. Brittle deformation appears to be dominant in the Phulbani block. All the above shear zones, in general, dip northwest or northward and are connected and genetically interlinked to the east−west trending MSZ and the N−S trending BTSZ. From these observations, it can be speculated that these shear zones may represent a stack of thrust sheets accreted together in the region. However, it is challenging to determine the roots of these shear zones and their vergence, which needs further focused multiscale structural studies.

3.3.2 TECTONIC DOMAINS/TERRANES

The EGMB-central can be divided into three tectonic domains/terranes considering the lithological assemblages, structural styles, metamorphism, magmatism, and the existence of boundary shear zones: (1) the Rajahmundry-Visakhapatnam-Koraput terrane; (2) the Berhampur-Bhubaneswar-Phulbani terrane; and (3) the Khariar-Bolangir terrane. All three terranes display a nearly triangular shape and are separated by crustal-scale shear zones (Fig. 3.22).

3.3.2.1 Rajahmundry-Visakhapatnam-Koraput terrane

The Rajahmundry-Visakhapatnam-Koraput terrane (RVKT) occurs in the southern part of the EGMB-central. It is bound by GR in the south, SSZ on the west, the east coast in the east, and NSVZ in the northeast (see Fig. 3.22). This terrane broadly coincides with the WKZ proposed by Ramakrishnan et al. (1998) and the tectonic domain II of Rickers et al. (2001). The RVKT, occurring to the north of the GR, is significantly different from that of the EGMB-south. The RVKT extends in NE−SW with a maximum width of 200 km that decreases gradually to about 50 km in the south. The major rock units in the southern part of the RVKT (N. Lat. $17°00'$ and $18°00'$ and E. Long. $81°00'$ and $83°00'$) comprise khondalitic group of rocks (garnetiferous sillimanite gneiss, quartzites, and calc−granulites) and a wide variety of charnockitic rocks (hypersthene-bearing granulites and gneisses) which were subjected to granulite facies metamorphism. The contact between the two major rock suites is mostly sheared and often migmatized. The pyroxene−free quartzo−feldspathic gneiss, often described as "leptynite" is the dominant variety of migmatite. In general, the rock units display metamorphic gneissosity strikingly parallel to the regional NE−SW trend and dip steeply to the east. The gneissosity, mostly parallel to relicts of compositional layering, is defined by alternations of quartz-rich layers and garnet-sillimanite−rich layers in the khondalitic suite, while it is defined by alternate felsic and mafic layers in calc−granulites and charnockitic suite. Large mafic intrusives and layered mafic−ultramafic−chromitite complexes are absent in this terrane, but the alkaline complexes, restricted to the SSZ, are located at the western margin. Saphirine bearing assemblages were reported from a few locations from RVKT, which yielded ages of 1000 Ma (Grew & Manton, 1986). The rocks were subjected to UHT metamorphism at some places in the terrane. This was followed by prolonged, near-isobaric cooling along

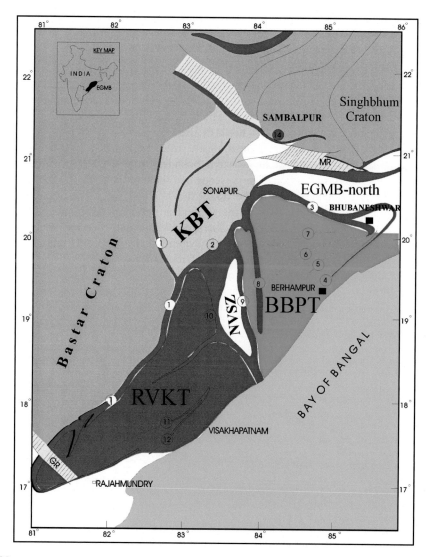

FIGURE 3.22

Map showing the delineation of major terranes separated by the crustal scale zones in the EGMB-central: *RVKT,* Rajahmundry-Visakhapatnam-Koraput terrane; *BBPT,* Berhampur-Bhubaneswar-Phulbani terrane; *KBT,* Khariar-Bolangir terrane. Important shear zones: 1. Sileru shear zone, 2. Koraput-Sonepur shear zone, 3. Mahanadi shear zone, 4. Chilkalake shear zone, 5. Digapahandi Shear zone, 6. Aska Taptapani Shear zone, 7. Banjanagar Shear zone, 8. Baligurha Tel Shear zone, 9. Vamsadhara shear zone, 10. Nagavali shear zone, 11. Narsipatnam shear zone, 12. Tuni-Eleshwaram shear zone, 13. Angul-Dhenkanal shear zone, and 14. Northern Boundary shear zone. *MR,* Mahanadi Rift; *GR,* Godavari Rift.

an anticlockwise P—T trajectory at lower crustal depths at c. 1.03—0.99 Ga (Bose, Dunkley, Dasgupta, Das, & Arima, 2011). Korhonen, Saw, Clark, Brown, and Bhattacharya (2011) also reported decompression-dominated retrograde history subsequent to a UHT peak metamorphism at c. 980 Ma. The UHT assemblages (high Mg—Al granulites) were ascribed to the emplacement of mafic melts resulting in the heating of lower crust through magmatic accretion during an early deformation at 1.4 Ga (Rickers, Raith, & Dasgupta, 2001a). Detailed geological and structural studiesin RVKT reveal many interesting inferences, which are described below.

Regional mapping of a critical part just north of the GR across the RVKT by using Landsat Thematic Mapper data shows a number of shear zones, structural trends, and structural domes and basins (Fig. 3.23). The SSZ occurs at the western boundary of the RVKT in contact with the Archean cratonic rocks of Bastar craton. The presence and the confinement of Kunavaram alkaline complex to the SSZ is a notable feature. The SSZ extends along the strike through the Elchuru alkaline complex in the south and the KAC in the north. The important rock units in the shear zone include granite gneisses, calc—granulites, charnockites, and khondalites that are extensively sheared and exhibit a spectrum of mylonites. The nature of asymmetrical folds, S—C fabrics, shear bands, and extensional cleavages suggest the dextral sense of movement. The SSZ in the terrane strikes NE—SW and dips due southeast with values around 45 degrees. The stretching lineations predominantly plunge SSW at 20—30 degrees. The shear movement was inferred to be a dominantly dextral strike-slip with a possible minor vertical slip component (Chetty & Murthy, 1993), though Nanda

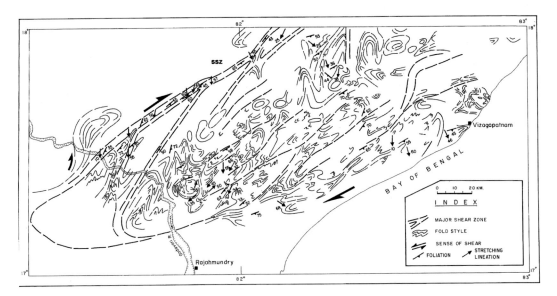

FIGURE 3.23

Simplified structural framework of the RVKT showing shear zones, structural domes and basins and a range of fold styles. *SSZ*, Sileru shear zone.

After Chetty, T.R.K. & Murthy, D.S.N. (1994). Collision tectonics in the Eastern Ghats Mobile Belt: Mesoscopic to Satellite scale structural observations. TERRA NOVA, 6, 72—81.

and Gupta (2012) suggest that the western khondalites were thrust onto the WCZ along the SSZ. The SSZ was believed to be Mesoproterozoic which has been reactivated through time and again as evidenced from the different alkaline magmatic events. The contrasting lithologies, metamorphism, and structural history of the rocks on either side suggest that the SSZ represents a Precambrian suture zone.

Besides the SSZ, there are also other important shear zones in the RVKT prominently observed at the contacts between the two major rock suites, which are relatively less known. Some of them are sinuous and often folded. These shear zones vary in their width from a few meters to even up to about a kilometer and are characterized by (1) a change in the character of the rocks from coarse-grained, folded, and banded gneisses to finer-grained, thinly banded quartzofeldspathic gneisses with dominant transposition fabrics, (2) the presence of migmatitic rocks, resulting in the emplacement of neosomes such as leptynites (pyroxene-free garnetiferous granitic rocks), quartzo-feldspathic gneisses in association with pegmatites, and vein quartz showing syn- or late-tectonic deformation, and (3) development of stretching lineations, boudinaged rocks, and occasional sheath folds. The notable among them are located around Narsipatnam, Burugubanda, and Eleswaram. The Narsipatnam shear zone is a relatively narrow belt ($\sim$1 km wide) and extends for about 200 km along NE−SW direction. It joins the tungsten bearing graphite deposits of Burugubanda-Tapasikonda in the south, restricted to a shear zone (Chetty & Murthy, 1992). In the north, it joins Tuni-Eleswaram shear zone and terminates near Vijayanagaram. The easterly dipping Tuni-Eleswaram shear zone apparently defines the eastern boundary of the RVKT possibly extending in southwest direction to join the SSZ in the form of a folded shear zone implying the refolding of large scale fold structures.

The folds vary in style from tight isoclinal, mostly asymmetrical in nature with attenuated limbs to open folds. The amplitudes of the folds vary from a few cm to few tens of kilometers. The amplitude is often highly exaggerated, particularly at the margins of the EGMB, indicating high strains demonstrated by the presence of incompetent rock types. The axial surface of the early folds is also often folded, giving rise to the refolded folds with near-coaxial deformation. The other distinct feature of the RVKT is the presence of closed structural forms of different orientation and sizes. Extensive 65 Ma old volcanism occurs to the east adjacent to the shear zone around Rajahmundry in the eastern part of the terrane.

A simplified tectonic framework of the RVKT shows the distribution of major rock units and associated structural trends (Fig. 3.24). Examination of general trends and dips of foliations, map pattern, and the distribution of rock types, geometry, and disposition of folds and ductile shear zones suggest an overall shortening of rock units by means of imbricate thrusting. Apart from the shear zone network, several closed structural forms of varying dimensions and NE−SW elongation in the form of structural domes and basins and are interpreted as possible sheath geometries (Chetty & Murthy, 1993). The presence of large scale refolded fold structures at the SW margin of the region and associated conformable structural fabrics indicate the presence of major fold closure, which may be genetically related to major low angle thrusting. The southwesterly fold closure together with the dominant plunge direction of stretching lineations and mesoscopic fold axes suggest the presence of antiformal fold closure formed due to a NW−verging major thrust, represented by the SSZ.

Field observations along two east−west trending traverses, separated by a distance of nearly 70 km, along (1) Upper Sileru-Visakhapatnam (SV), and (2) Chinturu-Eleswaram (CE), reveal the

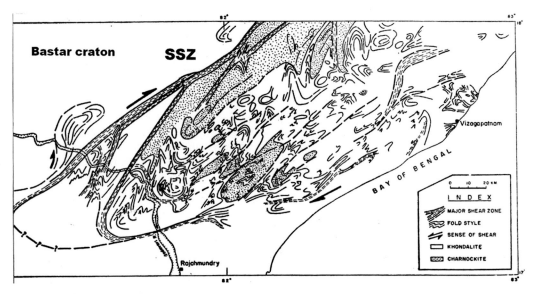

FIGURE 3.24

Simplified geological and tectonic map of the RVKT, where lithological units were superposed over the structural framework. *SSZ*, Sileru shear zone.

After Chetty, T.R.K. & Murthy, D.S.N. (1993). LANDSAT Thematic mapper data applied to structural studies of the Eastern Ghats granulite terrane in part of Andhra Pradesh. Journal Geological Society of India, 42, 373–391.

dominant presence of charnockites in the western part, while khondalites form the major rock unit in the eastern part. Quartzo—feldspathic migmatitic gneisses, often termed as leptynites, show intensely developed mylonitic fabrics at several places particularly at the contact zones of the two major rock suites. Profuse development of garnets and intense migmatization are well recorded in these rocks. The foliation fabrics trend NE—SW with moderate to steep dips to the east. The geological traverses show that the SSZ forms the boundary shear zone while the other shear zones are represented by the presence of migmatites with an intense deformation that occurs at the contact between the two major rock suites. The structures are interpreted in terms of major duplex structures and associated shear zones in the form of thrust zones. All the observations described above suggest NW-verging thrusts during D1 (Fig. 3.25). The steepness of shear zone fabrics (50—70 degrees) and the presence of oblique stretching lineations indicate subsequent transpressive deformation during D2. Pan-African U—Pb zircon age was obtained from zoned apatite—magnetite veins from Kasipatnam area, located west of Visakhapatnam (Kovach et al., 1998).

Structural mapping of regional shear zones through satellite data and aerial photographs is extremely useful in the EGMB, particularly in regions of inaccessibility and dense vegetation (Chetty & Murthy, 1993, 1994). It must also be noted that the field geologist's perspective is limited in identifying regional scale shear zones. The major shear zones, important fold forms, and the spatial distribution of rock types in the map (see Fig. 3.24) imply that the distribution of rock types is controlled by a tectonic fabric of the region. The charnockites, predominantly occurring along the western part, are generally bound by shear zones, suggesting that they could be allochthonous

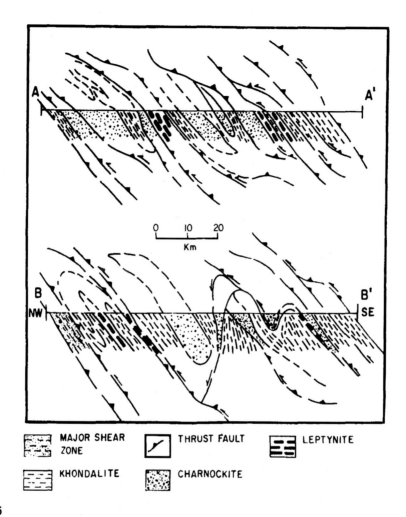

FIGURE 3.25

E−W trending structural cross sections across the RVKT showing major shear zones and west-verging thrusts often showing duplex structures.

After Chetty, T.R.K. & Murthy, D.S.N. (1994). Collision tectonics in the Eastern Ghats Mobile Belt: Mesoscopic to Satellite scale structural observations. TERRA NOVA, 6, 72–81.

blocks. The presence of sheath folds in the form of closed structural forms suggests that the rocks were subjected to high shear strains. Detailed structural mapping of one such large scale domal structure near Visakhapatnam indicates that they were formed during a single phase of deformation of early folding process (Natarajan & Nanda, 1981).

In summary, it is apparent that the structural style of the RVKT is typical of that of a thin-skinned fold-thrust terrane represented by westerly verging thrusts. The foliation fabrics and the stretching lineations of the RVKT suggest a predominant NE−SW structural trend, with steep dips to the east. The lineations plunge mostly to the south and southwestwards with gentle to moderate values. Broad fold closures to the southwest along with the associated lineations and the spatial disposition of different rock units indicate that the entire rock sequence of the RVKT may represent a regional antiformal fold structure (Chetty & Murthy, 1993). The structural architecture of the RVKT suggests an overall shortening of the rock units by means of imbricate thrusting. The northwestward verging thrusts associated with steeply east dipping shear zones (50−70 degrees) substantiate a compressive regime from the southeast during the evolution of the EGMB resulting from Precambrian collision tectonics (Chetty & Murthy, 1994).

3.3.2.1.1 Metamorphism

The petrological evolution of the rocks of the RVKT are largely constrained from aluminous and calc−silicate granulites (Dasgupta & Sengupta, 2003). The rocks are found to be polymetamorphic, with an early M1−UHT metamorphic event ($\sim 1000°C$ at 8−10 kbar) along a counterclockwise pressure-temperature (*P-T*) trajectory around 1400 Ma. During M2, the granulites were reworked with peak conditions estimated at 8−8.5 kbar and 850°C. The retrograde path of M2 is characterized by near isothermal decompression to ~ 5 kbar, which resulted in exhumation of the granulites to mid-crustal levels at around c. 1000−950 Ma. Recently, Das et al. (2011) determined a concordia age of 953 ± 6 Ma forM2 (U−Pb in zircon) from aluminous granulite from the RVKT. However, these granulite facies rocks were locally overprinted by amphibolite−facies (5 kbar, 600°C) M3 metamorphism, the age of which is constrained at c. 550−500 Ma (Mezger & Cosca, 1999). The c. 1000−900 Ma tectonothermal event in the EGMB has been correlated with that in the Rayner complex in East Antarctica implying that India−Antarctic fit forms a part of Rodinia (Harley, 2003). However, it may be noted that there have been conflicting petrologic and structural interpretations relating to the pressure (P)−temperature (T)−time (t) evolution of the RVKT (counterclockwise (CCW) versus clockwise (CW)) and difficulties in deciphering the geologic significance of ages that span a range from the late Mesoproterozoic to the early Neoproterozoic period (Bose et al., 2011).

It is now well known that the P−T trajectories are of diverse types. The early metamorphism M1 in the RVKT shows a general anticlockwise trajectory, in which the UHT metamorphic peak is followed by near isobaric cooling (Dasgupta & Sengupta, 2003). Contrary to this, a clockwise path with decompression following the UHT peak condition was suggested for the rocks in the Paderu area of the RVKT (Lal, Ackermand, & Upadhyay, 1987). Bhattacharya and Kar (2002) also proposed a clockwise path, in which peak P−T condition of ~ 10 kb and $\sim 1000°C$ was followed by high-temperature decompression and subsequent isobaric cooling. However, Pal and Bose (1997) reinterpreted the petrographic data of Lal et al. (1987) to make them consistent with a general anticlockwise path; the prograde path reached peak values of $P = 9.5$ kb, $T = \sim 1000°C$ and was followed by near isobaric cooling to $P = 9$ kb, $T = 900°C$. Further, some of the textural interpretations and thermodynamic calibrations were questioned. In the light of the controversies, Bhattacharya and Gupta (2001) suggested that the differences in the postpeak P−T paths in different areas may indicate the existence of partially disparate metamorphic sectors within the granulite belt. In the Koraput area, Nanda et al. (2008) demonstrated that an initial phase of granulite metamorphism was followed by intrusion of the KAC, after which the complex and the surrounding rocks were metamorphosed in a second phase of granulite facies metamorphism. They demonstrated that this second phase of granulite metamorphism also occurred along an anticlockwise P−T path, indicating heating followed by loading. This metamorphic imprint could be traced across the SSZ into the Archean charnockites of the WCZ (the Jeypore Province of Dobmeier & Raith, 2003), and correlated this with a phase of intracontinental orogenesis that occurred in the presence of locally derived carbon dioxide (Nanda & Gupta, 2012).

The RVKT hosts many localities of UHT rocks reported from Paderu, Araku, Anantagiri, Vizianagaram, Anakapalle, Chilka lake, Rayagada, Rajahmundry in the form of Mg−Al granulites. The UHT rocks occur as pods or lenses within the granulite supracrustals (Gupta, 2004). Often, they are seen as exotic blocks within mafic granulites. Among other features, UHT assemblages including Zn poor spinel + quartz, sapphirine + quartz, high alumina orthopyroxene ± sillimanite ± quartz, and high-temperature mesoperthite. Complex Fe−Ti−Al oxide solid solutions have been described from these rocks. Some workers suggested that the development of UHT minerals like sapphirine is syntectonically generated along with penetrative foliation during the early deformational events D1/D2 (Sengupta et al., 1990). In contrast, Dasgupta, Sanyal, Sengupta, and Fukuoka (1994) reported the growth of coronas and symplectite overprinting the metamorphic gneissosity implying their tectonic growth perhaps during D2. Gupta (2012) reported that the UHT metamorphism at the western boundary of the EGMB was after the D3-deformational event. The first report of c. 1.0 Ga granulite facies event was from the RVKT obtained from U−Pb and Th−Pb age of perrierite and zircon minerals derived from saphirine bearing granulites (Grew & Manton, 1986). However, the available geochronological data shows a broad range between 1250 and 1100 Ma age for the development of UHT metamorphism in the EGMB. Korhonen et al. (2011) provided further evidence for a major high-temperature metamorphic event in the EGMB between c. 1000 and 950 Ma, although prograde to peak metamorphism may have occurred as early as c. 1042 Ma.

In summary, it can be seen that the RVKT is characterized by the widespread occurrence of late-Mesoproterozoic/early-Neoproterozoic UHT metamorphism. The available metamorphic data from individual localities show that the evolution of the RVKT can be characterized by a broadly similar counterclockwise P−T path that provides an explanation for the spread of ages within the context of a single UHT metamorphic event characterized by slow cooling after the metamorphic peak (Korhonen, Clark, Brown, Bhattacharya, & Taylor, 2013). It is possible that the RVKT preserves a record of a single long-lived high-grade metamorphic evolution in the interval between 1130 and 930 Ma and the peak UHT metamorphism may be around 1000 Ma.

3.3.2.1.2 Tectonic synthesis

The regional synthesis of interpreted structures from the Landsat data in conjunction with the detailed structural analysis from the field measurements offers a new regional tectonic framework for the RVKT that suggests the following features: (1) the RVKT constitutes many large-scale recumbent fold structures, sheath folds, and a network of shear zones, (2) the SSZ forms a boundary between the granulite facies rocks of the EGMB and the granulite and amphibolite facies rocks of Bastar craton, and represents a Precambrian suture zone, (3) the juxtaposition between charnockites and khondalites is generally sheared and migmatized, (4) the broad fold closure to the southwest and the predominant stretching lineations suggest that the deformed rock sequence of the RVKT represents an antiformal structure, (5) two major tectonic events are identified: (i) the northwestward-verging fold styles, thrust systems, shear zones, and the predominant structural trends must have been developed during the first tectonic event synchronous with granulite facies metamorphism resulting from compressive regime from the east, and (ii) the rocks were subsequently subjected to a transpressional tectonic regime during the second tectonic event. The structural fabric suggests an essentially horizontal tectonic regime resulting in thrust systems and associated structures during the first event and high shear strains and migmatization during the second event. Features such as westward-verging thrusts, large-scale recumbent folds, major shear zones, structural domes and basins, indications of tectonic crustal shortening, extensive calc−alkali magmatism and widespread migmatization in the region are attributed to collisional processes during Proterozoic times (Chetty & Murthy, 1994).

The Bouguer anomaly map over the EGMB and adjoining Archean cratons (Fig. 3.26) shows a distinctive linear NE−SW trending gravity high following the major structural trend of the EGMB with a conspicuous break along the GR structure. The gravity high is flanked by a series of gravity lows to the west with a steep gradient to the east. Conspicuously, the gravity contours get crowded displaying steep gradient coinciding with the SSZ, which occurs along the contact of the Bastar craton and the EGMB. A chain of gravity lows over the eastern margin of the Bastar craton is followed eastward by a rise in the gravity field over the EGMB (Niraj Kumar, Singh, Gupta, & Mishra, 2004). The steep gravity gradient across the boundary of the Archean cratons/EGMB was explained in terms of two contrasting crustal domains with a density difference of 0.07−0.1 g/cm^3 (Subrahmanyam & Verma, 1986). In general, the gravity contours follow the trend of the shear zones and associated structural fabrics of the EGMB. Interestingly, the gravity contours strike near N−S and coincide with the trend of the NVSZ, in the central part of the EGMB. The gravity contours show a relative low over the NBSZ where the Singhbhum craton and the EGMB are juxtaposed.

3.3.2.2 Berhampur-Bhubaneswar-Phulbani Terrane

The Berhampur-Bhubaneswar-Phulbani Terrane (BBPT) occurs in the northeastern part of the EGMB and was described as Domain III of Rickers et al. (2001). The BBPT displays a triangular shape and is bordered by the NVSZ in the west, east coast in the east, and the MSZ in the north (see Fig. 3.22). The terrane is dominated by migmatitic rocks besides the major rock suites of khondalites and charnockites. The striking feature in the lithological assemblages of the BBPT is the occurrence of Chilka lake anorthositic complex at the east coast. The terrane constitutes a set of four important NE−SW trending shear zones emerging from the eastern end of the MSZ, centered around Bhubaneswar. Besides, the Baligurha-Tel shear zone also occurs at the western margin of the BBPT. In general, the rocks in BBPT trend NE−SW to near east−west with moderate to steep dips on either side. The trends significantly vary and get deflected to near east−west in the northern part around Phulbani and in the vicinity of MSZ. This part of the domain is described here as Phulbani-Daspalla domain.

The Phulbani-Daspalla domain forms an integral part of the BBPT in the north and is restricted by the MSZ in the north and the KSRSZ in the west (Chetty, 2010). Although the domain extends in an east−west direction, the eastern and western parts are distinctly different in their lithological and structural characteristics. The western part is dominated by charnockites, mafic granulites, and migmatites, and it displays near NE−SW trends with steep westerly dips, often associated with domal structures. However, the structural styles vary depending on the proximity to the marginal shear zones. A few km west of Phulbani, two small and isolated upper Gondwana basins are also mapped between two E−W striking parallel shear zones. The eastern part of the domain comprises mostly migmatitic rocks and megacrystic granitoids. Regional scale folds defined by the gneissic

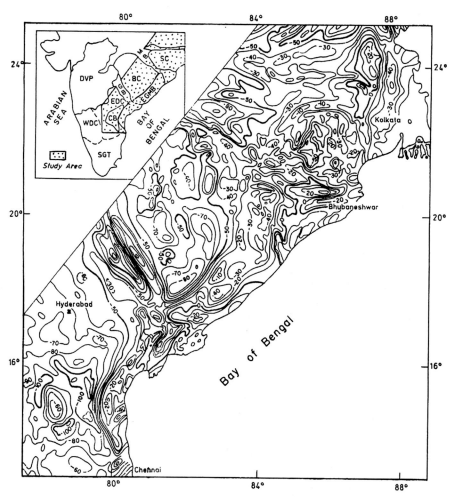

FIGURE 3.26

Bouguer gravity anamoly map of the EGMB and the adjacent cratonic regions exhibiting good correspondence with major shear zones and other geological boundaries.

After NGRI, 1978. NGRI/GHP-1 to 5 Gravity maps of India, Scale: 1:5,000,000. National Geophysical Research Institute, Hyderabad, India.

foliations are predominant with fold axes plunging to northwest with gentle to moderate values (Fig. 3.27). The stretching lineations plunge westward and gradually change to southwest. The deformational history reveals that the early folding seemed to be of near east−west trend with horizontal plunges, which has been later refolded on NW−SE axis. The early folds have been dragged, rotated, and stretched parallel to the MSZ, indicating dextral transpressional tectonics. Mappable curvilinear folds of sheath like geometry were also observed in the eastern part of the Phulbani-Daspalla domain.

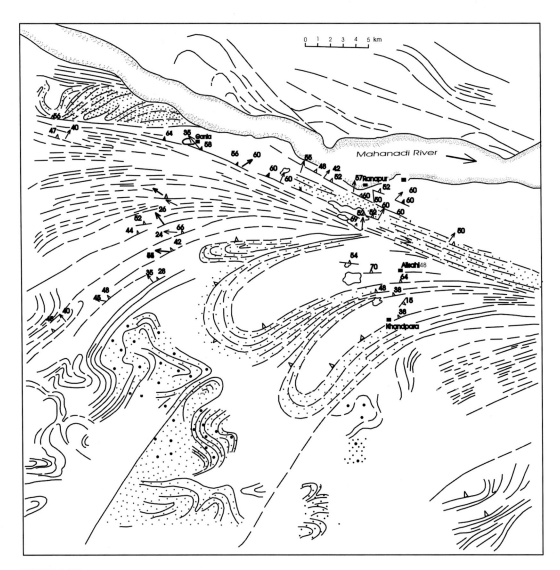

FIGURE 3.27

Structural map of a apart of the Mahanadi shear zone showing the deflection of structural fabrics in a dextral fashion with downward stretching lineations.

After Chetty, T.R.K. (2010). Structural architecture of the northern composite terrane, the Eastern Ghats Mobile Belt, India: Implications for Gondwana tectonics. Gondwana Research, 18, 565–582.

3.3.2.2.1 Chilka lake anorthosite complex

The Chilka lake anorthosite complex (CAC), located at the western margin of the Chilka lake near the east coast, constitutes one of the largest and important anorthosite massifs of the Indian shield

(Fig. 3.28). The CAC contains several variants including anorthosites, leuconorite, norite, jotunite, and quartz mangerite (Sarkar, Bhanumathi, & Balasubrahmanyan, 1981). They are well exposed in three localities: Banpur-Balugoan, Kallikota, and Rambha (Perraju, 1973). Inclusions of anorthosite within the leuconorite are a common feature in CAC. Field observations suggest that all the rocks have been syntectonically emplaced. The host granulite facies rocks around Chilka lake comprises para-, and orthogneisses and a migmatitic rock suite. Bands of two-pyroxene granulite/charnockites occur as mappable enclaves within quartzofeldspathic gneisses. An outcrop scale, similar nonplane, and noncylindrical fold structures have been mapped. The presence of sulfide veins containing

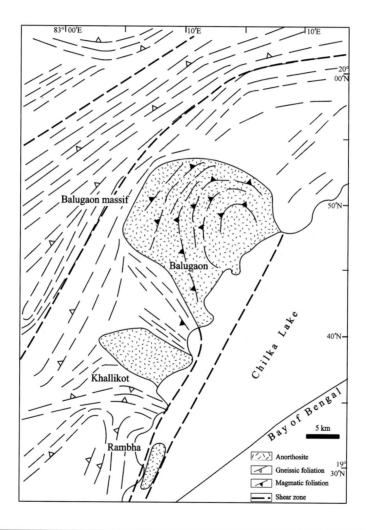

FIGURE 3.28

Structural map of Chilka lake anorthosite complex and the adjacent granulite facies rocks in Berhampur-Bhubaneswar-Phulbani terrane.

chalcopyrite and pyrrhotite are also recorded. Zircon U−Pb SHRIMP data suggest a metamorphic event at 781 ± 9 Ma, while, on the other hand, monazite U−Th−Pb EPMA data, yielded a group age of 988 ± 23 Ma for another metamorphic event. Some younger ages in the range of 550−500 Ma were also obtained (Bose, Das, Torimotoc, Arima, & Dunkley, 2016).

Banpur-Balugaon anorthosite was regarded as "Adirondack-type" emplaced into the cores of large fold structures. A tongue-like extension of anorthosite is observed at the western sheared contact. Large xenoliths of pyroxene granulites and other gneissic rocks are common in the Balugaon dome. Magmatic foliation is defined by mafic rich layers of leuconorite−norite and mafic poor layers of anorthosites. The layering is well preserved as flat-lying in the central parts. The plagioclase grains are as big as 18×10 cm, while the pyroxenes reach up to 10×8 cm. Ferrodiorite dykes more than 10 m wide and about 100 m long are commonly associated with meta-leuconorite and metanorite. The massif is also associated with rapakivi granites. Lenticular aggregates of bluish quartz occur at the margins of anorthosite and ferrodiorite.

Medium-grained anorthosites are recorded around Rambha in the form of a lensoid body dipping west and defining a lobate structure. The occurrence of wollastonite-bearing calc−silicate skarn rock as a xenolith in the Rambha unit of the complex bears testimony to contact metamorphism by the emplacement of anorthosite at high temperature (Leelanandam & Narasimha Reddy, 1998). Phase relations of the calc−silicate rocks in direct contact with the anorthosite unequivocally demonstrate an UHT ($>1000°C$) contact metamorphism caused by heat contributed by the hot anorthosite ($>1100°C$) (Sengupta, Dasgupta, Dutta, & Raith, 2008). Kallikota anorthosite occurs as small and isolated outcrops. The dominant rock type is the quartz mangerite. Small xenoliths of high-grade gneisses occur within the anorthosite. Peripheral zones of the anorthosite are intensely deformed. The well-developed gneissosity at the margins is conformable with the gneissic foliation in the host rocks. Varying geometric forms of the three outcrops with attendant variation in lithological character indicates that the CAC is a typical massif type complex.

The tectonic setting vis-à-vis the intrusion age of anorthosite magma around Chilka lake is still an issue for debate. The emplacement age was described at c. 983 Ma (Chatterjee, Crowley, Mukherjee, & Das, 2008) and that represents only one episode of anorthosite intrusion. The temperature of anorthosite magma was in excess of $\sim 1000°C$. However, no chilled margin is reported, possibly because the anorthosite magma intruded the country rocks that were already hot and had suffered deformation and granulite facies metamorphism prior to the emplacement of anorthosite. Diverse opinions exist about the timing of emplacement of the anorthosite with regard to the regional deformation and granulite facies metamorphism. The emplacement age of CAC was also considered to be synchronous with the emplacement of voluminous megacrystic granitoids in the vicinity. However, the metamorphism of anorthositic rocks was dated as c. 792 Ma (Krause, Dobmeier, Raith, & Mezger, 2001); Nagaraju (2007); Nasipuri, Bhadra (2013); Paul et al. (1990); Prasada Rao, Rao, and Murthy (1987). Finally, overprinting of Pan-African tectonic activity (0.69−0.47 Ga) in CAC was also documented. UHT metamorphism (1000°C) was recorded from Fe-Al−rich granulites and calc−silicate granulites along the contact of the anorthosite massif (Raith, Dobmeier, & Mouri, 2007).

Traditionally, anorthositic and alkaline magmatism was believed to be associated with extensional setting, but the CAC was interpreted to be generated in a compressional tectonic setting (Dobmeier, 2006). The structural evolution of the host high-grade gneisses around the Chilka Lake region was interpreted in terms of transpressional tectonics presumably resulting from oblique collision of India and East Antarctica craton at c. 690−660 Ma (Dobmeier & Simmat, 2002). The host rocks show a complex evolutionary history that commenced with a clockwise P−T path reaching peak HT/UHT conditions, resembling the evolution of the Rayner Complex of East Antarctica implying their correlation during the assembly of the supercontinent Rodinia.

The rocks in the BBPT show homogeneous Nd model ages of 1.8−2.2 Ga for both the orthogneisses and the meta-sediments and was dominated by reworking Grenvillian orogeny with more juvenile additions. According to Rickers et al. (2001a), both the terranes RVKT and BBPT on either side of the NVSZ are totally unrelated. This is further substantiated by the contrasting Neoproterozoic histories: the RVKT evolved through an early anticlockwise P−T path culminating in UHT metamorphism and the BBPT evolved through a clockwise P−T path (Bose et al., 2016). This distinction is vital in the interpretation that the BBPT was contiguous with the Rayner complex of East Antarctica that became a part of the EGMB during the assembly of Gondwana. Thus, the decompression event of c. 780 Ma in BBPT of the EGMB opens up new possibilities for interpreting the breakup of Rodinia. It is possible that the reworking in the BBPT brought the deeper crust to mid-crustal level at c. 780 Ma synchronously along with the initiation of the breakup of Rodinia.

3.3.2.3 Khariar-Bolangir Terrane

The Khariar-Bolangir Terrane (KBT) occurs in the north western part of the EGMB with the Bastar craton to the west and the SSZ/KSRSZ to the east. The KBT shows an arcuate shape with its convexity to northwest. The western margin is marked by a major shear zone defining the contact between the craton and the EGMB, which has been described as TBSZ or Lakhna shear zone (LSZ) by Biswal et al. (2003). Regionally, the LSZ can also be considered as a branch and an extension of the SSZ that got split near Bhawanipatna in the south. The LSZ is ~2 km wide and arcuate in nature with moderate dips to east finally becoming horizontal in the form of a listric thrust (Fig. 3.29). The listric geometry of the LSZ was well imaged by the seismic reflectors defining a detachment or decollement zone between the cratonic foreland and the overlying granulite facies rocks of the EGMB (Nayak, Choudhury, & Sarkar, 1998). Two major nappe structures, characterized by distinct lithological assemblages and deformational history, are well preserved and the KBT can be considered as a well-demarcated fold thrust belt. Gupta et al. (2000) demonstrated that granites in the cratonic foreland were thrust over by the EGMB thrust sheet, and suggested that a zone of "inverted metamorphism" could be observed in the craton, adjacent to the EGMB. Bhadra, Gupta, and Banerjee (2004) demonstrated that as a consequence of this "hot-over-cold" thrusting, cratonic foreland granites underwent sequential fabric formation. They also suggested that movement along the thrust contact may have occurred in two phases, the second operating along a slightly different movement vector. Proterozoic cratonic sedimentary rocks in the foreland (the Khariar sediments) were also deformed in the foreland of the thrust, probably during the second phase of movement (Bhadra & Gupta, 2016).

Deformation character varies along the LSZ with a thrust slip in the central part and strike-slip shear along the adjacent parts of the shear zone. The LSZ at the point of maximum curvature shows thrust character with a general strike of NNE−SSW and the dip varies from 45° to 60° toward the southeast. The shear zone comprises quartzofeldspathic mylonites in association with profusely developed down-dip stretching lineations. The kinematic indicators suggest the overthrusting of the granulites of the EGMB on to the craton. In an overall geometry of the LSZ, the high dip of the

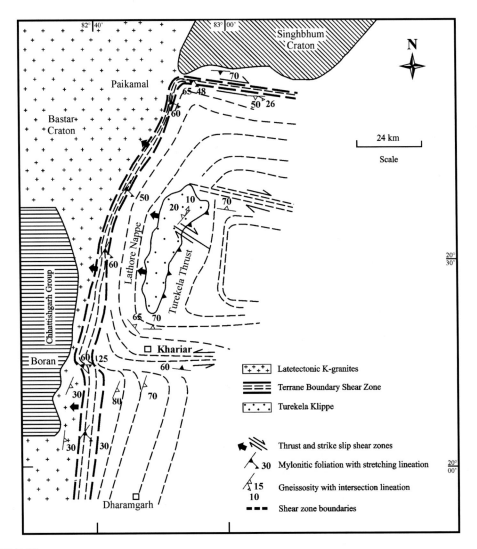

FIGURE 3.29

Map showing regional structure of Khariar-Bolangir-terrane (KBT) showing Lakhna shear zone and other associated thrust zones.

Modified after Biswal, T.K. & Sinha, S. (2003). Deformation history of the NW salient of the Eastern Ghats Mobile Belt, India. Journal of Asian Earth Sciences, 22, 157–169 (Biswal & Sinha, 2003).

zone can be considered as the frontal ramp of the decollement, with a calculated displacement of nearly 4.7 km (Biswal et al., 2000). Lateral ramps were delineated at Khariar in the south and Paikamal in the north. It was inferred that the cratonic granites must have served as the protolith of the mylonites and the basement behaved in a ductile manner during thrusting. The presence of amphibolites, derived from the retrogression of basic granulites during thrusting, suggests that the shear zone must have witnessed extensive retrogression.

The Lathore nappe, overlying the decollement, occurs as a huge tectonic slab (see Fig. 3.25) comprising charnockites, basic granulites, khondalites, and calc−granulites as important constituents with a well-developed gneissosity striking dominantly NE−SW. The rocks in the nappe preserve coaxial early fold structures (F1 and F2) developed along NNE−SSW axes often transformed into map scale sheath folds (Biswal, Sajeevan, & Nayak, 1998). The Lathore nappe close to the shear zone displays realignment of the folds parallel to the strike of the shear zone. Many alkali granite dykes have been synkinematically emplaced into the granulites along with thrusting in and around the Khariar lateral ramp. The Turekela nappe, overlying the Turekela thrust, is well preserved as a klippe surrounded by the rocks of the Lathore nappe. The Turekela nappe is dominated by khondalites and calc granulites with concordant leptynites within khondalites that witnessed multiple stages of folding giving rise to interference fold structures of different scales. Closely spaced NW−SE trending brittle dextral shear zones cut across both the nappe structures as well as the Turekela thrust. Regarding the age of thrusting, it has been earlier reported that the thrusting is coeval with the emplacement of nepheline syenite along the TBSZ near Khariar.

3.3.2.3.1 Khariar alkaline pluton

A 30-km long concordant nepheline syenite pluton occurs at Khariar, comprising syenite and nepheline syenite in addition to alkali granite, gabbro, essexite, malignite, shonkinite, and lamprophyre (Fig. 3.30). Smaller ultramafic bodies such as harzburgite, lherzolite, and pyroxenite are also observed in the close vicinity. The pluton displays well-preserved magmatic foliations, which are conformable with the foliations in host rocks suggesting synkinematic emplacement during thrusting. The rocks of the alkaline complex are interlayered with tholeiitic mafic rocks, which are interpreted to be remnants of an ocean floor (Upadhyay, Raith, Mezger, Bhattacharya, & Kinny, 2006). The nepheline syenite intrusive at Khariar yielded an U−Pb SHRIMP age of 520 Ma (Biswal et al., 2007) and the Mesoproterozoic ages (1480−1500 Ma) obtained were interpreted as the inherited ages of xenocrystic zircons of the older basement. However, Mahapatro, Nanda, and Tripathy (2008) suggested that Khariar nepheline syenites may represent two episodes of alkaline magmatism.

3.3.2.3.2 Anorthosite magmatism

The KBT comprises three important massif-type anorthosite complexes of Bolangir, Turkel, and Jugsaipatna. All the anorthosite complexes were described as a nappe sheet, and the SSZ as the root zone for the emergence of the nappe (Mahapatro et al., 2008). The Bolangir anorthosite complex (BAC) occurs in the north eastern part of the KBT and in the vicinity of the SSZ/KSRSZ (Fig. 3.31). The BAC, predominantly a massif type, is the largest of all the known anorthosites of India (~ 450 km^2) and is surrounded by a khondalitic group of rocks, basic granulites, and migmatitic gneisses including leptynites. The BAC includes anorthositic norite, norite anorthosite, and anorthosite. It contains enclaves of garnet−pyroxene−granulites. The BAC displays well-preserved magmatic layering with variable width from 1 cm to about 15 cm essentially in the form of platy schlieren, made up of biotite, ilmenite, and rarely garnet (Mukherjee, Bhattacharya, & Chakraborty, 1986). The country rocks are well foliated with metamorphic gneissosity. The magmatic foliations are nearly horizontal in the core while they are broadly dipping away at the margins suggesting domal geometry of the BAC. Several ferrodiorite and monzonite dikes of decimeter to meter-wide

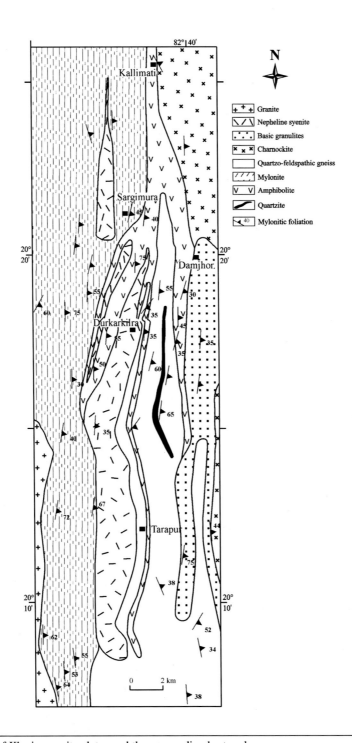

FIGURE 3.30

Geological map of Khariar syenite pluton and the surrounding host rocks.

Modified after Biswal, T.K., Ahuja, H., & Sahu, H.S. (2004). Emplacement kinematics of nepheline syenites from the Terrane Boundary Shear Zone of the Eastern Ghats Mobile Belt, west of Khariar, NW Orissa: Evidence from meso- and microstructures. Proceedings Indian Academy of Sciences (Earth and Planetary Sciences), 113, 785–793 (Biswal, Ahuja, & Sahu, 2004).

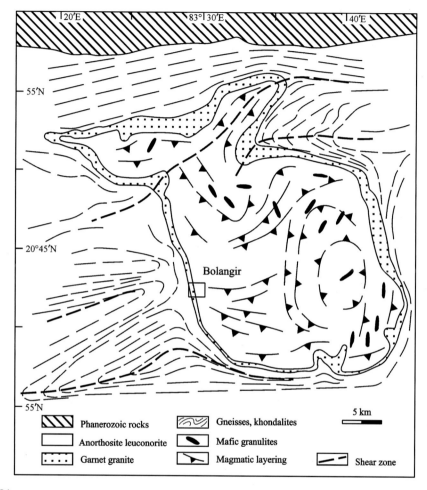

FIGURE 3.31

Map showing the geology and structure of Bolangir anorthosite complex.

Modified after Nasipuri, P., & Bhadra, S. (2013). Structural framework for the emplacement of the Bolangir anorthosite massif in the Eastern Ghats Granulite Belt, India: Implications for post-Rodinia pre-Gondwana tectonics. Minerology and Petrology, 107, 861-880.

occur in and outside the anorthosite (Bhattacharya, Raith, Hoernes, & Banerjee, 1998). The host rocks maintain conformable gneissic foliations with the flow layering at the contact zones. Further, the relationship of tectonic foliations and magmatic layering in the north western part suggests the forceful intrusion of anorthosite in the form of a tongue-like extension. These features imply that the intrusion of BAC must have been synchronous during the thrust-shear–dominated deformation of the country rocks. The parent magma of anorthosite is considered to be coeval with that of 1312 Ma old Chilka lake anorthosites in the EGMB at the east coast. However, it is interesting to note that the ferrodiorites of BAC yielded U−Pb upper intercept age of 933 ± 32 Ma with a lower intercept at 515 ± 20 Ma (Krause, Dobmeier, Raith, & Mezger, 2001).

The Turkel Anorthosite Complex (TAC) is located south of BAC and is confined between the Khariar/Lakhna shear zone and the SSZ within the KBT. The TAC comprises mostly anorthosite—leuconorite and is encircled by quartz—monzonite intrusions hosted within a widespread megacrystic K—feldspar granite. The host rocks around the TAC comprise granulite facies rock assemblages (Fig. 3.32) dominated by migmatitic quartzofeldspathic and metapelite gneisses along with mafic granulites, calc—silicate gneisses, and monzonite—charnockite—granitoids (Das, Nasipuri, Bhattacharya, & Swaminathan, 2008). The TAC occurs in the form of an elliptical pluton exposing anorthosite in the central part grading into leuconorite toward its margins. TAC is flanked by quartz—monzonite to monzonite in the west, north, and east (Maji, Bhattacharya, & Raith, 1997). Ferrodiorite occurs as discontinuous dykes, lenses, and veins at the contact between border-facies leuconorite and the quartz—monzonite. The rocks of the TAC look massif in the core and exhibit foliation fabrics that are increasingly distinct at the margins. The TAC shows sheared contacts with an assemblage of meta-sedimentary rocks, migmatites, and mafic granulites. The

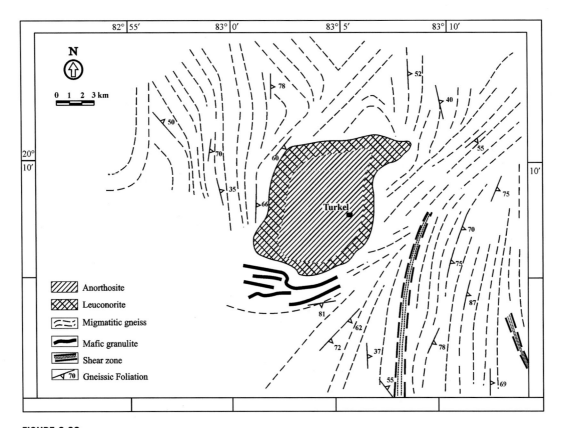

FIGURE 3.32

Map showing the geology and structure of Turkel anorthosite complex and the surrounding region.

Modified after Maji, A.K., Bhattacharya, A., & Raith, M. (1997). The Turkel anorthosite complex revisited. Proceedings of Indian National Academy of Sciences (Earth Planet Sciences), 106, *125–313.*

southern and the eastern margins of TAC are marked by the presence of the SSZ. The field observations suggest west-directed thrusting of the KBT on to the cratonic foreland during early Phanerozoic time (Biswal, Biswal, Mitra, & Moulik, 2002).

Zircon grains from TAC yielded the U−Pb isotope system crystallizion age at 980 ± 8 Ma implying that the leuconorite−anorthosite suite represents the oldest intrusive phase of the Turkel complex (Raith et al., 2014). The other components of TAC like the monzonite, quartz-monzonite, and ferrodiorite were emplaced later between 956 and 945 Ma. The rocks of the TAC underwent an early phase of metamorphism at around 901−879 Ma. A late-Neoproterozoic to early-Paleozoic Pan-African tectonothermal event was also recorded in TAC, which could be related to the development of the SSZ. Considering the age of c. 964−996 Ma for the surrounding megacrystic K−feldspar granitoids, it was suggested that TAC experienced a prolonged episode of slow cooling from the magmatic crystallization stage (Dharma Rao, Santosh, & Zhang, 2014), comparable to that of the BAC. The TAC was metamorphosed subsequently during the late Neoproterozoic nappe tectonics in the KBT. The TAC is interpreted to have originated from polybaric fractionation of mantle-derived high-Al gabbroic magmas (Maji et al., 1997).

The Jugsaipatna anorthosite complex (JAC) occurs at the southern extreme of the KBT in the close vicinity of the SSZ. It displays a circular form spreading over an area of 30 km^2. The JAC comprises anorthosite, leuconorite, troctolitic norite, gabbronorite, and eucritic gabbronorite as the major rock units and minor ferrodiorite and websterite (Fig. 3.33). Anorthosite−leuconorite−troctolitic norite in the core are surrounded by a rim of high-Al gabbronorite (HAG), (Mahapatro, 2010). Small dykes of ferrodiorite measuring up to 25 m in length and 20−50 cm in width intrude on the leuconorite. Interestingly, the mafic members such as HAG, troctolitic norite, and eucritic gabbronorite are not reported from the other anorthosite complexes of EGMB. Like many other plutons, the JAC also is massif type in the central part while the development of foliation fabrics at the margins is distinct. Charnockites, two pyroxene granulites, megacrystic granitoids, and migmatitic gneisses are the principal constituents of the host rocks. The geochemical characteristics of gabbronorite from the JAC are similar to those of high-Al gabbros reported elsewhere in the globe and suggest that these high-Al gabbros may be parental to the associated anorthositic rocks as mantle-derived melt produced by the differentiation of basaltic (pricritic) magma in upper mantle chambers (Mahapatro, Nanda, & Tripathy, 2010).

The recently obtained U−Pb zircon data indicate that the anorthosites, gabbros, and associated granitoids of JAC must have been emplaced contemporaneously at 918−996 Ma, which are comparable with similar ages of 983 ± 2.5 Ma for the anorthosites from Chilka Lake complex and the c. 930 Ma for Bolangir anorthosite in the EGMB. Further, these ages also coincide with the timing of UHT metamorphism recorded from different localities of the EGMB. The coeval nature of mantle-derived magmatism and UHT metamorphism in a collisional orogen following a prolonged subduction−accretion history was interpreted as asthenospheric upwelling, probably through a slab-window mechanism (Dharma Rao et al., 2014), which also coincides with the timing of final assembly of India within the Neoproterozoic supercontinent Rodinia.

From the field evidence, it is evident that the entire KBT constitutes a stack of northwesterly verging imbricate thrust sheets and nappes defining a fold thrust belt, juxtaposed with the EGMB. The fold thrust belt of the KBT comprises LSZ, Lathore nappe, Turekela klippe, and Turekela thrust reflecting the main architecture. The JAC forms a part of the fold thrust belt or a nappe in association with the other known anorthosites of TAC and BAC within the KBT. It is possible that

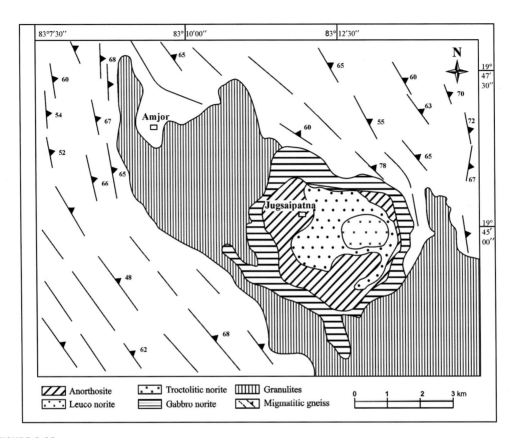

FIGURE 3.33

Geology of Jugsaipatna anorthosite complex.

Modified after Mahapatro, S.N., Nanda, J.K., & Tripathy, A.K. (2010). The Jugsaipatna Anorthosite Complex, Eastern Ghats Belt, India: Magmatic lineage and petrogenetic implications. Journal of Asian Earth Science, 38, 147–161.

the thrust tectonics, subsequent to continental collision, must have been the major cause for the juxtaposition of not only the KBT but also other terranes in the EGMB. The tectonic process may be similar to the thrusting in front of a hinterland indenter, as in case of the Himalayas. The mechanism is also comparable with the Caledonide type of fold-thrust belt where the thrusting is posttectonic to folding and metamorphism of the cover rocks and the basement was deformed in ductile manner during thrusting.

3.3.3 JEYPORE-BERHAMPUR CORRIDOR

Geological mapping along an east–west Jeypore-Berhampur corridor (300×5 km) in the central and the widest part of the EGMB (Nanda & Pati, 1989) reveals the distinct lithological associations separated by the major shear zones such as SSZ, NSZ, and VSZ (Fig. 3.34). Three dominant

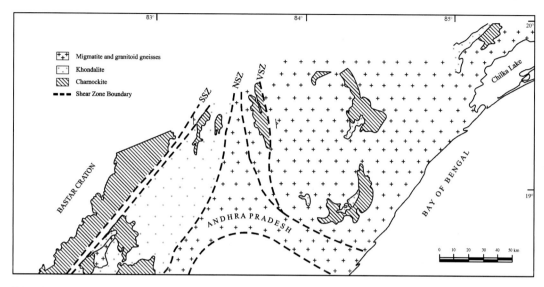

FIGURE 3.34

Geological map of Jeypore-Berhampur corridor. *SSZ*, Sileru shear zone; *NSZ*, Nagavali shear zone; *VSZ*, Vamsadhara shear zone.

Modified after Nanda, J.K. & Pati, U.C. (1989). Field relations and petrochemistry of granulites and associated rocks in the Ganjam-Koraput sector of the Eastern Ghats Belt. Indian Minerals, 43, *247–264.*

lithological associations from east to west can be delineated: (1) the eastern migmatite–granite zone with enclaves of khondalite, two-pyroxene granulite and charnockite; (2) the central khonda-lite zone with subordinate pyroxene granulite, charnockite, and granitoids; and (3) the western pyroxene granulite and charnockite zone with minor khondalitic slivers and patches of migmatite (Narayanaswami, 1975). The eastern zone lies between the NVSZ and the east coast that constitute a variety of migmatites with restites of quartzite, quartz–garnet–sillimanite gneisses and calc–silicate rocks, pyroxene granulites, and charnockites of varying proportions. They occur as bands, lenses, schlierens, and patches in a sea of leucosomes of dominantly adamellitic composi-tion. Mylonites and pseudotachylites are also common. Some of the large khondalitic restites occur in the form of domal structures and folds with variable axial trends.

Detailed geology of a part of the J–B corridor, confined to the region between the NSZ and the VSZ is presented (Fig. 3.35). The central zone of the J–B corridor ($\sim$15 km), occurring between the NSZ and the VSZ, comprises primarily a meta-sedimentary pile with subordinate and conform-able bands and lenses of charnockites and pyroxene granulites. The meta-sedimentary pile includes khondalites, quartzites, and calc–silicates representing psammitic, pelitic, and calcareous protoliths and exhibit conformable rhythmic interbanding. There appear to be structural repetitions implying isoclinal folding. The contact between the major rock groups of khondalitic and charnockitic rocks is mostly occupied by sheared migmatites. Khondalites often show minor concentrations of manga-nese oxides as well as graphite. Development of cordierite is also often recorded along the NVSZ. Megacrystic granites are the dominant rock type within the NSZ as described earlier.

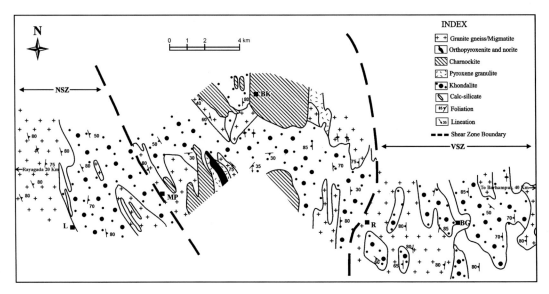

FIGURE 3.35

Regional geological map of the Nagavali shear zone (NSZ) and the Vamsadhara shear zone (VSZ) and the region between them.

Modified after Nanda, J.K. & Pati, U.C. (1989). Field relations and petrochemistry of granulites and associated rocks in the Ganjam-Koraput sector of the Eastern Ghats Belt. Indian Minerals, 43, *247–264.*

The zone between the SSZ and the Bastar craton is considered here as the third category with a width of ~20 km, which extends up to the Krishna river in the south. The dominant rock types in the zone are charnockite and pyroxene granulite and the latter occurs as enclaves within the former. These rocks are mostly massive and weakly foliated and were strongly retrogressed to amphibolites grade at the western part of this zone. The rocks, further west, are followed by intensely deformed cratonic rocks of Bastar craton.

3.4 EGMB-NORTH

The northern parts of the EGMB, covering the MR structure and the adjacent regions, form a composite terrane of high-grade metamorphic rocks, described here as EGMB-north (Chetty, 2010). In contrast to the regional NE–SW trend of the EGMB, the rocks in the EGMB-north trend east–west. The region demarcates a major block structure (120 × 50 km) with a composite network of shear zones (Fig. 3.36). The EGMB-north is separated from the 3.4 Ga old rocks of the Singhbhum craton in the north by a crustal-scale shear zone described here as Northern Boundary Shear Zone (NBSZ). The Koraput-Sonepur-Rairakhol shear zone (KSRSZ) marks the western boundary and MSZ marks the southern boundary. The other important shear zones of the composite terrane are

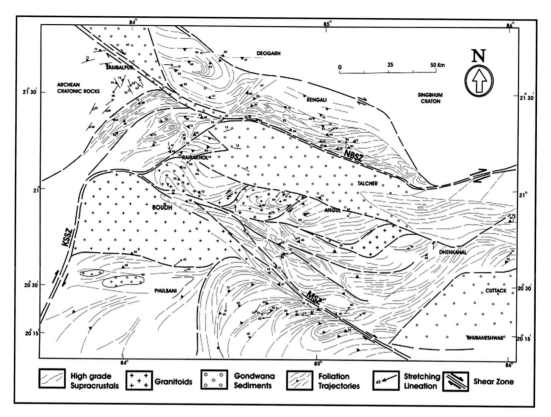

FIGURE 3.36

Map of the structural architecture of the composite terrane EGMB-north showing regional lithological distribution, shear zone network, distribution and trends of structural fabrics, and kinematic displacements. *KSSZ*, Koraput-Sonepur shear zone; *MSZ*, Mahanadi shear zone; *NBSZ*, Northern Boundary shear zone. Note the consistent dextral displacements along E—W shear zones as well as the NE—SW striking Riedel shear zones.

After Chetty, T.R.K. (2010). Structural architecture of the northern composite terrane, the Eastern Ghats Mobile Belt, India: Implications for Gondwana tectonics. Gondwana Research, 18, 565–582.

represented by domain boundaries as well as the oblique Riedel shear zones. The EGMB-north comprises high-grade orogen core and the Permo-Carboniferous Gondwana sedimentary basins. The major rock units include charnockitic and khondalitic gneisses, which have been variably migmatized and retrogressed.

3.4.1 SHEAR ZONES

The composite terrane of the EGMB-north is characterized by a network of major shear zones (see Fig. 3.36). Prominent among them are: Northern Boundary Shear Zone (NBSZ) in the north; the

MSZ in the south, and the Koraput-Sonepur-Rairakhol Shear Zone (KSRSZ) in the west. The other shear zones, described here as link shear zones, are spatially connected to the boundary shear zones. Protracted deformational history ranging from Neoarchean to Permo-Carboniferous/ Cretaceous is well preserved along the boundary shear zones and influenced the structural architecture of the region (Chetty, 2010). Structural trends are well reflected in satellite images and correspond to the shear zone network and compositional fabrics of various rock types defining the regional tectonic fabric of the region. The details of each shear zone are described below.

3.4.1.1 Northern boundary shear zone

The Northern Boundary Shear Zone (NBSZ) marks the northern boundary of the EGMB and separates the Singhbhum craton (3.2 Ga) to the north. The NBSZ has been named in various ways by earlier workers: Kerajang fault (Nash et al., 1996) and Northern Orissa Boundary fault (Mahalik, 1994). The NBSZ extends along near east−west direction with an over 250 km strike length and varying width (2−5 km), often extending to a maximum of ∼ 15 km in the central part (Fig. 3.37). The NBSZ is not uniform and exhibits complex architecture because of interdigitation of lithological assemblages of disparate age, metamorphism, and deformation. In the central part, the NBSZ strikes east−west and separates the Lower Gondwana basin (Permo-Carboniferous Talcher) to the south and the Rengali granulite facies rock domain to the north. The widened part at the northern margin of the NBSZ has been earlier described as Riamal fault, while the southern margin is described as the Kerajang fault (Crowe, Nash, Harris, Leeming, & Rankin, 2003). In the east, it extends in the form of Sukinda thrust hosting chromiferous−ophiolite dominated lithologies.

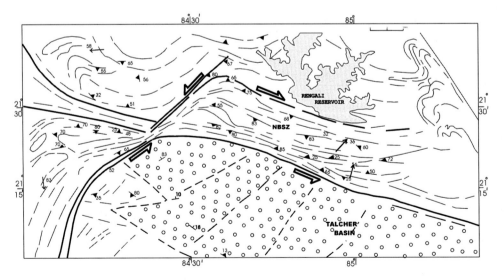

FIGURE 3.37

Map showing a part of the northern Boundary shear zone (NBSZ) with structural elements derived from the satellite images as well as from the field measurements suggesting dextral displacements.

A suite of ophiolitic differentiates in island arc environment that include gabbro—orthopyroxene—peridotite—dunite bodies with podiform to stratiform chromite occur along this suture zone (Banerjee, Mahakud, Bhattacharya, & Mohanty, 1987). In the west, the NBSZ extends in NW—SE direction and is closely associated with Lower Gondwana coal bearing sediments of Ib River. The NBSZ is characterized by linear and folded quartzite ridges for a strike length of tens of kilometers, which show intense deformational features. These ridges lie parallel to the regional foliation fabric in the country rocks. The NBSZ is also dominated by the presence of tectonic slivers of cratonic basement gneisses along with high-grade rock assemblages including quartzites, which have been intensely deformed and retrogressed to lower amphibolite to greenschist facies conditions. The quartzites are, in general, cherty type and are closely associated with sericite schists. The quartzites exhibit mylonitization and intense fracturing while intense folding and puckers are reflected in sericite schist. Deformed gneissic nepheline syenite bodies ($\sim$1.45 Ga) intruded these quartzites (Mahapatro, Tripathy, Nanda, & Rath, 2013). Micaceous pegmatites are also common along the NBSZ. While the gneisses show gentle to moderate dips to the south, mylonitic fabrics display steep southerly dips. North-verging tight to isoclinal folding with gentle plunges on either side is common. The field observations are consistent with the reverse-upward trajectories in high-grade rocks that were transported and juxtaposed against the low-grade rocks of the Singhbhum craton through high-angle thrusting. At deeper levels, the NBSZ obtains listric geometry to south. Misra and Gupta (2014) gave an alternative interpretation to the contact between the Talchir basin and the Rengali Province to be northward and described the contact as a subvertical strike-slip shear zone with dextral sense that formed a dilational step-over.

Lower metamorphic grade and reworking relationships indicate that the NBSZ is a retrograde shear zone that continued to deform later in transpressional orogenesis into the brittle environment. Discrete brittle shear zones are also common in the widened parts of the NBSZ and are associated with fault gauges and intense brittle fracturing in quartzite horizons. The association of Gondwana basins with the NBSZ was attributed to dextral strike-slip displacement kinematics resulting in the development of pull-apart—type basins (Chetty, 1995b). The NBSZ was interpreted to be a lateral ramp allowing the granulite facies rocks of the EGMB to be transported northward on to the Singhbhum craton giving rise to Rengali domain. The presence of K—feldspar megacrystic granite along the NBSZ probably indicate the initiation of the NBSZ at c. 980 Ma (Crowe et al., 2003) as a dextral transpressional shear zone. A reactivation deformational event along the NBSZ was inferred to be during 490—470 Ma based on ^{40}Ar-^{39}Ar plateau ages for white mica and biotite derived from pervasively retrogressed mylonites (Crowe, Cosca, & Harris, 2001). This event of brittle reactivation of the NBSZ during the Permo-Carboniferous period must have been associated with continued transcurrent dextral shearing, subsequent faulting, and displacement in coal deposits of Lower Gondwana basins of the Talcher and Ib River. Further, available AFT ages of zircons obtained from the NBSZ suggest a long residence time and a slow cooling of the metamorphic basement rocks during 400—300 Ma (Lisker & Fachmann, 2001).

3.4.1.2 Mahanadi shear zone

The MSZ is a first-order tectonic feature representing the southern boundary of the composite orogen of the EGMB-north. The MSZ trends WNW—ESE and extends for over 150 km along strike, with width varying from 2 to 8 km. It occurs all along the Mahanadi River course and lies subparallel to NBSZ and separates the Tikarapara domain to the north and Phulbani-Daspalla domain to the south (Chetty, 2010). The MSZ is characterized by a series of parallel strike ridges with predominant rock types represented by khondalites and charnockites, which are invariably, retrogressed, migmatized, and mylonitized (Fig. 3.38). The foliations in MSZ strike WNW−ESE with varying northerly dips (30−50 degrees). The well developed stretching lineations are predominantly down-dip in nature. Spectacular outcrops of ultramylonites in the form of strike ridges trending for a length of over 12 km and a width of ∼250 meters are the striking features of the MSZ. Pseudotachylite veins of ∼10 cm wide criss-cross the outcrops of ultramylonites. The orientation and geometry of fabrics on either side of the MSZ change dramatically. The marked deflection of foliation trajectories and other kinematic features strongly indicate large strains of dextral displacements.

The MSZ, in the western part, splits into several splay shear zones: the major one being that along the NW−SE direction while the others extend towards west, delineating the southern

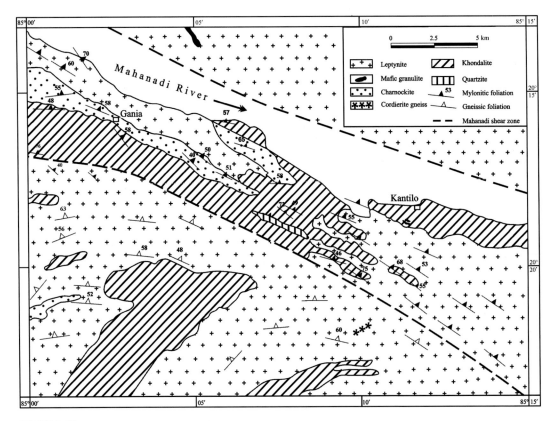

FIGURE 3.38

Geological map of a part of the Mahanadi shear zone around Gania-Kantilo region.

Modified after Mahapathro,S.N., tripathy, A.K., Nanda, J.K., & Abhinaba Roy (2009). Coexisting ultramylonite and Pseudotachylyte from the eastern segment of the Mahanadi shear zone, eastern Ghats Mobile Belt. Journal Geological Society of India *74, 679−689.*

boundary of a large megacrystic granitoid body at Sonepur. The latter shear zone has been described as the Ranipathar shear zone (Nash et al., 1996). The Sonepur granite body appears more massive and internally less deformed and envelopes all the splay shear zones that emerge from the MSZ and yielded a SHRIMP Zircon age of c. 980 Ma (Crowe et al., 2003). This age is consistent with similar ages for megacrystic granites elsewhere in the EGMB (Kovach et al., 1998). Intense overprinting reactivation was recorded along the Ranipathar shear zone as indicated by biotite $^{40}Ar/^{39}Ar$ date of 504 Ma (Crowe et al., 2001). The MSZ also constitutes megacrystic granitoids at several places and the feldspar phenocrysts have been reworked and recrystallized to give rise to myrmekites, granulation of feldspar grains, and a variety of δ-type and sigmoidal kinematic indicators. This is further substantiated by microscopic kinematic indicators of a thin section made from an ultramylonite sample collected from Ranipathar shear zone (Fig. 3.39). However, it may be noted that all the branches or splays of the MSZ coalesce finally with the KSRSZ in the west. In the east, the MSZ is relatively narrow (∼2 km) and well defined for a strike length of ∼50 km extending up to the east coast. The important rock units are interbanded sequence of khondalites and charnockites, migmatitic gneisses, and mafic granulites (Fig. 3.38). A few NE-SW−trending oblique shear zones emerge in the east apparently controlling the development of rhombohedral shaped upper Gondwana Athgarh basin. Interestingly, a 117 Ma old mafic dyke located at Naraj intrudes these upper Gondwana sediments (Lisker & Fachmann, 2001).

3.4.1.3 Koraput-Sonepur-Rairakhol shear zone

The Koraput-Sonepur-Rairakhol Shear Zone (KSRSZ) forms the northern extension of the SSZ and marks the western boundary of the EGMB-north (see Fig. 3.36). The KSRSZ extends for over 70 km in NNE−SSW direction with a width of 1−3 km and traverses further through Rairakhol. The shear zone is characterized by mylonitic fabrics and megacrystic gneisses. The shear zone, in general, dips at moderate to steep angles to southeast with oblique stretching lineations. A few small nepheline syenite bodies (1413 Ma, Rb−Sr) associated with nonfeldspathoidal syenites are also located in the proximity of KSRSZ around Rairakhol (Panda, Patra, Parta, & Nana, 1993). The country rocks around these bodies include the interbanded sequence of quartzites, khondalites and charnockites. A few subparallel and sympathetic shear zones are also developed west of Rairakhol, in the region of cratonic area south of Sambalpur overlain by granulite facies rocks, similar to that of the Rengali province or the KBT. The KSRSZ also displays an eastward dextral shift of ∼10 km and restricts the western extension of the Sonepur megacrystic granite as well as the western contact of Talcher basin. The rocks to the west of KSRSZ are termed as Badarama complex, which show similar characteristics to those of the granulite facies rocks of Rengali Province. The western margin of Badarama complex is also marked by another splay of KSRSZ that trends near north south and coalesce with the NBSZ (see Fig. 3.36). To the east of the KSRSZ, structural fabrics trend E−W, becoming NNE−SSW toward the west.

3.4.1.4 Link shear zones

The domain boundary shear zones, a distinct oblique set of NE-SW−trending Riedel shear zones, are interlinked together and are described here as link shear zones that occur between the boundary shear zones of NBSZ and MSZ (see Fig. 3.36). The domain boundaries are characterized by intense shearing and migmatization. The prominent among the link shear zones is the Angul-Dhenkanal shear zone, which occurs at the southern contact of the Talcher basin. The shear zone is 2−3 km wide

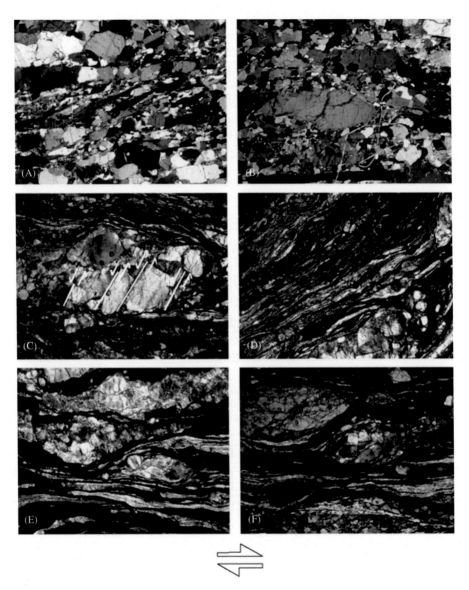

FIGURE 3.39

Microphotographs showing (A) C-type shear band structures, (B) hornblende crystal in the form of a "fish" structure surrounded by mylonitic foliation constituting recrystallized quartz and feldspar grain, (C) large porphyroclasts of feldspar showing "book shelf"–type of structure with antithetic relative displacements along microfaults, (D) zone of ultramylonite with straight internal layering (upper part) in a coarse grained host rock composed of quartz, feldspar, and biotite, (E) large number of rigid elongate crystals in a porphyroclast showing imbrication with antithetic microfaults, and (F) cluster of fragments of mantled porphyroclasts surrounded by ultramylonite layers. All the pictures are enlarged to XPL X 4X, consistent with dextral sense of shear.

After Chetty, T.R.K. (2010). Structural architecture of the northern composite terrane, the Eastern Ghats Mobile Belt, India: Implications for Gondwana tectonics. Gondwana Research, 18, *565–582.*

and ~200 km long with broad southerly steep dips hosting many satellite bodies of alkaline rocks (Panda et al., 1993). Reactivated basement lithologies and a few preferably located dykes are also exposed along the zone. The shear zone has accommodated many granitic plutons and the time of granitic intrusive activity is around 850 Ma (Rb−Sr muscovite mineral age, Halden, Bowes, & Dash, 1982).

The other link shear zones are relatively narrow, and strain partitioning is well reflected in the form of distinct sigmoidal foliation geometries, consistent with the dextral displacements (see Fig. 3.36). Interestingly, the shear zones within the Tikarapara domain coalesce and merge with the MSZ and KSRSZ in the west. They also often get deflected towards NE converging with the geometry of Riedel shear zones and finally terminate against the NBSZ in the east. While the shear zones at the domain boundaries are steep, the shear zones within domains like Tikarapara show moderate to steep dips to north. These are characterized by well-developed early high-grade ductile mylonitic fabrics with quartz and feldspar aggregate ribbons. These are superimposed by brittle fracturing at several places often displaying fault gauge and fault breccia. All these observations suggest that repeated reworking had been a common feature along the link shear zones with varied deformational characteristics. The geometric relationships of foliation trajectories, sigmoidal shaped enclaves, and other kinematics along the shear zones indicate dextral shearing.

3.4.2 STRUCTURAL DOMAINS

The EGMB-north constitutes many lithostructural domains of granulite facies rocks separated by east−west trending shear zones described in the foregoing. From north to south, the distinct domains include: Rengali domain, Angul domain, and Tikarapara domain (see Fig. 3.36). The geology of each domain is detailed below.

3.4.2.1 Rengali domain

The Rengali domain, a triangular-shaped region north of the NBSZ, is bound by shear zones on all sides. It comprises granulite facies metamorphic rocks, namely charnockites, migmatitic gneisses, khondalites, quartzites, and a few small occurrences of isolated basic granulites. The rocks, in general, exhibit E-W−trending tight to isoclinal folds with gentle to moderate dips to south. The pervasive metamorphic gneissic foliation in the rocks has been superimposed locally by mylonitic secondary foliations recording lower amphibolite facies metamorphism. Petrological studies reveal widespread biotite melting at peak granulite facies metamorphic conditions of 7.8 ± 0.13 kbar, $849 \pm 31°C$ and subsequent melt extraction, producing a mixture of residual granulites and melts in the Rengali Province.

The identical Pb−Pb zircon age of 2802 ± 3 Ma and SHRIMP age of 2801 ± 10 Ma, obtained for hornblende-bearing orthogneiss were interpreted to indicate crystallization of precursor granitoids suggesting the extensive magmatism at 2.8 Ga within the Rengali domain (Crowe et al., 2003). The Ar−Ar plateau age for hornblende indicates cooling below 500°C at 699 ± 8 Ma, indicating retrogression to amphibolite facies (Crowe et al., 2001). Electron microprobe geochronology of texturally well constrained monazites indicates the timing of peak granulite metamorphism at 3057 ± 17 Ma and its metamorphic reheating at 2781 ± 16 Ma (Mahapatro, Pant, Bhowmik, Tripathy, & Nanda, 2012). The available age data shows that the protoliths of Rengali domain yielded Archean ages (~3.0 Ga) and the event at 2.8 Ga denotes a major tectonothermal event, followed by isolated mafic dyke activity around 0.8 Ga. The 2.8 Ga metamorphism had a peak P—T regime of 9.5 Kb at 950°C followed by postpeak isobaric cooling through 200—350°C at 8.5—9.5Kb (Kar, 2007). Fission Track ages between 270 ± 13 and 252 ± 9 Ma suggest that Rengali domain entered the partial annealing zone for apatite (120—160°C) during the Late Permian (Lisker & Fachmann, 2001).

The status of Rengali domain vis-à-vis the EGMB, the following observations are pertinent: (1) the disposition of granulite facies rocks and the association of 2.8 Ga gneissic fabrics of the Rengali domain; (2) the presence of shear zones all around during Neoproterozoic reworking; (3) the occurrence of flat fabrics; and (4) the close association of the suture zone in the form of NBSZ reveals that the granulite facies rocks of the Rengali domain indicate that they were originally a part of the EGMB, which have been transported onto the Singhbhum craton through high angle thrusting. This hypothesis finds further support from evidence such as (1) the rocks of both the EGMB and the Rengali province have undergone similar cooling history, inferred on the basis of $^{40}Ar/^{39}Ar$ data (Crowe et al., 2001); (2) consistency of similar Hb ages (699 Ma) for the Rengali province as well as those from the EGMB (650 Ma); and (3) continuous cooling history in the northern parts of Rengali province while perturbed thermal history prevailed in the southern parts. The Angul and Rengali domains show that the two were juxtaposed at similar crustal levels around 700 Ma, but the former was at deeper level during the Grenvillian, implying that the Angul domain must have been tectonically uplifted because of transpressional tectonics during the Pan-African period. In the light of above discussion and the similar tectonic history at the western margin of the EGMB, the argument that the Rengali province represents a part of Bastar craton or Singhbhum craton may not be correct. Considering the structural architecture, the geometry of shear zones, and the involvement of transpressional and resurgent tectonics, the hypothesis that the Rengali domain constituting the high-grade granulite facies rocks must have been originally a part of the EGMB, which was upthrusted and juxtaposed onto the Archean Singhbhum craton through the north-verging NBSZ, is preferred. The model of thrusting of hot granulites over the cold craton and their cooling and decompression seem to be the convincing reasons for the hypothesis regarding the status of the Rengali domain described above.

3.4.2.2 Angul domain

The Angul Domain, bound by shear zones on either side, occurs to the south of Gondwana sediments of Talcher basin and extends in an east—west direction. The host rocks include khondalites, charnockites, and basic granulites accompanied by extensive migmatization. The early L—S fabrics of 2.8 Ga are best preserved locally in high-grade assemblages where coaxially refolded compositional layering is evident. The initial subhorizontal foliations are largely obliterated by intense transposition and folding into steeper mylonitic fabrics. A distinct oblique set of NE-SW—trending shear zones, akin to the Riedel shear zones, links the domain boundary shear zones defining dextral regional scale lozenge-shaped geometries, which are often occupied by coarse-grained granitoid bodies. A few narrow (1×10 m) mafic dykes are also recorded within the domain. Megacrystic granitic gneisses with enclaves of charnockitic and khondalitic rocks displaying dextral transpressional kinematics are also common. Structural and metamorphic studies by Sarkar, Gupta, and Panigrahi (2007) identified five deformation events, with two major cycles of compressional tectonics separated by a phase of extension and uplift. The first cycle accompanied granulite facies metamorphism, while the second cycle was a phase of reworking represented by amphibolite facies metamorphism, although dehydration reactions also occurred during this event in zones deficient in aqueous fluids.

Some of the shear zones are mostly marked by pegmatites and associated migmatites. Muscovite from a pegmatite yielded Rb−Sr age of 854 ± 6 Ma (Halden et al., 1982) and amphiboles revealed ^{40}Ar-^{39}Ar age of 854 ± 4 Ma (Lisker & Fachmann, 2001), which can be related to granite magmatism and migmatization in the domain. The subsequent Pan-African event is represented by the development of 512 Ma old pseudotachylites. This event, which is recorded at several places in the entire EGMB, may represent the imprints of Pan-African orogeny preserved mostly along the regional shear zones.

3.4.2.3 Tikarapara domain

This domain occurs between the Angul domain and the MSZ. The major rock types include khondalites, charnockites, basic granulites, quartzites and calc−silicate rocks with occasional pegmatites and quartz veins. Pseudotachylites of the domain yielded ages of 550 Ma representing large scale transcurrent shearing and reactivation. Unlike in other domains, the foliations in this domain dip toward north at moderate to steep angles. Stretching lineations and mesoscopic fold axes consistently show gentle to moderate plunges to east. Reorientation and reorganization of foliation fabrics into sigmoidal geometry are the characteristic features indicating consistent dextral displacement history in the domain. Broadly, the internal fabrics and the boundary shear zones of the Tikarapara Domain define a south verging "half flower structure." The presence of upper Gondwana Athgarh basin in the eastern part of the domain is striking. The basin seems to be strongly controlled by NW−SE trending MSZ and NE−SW trending Riedel shear zones. The AFT ages from the eastern part of the domain yielded 318 + 18 Ma and 315 Ma pointing to a continuous slow cooling without any major influence of Gondwana rifting/basin formation (Lisker & Fachmann, 2001).

3.4.3 CRUSTAL CROSS SECTION

A north−south crustal cross section constructed across the composite terrane of the EGMB-north (Fig. 3.40) shows the nature, disposition and the geometry of shear zones and structural domains displaying the present structural architecture. From north to south, the Rengali domain, NBSZ, Angul domain, Tikarapara domain, and the MSZ represent distinct geological units. North-verging nappes and fold structures are common in the Rengali domain with horizontal to gently dipping structures to south. Considering the general structure and the presence of shear zones on all sides, the Rengali domain was inferred to have been transported and thrust from south over the Singhbhum craton rooted through the north-verging NBSZ (Chetty, 2010).

The NBSZ is a northern boundary shear zone and a suture zone with complexities along the strike. It shows maximum thickness in the central part and dips steeply to the south and can be described as lateral ramp geometry. Inward-dipping normal faults mark the Talcher basin, which is closely associated with the NBSZ, suggesting a rifting process in phases not only during the deposition but also subsequently. The Angul domain forms the central part of the cross section and is characterized by steep dips and intense migmatization intruded by 980 Ma old megacrystic granitoids. The basement seems to have been reworked giving rise to syntectonically emplaced and deformed oval shaped granitoids displaying dextral kinematics.

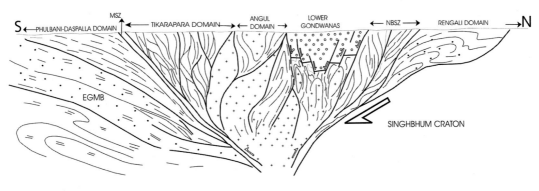

FIGURE 3.40

An interpretative north—south structural cross-section across the composite terrane of the EGMB-north showing the geometry, disposition, and nature of shear zones suggesting a positive "flower structure." MSZ, Mahanadi shear zone; NBSZ, Northern Boundary shear zone.

The Tikarapara domain is characterized by intricately folded and intensely imbricated south-verging structures with overall northerly dips. This domain is bound by south-verging thrust following the MSZ. Considering the geometry and disposition of NBSZ and the MSZ and other complementary shear zones and their temporally consistent dextral kinematics, it can be inferred that the EGMB-north represents a typical crustal-scale "flower structure" (Chetty, 2010). The EGMB-north was interpreted to have been reactivated in a dextral transpressional tectonic regime possibly during Pan-African times. Subsequently, the first phase of rifting must have developed giving rise to progressive evolution of Gondwana basins.

The south-dipping MSZ and north-verging NBSZ merge together at depth in the form of a detachment zone, giving rise to extensive granitoids in the axial zone of the composite terrane of EGMB-north. The MSZ may have been originated initially as a south-verging complementary thrust emanating from the crustal-scale south-dipping NBSZ. Both these shear zones were possibly subjected to dextral transpressional tectonics during the Grenvillian and up to the Pan-African period. It is also emphasized that NBSZ and the MSZ represent the boundaries of the MR where the rifting processes were superimposed giving rise to the development of normal faulting, sedimentation, deposition, and magmatism.

3.4.4 TECTONIC SYNTHESIS

The composite terrane of high-grade metamorphic rocks of the EGMB-north, strikes east—west in contrast to the regional NE—SW trend of the EGMB. The boundary shear zones and the link shear zones are spatially and genetically connected affecting the composite terrane spatially and temporally. The transpressive nature of shear zones, the intrusion of granitoids in the axial zone, high-angle thrusting, the geometry of shear zones exhibiting "flower structure"—all combined together to make the terrane a classic orogen of oblique convergence during the Neoproterozoic.

Multiscale field-based structural observations along the Northern Boundary Shear Zone (NBSZ) as well as the MSZ and Koraput-Sonepur-Rairakhol shear zone (KSRSZ) reveal that the rocks in the EGMB-north witnessed dextral kinematic displacements. The pervasive first formed gneissic fabrics were continuously reworked and partitioned into a series of east—west crustal-scale shear zones. A north—south structural cross section illustrates different domains displaying distinctive internal structures with widely varying different geological evolution history and strain partitioning, separated by crustal-scale shear zones, cumulatively presenting a crustal-scale "flower structure" across the composite terrane (Fig. 3.41).

The process of kinematic changes took place in the EGMB-north gradually after the cessation of Neoproterozoic tectonics and the associated cooling and exhumation history of granulite facies rocks together with the initial stages of the dispersal of Gondwana fragments. The transpressional tectonic stresses ceased and a transtensional tectonic regime was initiated at the end of the Neoproterozoic ($\sim$ 400 Ma) giving the way for the rifting in the region. The rifting took advantage of the preexisting shear zones in the basement and resulted in the development of normal faults. This was simultaneously followed by glacial sedimentation and deposition of Lower Gondwana

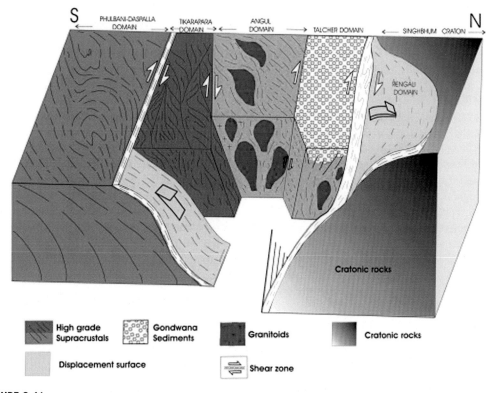

FIGURE 3.41

A 3D schematic block diagram showing the disposition and geometry of distinct domains separated by major shear zones and structural fabrics of different blocks for the EGMB-north showing positive "flower structure."

After Chetty, T.R.K. (2010). Structural architecture of the northern composite terrane, the Eastern Ghats Mobile Belt, India: Implications for Gondwana tectonics. Gondwana Research, 18, 565–582.

sediments during the Permian. Widening of the basins and synsedimentary growth faulting were the characteristic features of this event. This event was well constrained by the fission track ages obtained between 300 and 250 Ma, marking the partial annealing zone for apatite at the latest in Late Permian times (Lisker & Fachmann, 2001). Dispersal of the Gondwana fragments continued, and another thermal event around 140 Ma resulted in a renewed rifting process causing tilting and rotation of the inward-dipping normal faults in the coal seams of the Lower Gondwana basin and the development of new isolated upper Gondwana basins at both ends of the MSZ. These are evidently associated with major E−W trending fractures, south of the MSZ. These fractures make an acute angle with the MSZ, implying that the rifting has resulted not only in the reactivation of pre-existing shear zones essentially but also in locally generating a new set of fracture zones. This assumption is further supported by the presence of a small isolated shallow Gondwana basins closely associated with E−W trending fracture zones further south with in the EGMB near Srikakulam. The reactivation tectonics seemed to have occurred along the major shear zones taking advantage of their steep geometry associated with mylonitic fabrics. They also aided in accommodating progressive transformation from transpression to transtension. However, transpressional orogenesis seems to be a crucial common factor for later reactivation by rifting during the Gondwana dispersal (Goscombe & Gray, 2008). The kinematic changes from a long-lived transpressional tectonics to a transtensional regime are consistent with the change in global geodynamics and pattern of mantle convection, which is responsible for the reconfiguration of Rodinia and its transformation into Gondwana (Brown, 2007). Some of the small and narrow dykes recorded along the shear zones in EGMB-north indicate a localized thermal overprint at around 140−110 Ma that reflects a reactivation event along shear zones during the Cretaceous. This is probably coeval with the onset of sea floor production in the Bay of Bengal (117 Ma; Lisker & Fachmann, 2001).

The Neoproterozoic regional dextral transpressional tectonics along the shear zones and their repeated reactivation could be responsible for the initiation and successive evolution of Gondwana basins and different episodes of sedimentation in the EGMB-north (Chetty, 1995b). Available geochronological data shows that the structural architecture of the region is post-Grenvillian, which has been repeatedly reactivated through long-lived transpressional tectonics. The composite terrane is characterized by all the typical features of an oblique convergent orogen with transpressional kinematics in the middle to lower crust. The extension of the MR can be observed in East Antarctica in the form of the Lambert Rift on a supercontinent scale. The kinematic changes from transpression to transtensional stresses were found to be associated with global geodynamics related to the transformation from Rodinia to Gondwana configuration.

3.5 TECTONIC SYNTHESIS−EGMB

The EGMB must have witnessed several deformational and metamorphic events, but three events are considered as the most significant. The earliest deformational event that could be recognized is the development of the compositional fabric or the gneissic banding, which can be related to a 2.8 Ga deformational event during the Neoarchean (e.g., Ramakrishnan & Vaidyanathan, 2008; Sarkar et al., 2000). This event must have been associated with northwest−directed thrusting generating NE-SW−trending fabrics with gentle to moderate dips at the western margin juxtaposing the

Bastar craton and ENE—WSW fabrics at the northern margin juxtaposing the Singhbhum craton (Fig. 3.42). The event is also marked by northwest-verging thrusts, giving rise to large-scale recumbent fold structures and other flat lying fabrics (Fig. 3.43) associated with crustal thickening and granulite facies metamorphism during 3000—2600 Ma (Chetty & Murthy, 1994). This is consistent with the existence of several thrust zones that were demarcated all over the EGMB. The other most significant thermal event recognized in the evolution of the EGMB is the Grenvillian orogenic event (1100—800 Ma), which was marked by the development of shear zones and associated metamorphism and migmatization (Chetty & Murthy, 1994; Aftalion, Bowes, Dash, & Dempster, 1988). The preexisting 2.8 Ga gneissic fabrics were superimposed by 1Ga mylonitic fabrics during this event. As described earlier, the emplacement of megacrystic granitoid bodies along NBSZ as well as MSZ and the metamorphic events at several other places throughout provide indisputable evidence for the Grenvillian orogenic event in the EGMB. Dextral transpressional kinematics and intense strain partitioning during this period seem to be the dominant tectonic scenario continued until the end of the Neoproterozoic up to 500 Ma. This deformational event was related to the Mesoproterozoic oblique convergence between the Indian shield and the East Antarctica, broadly coinciding with the formation of the Rodinia supercontinent (Yoshida et al., 1996). The other major event was the thrusting between 540 and 500 Ma, which was posttectonic to folding and granulite metamorphism of the cover rocks. This is based on U—Pb Zircon SHRIMP ages (550—500 Ma) from a syntectonically emplaced nepheline syenite overlying the tectonic contact with basement granite—gneiss in KBT (Biswal et al., 2007). However, the ages and the nature of deformational history with respect to the events is still a matter of debate, for instance, the time of thrusting and whether it is synchronous or posttectonic with the granulite facies metamorphism. Future research can resolve these critical and controversial issues in complex terranes like the EGMB.

The Pan-African event, recorded around 500 Ma, could possibly represent a third deformation event, most probably restricted to shear zones. Both the Grenvillian and the Pan-African orogenic events were restricted to the granulite facies rocks of the EGMB and failed to significantly affect the neighboring marginal cratonic regions. But the Neoproterozoic history has been well recorded in different constituents of East Gondwana Land that include Prydz Bay and Denman glacier regions of East Antarctica and the Leeuwin complex in southwest Australia (e.g., Fitzsimons, 2003). The kinematic changes from a long-lived transpressional tectonics to a transtensional regime are consistent with the change in global geodynamics and pattern of mantle convection, responsible for the reconfiguration of Rodinia and its transformation into Gondwana (Brown, 2007).

Shear zones in the Proterozoic orogens are the most complex geological structures developed in mid to lower crustal levels. Heterogeneity in structural fabrics, wide variation in strain patterns, multiple pulses of magmatism, metamorphism, and tectonics are the characteristic features of deep crustal shear zones of the EGMB. Therefore, a determination of the age of shearing is a challenging task. The timing of shearing is estimated mostly by an indirect means by dating the spatially and genetically connected igneous plutons and associated magmatic enclaves within the shear zones. The only direct method of knowing the age of shearing is to date the minerals that are associated with the shear zone fabrics, which invariably yields the latest deformational event. Dating the last phase of recrystallization can be done by dating micas or the rims of zircons. All the geochronological data set from the major boundary shear zones such as SSZ, NBSZ, MSZ and the NVSZ in the EGMB ranges from 2800 to 450 Ma focused during three major deformational/shearing events. The first and the foremost is the shearing associated with the earliest thrusting around 2.8 Ga;

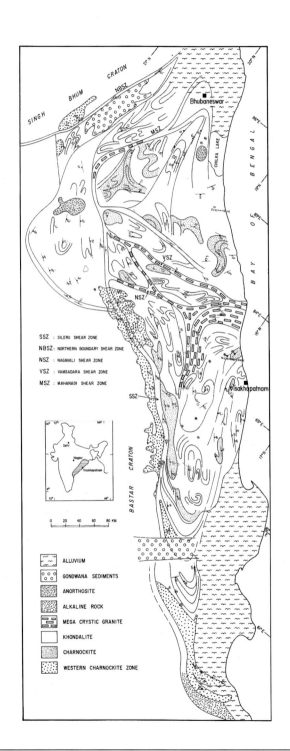

FIGURE 3.42

Map showing comprehensive tectonic map of the Eastern Ghats Mobile Belt showing major shear zone network and important lithologies.

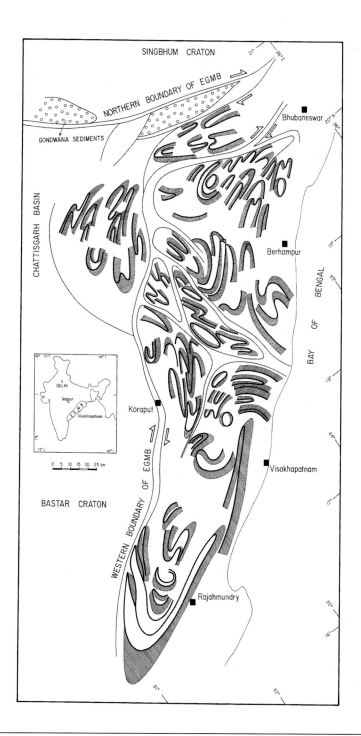

FIGURE 3.43

Regional network of shear zones and the varied complex fold styles in different domains exhibiting various trends of their axial planes.

the second is the prolonged Grenvillian event (1450−950 Ma) mostly along the preexisting thrust planes and the third is the Pan-African shearing (600−400 Ma), restricted to shear zones, culminating in the final collision and assembly of east Gondwana.

Two contrasting data sets exist regarding the P−T−t history of the granulites of the EGMB. One, based on isobaric cooling paths that show anticlockwise trajectories, suggesting magmatic underplating (e.g., Sengupta et al., 1990), and the other, decompression paths showing clockwise trajectories implying models of collision tectonics (e.g., Lal et al., 1987; Mohan, Tripathi, & Motoyoshi, 1997). However, the precise locations of those analyzed rock samples for the P−T−t histories with reference to shear zones are not clear. Further, assuming that those divergent opinions of P−T−t histories may be factual, it is possible that they may represent different thrust sheets or terranes. It may also be noted that there can be contrasting P−T trajectories preserved in the same orogenic belt. It has been well established that the rocks from the shear zones are affected by several processes such as partial melting, mixing of fluids that come from deeper levels and strain variations. Geochemical and mineralogical changes are inevitable (Beach, 1976), and distinct differences may emerge between the sheared and nonsheared rock domains. In the light of the above, a collection of rock samples without knowing their geological and structural environment may lead to erroneous interpretations.

Satellite image interpretation supported by multiscale field observations indicate that the fragments of granulite facies rocks seen on the craton beyond the SSZ and the NBSZ suggest that they are upthrusted, dismembered, and transported masses of granulites. These shear zones also represent sole thrusts and are considered as the sites of suture zones across which the EGMB was accreted and juxtaposed with India. The presence of thrust systems and associated structures, large recumbent folds, and the network of shear zone systems suggest that the tectonic history of the EGMB is similar to that described for continent−continent collisional belts (Chetty & Murthy, 1994). Geochemical data for the protolith of mafic granulites from the EGGB suggest an island-arc environment (Rao & Rao, 1992).

The presence of alkaline plutons is restricted to the SSZ all over its length and their emplacement history of these plutons has been widely debated. Some workers (e.g., Biswal et al., 2007; Chetty & Murthy, 1994, 1998a) believe that the alkaline plutons are syntectonically emplaced, while others describe postemplacement deformation and shearing resulting in development of solid-state metamorphic fabrics (e.g., Gupta, 2004; Upadhyay & Raith, 2006a). Vijaya Kumar et al. (2007) reported that the parental magma of these rocks must have been derived from partial melting of the enriched mantle sources. The earlier geochronological data on the alkaline rocks was sparse mostly with whole rock Rb−Sr system indicating a wide range of "intrusion" ages between 1446 and 856 Ma (Sarkar & Paul, 1998). The recently available and more reliable U−Th−Pb systematics in zircons (using SHRIMP and/or TIMS) on all alkaline complexes all along the SSZ point to their emplacement in a relatively narrow time frame between 1262 and 1480 Ma (Upadhyay, 2008). According to him, the SSZ may represent the location of a palaeo-rift and the alkaline complexes must have been later deformed during the collisional event associated with the thrusting of the EGMB. Further, the alkaline complexes of the SSZ resemble the deformed alkaline rocks and carbonatites (DARC) of southeastern Africa for which Burke, Ashwal, and Webb (2003) proposed a model of emplacement during rifting and basin formation and deformation during closure and basin inversion.

Geological and geophysical characteristics suggest that the SSZ could be a Precambrian suture zone extending to lithospheric−mantle depths. The available ages of different alkaline plutons indicate that there must have been multiple events of accretion, which is a common feature of many ancient accretionary systems. However, the present structural architecture of the EGMB represents final amalgamation history related to the assembly of the Gondwana supercontinent during the Neoproterozoic-Cambrian period (Chetty & Santosh, 2013).

The EGMB was first interpreted as a collage of juxtaposed terranes on the basis of network of shear zones, fold styles, and stretching lineations and broad lithologies (Chetty, 2001). The mineral-stretching lineations, mostly defined by the elongation of mineral grains such as mica, hornblende, feldspar, sillimanite, quartz, etc., display a systematic variation over distances across a few tens of kilometers (Fig. 3.44). However, distinct orientations of gneissic foliation exhibiting different fold styles, axial surfaces of early formed folds F1/F2, and a regional consistency in the orientation of stretching lineations that characterize the individual tectonic domains separated by the shear zones are well recognized. Subsequently, different crustal domains with distinct isotopic characteristics were identified (Rickers et al., 2001), which broadly coincide with the earlier classification of tectonic domains/terranes. The divergent P−T−t histories, the broad spectrum of geochronological data, and the observations on contrasting structural orientations from different domains support the applicability of the terrane concept to the EGMB. The terranes in the EGMB are characterized by internal homogeneity and continuity of structure and independent tectonic style and history. The boundaries between terranes in the form of shear zones provide linkages constrain the age of terrane amalgamation. The characterization and description of these terranes/lithostructural domains/crustal domains and their amalgamation history have been reviewed by several workers (Dobmeier & Raith, 2003; Gupta, 2012). It is possible that different terranes could represent different thrust nappes as allochthonous tectonic sheets representing a tectonostratigraphic terrane. The marginal shear zones of the EGMB appear analogous in geometry to higher level structures developed in a thin-skinned thrust tectonic regime (Chetty & Murthy, 1998a). Many different geological and geophysical approaches are required to prove or to disprove the applicability of the terrane concept to the EGMB and the answers remain more in the field and in experimental laboratories.

Two important tectonic models for the evolution of the EGMB were proposed based on geological, geophysical, and geochronological data sets. Although there has been ambiguity in the subduction polarity, the plate tectonic processes of subduction−accretion and collision have gained increasing importance in all the models of recent times. The model proposed by Vijaya Kumar and Leelanandam (2008) involves two episodes of convergence with an initial onset of continental rifting (Fig. 3.45). While the first rifting was initiated at ∼2.0 Ga, the culmination took place through continent−continent collision at 1.55 Ga. The second rifting was at 1.5−1.35 Ga, which facilitated the emplacement of alkaline rocks and carbonatites (ARCs). They suggested that the polarity of subduction changed from the earlier westward to eastward at around 1.5 Ga and that the emplacement of Kandra ophiolite (1.85 Ga) and KOM (1.35 Ga) may be related to the former westward subduction. It was described earlier in this chapter that the Pan-African tectonics was mostly restricted to major shear zones in the EGMB. There is also a strong opinion that the final amalgamation of the EGMB with the Indian shield took place along the SSZ during the Pan-African period through eastward subduction (Das et al., 2008). This is well supported by the present day structural architecture together with the geometry of the shear zones of the EGMB suggesting eastward subduction polarity. Based on the seismic receiver function analysis, Ramesh, Bianchi, and Das Sharma (2010) identified westerly dipping interface between 160 and 200 km reflecting a subduction-related relict association with westward subduction polarity.

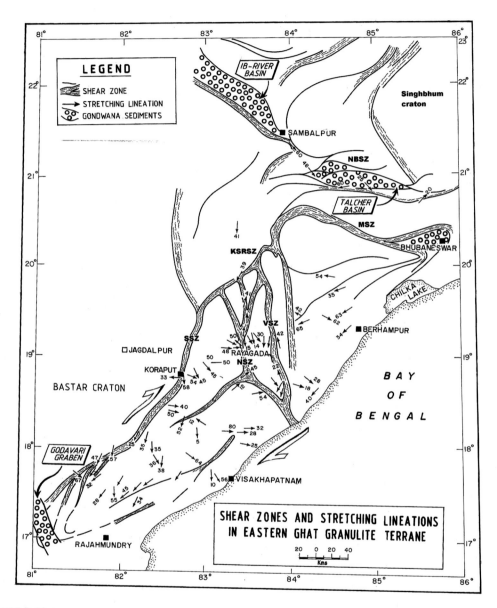

FIGURE 3.44

A network of deep crustal shear zones and the orientation of stretching lineations with in the EGMB. Note the consistent behavior of stretching lineations in each of the lithostructural domain.

After Chetty, T.R.K. (2001). The Eastern Ghats Mobile Belt, India: A collage of juxtaposed terranes (?). Gondwana Research, 4, 319–328.

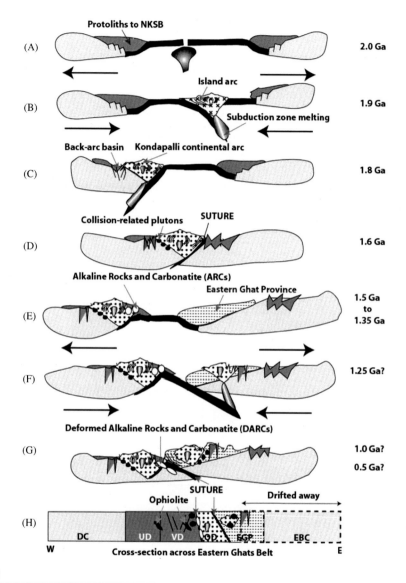

FIGURE 3.45

A schematic tectonic model illustrating the different stages of evolution of the EGMB showing changes in subduction polarity (after Vijaya Kumar & Leelanandam, 2008). The chronological sequence includes the following: (A) continental rifting and the development of Atlantic-type ocean margin, (B) oceanic crust decoupling and the eastward subduction leading to oceanic island arc, (C) accretion of island arc and switching over to westward subduction, (D) culmination of continent–continent collision generating granitic magmas (Vinukonda pluton) and the development of SSZ suture zone at the eastern margins of Dharwar and Bastar cratons, (E) further second cycle of rifting along the preexisting suture zone in association with alkaline magmatism, (F) eastward subduction followed by convergent tectonics and the major episode of mafic and felsic magmatism, (G) continued eastward subduction and deformation of alkaline rocks converting them into DARCS, and (H) juxtaposition of the Dharwar craton and the Antarctica with the EGMB in between constituting different tectonic domains. (For more details see Vijaya Kumar & Leelanandam, 2008.)

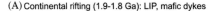
(A) Continental rifting (1.9-1.8 Ga): LIP, mafic dykes

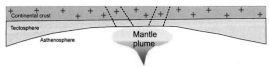
(B) Birth of oceanic crust (ophiolite formation)

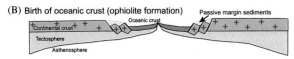
(C) Initiation of subduction, arc magmatism and accretion (Pacific-type)

(D) Ocean closure (arc accretion)

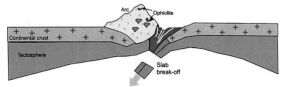
(E) Continental collision (Himalayan-type): c. 1.0 Ga

(F) Detailed anatomy of the collisional suture

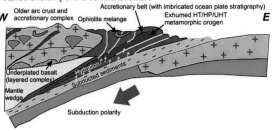

FIGURE 3.46

A tectonic model with consistent westward subduction throughout the Proterozoic history and the development of the EGMB: (A) initiation with continental rifting; (B) development of oceanic crust; (C) westward subduction, followed by arc magmatism and the formation of ophiolites; (D) ocean closure, arc accretion followed by slab break-off; (E) culmination of Himalayan style collision tectonics; and (F) anatomy of accretionary tectonics, ophiolite mélanges, exhumation of high-pressure and UHT assemblages. (For more details see Dharma Rao et al., 2011.)

The westward subduction throughout the geologic history after 1.9 Ga without any change in polarity was contemplated, and accretion tectonics were proposed in another model by Dharma Rao et al. (2011) (Fig. 3.46). The possible sequence of different chronological events in the model during the evolution of the EGMB is as follows: (1) development of a Paleoproterozoic NE–SW trending rift along the eastern margin of proto-India during the breakup of Columbia; (2) generation of an intraoceanic island arc due to subduction of oceanic lithosphere at 1.9 Ga; (3) opening of the rift and the formation of an ocean basin at around 1.9–1.8 Ga, between eastern India and East Antarctica where the sedimentary sequences of the EGMB were deposited; (4) accretion of intraoceanic island crust to the rifted margin of southeast India leading to the development of a continental magmatic arc, which can be correlated with accretion of the Napier complex to proto-India, (5) ocean spreading, initiation of subduction, accretion of ophiolites of Kanigiri and Kandra starting from 1.8 up to 1.3 Ga; (6) culmination of convergent forces into a continent–continent collision with associated mafic and felsic magmatism like Vinukonda pluton; and (7) a major episode of postcollision alalkaline magmatism preserved in nepheline syenites and other related rocks preserved along the SSZ. The EGMB must have possibly witnessed a Pacific-type orogeny during the Paleoproterozoic and Mesoproterozoic culminating in Himalayan-style collision during the early Neoproterozoic, probably associated with the assembly of the Rodinia supercontinent (Dharma Rao et al., 2011). This hypothesis is further supported by the contiguity of the EGMB and the SGT indicating similar orogenic processes at least during the Neoproterozoic period (Chetty & Santosh, 2013). However, the complexity and the ambiguity regarding the subduction polarity can be resolved only by further detailed geological and geophysical studies.

3.6 CORRELATION WITH THE SGT

The orogens of the EGMB and the SGT expose a deep Precambrian continental crust consisting of complexly deformed Archean to Neoproterozoic high-grade metamorphic and magmatic assemblages. They constitute discrete crustal blocks or terranes separated by major shear zones often with boundaries hitherto unsuspected both within and around the margins of these orogenic belts, with plate tectonic history analogous to that of modern orogenic belts (Chetty & Santosh, 2013). The lithological assemblages in both the orogens are high-grade metamorphic rocks comprising charnockites, khondalites, leptynites, and migmatites as the dominant units, invaded by a variety of younger intrusive of alkaline, anorthositic, and granitic rocks. However, the continuity of these two orogens on the continental surface is broken over a distance of ~400 km and disappears in the Bay of Bengal near Ongole at the east coast and again reappears at Chennai. Therefore, some opine that these two EGMB–SGT orogens were not contiguous and that they belong to two independent orogenies. Broad tectonic trends and structural framework of EGMB–SGT displaying major structural elements such as suture zones, shear zones, fold patterns, thrust zones, and different crustal blocks and other major tectonic trends are well depicted (Fig. 3.47). Two distinct geologic domains could be recognized from the map: the Achaean Dharwar Craton in the center, and the Proterozoic orogens skirting the three cratons of Dharwar, Bastar, and Singhbhum. The interface between the cratons and the orogenic belts is a curvilinear high-strain zone, which has been well established as a Precambrian mega-suture zone (PMSZ) with a total strike length of ~1500 km and width ranging

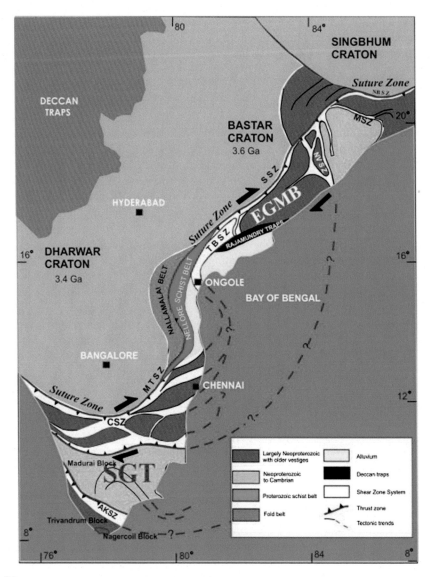

FIGURE 3.47

Tectonic framework of the Proterozoic orogenic belts of southern India showing major crustal blocks and tectonic trends and features like suture zone, thrust zones, shear zone system, major geological units and extended trend lines. *EGMB*, Eastern Ghats Mobile Belt; *SGT*, Southern Granulite Terrane; *CSZ*, Cauvery suture zone; *AKSZ*, Achankovil shear zone; *TBSZ*, Terrane Boundary shear zone; *NVSZ*, Nagavali-Vamsadhara shear zone; *NBSZ*, Northern Boundary shear zone; *MTSZ*, Mettur Shear Zone; *SSZ*, Sileru Shear Zone.

After Chetty, T.R.K. & Santosh, M. (2013). Proterozoic orogens in southern Peninsular India: Contiguities and complexities. Journal of Asian Earth Sciences, 78, 39–53.

from 3 to about 60 km. The PMSZ is composed of not only high-grade supracrustals but also other rock units such as the intensely deformed Nallamalai fold-thrust belt and the adjacent NSB. Both the units occur as highly stretched domains displaying dextral kinematics (Chetty, 2001).

The E–W trending CSZ extends from the west coast to the east coast and disappears in the Bay of Bengal and continues further along northeast and swings northward to join the NVSZ of the EGMB (Chetty, 1995b). Another important branch of the CSZ deviates from Bhavani toward the northeast and extends through the Mettur shear zone (MTSZ), hosting c. 780 Ma old alkaline magmatism. The MTSZ extends further north and coincides with the western boundary of the Proterozoic Nallamalai fold belt. There are also some other branches of CSZ that lie subparallel to MTSZ further east, which merge with the margins of the Nallamalai fold belt and the NSB in the form of bounding shear zones. Further northward, these zones are marked by the presence of the SSZ at the western margin and the NBSZ at the northern margin of the EGMB.

Another major shear zone, known as the Salem-Attur Shear Zone, the northern marginal zone of the CSZ, extends eastward into the Bay of Bengal and deflects to the north and northwest to join the Terrane boundary shear zone (TBSZ) as defined by Nagaraju et al. (2008) that lies between the eastern margin of the Cuddapah basin and the EGMB encompassing the Nelolore schist belt as a tectonic sliver or an accretionary belt. The dextral transpressional tectonics (Neoproterozoic) seem to have played a dominant role in the development of PMSZ as well as the orogens of EGMB–SSGT, which are responsible for the present crustal architecture of the southern Peninsular India (Chetty & Santosh, 2013). Transpressional regimes were widespread in both the orogens that gave rise to spatially varying complexity in structural fabrics and strain patterns resulting in structural heterogeneity. In summary, the EGMB–SGT represents a classic transpressional orogen that provides a strong piercing marker for the reconstruction models of Gondwana. Further, the orientation and complex geology of the composite terrane of the EGMB-north were ascribed to (1) the obliquity of convergence and (2) the juxtaposition of the three cratons namely Bastar, Singhbhum, and the Antarctica.

The most significant development in understanding the evolution of the EGMB–SGT in recent years is the findings of several Precambrian ophiolites all over the orogen with ages ranging from 1.85 to 0.78 Ma (Fig. 3.48). From north to south, the recently reported occurrences of ophiolitic complexes include: Sukinda, Kondapalli, Kanigiri, Kandra, Devanur, and Manamedu. All the known ophiolitic rocks are characterized by thrust-nappe emplacement history and suprasubduction zone tectonic setting. The subduction, accompanied by the closure of a large ocean, would develop into a sequence of accretionary complexes presenting high-temperature and high-pressure regional metamorphic belts. It has been postulated that the Gondwana-forming orogens had consumed the intervening Mozambique Ocean during the Neoproterozoic.

In general, the EGMB–SGT reveal the broad geometry and kinematics of shear zones and the mechanisms of accommodating three-dimensional transpressive deformation at deep to mid-crustal levels with large orogen-parallel movements and shortening during continental convergence. Broadly, the following salient features in EGMB–SGT can be identified: (1) the occurrence of high-grade granulite facies rock assemblages with continuing structural fabrics and shear zones and similar structural styles of fold-thrust tectonics; (2) constitution of different crustal blocks/terranes of distinct geologic history separated by a network of mid-crustal shear zones and associated episodic reactivation tectonics; (3) the presence of high pressure granulites and UHT metamorphic mineral assemblages of different ages; (4) the existence of crustal scale "flower structures"; (5) the

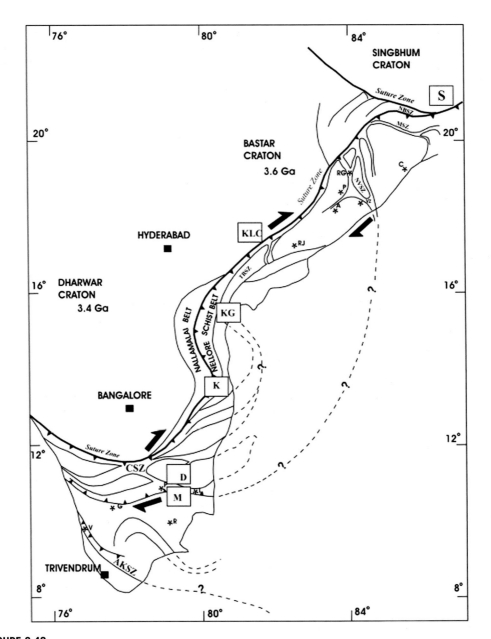

FIGURE 3.48

Map showing the tectonic framework of Proterozoic Orogenic Belts of Southern India (POSI) displaying the recently reported locations of UHT assemblages and Precambrian ophiolitic complexes. *CSZ*, Cauvery suture zone; *AKSZ*, Achankovil shear zone; *TBSZ*, Terrane Boundary shear zone; *NVSZ*, Nagavali-Vamsadhara shear zone; *NBSZ*, Northern Boundary shear zone. Ophiolite Complexes: *M*, Manamedu; *D*, Devanur; *K*, Kandra; *KG*, Kanigiri; *KLC*, Kondapalli layered complex; *S*, Sukinda. UHT locations: *V*, Vadavathur; *R*, Rajapalayam; *G*, Ganguvaripatti; *PD*, Pangidi; *L*, Laxmanapatti; *RJ*, Rajahmundry; *A*, Araku; *P*, Paderu; *VZ*, Vizianagaram; *RG*, Rayagada.

After Chetty, T.R.K. & Santosh, M. (2013). Proterozoic orogens in southern Peninsular India: Contiguities and complexities. Journal of Asian Earth Sciences, 78, *39–53.*

emplacement of Neoproterozoic anorthosites, alkaline rocks, and granitoids; (6) the common occurrence of Precambrian ophiolitic complexes with ages ranging from 3.12, 2.5, to 0.8 Ga; and (7) a wide spectrum of ages from magmatic and metamorphic rocks ranging from Neoarchean (2.5 Ga) to Neoproterozoic (0.5 Ga). The orogenic evolution of the EGMB−SGT was involved in subduction−accretion−collision tectonics during at least three major phases; Neoarchean, Meso- to Neoproterozoic, and end Neoproterozoic-Cambrian. The amalgamation of these orogens with the proto-Indian continent seems to be complex and has occurred at different places during different times. The arcuate suture/shear zones associated with the EGMB−SGT around the Dharwar-Bastar-Singhbhum cratons are interpreted as interference shear systems accommodating the crustal flow and that the cratons as hot orogeny developed with weak a lithosphere that enhanced the lateral flow in response to accretionary history of the EGMB−SGT (Chardon, Jayananda, Chetty, & Peaucat, 2008).

The present structural architecture seems to be the end result of a final amalgamation history of the EGMB−SGT with the proto-India during the Neoproterozoic-Cambrian period, related to the assembly of the Gondwana supercontinent. The EGMB−SGT witnessed Phanerozoic-style plate tectonic models, and can be considered as typical ancient analogs of modern orogens. The geometry of the shear zones show outward dipping from the Dharwar Craton, suggesting the polarity of subduction to be away from the Dharwar Craton at least during the Neoproterozoic. The combined accretion and collision history suggest a possible convergence of different cratons and a controlled strain partitioning at the scale of the supercontinents of Rodinia and Gondwana.

LIST OF ABBREVIATIONS

AKSZ	Achankovil shear zone
ARCs	alkaline rocks and carbonatites
BAC	Bolangir anorthosite complex
BBPT	Berhampur-Bhubaneswar-Phulbani Terrane
BTSZ	Baligurha Tel shear zone
CAC	Chilka lake anorthosite complex
CMP	Chimakurti mafic-ultramafic pluton
CMZ	central migmatitic zone
CSZ	Cauvery shear/suture zone
DARCs	deformed alkaline rocks and carbonatites
EGMB	Eastern Ghats Mobile Belt
EKZ	Eastern khondalitic zone
GR	Godavari Rift
JAC	Jugsaipatna anorthosite complex
KAC	Koraput alkaline complex
KAP	Kunavaram alkaline pluton
KBT	Khariar-Bolangir Terrane
KLC	Kondapalli layered complex
KOM	Kanigiri ophiolite mélange
KSRSZ	Koraput-Sonepur-Rairakhol shear zone
KSSZ	Koraput-Sonepur shear zone

LSZ	Lakhna shear zone
MR	Mahanadi Rift
MSZ	Mahanadi shear zone
MTSZ	Mettur shear zone
NBSZ	Northern boundary shear zone
NSB	Nellore Schist Belt
NSZ	Nagavali shear zone
NVSZ	Nagavali-Vamsadhara shear zones
PdGP	Podili granite pluton
PGP	Pasupugallu Gabbro Pluton
PMSZ	Precambrian mega-suture zone
RVKT	Rajahmundry-Visakhapatnam-Koraput Terrane
SGT	Southern Granulite Terrane
SSZ	Sileru shear zone
TAC	Turkel Anorthosite Complex
TBSZ	Terrane Boundary shear zone
VGP	Vinukonda granite pluton
VSZ	Vamsadhara shear zone
WCZ	Western charnockitic zone
WKZ	Western khondalitic zone

REFERENCES

Aftalion, M., Bowes, D. R., Dash, B., & Dempster, T. J. (1988). Late Proterozoic charnockites of Orissa, India-a U-Pb and Rb-Sr isotopic study. *Journal Geology, 96*, 663–676.

Aftalion, M., Bowes, D. R., Dash, B., & Fallick, A. E. (2000). *Late Pan-African thermal history in the Eastern Ghats Terrane, India, from U−Pb and K−Ar isotopic study of the Mid-Proterozoic Khariar alkali syenite, Orissa* (57, pp. 26–33). Geological Survey of India, .

Banerjee, P. K., Mahakud, S. P., Bhattacharya, A. K., & Mohanty, A. K. (1987). *On the northern margin of the Eastern Ghats in Orissa. Records* (118, pp. 1–8). Geological Survey of India, .

Beach, A. (1976). The interrelations of fluid transport, deformation, geochemistry and heat flowing early Proterozoic shear zones in Lewisian Complex. *Philosophical Transactions of Royal Society, London A, 280*, 569–604.

Bhadra, S., & Gupta, S. (2016). Reworking of a basement-cover interface during terrane boundary shearing: An example from the Khariar basin, Bastar craton, India. In S. Mukherjee, & K. F. Mulchrone (Eds.), *Ductile Shear Zones: From Micro- to Macro-scales* (pp. 164–181). Chichester: John Wiley and Sons Ltd.

Bhadra, S., Gupta, S., & Banerjee, M. (2004). Structural evolution across the Eastern Ghats Mobile Belt — Bastar craton boundary: Hot over cold thrusting in an ancient collision zone. *Journal of Structural Geology, 26*, 233–245.

Bhattacharya, A., & Gupta, S. (2001). A reapprisal of polymetamorphism in the Eastern Ghat Belt: A view from the north of the Godavari rift. *Proceedings of Indian Academy (Earth and Planetary Sciences), 110*, 369–383.

Bhattacharya, A., Raith, M., Hoernes, S., & Banerjee, D. (1998). Geochemical evolution of the massif-type anorthosite complex at Bolangir in the Eastern Ghats Belt of India. *Journal of Petrology, 39*, 1169–1195.

Bhattacharya, S., & Kar, R. (2002). High-temperature dehydration melting and decompressive P-T path in a granulite complex from the Eastern Ghats, India. *Contributions to Mineralogy and Petrology, 143*, 175–191.

Bhui, U., Sengupta, P., & Sengupta, P. (2007). Phase relations in mafic dykes and their host rocks from Kondapalle, Andhra Pradesh, India: Implications for the time−depth trajectory of the Palaeoproterozoic (late Archean?) granulites from southern eastern Ghats Belt. *Precambrian Research, 156*, 153−174.

Biswal, T. K., & Sinha, S. (2003). Deformation history of the NW salient of the Eastern Ghats Mobile Belt, India. *Journal of Asian Earth Sciences, 22*, 157−169.

Biswal, T. K., Biswal, B., Mitra, S., & Moulik, M. R. (2002). Deformation pattern of the NW terrane boundary of the Eastern Ghats Mobile Belt, India: A Tectonic Model and Correlation with Antarctica. *Gondwana Research, 51*, 45−52.

Biswal, T. K., Sajeevan, G., & Nayak, B. P. (1998). Deformational history of Eastern Ghats Mobile Belt around Lathore, Balangir district, Orissa. *Journal Geological Society of India, 51*, 219−225.

Biswal, T. K., Ahuja, H., & Sahu, H. S. (2004). Emplacement kinematics of nepheline syenites from the Terrane Boundary Shear Zone of the Eastern Ghats Mobile Belt, west of Khariar, NW Orissa: Evidence from meso- and microstructures. *Proceedings Indian Academy of Sciences (Earth and Planetary Sciences), 113*, 785−793.

Biswal, T. K., Jena, S. K., Datta, S., Das, R., & Khan, K. (2000). Deformation of the terrane boundary shear zone (Lakhna Shear Zone) between the Eastern Ghats Mobile belt and the Bastar craton, in Balangir and Kalahandi districts of Orissa. *Journal Geological Society of India, 55*, 367−380.

Biswal, T. K., Waele, B. D., & Ahuja, H. (2007). Timing and geodynamics of the juxtaposition of the Eastern Ghats Mobile Belt against the Bhandara craton, India: A structural and Zircon U-Pb SHRIMP study of the fold thrust belt and associated nepheline syenite plutons. *Tectonics, 26*, Tc 4006.

Bose, M. K. (1970). Petrology of the intrusive alkaline suite of Koraput, Orissa. *Journal Geological Society of India, 11*, 99−126.

Bose, S., Dunkley, D. J., Dasgupta, S., Das, K., & Arima, M. (2011). India−Antarctica Australia−Laurentia connection in the Paleoproterozoic−Mesoproterozoic revisited: Evidence from new zircon U−Pb and mon-azite chemical age data from the eastern Ghats Belt, India. *Geological Society of America Bulletin, 123*, 2031−2049.

Bose, S., Das, K., Torimotoc, J., Arima, M., & Dunkley, D. J. (2016). Evolution of the Chilka Lake granulite complex, northern Eastern Ghats Belt, India: First evidence of ∼780 Ma decompression of the deep crust and its implication on the India−Antarctica correlation. *Lithos*. Available from http://dx.doi.org/10.1016/j.lithos.2016.01.017.

Brown, M. (2007). Metamorphic conditions in orogenic belts: A record of secular change. *International Geology Review, 49*, 193−234.

Burke, K., Ashwal, L. D., & Webb, S. J. (2003). New way to map old sutures using deformed alkaline rocks and carbonatites. *Geology, 31*, 391−394.

Chardon, D., Jayananda, M., Chetty, T. R. K., & Peaucat, J. J. (2008). Precambrian continental strain and shear zone patterns: The south Indian craton. *Journal Geophysical Research, 113*, B08402. Available from http://dx.doi.org/10.1029/2007JB005299.

Chatterjee, N., Crowley, J. L., Mukherjee, A., & Das, S. (2008). Geochronology of the 983-Ma Chilka Lake Anorthosite, Eastern Ghats Belt India: Implications for Pre-Gondwana tectonics. *Journal of Geology, 116*, 105−118.

Chetty, T. R. K., & Murthy, D. S. N. (1992). Structural controls of graphite-tungsten mineralization in Eastern Ghat Mobile Belt. In S. C. Sarkar (Ed.), *Metallogeny related to the tectonics of the Proterozoic mobile belts* (pp. 153−162). New Delhi: Oxford IBH publishing Co. Pty. Ltd.

Chetty, T. R. K., & Murthy, D. S. N. (1993). LANDSAT Thematic mapper data applied to structural studies of the Eastern Ghats granulite terrane in part of Andhra Pradesh. *Journal Geological Society of India, 42*, 373−391.

Chetty, T. R. K., & Murthy, D. S. N. (1994). Collision tectonics in the Eastern Ghats Mobile Belt: Mesoscopic to Satellite scale structural observations. *TERRA NOVA, 6*, 72−81.

Chetty, T. R. K., & Murthy, D. S. N. (1998a). *Regional tectonic framework of the Eastern Ghats Mobile Belt: A new interpretation*, . *Proceedings of Workshop on Eastern Ghats Mobile Belt* (44, pp. 39−50). Geological Survey of India, .

Chetty, T. R. K., & Murthy, D. S. N. (1998b). Elchuru-Kunavaram-Koraput (EKK) shear zone, Eastern Ghats Granulite Terrane, India: A possible Precambrian suture zone. In A. T. Rao, S. R. Divi, & M. Yoshida (Eds.), *Precambrian crustal processes in East coast granulite-greenstone regions of India and Antarctica with East Gondwana* (pp. 37−48). Gondwana Research Group, Memoir No. 4.

Chetty, T. R. K. (1995a). A correlation of Proterozoic shear zones between Eastern Ghats, India and Enderby Land, East Antarctica. In M. Santosh, & M. Yoshida (Eds.), *India and Antarctica during the Precambrian* (pp. 205−220). Memoir, Geological Society of India 34.

Chetty, T. R. K. (1995b). *Strike-slip tectonics and the evolution of Gondwana basins of eastern India*. *Proceedings of 9th International Gondwana Symposium*. Geological Survey of India.

Chetty, T. R. K. (2001). The Eastern Ghats Mobile Belt, India: A collage of juxtaposed terranes (?). *Gondwana Research*, *4*, 319−328.

Chetty, T. R. K. (2010). Structural architecture of the northern composite terrane, the Eastern Ghats Mobile Belt, India: Implications for Gondwana tectonics. *Gondwana Research*, *18*, 565−582.

Chetty, T. R. K., Vijaya, P., Narayana, N. L., & Giridhar, G. V. (2003). Structure of the Nagavali shear zone, Eastern Ghats Mobile Belt, India: Correlation in the East Gondwana reconstruction. *Gondwana Research*, *6*, 215−229.

Chetty, T. R. K., & Santosh, M. (2013). Proterozoic orogens in southern Peninsular India: Contiguities and complexities. *Journal of Asian Earth Sciences*, *78*, 39−53.

Clark, G. S., & Subba Rao, K. V. (1971). Rb-Sr isotopic age of the Kunavaram Series, a group of alkaline rocks from India. *Canadian Journal of Earth Sciences*, *8*, 1597−1602.

Crowe, W. A., Cosca, M. A., & Harris, L. B. (2001). 40Ar/39Ar geochronology and Neopro-terozoic tectonics along the northern margin of the Eastern Ghats Belt in north Orissa, India. *Precambrian Research*, *108*, 237−266.

Crowe, W. A., Nash, C. R., Harris, L. B., Leeming, P. M., & Rankin, L. R. (2003). The geology of the Rengali province: Implications for the tectonic development of northern Orissa, India. *Journal of Asian Earth sciences*, *21*, 697−710.

Cruden, A. R. (1998). On the emplacement of tabular granites. *Journal Geological Society, London*, *155*, 853−862.

Das, K., Bose, S., Karmakar, S., Dunkley, D. J., & Dasgupta, S. (2011). Multiple tectonometamorphic imprints in the lower crust: First evidence of ca. 950 Ma (zirconU−Pb SHRIMP) compressional reworking of UHT aluminous granulites from the Eastern Ghats Belt, India. *Geological Journal*, *46*, 217−239.

Das, S., Nasipuri, P., Bhattacharya, A., & Swaminathan, S. (2008). The thrust-contact between the Eastern Ghats Belt and the adjoining Bastar craton (Eastern India):evidence from mafic granulites and tectonic implications. *Precambrian Research*, *162*, 70−85.

Dasgupta, S., Bose, S., & Das, K. (2013). Tectonic evolution of the Eastern Ghats Belt, India. *Precambrian Research*, *227*, 247−258.

Dasgupta, S., Ehl, J., Raith, M. M., Sengupta, P., & Sengupta, P. (1997). Mid-crustal contact metamorphism around the Chimakurthy mafic−ultramafic complex, Eastern Ghats Belt, India. *Contributions to Mineralogy and Petrology*, *129*, 182−197.

Dasgupta, S., Sanyal, S., Sengupta, P., & Fukuoka, M. (1994). Petrology of granulites from Anakapalle—evidence for Proterozoic decompression in the Eastern Ghats, India. *Journal of Petrology*, *35*, 433−459.

Dasgupta, S., & Sengupta, P. (2003). Indo-Antarctic correlation: A perspective from the Eastern Ghats granulite belt, IndiaIn M. Yoshida, B. F. Windley, & S. Dasgupta (Eds.), *Proterozoic East Gondwana: Supercontinent Assembly and Breakup* (206, pp. 131−143). Geological Society of London, Special Publication.

Dharma Rao, C., & Reddy, U. V. B. (2009). Petrological and geochemical characterization of proterozoic ophiolitic mélange, Nellore–Khammam Schist Belt, SE India. *Journal of Asian Earth Sciences, 36,* 261–276.

Dharma Rao, C., Santosh, M., & Yuan, B. W. (2011). Mesoproterozoic ophiolitic mélange from these periphery of the Indian Plate: U–Pb zircon ages and tectonic implications. *Gondwana Research, 19,* 384–401.

Dharma Rao, C. V., Santosh, M., & Dong, Y. (2012). U–Pb zircon chronology of the Pangidi–Kondapalle layered intrusion, Eastern Ghats belt, India: Constraints on Mesoproterozoic arc magmatism in a convergent margin setting. *Journal of Asian Earth Sciences, 49,* 362–375.

Dharma Rao, C. V., Santosh, M., & Zhang, S.-H. (2014). Neoproterozoic massif-type anorthosites and related magmatic suites from the Eastern Ghats Belt, India: Implications for slab window magmatism at the terminal stage of collisional orogeny. *Precambrian Research, 240,* 60–78.

Dharma Rao, C. V., Santosh, M., Zhang, Z., & Tsunogae, T. (2013). Mesoproterozoic arc magmatism in SE India: Petrology, zircon U–Pb geochronology and Hf isotopes of the Bopudi felsic suite from Eastern Ghats Belt. *Journal of Asian Earth Sciences, 75,* 183–201.

Dobmeier, C. (2006). Emplacement of Proterozoic massif-type anorthosite during regional shortening: Evidence from the Bolangir anorthosite complex (EGP, India). *International Journal of Earth Sciences, 95,* 543–555.

Dobmeier, C., & Raith, M. M. (2003). Crustal architecture and evolution of the Eastern Ghats Belt and adjacent regions of IndiaIn M. Yoshida, B. F. Windley, & S. Dasgupta (Eds.), *Proterozoic east Gondwana: Supercontinent assembly and breakup* (206, pp. 169–202). Geological Society of London, Special Publication.

Dobmeier, C., & Simmat, R. (2002). Post-Grenvillean transpression in the Chilka Lake area, Eastern Ghats Belt–implications for the geological evolution of peninsular India. *Precambrian Research, 113,* 243–268.

Dobmeier, C. J., Lütke, S., Hammerschmidt, K., & Mezger, K. (2006). Emplacement and deformation of the Vinukonda meta-granite (Eastern Ghats, India) –Implications for the geological evolution of Peninsular India and for Rodinia reconstructions. *Precambrian Research, 146,* 165–178.

Fitzsimons, I. C. W. (2003). Proterozoic basement provinces of southern and southwestern Australia and their correlation with Antarctica. In: Proterozoic East Gondwana: Supercontinental assembly and breakupIn M. Yoshida, B. F. Windley, & S. Dasgupta (Eds.), *Proterozoic East Gondwana: Supercontinent Assembly and Breakup* (206, pp. 93–130). Geological Society of London, Special Publication.

Goscombe, B. D., & Gray, D. R. (2008). Structure and strain variation at midcrustal levels in a transpressional orogen: A review of Kaoko Belt structure and the character of west Gondwana amalgamation and dispersal. *Gondwana Research, 13,* 45–85.

Grew, E. S., & Manton, W. I. (1986). A new correlation of saphirine granulites in the Indo-Antarctic metamorphism terrane: Late Proterozoic dates from the Eastern Ghats Province of India. *Precambrian Research, 33,* 123–137.

Gupta, J. N., Pandey, B. K., Chabria, T., Banerjee, D. C., & Jayaram, K. M. (1984). Rb-SrGeochronologic studies on the granites of Vinukonda and Kanigiri Prakasam district, Andhra Pradesh, India. *Precambrian Research, 26,* 105–109.

Gupta, S. (2004). *The Eastern Ghats Belt, India- a new look at an old orogen* (84, pp. 75–100). Geological Survey of India, .

Gupta, S. (2012). Strain localization, granulite formation and geodynamic setting of 'hot orogens': Case study from the Eastern Ghats Province, India. *Geological Journal, 47,* 334–351.

Gupta, S., Bhattacharya, A., Raith, M., & Nanda, J. K. (2000). Contrasting pressure-temperature deformational history across a vestigial craton-mobile belt assembly: The western margin of the Eastern Ghats belt at Deobhog, India. *Journal of Metamorphic Geology, 18,* 683–697.

Gupta, S., & Bose, S. (2004). Deformation history of the Kunavaram complex, Eastern Ghats belt, India: Implication for alkaline magmatism along the Indo-Antarctica suture. *Gondwana Research, 7,* 1228–1335.

Gupta, S., Nanda, J., Mukherjee, S. K., & Santra, M. (2005). Alkaline magmatism versus collision tectonics in the Eastern Ghats Belt, India: Constraints from structural studies in the Koraput complex. *Gondwana Research*, *8*, 403−419.

Halden, N. M., Bowes, D. R., & Dash, B. (1982). Structural evolution of migmatites in granulite facies terrain-Precambrian crystalline complex of Angul, Orissa, India. *Transactions of the Royal Society of Edinburgh: Earth Sciences*, *73*, 109−118.

Harley, S. L. (2003). Archean−Cambrian crustal development of East Antarctica: Metamorphic characteristics and tectonic implicationsIn M. Yoshida, B. F. Windley, & S. Dasgupta (Eds.), *Proterozoic East Gondwana: Supercontinent assembly and breakup* (206, pp. 203−230). Geological Society London, Special Publication.

Henderson, B., Collins, A. S., Payne, J., Forbes, C., & Saha, D. (2014). Geologically constraining India in Columbia: The age, isotopic provenance and geochemistry of the protoliths of the Ongole Domain, Southern Eastern Ghats, India. *Gondwana Research*, *26*, 888−906.

Kaila, K. L., & Bhatia, S. C. (1982). Gravity study along Kavali−Udipi deep seismic sounding profile in the Indian peninsular shield: Some inferences about origin of anorthosites and Eastern Ghat orogeny. *Tectonophysics*, *79*, 129−143.

Kaila, K. L., Tewari, H. C., Roy Chowdhury, K., Rao, V. K., Sridhar, A. R., & Mall, D. M. (1987). Crustal structure of the northern part of the Proterozoic Cuddapah basin of India from deep seismic soundings and gravity data. *Tectonophysics*, *140*, 1−12.

Kar, R. (2007). Dominal fabric development associated microstructures and P-T records attesting to polymetamorphism in a granulite complex of the Eastern Granulites Belt, India. *Journal of Earth System Science*, *116*, 21−35.

Korhonen, F. J., Clark, C., Brown, M., Bhattacharya, S., & Taylor, R. (2013). How long lived is ultrahigh temperature (UHT) metamorphism? Constraints from zircon and monazite geochronology in the Eastern Ghats orogenic belt, India. *Precambrian Research*, *234*, 322−350.

Korhonen, F. J., Saw, A. K., Clark, C., Brown, M., & Bhattacharya, S. (2011). New constraints on UHT metamorphism in the Eastern Ghats Province through the application of phase equilibria modelling and in situ geochronology. *Gondwana Research*, *20*, 764−781.

Kovach, V. P., Berezhnaya, N. G., Salnikova, E. B., Narayana, B. L., Divakar Rao, V., Yoshida, M., & Kotov, A. B. (1998). U-Pb zircon age and Nd isotope systematics of megacrystic charnockites in the Eastern Ghats Granulite Belt, India, and their implication for East Gondwana reconstruction. *Journal African Earth Sciences*, *27*, 125−127.

Kovach, V. P., Simmat, R., Rickers, K., Berezhnaya, N. G., Salnikova, E. B., Dobmeier, C., ... Kotov, A. B. (2001). The western charnockite zone of the eastern Ghats Belt, India — an independent crustal province of late Achaean (2.8 Ga) and Palaeoproterozoic (1.7−1.6 Ga) terrains. *Gondwana Research*, *4*, 666−667.

Krause, O., Dobmeier, C., Raith, M. M., & Mezger, K. (2001). Age of emplacement of massif-type anorthosites in the Eastern Ghats Belt, India: Constraints from U−Pb Zircon dating and structural studies. *Precambrian Research*, *109*, 25−38.

Krause, O., Dobmeier, C., Raith, M. M., & Mezger, K. (2001). Age of emplacement of massif-type anorthosites in the Eastern Ghats Belt, India: Constraints from U−Pb zircon dating and structural studies. *Precambrian Research*, *109*, 25−38.

Krishnan, M. S. (1961). *The structure and tectonic history of India*. Memoir, Geological Survey of India, 87, 137p.

Lal, R. K., Ackermand, D., & Upadhyay, H. (1987). P−T−X relationships deduced from corona textures in sapphirine−spinel−quartz assemblages from Paderu, southern India. *Journal of Petrology*, *28*, 1139−1168.

Leelanandam, C. (1972). An anorthositic layered complex near Kondapalli, Andhra Pradesh. *Quarterly Journal Geology Mining Metallurgical Society of India*, *44*, 105−107.

Leelanandam, C. (1989). The prakasam alkaline province in Andhra Pradesh, India. *Journal of Geological Society of India, 34,* 25−45.

Leelanandam, C. (1997). The Kondapalli layered complex, Andhra Pradesh, India: A synoptic overview. *Gondwana Research, 1,* 95−114.

Leelanandam, C. (2015). Chromite from the Kondapalli Layered Complex. *Journal of economic geology & Geo resource management Prof. Calamur Mahadevan Commemorative Special, Volume 10,* 1−14.

Leelanandam, C., Burke, K., Ashwal, L. D., & Webb, S. J. (2006). Proterozoic mountain building in Peninsular India: An analysis based primary on alkaline rock Distribution. *Geological magazine, 143,* 195−212.

Leelanandam, C., & Narasimha Reddy, M. (1998). *Precambrian anorthosites from Peninsular India-problems and perspectives* (44, pp. 152−169). Geological Survey of India, .

Lisker, F., & Fachmann, S. (2001). The Phanerozoic history of the Mahanadi region, India. *Journal Geophysical Research, B: Solid Earth, 106,* 22027−22050.

Madhavan, V., & Leelanandam, C. (1977). A study of the relative proportions of mafic and felsic rocks in the Elchuru alkaline massif, Andhra Pradesh. *Chayanica Geologica, 3,* 122−132.

Mahalik, N. K. (1994). Geology of the contact between the Eastern Ghats Belt and the North Orissa craton. *India. Journal Geological Society of India, 44,* 41−51.

Mahapatro, S. N., Nanda, J. K., & Tripathy, A. K. (2008). Anorthosite-alkaline magmatism and granulite nappe over Bastar craton in the northwestern part of the Eastern Ghats Mobile Belt. *IAGR (Mumbai) Abstract Series, 5,* 162−164.

Mahapatro, S. N., Nanda, J. K., & Tripathy, A. K. (2010). The Jugsaipatna Anorthosite Complex, Eastern Ghats Belt, India: Magmatic lineage and petrogenetic implications. *Journal of Asian Earth Science, 38,* 147−161.

Mahapatro, S. N., Pant, N. C., Bhowmik, S. K., Tripathy, A. K., & Nanda, J. K. (2012). Archean granulite facies metamorphism at Singhbhum Craton-Eastern Ghats Mobile belt interface: Implication of the Ur supercontinent assembly. *Geological Journal, 47,* 312−333.

Mahapatro, S. N., Tripathy, A. K., Nanda, J. K., & Rath, S. C. (2013). Petrology of the Udayagiri Anorthosite Complex, Eastern Ghats Belt, India. *Journal Geological Society of India, 82,* 319−329.

Maji, A. K., Bhattacharya, A., & Raith, M. (1997). The Turkel anorthosite complex revisited. *Proceedings of Indian National Academy of Sciences (Earth Planet Sciences), 106,* 125−313.

Mezger, K., & Cosca, M. (1999). The thermal history of the Eastern Ghats Belt (India) as revealed by U-Pb and 40Ar/39Ar dating of metamorphic and magmatic minerals: Implications for the SWEAT correlation. *Precambrian Research, 94,* 251−271.

Misra, S., & Gupta, S. (2014). Superposed deformation and inherited structures in an ancient dilational step-over zone: post-mortem of the Rengali Province, India. *Journal of Structural Geology, 59,* 1−17.

Mohan, A., Tripathi, P., & Motoyoshi, Y. (1997). Reaction history of sapphirine granulites and a decompressional P−T path in a granulite complex from the Eastern Ghats. *Proceedings, Indian Academy of Sciences (Earth Planetary Sciences), 106,* 115−129.

Mukherjee, A., Bhattacharya, A., & Chakraborty, S. C. (1986). Convergent phase equilibria at the massif anorthosite-granulite interface near Bolangir, Orissa, India, and the thermal evolution of a part of the Indian Shield. *Precambrian Research, 34,* 69−104.

Mukhopadhyay, D., & Basak, K. (2009). The Eastern Ghats Belt- A polycyclic Granulite terrain. *Journal Geological Society of India, 73,* 489−518.

Nagaraju, J. (2007). *Structure and tectonics of the transition zone, southern Eastern Ghats, in parts of Prakasam district, Andhra Pradesh, India. unpublished Ph.D thesis.* Hyderabad: Osmania University.

Nagaraju, J., & Chetty, T. R. K. (2005). Emplacement history of Pasupugallu Gabbro Pluton, Eastern Ghats Belt, India: A structural study. *Gondwana Research, 8,* 87−100.

Nagaraju, J., Chetty, T. R. K., Vara Prasad, G. S., & Patil, S. K. (2008). Compressional tectonic setting during the emplacement of Pasupugallu Gabbro Pluton, western margin of the Eastern Ghats Mobile Belt, India: Evidence from AMS fabrics. *Precambrian Research, 162*, 86−101.

Nanda, J., & Gupta, S. (2012). Intracontinental orogenesis in an ancient continent-continent collision zone: Evidence from structure, metamorphism and P-T paths across a suspected suture zone within the Eastern Ghats Belt, India. *Journal of Asian Earth Sciences, 49*, 376−395.

Nanda, J., Gupta, S., & Dobmeier, C. J. (2008). Metamorphism of the Koraput Alkaline Complex, Eastern Ghats Province, India−evidence for reworking of a granulite terrane. *Precambrian Research, 165*, 153−168.

Nanda, J. K. (2008). Tectonic framework of Eastern Ghats Mobile Belt: An overview. *Memoir, Geological society of India, 74*, 63−87.

Nanda, J. K., & Pati, U. C. (1989). Field relations and petrochemistry of granulites and associated rocks in the Ganjam-Koraput sector of the Eastern Ghats Belt. *Indian Minerals, 43*, 247−264.

Narayanaswami, S. (1975). *Proposal for Charnockite-Khondalite System in the Archean shield of Peninsular India* (23, pp. 1−16). Geological Survey of India, Miscellaneous Publication.

Nash, C. R., Rankin, L. R., Leeming, P. M., & Harris, L. B. (1996). Delineation of lithostructural domains in northern Orissa (India) from Landsat Thematic Mapper imagery. *Tectonophysics, 260*, 245−257.

Nasipuri, P., & Bhadra, S. (2013). Structural framework for the emplacement of the Bolangir anorthosite massif in the Eastern Ghats Granulite Belt, India: Implications for post-Rodinia pre-Gondwana tectonics. *Mineralogy and Petrology, 107*, 861−880.

Natarajan, V., & Nanda, J. K. (1981). Large scale basin and dome structures in the high grade metamorphics near Visakhapatnam, South India. *Journal Geological Society of India, 22*, 584−592.

Nayak, P. N., Choudhury, K., & Sarkar, B. (1998). *A review of Geophysical Studies of the Eastern Ghats Mobile Belt. Proc. of workshop on Eastern Ghats Mobile Belt* (44, pp. 87−94). Geological Survey of India.

Niraj Kumar, Singh, A. P., Gupta, S. B., & Mishra, D. C. (2004). Gravity Signature, Crustal Architecture and Collision Tectonics of the Eastern Ghats Mobile Belt. *Journal of Indian Geophysical Union, 8*(2), 1−10.

Pal, S., & Bose, S. (1997). Mineral reactions and geothermobarometry in a suite of granulite facies rocks from Paderu, Eastern Ghats granulite belt: A reappraisal of the P−T trajectory. *Proc. Indian Academy of Sciences (Earth Planetary Sciences), 106*, 77−89.

Panda, P. K., Patra, P. C., Parta, R. N., & Nana, J. K. (1993). Nepheline syenites from Rairakhol, Sambalpur district, Orissa. *Journal Geological Society of India, 41*, 144−151.

Paul, D. K., Ray Barman, T. K., McNaughton, N. J., Fletcher, I. R., Potts, P. J., Ramakrishnan, M., & Augustine, P. F. (1990). Archean-Proterozoic evolution of Indian charnockites: Isotopic and geochemical evidence from granulites of the Eastern Ghats belt. *Journal Geology, 98*, 253−263.

Perraju, P. (1973). *Anorthosites of Puri district, Orissa, Records* (105, pp. 102−116). Geological Survey of India, .

Prasada Rao, A. D., Rao, K. N., & Murthy, Y. G. K. (1987). *Gabbro-anorthosite-pyroxenite complexes and alkaline rocks of Chimakurti-Elchuru area, Prakasam district, A. P: Records* (116, pp. 1−20). Geological Survey of India.

Raith, M. M., Dobmeier, C., & Mouri, H. (2007). Origin and evolution of Fe−Al granulites in the thermal aureole of the Chilka Lake anorthosite, Eastern Ghats Province, India. *Proceedings of Geological Association, 118*, 87−100.

Raith, M. M., Mahapatro, S. N., Dewashish Upadhyaya, D., Berndt, J., Mezgere, K., & Nanda, J. K. (2014). Age and P−T evolution of the Neoproterozoic Turkel Anorthosite Complex, Eastern Ghats Province, India. *Precambrian Research, 254*, 87−113.

Ramakrishnan, M., Nanda, J. K., & Augustine, P. F. (1998). *Geological evolution of the Proterozoic Eastern Ghat Mobile Belt, . Proceedings of Workshop on Eastern Ghats Mobile Belt* (44, pp. 1−21). Geological Survey of India, .

Ramakrishnan, M., & Vaidyanathan, R. (2008). *Geology of India* (volume 1Geological Society of India, .

Ramesh, D. S., Bianchi, M. B., & Das Sharma, S. (2010). Images of possible fossil collision structures beneath the Eastern Ghats belt, India, from P and S receiver functions. *Lithosphere*, 2, 84−92.

Rao, A. T., & Rao, J. U. (1992). Basic pyroxene granulites from Visakhapatnam region in the Eastern Ghats, India. In S. S. Augustithis (Ed.), *High-Grade Metamorphics* (pp. 123−140). Theophrastus Publications.

Ratnakar, J., & Vijaya Kumar, K. (1995). Petrogenesis of quartz-bearing syenite occurring within nepheline syenite of the Elchuru alkaline complex, Prakasam Province, Andhra Pradesh. *Journal of the Geological Society of India*, 46, 611−618.

Rickers, K., Mezger, K., & Raith, M. M. (2001). Evolution of the continental crust in the Proterozoic Eastern Ghats Belt, India and new constraints for Rodinia reconstruction: Implications from Sm-Nd, Rb-Sr and Pb-Pb isotopes. *Precambrian Research*, *112*, 183−212.

Rickers, K., Raith, M., & Dasgupta, S. (2001a). Multistage reaction textures in xenolithic high-MgAl granulites at Anakapalle, Eastern Ghats Belt, India: Examples of contact poly metamorphism and infiltration-driven metasomatism. *Journal of Metamorphic Geology*, 19, 563−582.

Sarkar, A., & Paul, D. K. (1998). *Geochronology of the Eastern Ghats Precambrian MobileBelt−A review* (44, pp. 51−86). Geological Survey of India.

Sarkar, A., Bhanumathi, L., & Balasubrahmanyan, M. N. (1981). Petrology, geochemistry and geochronology of the Chilka Lake igneous complex, Orissa State, India. *Lithos14*, 93−111.

Sarkar, A., Nanda, J. K., Paul, D. K., Bishui, P. K., & Gupta, S. N. (1989). Late Proterozoic alkaline magmatism in the Eastern Ghats Belt: Rb-Sr isotopic study on the Koraput complex, Orissa. *Indian Minerals*, 43, 265−272.

Sarkar, A., Pati, U. C., Panda, P. K., Patro, P. C., Kundu, H. K., & Ghosh, S. (2000). *Late-Archean charnockitic rocks from the northern marginal zones of the Eastern Ghats Belt: A geochronological study* (57, pp. 171−179). Geological Survey of India.

Sarkar, M., Gupta, S., & Panigrahi, M. K. (2007). Disentangling tectonic cycles along a multiply deformed terrane margin: Structural and metamorphic evidence for mid-crustal reworking of the Angul granulite complex, Eastern Ghats Belt, India. *Journal of Structural Geology*, 29, 802−818.

Sarkar, T., Schenk, V., & Berndt, J. (2015). Formation and evolution of a Proterozoic magmatic arc: Geochemical and geochronological constraints from metaigneous rocks of the Ongole domain, Eastern Ghats Belt, India. *Contributions to Mineralogy and Petrology*, 169, 5. Available from http://dx.doi.org/10.1007/s00410-014-1096-1.

Sengupta, P., Dasgupta, S., Bhattacharya, P. K., Fukoka, M., Chakraborti, S., & Bhowmick, S. (1990). Petrotectonic imprints in the saphirine granulites from Anantagiri, Eastern Ghats Mobile Belt, India. *Journal of Petrology*, 31, 971−996.

Sengupta, P., Dasgupta, S., Dutta, N., & Raith, M. M. (2008). Petrology across a calc−silicate−anorthosite interface from the Chilka Lake Complex, Orissa: Implications for Neoproterozoic crustal evolution of the Eastern Ghats Belt. *Precambrian Research*, 162, 40−58.

Sesha Sai, V. V. (2004). *Petrographic and petrochemical charecterisation of Proterozoic granites in Nellore schist belt and northeastern fringes of Cuddapah basin* (137, pp. 184−188). Records Geological Survey of India.

Sesha Sai, V. V. (2013). Proterozoic Granite Magmatism along the Terrane Boundary Tectonic Zone to the East of Cuddapah Basin, Andhra Pradesh − Petrotectonic Implications for Precambrian Crustal Growth in Nellore Schist Belt of Eastern Dharwar Craton. *Journal Geological Society of India*, 81, 167−182.

Shaw, R. K., & Arima, M. (1996). Mineral chemistry, reaction textures, thermobarometry, and P-T path from orthopyroxene granulites of Rayagada, Eastern Ghats, India. *Journal Southeast Asian Earth Sciences*, 14, 175−184.

Shaw, R. K., Arima, M., Kagami, H., Fanning, C. M., Shairashi, K., & Motoyashi, Y. (1997). Proterozoic events in the Eastern Ghats granulite belt, India: Evidence from Rb-Sr, Sm-Nd systematics and SHRIMP dating. *Journal of Geology*, 105, 645−658.

Simmat, R., & Raith, M. M. (2008). U−Th−Pb monazite geochronometry of the Eastern Ghats Belt, India: Timing and spatial disposition of poly-metamorphism. *Precambrian Research*, 162, 16−39.

Singh, A. P., Mishra, D. C., Gupta, S. B., & Rao, M. R. K. P. (2004). Crustal structure and domain tectonics of the Dharwar Craton (India):insight from new gravity data. *Journal of Asian Earth Sciences*, 23, 141−152.

Subba Rao, K. V. (1971). The Kunavaram Series-a group of alkaline rocks, Khammam District, Andhra Pradesh, India. *Journal of Petrology, 12*, 621−641.

Subba Rao, T. V., Bhaskar Rao, Y. J., Siva Raman, T. V., & Gopalan, K. (1989). *Rb-Sr age and petrology of the Elchuru alkaline complex: Implications to alkaline magmatism in the Eastern Ghats mobile belt* (pp. 207−223). Memoir. Geological Society of India 15.

Subrahmanyam, C. (1978). On the relation of the gravity anomalies to geotectonics of the Precambrian terrains of the South Indian shield. *Journal Geological Society of India, 19*, 251−263.

Subrahmanyam, C., & Verma, R. K. (1986). Gravity field, structure and tectonics of the Eastern Ghats. *Tectonophysics, 126*, 195−212.

Subramanyama, K.S.V., Santosh, M., Qiong-Yan, Y., Ze-Ming, Z., Balaram, V., Reddy, U.V.B., 2016. Mesoproterozoic island arc magmatism along the south eastern margin of the Indian Plate: Evidence from geochemistry and zircon U-Pb ages of mafic plutonic complexes. http://dx.doi.org/10.1016/j.jseaes.2016.07.027.

Upadhyay, D. (2008). Alkaline magmatism along the southeastern margin of the Indian shield: Implications for regional geodynamics and constraints on craton−Eastern Ghats Belt suturing. *Precambrian Research, 162*, 59−69.

Upadhyay, D., & Raith, M. M. (2006a). Petrogenesis of the Kunavaram alkaline complex: Implications for the evolution of the craton-Eastern Ghats Belt suture, SE, India. *Precambrian Research, 150*, 73−94.

Upadhyay, D., Raith, M. M., Mezger, K., Bhattacharya, A., & Kinny, P. D. (2006). Mesoproterozoic rifting and Pan-African continental collision in SE India: Evidence from the Khariar alkaline complex. *Contributions to Mineralogy and Petrology, 151*, 434−456.

Vijaya Kumar, K., & Leelanandam, C. (2008). Evolution of the Eastern Ghats Belt, India : A plate tectonic perspective. *Journal Geological Society of India, 72*, 720−749.

Vijaya Kumar, K., Carol, D. F., Ronald Frost, B., & Kevin, R. C. (2007). The Chimakurti, Errakonda, and Upplapadu plutons, Eastern Ghats Belt, India: An unusual association of tholeiitic and alkaline magmatism. *Lithos, 97*, 30−57.

Vijaya Kumar, K., Leelanandam, C., & Ernst, W. G. (2011). Formation and fragmentation of the Paleoproterozoic supercontinent Columbia: Evidence from the Eastern Ghats Granulite Belt, SE India. *International Geology Review, 53*, 1297−1311.

Vijaya Kumar, K., & Ratnakar, J. (2001). Petrogenesis of Ravipadu gabbro pluton, Prakasam province, Andhra Pradesh. *Journal of Geological Society of India, 57*, 113−140.

Yoshida, M., Bindu, R. S., Kagami, H., Rajesham, T., Santosh, M., & Shirahata, H. (1996). Geochronologic constraints of granulite terranes of India and their implications for the Precambrian assembly of Gondwana. *Journal Asian Earth Sciences, 14*, 137−147.

FURTHER READING

Biswal, T. K. (2000). Fold-thrust belt geometry of Eastern Ghats mobile belt, a structural study from its western margin, Orissa, India. *Journal of African Earth Sciences, 31*, 25−33, Special Publication.

Prasada Rao, G.H.S.V., Murty, Y.G.K. Dikshitulu, M., 1964. Stratigraphic relation of Precambrian Iron formation and associated sedimentary sequence in parts of Keonjhar, Cuttack, Dhenkanal and Sundergarh districts of Orissa, International Geological congress, x, 72−87.

Upadhyay, D., & Raith, M. M. (2006b). Intrusion age, geochemistry and metamorphic conditions of a quartz monzonite intrusion at the craton-Eastern Ghats Belt contact near Jojuru, India. *Gondwana Research, 10*, 267−276.

Vijaya Kumar, K., Ernst, W. G., Leelanandam, C., Wooden, J. L., & Grove, M. J. (2010). First Paleoproterozoic ophiolite from Gondwana: Geochronologic−geochemical documentation of ancient oceanic crust from Kandra, SE India. *Tectonophysics, 487*, 22−32.

THE CENTRAL INDIAN TECTONIC ZONE

4

4.1 INTRODUCTION

The Central Indian Tectonic Zone (CITZ) is a conspicuous zone of deformation in central India that extends from the west coast up to Meghalaya plateau in the east (Fig. 4.1). The CITZ, also regarded as Satpura Mobile Belt, merges with the Aravalli-Delhi orogen in the west, while it swerves around the nucleus of Singhbhum craton (SC) and joins the Eastern Ghats Mobile belt in the east. It extends for about 1500 km with a width of about ∼180 km and divides the Indian subcontinent into two distinct crustal provinces: the Bundelkhand craton (BKC) surrounded by Vindhyan sediments to the north, and the Deccan province constituting Bastar, Singhbhum, and Dharwar cratons to the south (Ramakrishnan & Vaidyanathan, 2008). A large part of the CITZ is covered by younger sequence comprising Gondwana sediments, Deccan basalts, and Quaternary alluvium. The CITZ is bounded by a structural belt, popularly known as the Narmada-Son lineament (NSL) in the north and the Central Indian shear zone (CISZ) in the south. The NSL juxtaposes with the Vindhyan basin, located to the south of BKC. The major rock types in the CITZ are gneisses and granites, supracrustal rock units, and boudin-type granulite facies rocks. The widespread gneisses are overlain by three important

Proterozoic Orogens of India. DOI: http://dx.doi.org/10.1016/B978-0-12-804441-4.00004-3

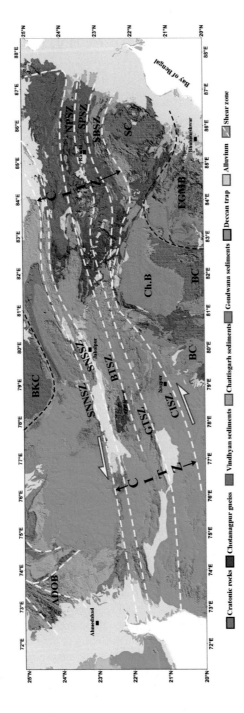

FIGURE 4.1

Geological and tectonic map of the Central Indian tectonic zone (CITZ) showing the broad lithological assemblages and tectonic frame work (geology superimposed on the DEM/ interpreted tectonic map of DEM). *BKC*, Bundelkhand craton; *SNNSZ*, Son-Narmada north shear zone; *SNSSZ*, Son-Narmada south shear zone; *BTSZ*, Balrampur-Tattapani shear zone; *GTSZ*, Gavilgarh-Tan shear zone; *CISZ*, Central Indian shear zone; *NPSZ*, North Purulia shear zone; *SPSZ*, South Purulia shear zone; *SBSZ*, Singhbhum shear zone; *SC*, Singhbhum craton; *BC*, Bastar craton; *Ch. B*, Chattisgarh Basin; *EGMB*, Eastern Ghats mobile belt; *ADOB*, Aravalli-Delhi orogenic belt.

east-west−trending, subparallel linear supracrustal belts, which are separated by crustal−scale ductile shear zones. The shear zones are also closely associated with a chain of granulite boudins (Yedekar, Jain, Nair, & Dutta, 1990). For the sake of brevity and clarity, the CITZ has been divided into west, central, and eastern sectors for the purpose of geological description despite the continuation of structural features and lithological assemblages among them following Acharyya (2003). The western sector is completely covered by Deccan traps, the central sector is partly covered by Deccan traps, and the Gondwana sediments and the eastern sector is broadly exposed. Since the western sector is covered by younger sequences like the Deccan traps and Gondwana sediments, only extended structural elements could be inferred and were considered for tectonic synthesis. The detailed description of central and eastern sectors is provided below.

4.2 CITZ: CENTRAL SECTOR

4.2.1 INTRODUCTION

The central sector of the CITZ is defined here as the region that lies between 76 and 84°E. Long. and 21 and 26°N. Lat. (Fig. 4.2). The region is dissected by a number of subparallel E−W trending shear zones. The marginal shear zones on either side of the NSL mark the northern part of the CITZ and the CISZ defines the southern boundary. Three prominent supracrustal belts occur in the regions that include: (1) the Mahakoshal belt in the north, (2) the Betul belt in the center, and (3) the Sausar belt in the south. Three significant granulite belts are also mapped in the region, which are described as (1) the Makrohar granulite belt (MGB) in the north, (2) the Ramakona-Katangi granulite belt (RKGB) in the middle, and (3) the Balaghat-Bhandara granulite belt (BBGB) in the south. Detailed characteristics of each of these geological units are described below.

4.2.2 SHEAR ZONES

4.2.2.1 Introduction

The region is traversed by a number of ENE-WSW trending important shear zones (see Fig. 4.2). Among them, the prominent ones, from north to south, include: (1) the Son-Narmada north shear zone (SNNSZ), (2) the Son-Narmada south shear zone (SNSSZ), (3) the Balarampur-Tattapani shear zone (BTSZ), (4) the Gavilgarh-Tan shear zone (GTSZ), and (5) the CISZ. They occur mostly at the boundaries of distinct lithological assemblages, separating the supracrustals, gneisses, and granulite belts. The characteristics of each shear zone are described below.

4.2.2.2 Son-Narmada north shear zone

The SNNSZ occurs at the northern margin of the NSL and separates the Mahakoshal supracrustal belt (MSB) from the Proterozoic Vindhyan sediments and some linear patches of Neoarchean gneisses to the north. The SNNSZ is a southerly dipping ductile shear zone with ENE−WSW trends in association with steep (70−80 degrees) southerly dips. It is characterized by subparallel isoclinal folds with south-dipping axial planes. Funnel shaped syenite plutons in association with trachytic dykes, lamprophyres, carbonatites, and barite veins are common. Ultramafic−mafic intrusions are also reported along the shear zone along with numerous quartz veins associated with gold

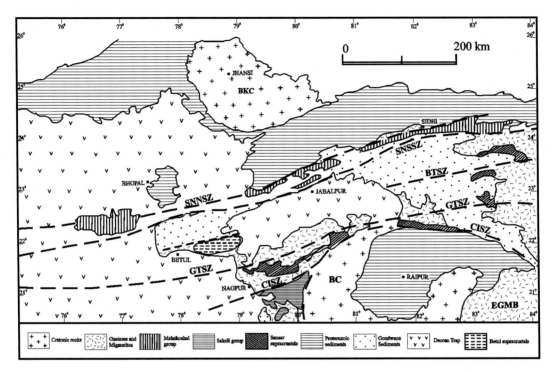

FIGURE 4.2

Geological and tectonic map of the CITZ (central sector). *BKC*, Bundelkhand craton; *SNNSZ*, Son-Narmada north shear zone; *SNSSZ*, Son-Narmada south shear zone; *BTSZ*, Balrampur-Tattapani shear zone; *GTSZ*, Gavilgarh-Tan shear zone; *CISZ*, Central Indian shear zone; *BC*, Bastar craton; *EGMB*, Eastern Ghats mobile belt.

Modified after Roy, A., Ramachandra, H.M., & Bandopadhyay, B.K. (2000). Supracrustal belts and their significance in the crustal evolution of Central India. Geological Survey of India, Special Publication, 55, 361–380.

mineralization. Many dolerite dykes cut across the region. The syenite plutons emplaced along the shear zone are dated at ~1800 Ma. Deep Seismic Sounding (DSS) studies suggest that the shear zone reaches to mantle depths (Kaila, Reddy, Dixit, & KoteswaraRao, 1985). In summary, it can be mentioned that it is a fundamental crustal scale feature which was active since the late Neoarchean to Neoproterozoic.

4.2.2.3 Son-Narmada south shear zone

The SNSSZ, also a south dipping ductile shear zone, occurs at the southern margin of the NSL and extends in a near east–west fashion for about 700 km along the strike. It hosts a chain of calk–alkaline and granite gneisses with well-developed mylonitic foliation, often associated with well-developed, down-dip stretching lineations (Roy & Hanuma Prasad, 2003). Similar fabrics are common both in supracrustal sequence as well as in sheet like granitoids in the vicinity of the shear zone. The granitoids show both magmatic and solid state deformation fabrics suggesting the syntectonic emplacement. They are defined by the alignment of K–feldspar megacrysts, mafic schlierens, and microgranitic enclaves. The granitoids also contain slivers of migmatitic gneiss, sillimanite–corrundum schist, mafic granulites, and anorthosites (Pichai Muthu, 1990), which are comparable to Chotanagpur granulite-gneissic complex (CGGC), which exist in the eastern extensions of the region, suggesting a mantle-crust interaction. The granitoids with reported Rb–Sr ages of 1.5–1.8 Ga extend further south. The kinematic indicators suggest largely thrust-sense with minor sinistral strike-slip movements (Roy, Sarkar, Jeyakumar, Aggarwal, & Ebihara, 2002). DSS studies revealed that the marginal shear zones of NSL reach up to mantle depths. In summary, both SNSSZ and SNNSZ together can be regarded as Paleoproterozoic sinistral transpressional tectonic zone.

4.2.2.4 Balarampur-Tattapani shear zone

The BTSZ is an important shear zone that traverses all along the CITZ in the central part (Fig. 4.3) and can be traced from SE of Shahdol to 10 km south of Tattapani with steep northerly dips at places. The BTSZ coincides with the northern margin of Betul-Chindwara supracrustal belt and extends in an ENE–WSW direction up to Balarampur. The shear zone displaces the Sausar metasedimentary belts. Dolerite dykes and sills are common along the shear zone. Further east of Balarampur, the shear zone emerges into many branches and extend in an east–west direction coinciding with the Damodar graben in the eastern sector of the CITZ. In summary, it can be mentioned that the BTSZ is more prominently reflected in satellite images. The shear zone also hosts Pb–Zn and W mineralization.

4.2.2.5 Gavilgarh-Tan shear zone

The significant crustal–scale shear zone in the central part of the CITZ has been commonly described as the Tan shear zone. The shear zone also extends to the west in the form of Gavilgarh fault and hence it is appropriate to term this as the GTSZ, following Golani, Bandyopadhyay, and Gupta (2001). The GTSZ trends ENE–WSW along a total strike length of about 500 km through Gavilgarh-Seoni-Ambikapur (see Fig. 4.3). The width varies from less than 2 km in the west to over 15 km in the east. It separates the base metal–enriched Betul supracrustal belt (BSB) to the north and biotite-gneiss-migmatite–bearing Sausar supracrustal belt (SSB) to the south. The GTSZ occurs at the southern slopes of Satpura hills.

In the western part it is mostly covered by Gondwana sediments, but the exposed part is characterized by mylonites, hot springs, and highly disturbed Gondwana sediments. The segment along the Kanhan river typically exposes a spectrum of mylonites derived from quartzo–feldspathic gneisses, pink to brick red colored granites, pegmatites, aplites, and quartz veins (Fig. 4.4). Gray orthogneisses and megacrystic biotite granites are abundant. Stretched lenses of amphibolites with pyrite desseminations are also common. The shear zone in this segment trends N60°E–S60°W with steep dips to south. Ultramylonites are well developed for a strike length of over 15 km and the width varies from a few meters to 200 m. The stretching lineations defined by the alignment of quartz and mica are well developed and they plunge mostly with gentle to moderate values to ENE. A few plunges to the WNW are also common. Small scale structural features such as highly asymmetric "S" shaped profiles, S–C fabrics and other kinematic features defined by porphyroclasts of K–feldspar show consistent sinistral sense of shearing (Chatopadhyay & Khasdeo, 2011; Golani et al., 2001). Stretching lineations are well developed showing plunges dominantly to east. GTSZ, in general, is dominantly a strike-slip ductile shear zone, but dip-slip movements noticed at a few places imply sinistral transpressional kinematics. Chattopadhyay, Khasdeo, Holdsworth, and Smith (2008) identified two sets of pseudotachylyte veins in this shear zone. From a detailed

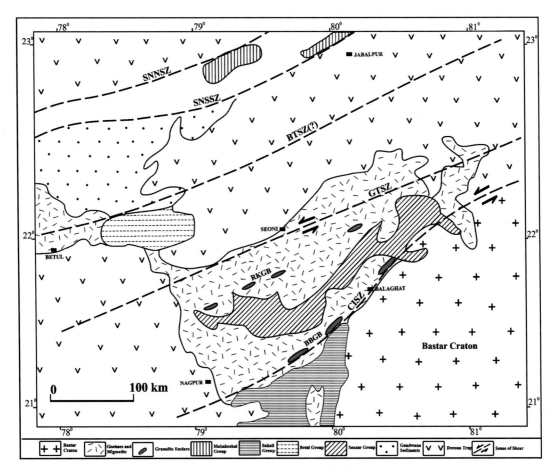

FIGURE 4.3

Geological and tectonic map of the high-grade granulite belts in the central sector of the CITZ. *SNNSZ*, Son-Narmada north shear zone; *SNSSZ*, Son-Narmada south shear zone; *BTSZ*, Balarampur-Tattapani shear zone; *GTSZ*, Gavilgarh-Tan shear zone; *CISZ*, Central Indian shear zone; *RKGB*, Ramakona-Katangi granulite belt; *BBGB*, Bhandara-Balaghat granulite belt; *BC*, Bastar craton.

Modified after Chattopadhyay, A. & Khasdeo, L. (2011). Structural evolution of Gavilgarh-Tan Shear Zone, central India: A possible case of partitioned transpression during Mesoproterozoic oblique collision within Central Indian Tectonic Zone. Precambrian Research, 186, 70–88.

microscopic and field study, they inferred that at least two phases of late brittle-ductile and/or brittle shearing has affected the ductile shear fabric. Kinematic analyses show that the brittle-ductile event was dextral in nature and the latest brittle deformation was again sinistral. This was interpreted by the authors as two phases of discrete fault reactivation in GTSZ with gradual exhumation of the deep-seated ductile shear zone. Later Chattopadhyay et al. (2015) attempted to date these events by $^{40}Ar/^{39}Ar$ dating of mylonites and pseudotachylytes and suggested that the sinistral

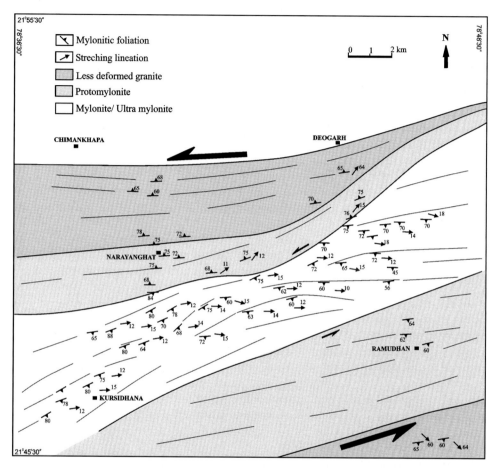

FIGURE 4.4

Structural map of a part of GTSZ displaying foliation trajectories and stretching lineations (modified after Chattopadhyay, A. & Khasdeo, L. (2011). Structural evolution of Gavilgarh-Tan Shear Zone, central India: A possible case of partitioned transpression during Mesoproterozoic oblique collision within Central Indian Tectonic Zone. *Precambrian Research*, 186, 70–88; Golani, P.R., Bandyopadhyay, B.K., &Gupta, A. (2001). Gavilgarh-Tan Shear: Prominent ductile shear zone in Central India with multiple reactivation history. *Geological Survey of India Special Publication, 64*, 265–272).

The lineations display gentle plunges in the central part of the shear zone, but are steep and down dip in nature in the south eastern sector of the map, possibly due to strain variations.

ductile shearing in GTSZ took place around 880 Ma, the first fault reactivation (brittle-ductile) occurred at c. 670 Ma, and the next brittle reactivation occurred at c. 459 Ma.

The eastern segment of the GTSZ is characterized by the presence of a variety of granitic gneisses including monzonites and other granitoids displaying both semisolid and solid state deformation fabrics that were transformed into protomylonites, mylonites, and ultramylonites. This part of the GTSZ, extending for more than 900 km in the eastern sector of the CITZ, is popularly known as the North Purulia shear zone (NPSZ). The general strike of the GTSZ is ENE−WSW and is continuous from the western part. The shear zone is covered by sedimentary/volcanic sequence at several places. The width of the shear zone between Seoni and Nainpur varies from 1 to 4 km with a total length of ∼100 km. The granite gneisses contain slivers of calc−silicates and include various phases of porphyritic to homophanous granites. They also comprise amphibole and biotite as major mafic minerals apart from quartz and feldspar. Repeated reactivation of GTSZ during Gondwana and the post-Deccan trap period reflects its geodynamic significance from Neoarchean to Quaternary time. Recently Bhattacharjee, Vikrant Jain, Chattopadhyay, Biswas, and Singhvi (2016) has shown that the western brittle part of GTSZ (called the Gavilgarh Fault Zone: GFZ) records evidences of a number of Quaternary fault reactivation events. The Quaternary fault movements along the GFZ were dated by luminescence dating (OSL/IRSL dating) of river terrace sediments. At least three distinct movements at c. 65−80 ka, c. 50 ka, c. 30−40 ka, and c. 14 ka were identified by these authors.

Geochemically, the rocks along the GTSZ show a distinct calc−alkaline affinity and peraluminous indicating a crustal source, continental collision, and an arc type implying orogenic character (Chattopadhyay & Khasdeo, 2011). Amphibole consists of up to 20% in quartz monzo−diorite suggesting their possible generation from mantle source. These rocks yielded Rb−Sr age of 1147 ± 16 Ma, with initial Sr87/Sr86 ratio of 0.7096 (Pandey, Krishna, & Chabria, 1998). Charnockite−mangerite association are well exposed around Mandla exhibiting similar fabrics as those of granitic rocks in the adjoining regions implying their coeval nature. In general, the GTSZ can be categorized as a significant Proterozoic sinistral transpressional shear zone. It may also represent a possible "collisional suture" and possibly accommodate the transpressional strain during oblique convergence of the Bundelkhand and Bastar cratonic provinces (Chattopadhyay & Khasdeo, 2011). The role of GTSZ in understanding the structural architecture and the evolution history of the CITZ is very significant, which warrants further detailed investigations.

4.2.2.6 Central Indian shear/suture zone

The Central Indian shear/suture zone (CISZ) is the most significant ductile shear zone representing the southern boundary of the CITZ (see Fig. 4.3), which is well established as a suture zone (Yedekar et al., 1990). It trends NE−SW in the western part and ENE−WSW in the eastern part with a general steep dip to north. The CISZ extends for over 1200 km in an east−west trend from the west coast to the Bay of Bengal in the east, largely covered by younger sequences at several places. The CISZ is well exposed between SE of Nagpur and to south of Korba. The width varies from a few tens of meters to a few kilometers. It is characterized by intense mylonitization involving a variety of lithologies like granitic gneisses and other supracrustal rocks. Narrow lenses of basic granulites, Mg−Al pelites, Al−rich (kyanite) pelites, BIF, and calc silicates associated with boudin-type charnockites are recorded all along between Bhandara and Balaghat. The central part of the CISZ is marked by silicification and brecciation to the east of Balaghat striking ENE−WSW with steep northerly dips. Mylonites were differently mapped as schists, phyllites, and conglomerates of SSB particularly around the Balaghat area. However, detailed studies established them to be phyllonites with well-developed streaky quartz, convolutedly rolled quartz veins forming pebbles, and excellent banded structure. Relics of high-strain gneisses occur as pods with in the CISZ with a complete gradation to phyllonite (Jain, Nair, & Yedekar, 1995). Often, the mylonites were derived from migmatitic gneisses, metasediments, and volcanic rocks, relics of which are still preserved in mylonite zones. Mesoscopic shear kinematics established the sinistral sense of displacements along the CISZ.

Ultramafic and mafic rocks (often serpentinised) along with anorthositic intrusives occur within "Tirodi-biotite gneisses" (TBG) comprising a tonalite—granodiorite suite of granitoids. Geochronological and petrological studies reveal contrasting characteristics of rocks on either side of the CISZ implying that the blocks to the north and south must have been amalgamated by an arc-continent collision during the Late Mesoproterozoic—Early Neoproterozoic Sausar Orogeny (Bhowmik, Wilde, & Bhandari, 2011). The high-grade Bhandara-Balaghat granulite belt surrounded by Tirodi biotite gneisses occurs to the north of the CISZ. The migmatization has yielded Rb—Sr ages of 1525 ± 70 Ma. Concordant and retrogressed charnockite and pyroxene granulite lenses (0.5—0.75 km wide and 0.5—2.0 km long) also occur along the CISZ. To the south of CISZ, the N-S—trending Kotri-Dongargarh, Sakoli, and Sonakhan belts occur in association with basement migmatitic gneisses ("Amgaon" gneisses), which got deflected and merged with CISZ. The basement is intruded by tonalitic and monzonitic plutons. The conformable low-grade volcanic and vocano-clastic sequences overlying the basement gneisses. were metamorphosed to upper amphibolite facies and the volcanic rocks yielded Rb—Sr isochron ages between 2200 and 2500 Ma. The Malanjkhand and Dongargarh granitoids intruded these gneisses around 2300 Ma. Diabase dyke swarms also occur concordantly with the gneisses and show Island arc tholeite signatures developed at plate margins (Yedekar et al., 1990). Recent metamorphic and geochronologic studies also indicate that the final amalgamation of the different cratonic blocks on either side of the CISZ had taken place during the 1.0 Ga continent-continent collisional orogenesis (Bhowmik, Bernhardt, & Dasgupta, 2010).

The CISZ extends eastwards along the northern fringes of Chattisgarh basin, coinciding with the southern contact of Bilaspur-Raigarh belt. The shear zone has a width of 2—5 km with a complete spectrum of mylonites ranging from proto-, mylonite-, and ultramylonites including phyllonites. The CISZ extends further along WNW—ESE direction for about 100 km with a well-exposed mylonite zone in association with two-pyroxene granulites. There is some ambiguity about the further continuation of the CISZ in the east. It is evident from the available satellite data and available literature that the CISZ splits into two shear zones: the northern branch is termed as the South Purulia shear zone (SPSZ), and the southern branch that coincides with the Singhbhum shear zone (SBSZ) via Jharsuguda-Rourkela-Jamshedpur. While some workers consider the SPSZ as an extension of the CISZ and form the southern boundary of the CITZ (Ramakrishnan & Vaidyanathan, 2008), some others opined that the CISZ extends through the basement of the Gondwana cover and coincides with the SBSZ (Jain et al., 1995; Rekha et al., 2011). It is more likely that the CISZ is represented further east in the form of the Singhbhum fold belt (SFB), bounded by SPSZ in the north and the SBSZ in the south. In all probability, the SBSZ may form the southern boundary of the CITZ in the eastern sector. However, it requires further detailed multidisciplinary investigations to establish the extensions of the CISZ and the southern boundary of the CITZ.

4.2.3 SUPRACRUSTAL BELTS

4.2.3.1 Introduction

Based on the spatial distribution, the important supracrustal belts from north to south have been distinguished as (1) the MSB, (2) the BSB, and (3) the SSB. All the belts expose basement gneisses comprising migmatites, granites, and granodiorites with enclaves of metavolcano sedimentary assemblages often metamorphosed to amphibolites to granulite grade. The other supracrustal associations structurally overlie the basement. All the belts are bounded by major shear zones and the major lithologies show tectonic contacts. Although there are many similarities among the supracrustal belts, the lithostratigraphic association is different in different areas and hence their correlation is challenging. For instance, BSB has more mafic−felsic volcanic, MSB has mafic volcanic, and the SSB has little or no volcanic. Such subtle differences broadly help in distinguishing the supracrustal belts as separate entities. The details of each belt are described below.

4.2.3.2 Mahakoshal supracrustal belt

The MSB occurs as a distinct linear belt, popularly known as NSL in the classic description of geology of central India. The MSB, restricted to the northern margin of the CITZ, trends ENE−WSW and is bounded by crustal−scale shear zones viz., the SNNSZ in the north and SNSSZ in the south (Fig. 4.5). The MSB is about 600 km long with a width of ~20 km, exposed between Narsinghpur in the west and Govindvallabh Pant sagar in the east. The MSB juxtaposes the Proterozoic Vindhyan sediments and the Neoarchean Sidhi gneissic complex to the north, while Proterozoic granites, gneisses mostly covered by the cover sequence, occur to the south.

The MSB constitutes the basement gneisses and the overlying supracrustal sequence of metavolcanic and metasediments (Jain et al., 1995). The basement is represented mostly by hornblende granodiorite gneiss with enclaves of quartz−mica schist, quartz−sillimanite schist, crystalline lime stone, banded-iron formation (BIF), amphibolites, and talc−chlorite schist. The basement gneiss is overlain by quartzites, stromatoliic carbonates, BIF and phyllites, pillowed basalt, manganiferous chert, and ultramafic rocks, followed by polymict conglomerate at the top. The conglomerate comprises pebbles of BIF, chert, jasper, quartzite, etc. There is another supracrustal sequence that overlies the conglomerate that consists of graywacke−argyllite, carbonate−phyllite, feldspathic quartzite, and alternate sequences of BIF−phyllite. The volcanic sequence and the sediments are, in general, interlayered and the former represents platformal volcanism in response to crustal

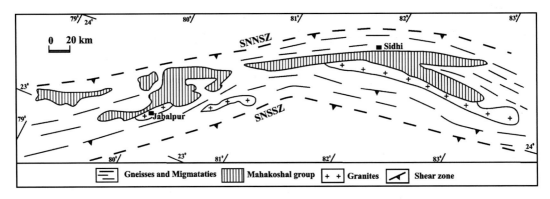

FIGURE 4.5

Simplified geological map of Mahakoshal supracrustal group of rocks bound by south dipping shear zones. *SNNSZ*, Son-Narmada north shear zone; *SNSSZ*, Son-Narmada south shear zone.

Modified after Roy, A. & Devarajan, M.K. (2000). A reappraisal of the Stratigraphy and tectonics of the palaeo-proterozoic Mahakoshal supracrustal belt, Central India. Geological Survey of India Special Publication, 57, 79−97 (Roy & Devarajan, 2000).

thinning or rifting. The basic volcanic rocks are predominantly tholeitic in composition and occur in association with gabbros, serpentinised ultramafic bodies of dunites and pyroxenites (Roy, Ramachandra, & Bandopadhyay, 2000).

The rocks of MSB were subjected to two major events of deformation. During the first event, upright to inclined folds with southward dipping axial planes, weakly noncylindrical with gentle plunges and pervasive foliations were developed. The second event of deformation resulted in E−W trending folds with steep south dipping intense fabrics with well-developed stretching lineations. Noncoaxial deformation is marked by complete transposition of earlier foliation fabrics during this event. The rocks are characterized by green schist facies metamorphism with peak P−T metamorphic conditions of 3−3.5 kbar and at 550°C.

4.2.3.2.1 Geophysical signatures

The regional gravity anomaly map shows a broad region of ENE-WSW trending "gravity high" in which the NSL appears as a narrow "low" in the Bouguer anomaly map of central India (Qureshy, 1982). The NSL is characterized by high amplitude elongated gravity high. The DSS studies across the NSL indicated the existence of a thick pile of high density mafic−ultramafic rocks beneath the NSL, indicating the emplacement of mantle products at a shallow crustal level of 2−3 km depth (Kaila, 1988). The crustal thickness varying between 38 and 45 km is higher compared to the average thickness of 35 km assumed for the regional Indian crust. The thickening of the crust must have taken place during Neoarchean-Paleoproterozoic time due to underplating of mantle derived high density material in the lower crust. An integrated geophysical study including DSS, gravity, heat flow, and geomagnetic data delineated the presence of serpentinite body above the Moho (at 41 km), which is well reflected in the form of a 5km-thick, low-velocity zone (Reddy et al., 1995).

The density model suggests an upwarp of Moho with in the NSL, bounded by crustal−scale shear zones (Venkat Rao, Srirama, & Ramasastry, 1990). The Magnetotelluric results across the NSL show mantle conductivity at Moho depth north of SNSSZ, suggesting an upwarp of the Moho, which corroborates with the 2-D gravity model (Das et al., 2007). The results derived from the 2-D inversion of a Magnetotelluric data along a 270 km long N−S profile across the NSL zone revealed the highly conductive nature of the mid-lower crust, which could be due to partial melting of subsurface rocks caused by mantle upwarping in the region (Dhanajayanaidu, Veeraswamy, & Harinarayana, 2011). The receiver function modeling studies show a distinct (15 km) crustal thickening and the complex nature of the Moho, with substantial evidence for postrifting magmatic underplating in the lower crust. The study also reveals a duplex nature of the Moho with the crust being distinctly thicker (> 52 km) within the NSL compared to that on either side (40 km) (Kumar, Singh, Kumar, & Sarkar, 2015).

4.2.3.2.2 Tectonic synthesis

The NSL, constituting the MSB, is restricted between the two major boundary shear zones, the SNNSZ and the SNSSZ, reaching the mantle. The MSB represents a sediment-dominated, volcanosedimentary association, which was possibly evolved from an intracratonic mantle activated by an aborted rift (Roy & Bandopadhyay, 1990). The existence of calc−alkaline magmatism in association with contractional tectonic setting and low-pressure metamorphism favors continental margin-arc tectonic setting at around 1.8 Ga. It can also be inferred that the MSB was initiated as a back-arc rift in the basement (?). Initially, the MSB must have been initiated as pericratonic basin, in which shallow marine quartzite—carbonate—pelite—chert, and BIF were deposited. Subsequent rifting and thermal doming resulted in the emplacement of tholeitic magma; accompanied by cogenetic ultramafic intrusions and basic volcanics. Further upliftment of rifted margins led to the deposition of the debris flow that culminated in the formation of argillite—BIF. Geochemical characteristics of mafic volcanic rocks indicate a high degree of melting of the shallow mantle source in a rift environment (Chaudhuri & Basu, 1990). Geochemically, the basalts plot in the fields of MORB, OIB, or continental rift and show ambiguous signatures. However, it is also not clear whether the MSB represents an ophiolite assemblage, which warrants more detailed investigations.

The MSB is characterized by north-verging thrusts related to convergent margin tectonics and can be regarded as a Paleoproterozoic sinistral transpressional tectonic zone. However, the resurgent tectonics were relatively more pronounced and protracted at the southern boundary compared to the northern boundary. Many splays emanate from both the boundary shear zones but they are more diffused in the eastern sector of the CITZ. The presence of granulite enclaves and boudins in the gneisses at the southern margin of the MSB near MGB supports the convergent tectonics in the evolution of the MSB associated with possible Paleoproterozoic sinistral transpressional tectonics.

Widely differing views prevail over the nature of the NSL. While some consider it as a rift valley, some others have described it as a fault zone, a major thrust zone, a horst, or a rift valley with boundary faults (Yellur, 1968). Some workers interpreted this as being due to the collision of the Indian plate with the Eurasian plate and as a suture zone of collision of the Bundelkhand protocontinent in the north and the Dharwar protocontinent in the south (Mishra, 1977). Based on geophysical information, the NSL was interpreted in terms of a typical rift structure extending into the Arabian Sea up to the Murray ridge, and into Madagascar, Somalia, and Ethiopia. The stratigraphy of the region is akin to a horst-type feature, delimited by the boundary shear zones. The nature of NSL must be understood in the context of broader perspective of the geodynamics of the CITZ in the realm of plate tectonics. It also becomes relevant to examine the NSL with a broader perspective of the geodynamics of the CITZ and the other Proterozoic orogens of India.

4.2.3.3 Betul Supracrustal Belt

The BSB, occurs in the central part of the CITZ between MSB and SSB (see Fig. 4.2). The exposed BSB trends near east—west with a strike length of ~135 km and a width of about 25 km. The region around BSB is largely covered by Deccan traps in the south and Gondwana sediments in the north. The BSB is bounded and traversed by a series of subparallel shear zones (Fig. 4.6). While the northern margin is marked by BTSZ, the Tapti shear zone demarcates the southern margin. The GTSZ separates both the BSB and the SSB. Despite several attempts to correlate all the supracrustal belts of the CITZ, no firm conclusion could be drawn due to a lack of adequate understanding of geological characteristics and detailed maps.

The BSB comprises primarily gneisses, granitoids, and other supracrustals. All the rocks of BSB are intensely deformed and sheared with varying degrees of metamorphism. The general sequence of BSB from the bottom includes: Amla gneiss and Golighat supracrustals; Bargoan group of volcano-sedimentary sequence in association with older granitoids; Bhopali metasedimentary rocks, and a suite of mafic-ultramafic rocks and younger granitoids (Roy & Hanuma Prasad, 2003). Series of arcuate/sigmoidal regional ductile shear zones of NE—SW to ENE—WSW trends are distinct in the region separating various rock units as different tectonic blocks. The foliation trajectories show that many NE-SW trending shear zones deflect to a near east—west direction.

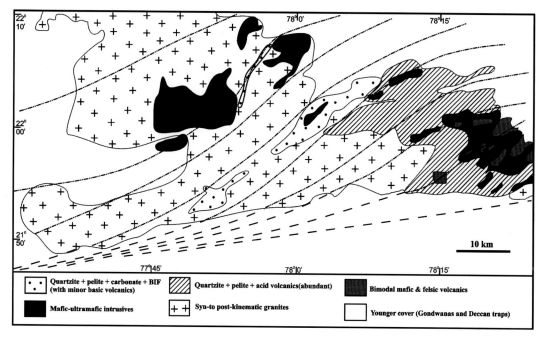

FIGURE 4.6

Geological and structural map of Betul supracrustal belt bounded by near E—W trending major shear zones. Sigmoidal shear zones separating distinct lithologies occur in between linking the bounding shear zones.

Modified after Roy, A. & Hanuma Prasad, M. (2003). Tectonothermal events in Central Indian Tectonic Zone (CITZ) and its implications in Rodinian crustal assembly. Journal of Asian Earth Science, 22, 115—129.

The rocks are isoclinally folded with axial planes dipping northwards subparallel to the shear zones. They show steep northerly dipping foliations and moderately plunging stretching lineations implying obique slip displacements. These shear zones seem to be genetically related to the regional bounding shear zones on either side of the BSB. Dykes, carbonate veins, quartz veins, and granites were emplaced along these shear zones. Granites were emplaced along the shear zones during and post shearing events that yielded Rb—Sr age of 850 ± 15 Ma. These structural features jointly with lithological characteristics can be interpreted in terms of duplex structures, possibly bound by roof and floor thrusts in map view (see Fig. 4.6), which are characteristic features of oceanic crust. Sulfide-rich zones including native gold and platinoids, associated with Fe—Ni—Cu—Pb—Zn sulfides and tungsten-bearing phases in the gabbroic intrusions, are extensively described in BSB (Sarkar & Gupta, 2012). Such an interpretation is consistent with the existence of Zn—Pb—Cu mineralization where ore bodies formed either as exhalations on the sea floor or as subsea-floor replacements or one involving both the processes (Ghosh & Praveen, 2008). It may also be worth to understand whether the mafic—ultramafic rocks of BSB represent analogs of ophiolite complex.

Amla gneisses (~ 1500 Ma), also known as the "Betul gneissic complex," occur around Betul and represent the basement comprising mainly gneisses, migmatites, basic schists, and amphibolites in association with granitic and basic intrusives. The basement gneisses contain older metamorphic enclaves (described as the Golighat group) of quartzite, quartz—mica schist, dolomitic marble, calc—silicate rock, tremolite—actinolite schist (altered ultramafic rock), and amphibolites in association with granulite-grade BIF. These enclaves are dominant in the western and northern parts of BSB. The enclaves of dolomitic marble and calc—silicates occur as small lensoid bodies in the vicinity of shear zones, which are considered as tectonic slices.

The Bargoan and Bhopali group of supracrustals that overlie the basement comprise Amla gneisses and display the tectonic contact. Bargoan group of mafic—ultramafic rocks, comprising mostly tholeitic metabasalts, magnesian basalts, felsic volcanic, and pyriteferous cherts are dominant in the eastern part of BSB (Fig. 4.7). The metabasalts are, in general, pillowed and consist of hornblende and andesine. Geochemically, they are low K—tholeites and high Mg—lavas. The tuffs are commonly associated with unusual sedimentary sequence. Biotite—garnet—staurolite schists with gahnite are the other sedimentary assemblage hosting Zn—Cu—Pb mineralization. Younger granites (850 ± 15 Ma) in Bargoan group were subjected to intense shearing, giving rise to augen gneisses and mylonitic fabrics supporting the presence of shear zones.

Bhopali group of younger sediments occur in the western and central parts of BSB, which are generally bound by shear zones. The sediments include phyllites intercalated with graded and current bedded quartzites, dolomitic lime stones, calc—phyllites, BIF, and metabasalts. Bhopali group of rocks, in general, show unconformable relationship with the basement Amla gneisses, but the presence of tectonic contact with Golighat group in the south and Gondwana sediments in the north are also not uncommon.

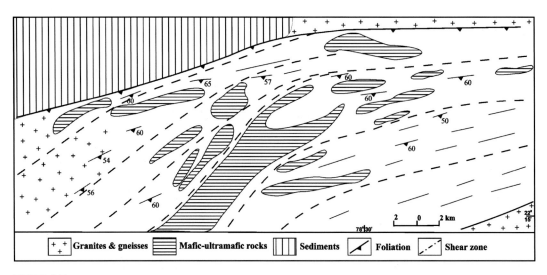

FIGURE 4.7

Structural map of a small segment in the eastern part of Betul supracrustal belt displaying sigmoidal shear zones separating distinct lithologies and associated mineralization.

Modified after Ghosh B. and Praveen M.N., Indicator minerals as guides to base metal sulphide mineralisation in Betul Belt, Central India, Journal of Earth System Science 117, 2008, 521–536.

While the volcano-sedimentary litho-associations dominate in the eastern and central parts of the belt, the mafic–ultramafic complex occurs in the western and northwestern parts as large individual intrusive bodies consisting of pyroxenite, gabbro, diorite, and foliated ultramafics exposed around Padhar, Gajpur and several other locations (Ghosh, Raj, & Nandy, 1998), locally known as the "Padhar mafic complex." They occur over 160 km^2 comprising essentially both unaltered olivine websterite, clinopyroxenite, hornblendite, gabbro, norite, and hornblende gabbro along with their altered equivalents viz. serpentinite, talc–antigorite schist, and chlorite schist (Roy et al., 2004). They are mostly undeformed but broadly oriented along ENE–WSW direction with enclaves of granulite grade BIF, calc–silicate rock, and amphibolites. The rocks are metamorphosed to middle to upper amphibolite facies. The rocks show enrichment in Rb, Ba, Th, and Pb and a depletion of Nb, Hf, and Zr. Major and trace element data along with REE, confirms the presence of a metasomatized mantle above a Mesoproterozoic subduction zone in the region. Differential petrological evolution, corresponding to the different magma batches, as reflected in almost all of the binary element/oxide variation diagrams, testifies to small scale metasomatic heterogeneity in the underlying suprasubduction zone mantle wedge (Chakraborty & Roy, 2012). From a detailed field and petrographic study corroborated by geochemical and metamorphic characterizations, a continental margin arc–type tectonic setting was also suggested for the Betul Belt.

4.2.3.3.1 Tectonic synthesis

The BSB forms a conspicuous lithotectonic unit of the CITZ between MSB and SSB. Quartzite, meta-pelite, bimodal volcanic rocks (basalt-rhyolite), meta-exhalites, calc–silicate rocks and BIFs constitute the supracrustal rocks of this belt. The belt is traversed by several ENE–WSW trending ductile shear zones having subvertical-to-steep dips, which were developed during deformation (Roy & Hanuma Prasad, 2001, 2003). The Betul belt represents a unique litho-package in the CITZ comprising a prominent bimodal volcanic sequence in which felsic volcanic rocks are dominant (Fig. 4.1). Sulfide mineralization is hosted by the felsic volcanic rocks similar to the volcanic-hosted massive sulfide deposits in other parts of the world.

The BSB is a complex mosaic of distinct crustal units, which got accreted together during the Neoarchean-Paleoproterozoic period. Several ductile shear zones showing evidences of oblique slip movement have brought slivers of supracrustal rocks of varying metamorphic grade and syntectonically emplaced granitic rocks in tectonic juxtaposition (Roy et al., 2004). The contact zone between the gneisses and the metasediments is a well-defined tectonic contact.

Geochemical data of mafic–ultramafic rocks suggest that they are derived from an enriched mantle source with subsequent contamination of continental crust. The existence of low k–tholeites, bimodal volcanics, and calk–alkaline granitoids favor continental margin arc setting for the BSB. The presence of ~1500 Ma syntectonic and older granites and 850 Ma old younger granites suggest the contemporaneity of BSB and the MSB with the later representing back-arc rift at the old continental margin. However, further studies are warranted to understand the timing and tectonic setting of the BSB.

4.2.3.4 Sausar supracrustal belt

The SSB is an arcuate fold belt trending ENE–WSW direction with convexity to south (Fig. 4.8). The SSB is in close association with the adjacent high-grade granulite belts on either side: RKGB at the northern margin and Balaghat-Bhandara granulite belt at the southern margin. The SSB and

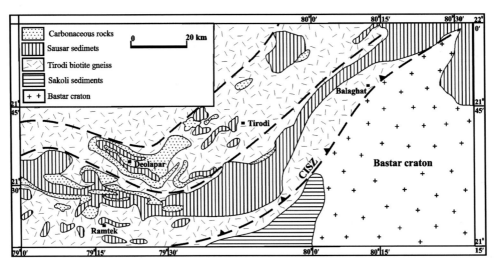

FIGURE 4.8

Simplified geological and structural map of Sausar supracrustal belt, which is bound by the Central Indian shear zone in the south.

Modified after Narayanaswamy, S., Chakravarthy, S.C., Vemban, N.A., Shukla, K.D., Subramaniam, M.R., Venkatesh, V., & Nagarajaiah, R.A. (1963). The geology and manganese ore deposits of the manganese belt in Madhya Pradesh and adjoining parts of Maharashtra. Part I: General introduction. Bulletin Geological Survey of India A-22 (I).

the granulite belts are inseparable and are together bound by crustal−scale shear zones namely the GTSZ in the north and the CISZ in the south. Hence, all the units of supracrustals, granulite belts, and the shear zones are cumulatively described as Sausar Mobile Belt (SMB) (Acharyya & Roy, 2000), which extends for about 350 km with a width of about 70 km.

The SSB consists of two major lithotectonic units: Tirodi biotite gneiss (TBG) and supracrustal sedimentary assemblages. The TBG comprises a wide range of rocks including biotite gneiss, felsic migmatite gneiss, TTG suite of rocks with enclaves of amphibolites, and metamorphosed ultramafic rocks−all together considered as the basement gneiss. (Chattopadhyay, Bandyopadhyay, & Khan, 2001). Tonalitic migmatite gneisses often show thin, extremely continuous, subparallel bands, centimeters to meters in thickness with alternating amphibolitic and quartzo−feldspathic material. The younger phase of the TBG includes a suite of pink, foliated granites and augen gneisses, which occur as sheet-like bodies at the contact zone between the migmatitic gneisses and the structurally overlying Sausar metasedimentary rocks implying tectonic nature of the contact. The supracrustal rocks consist of stable platform sequence of quartzite−pelite−carbonate in association with Mn−oxide formations (Narayanaswamy et al., 1963). At a few places, the presence of a polymict conglomerate with pebbles of gneisses and granites has also been recorded overlying the basement TBG. Therefore, the lateral continuity of a majority of supracrustal units is obscured and disrupted making the chronological sequence a challenge. There is also another view that the Sausar supracrustals were metamorphosed to upper amphibolite to granulite facies conditions, which were extensively migmatized giving rise to high-grade migmatitic gneisses, described as Tirodi biotite gneisses (TBG). The TBG is a multicomponent gneissic rock that recorded evidences of high-grade (granulite facies) metamorphism with a clockwise P−T path, and migmatization, which is totally absent in the meta-sedimentary rocks of the Sausar Group (Bhowmik, Pal, Roy, & Pant, 1999). A cooling age of 980 Ma from Ar/Ar systematics for Mn minerals in the Sausar supracrustals was obtained by Lippolt and Hautmann (1994). As the Tirodi gneiss is considered to be older to the Sausar (Narayanaswamy et al., 1963), it is possible that the younger ages mentioned above may be directly related to the evolution of the Sausar supracrustals.

Both supracrustal sediments and the basement gneisses jointly witnessed intense deformation involving thrusting, folding, and tectonic slicing. The deformational history suggests the involvement of low-angle thrusting and tectonic interleaving of basement and supracrustal rocks resulting in E-W−trending tight isoclinal folds, recumbent to reclined folds during the first event of deformation. This was followed by the development of steeply inclined noncylindrical folds during the second event of deformation. The nature of deformation is correlatable with a fold-thrust mechanism with south-verging thrusting (Roy & Hanuma Prasad, 2003). This concept is further supported by detailed structural mapping in the area around Deolapar in the north western part of the SSB. Details of the Deolapar nappe are described below.

4.2.3.4.1 Deolapar nappe

Existence of nappe structure in SSB around Deolapar was first advocated by West (1936) considering the stratigraphic inversion and omission, which was challenged by subsequent workers. However, the presence of thrust-nappes in SSB was emphasized after detailed structural mapping in Deolapar area (Chattopadhyay, Huin, & Khan, 2003). The northern part of the SSB exposes a curvilinear belt of calc−silicate rock and calcitic marble of the SSB. These rocks are folded into a series of large-scale alternate antiforms and synforms with gentle plunges (Fig. 4.9) and the TBG occurs on either side of the fold belt. Field observations suggest the existence of the older (pre-Sausar) discrete shear zones in the TBG, which are found to terminate abruptly at the TBG−SSB contact. A typical unconformable basement-cover relation between the TBG and SSB was lacking in the region, and the contact is often sheared possibly due to the allochthonous nature of the supracrustals. However, an isolated patch of polymict conglomerate at the SSB−TBG contact was reported from the eastern part of the Sausar fold belt, near Balaghat (Chattopadhyay et al., 2001).

Several low-angle thrusts were documented all along the northern margin of the SSB, which were responsible for the development of E-W trending tight to isoclinal and recumbent fold structures and intimate tectonic interleaving of tectonic slices of basement, disturbing the original stratigraphic sequence. Further deformation of thrust sheets and associated recumbent folds gave rise to map scale upright to steep folds plunging mostly to the east and southeast during the second deformational event. A spectrum of mylonites occurs in the central part of TBG with very strong down-dip−stretching lineations. Development of sheath folds is also common. The features such as intense thrust imbrications, development of mixed tectonic slices in the northern margin, and south-verging, low-angle thrust zones toward the foreland (Fig. 4.10) suggest the north ward subduction (Chattopadhyay et al., 2001), which is contradictory to the southward subduction invoked by Yedekar et al. (1990).

Detailed investigations around the Deolapar nappe reveal the presence of gently plunging alternate antiforms and synformal fold structures in the SSB, surrounded by the basement TBG

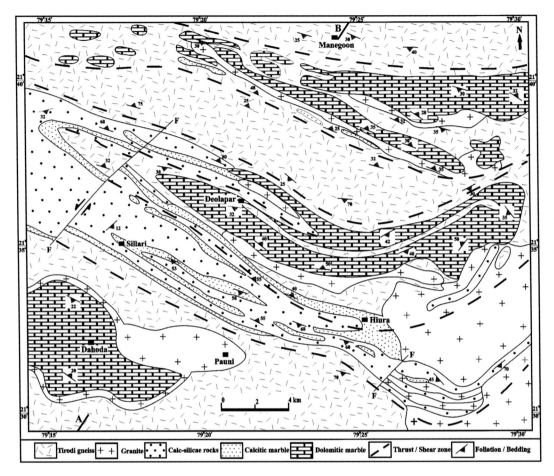

FIGURE 4.9

Detailed geological and structural map of the southern part of Sausar supracrustal group of rocks (modified after Chattopadhyay, A., Huin, A.K., & Khan, A.S. (2003). Structural framework of Deolapar area, central India and its implications for Proterozoic nappe tectonics. *Gondwana Research, 6*, 107−117). Note the presence of ductile shear zones at the lithological boundaries and NE−SW trending transverse dextral brittle shear zones. A−B represents the structural cross section depicted in Fig. 4.10.

(Fig. 4.11). The TBG is a multicomponent gneissic rock that records the evidences of high-grade granulite facies metamorphism with a clockwise P−T path (Bhowmik et al., 1999). This implies that the TBG witnessed an older thermal history, which is also supported by the presence of a number of discrete shear zones that abruptly end against the TBG−SSB contact. Small scale sinistral shear sense indicators were recorded.

Thin sheets of dolomite marble rest directly over the TBG. The presence of polymict conglomerate at the contact zone between TBG and SSB is lacking in this part of the region. The contact

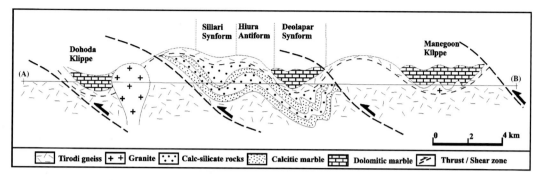

FIGURE 4.10

Interpreted structural cross section across the Nappe complex of Deolapar region displaying a series of south verging thrust planes and imbricate structures (modified after Chattopadhyay, A., Huin, A.K., & Khan, A.S. (2003). Structural framework of Deolapar area, central India and its implications for Proterozoic nappe tectonics. *Gondwana Research*, 6, 107−117). A−B represents the structural cross section derived from Fig. 4.9.

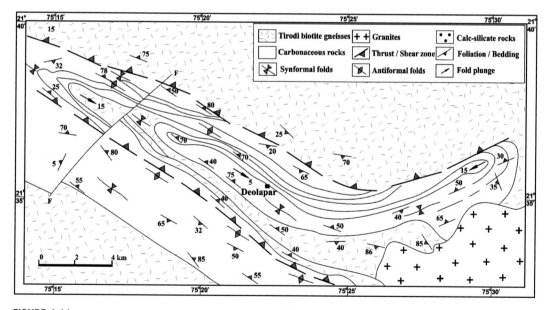

FIGURE 4.11

Detailed structural map of Deolapar nappe complex (modified after Chattopadhyay, A., Huin, A.K., & Khan, A.S. (2003). Structural framework of Deolapar area, central India and its implications for Proterozoic nappe tectonics. *Gondwana Research*, 6, 107−117). Notice the lithology-bound ductile shear zones inward-fold plunges of the fold structure in the nappe complex and the presence of brittle shear zone.

zone shows intensely sheared granite often containing quartz—fibrolite "tabloids" suggesting the allochthonous nature of SSB (Chattopadhyay et al., 2003). The whole allochthonous unit shows a doubly plunging synformal structure that closes at both the ends with inward plunges (see Fig. 4.11), which may also represent deformed sheath fold. The gneisses show different characteristics away from the contact zone with the synformal structure. A few outcrops of TBG containing enclaves of metabasic rocks occur in the core of the synformal fold structure.

The following features were cited as favorable evidence to characterize this rock unit as the Deolapar nappe (Chattopadhyay et al., 2003): (1) the occurrence of TBG in the core of Deolapar synformal structure of SSB, suggesting the regional inversion of the SSB by large scale folding; (2) there is no evidence of stratigraphic repetition on either side of TBG in the core; (3) the existence of Klippen structures in the adjacent region; (4) the occurrence of south-verging asymmetrical folds emanating from south verging thrusts; (5) the association of small-scale recumbent folds; (6) profuse development of south-verging thrusts in the northern margin of the synnclinorium; and (7) a decrease in the intensity of deformation from north to south. All these lines of evidence suggest that the Deolapar nappe must have been transported from north to south. According to Chattopadhyay et al. (2003), the Deolapar nappe represents a large thrust sheet comprising tectonic slices of both TBG as well as SSB transported along a low dipping basal thrust plane without the involvement of large-scale recumbent folding. The thrust sheet extends further showing up "klippen structures" in the region (see Fig. 4.11). However, the exact timing of thrusting is not clearly known. It is likely that the thrusting might have taken place during the earliest phase of deformation around ~950 Ma.

Several horizons of marbles, pelites, and Mn—formations with variable lithological characteristics are recorded in SSB. A polymict, granite clast conglomerate having micaceous matrix occurs at the base of the SSB. The SSB has been found to extend up to the northern fringes of Chattisgarh formations and continue further east as inliers within the Mahanadi Gondwana basins. The southern and northern parts of this belt are separated by CISZ passing through the area.

There have been varied opinions about the tectonic setting of the SSB. One view is that the sediments of the SSB were deposited in a shallow marine environment, and the other view is that the deposition was on the passive margin implying its source from north. Structural observations around Deolapar led to a conclusion that the Sausar supracrustals display a thrust-nappe structure and the SSB was subjected to Proterozoic nappe tectonics (Chattopadhyay et al., 2003). Some workers also argued that the basement gneisses of TBG (1.5 Ga old) represent migmatized equivalents of supracrustal rocks (Yedekar et al., 1990). Based on petrological characteristics, Bhowmik et al. (1999) opined that they may be related to high-grade gneisses of RKGB that preceded the deposition of SSB. According to them, the sedimentation of SSB postdates the high-pressure granulite facies metamorphism (8—10 kb and 800°C), which is interpreted to represent the main suturing event along the CITZ. The granulites must have been reworked to amphibolites facies conditions subsequently. Another extreme possibility is that the TBG may represent retrogressed granulites. Sarkar, Trivedi, and Gopalan (1986) reported Rb—Sr whole isochron age of 1525 ± 70 Ma and a mineral isochron age of 860 Ma, suggesting its early peak metamorphic event and the subsequent retrogression event of metamorphism. In such a scenario as described above, the concept of southerly subduction may not be valid.

The TBG shows large geochemical variations with contrasting geochemical and isotopic signatures at different places exhibiting the heterogeneous character possibly reflecting the nature of protoliths. It was suggested that the TBG might have been derived from partial melting of anhydrous, lower continental crust due to possible heating by ponded basaltic magmas (underplated) at the base of the crust (Subba Rao, Narayana, Divakara Rao, & Reddy, 1999). Geochronological constraints indicated that the TBG could be the resultant product of crustal anatexis during the ~1.6 Ga ultrahigh-temperature (UHT) metamorphism (Bhowmik et al., 2011). The c. 1.62 Ga age of magmatic emplacement obtained by Bhandari, Pant, Bhowmik, and Goswami (2011) is also identical with the age of 1.6 Ga for the UHT metamorphic event. These discrete magmatic and metamorphic events can be correlated with the formation of an early Mesoproterozoic accretionary orogen in the CITZ. A hot orogen model was postulated with a magmatic arc setting for the tectonothermal evolution of the TBG. Syntectonic and posttectonic granites intrusive into the Sausar Group metasedimentary assemblage has recently been dated to give ~955 Ma and ~928 Ma, respectively. The ~955 Ma event is suggested as the main metamorphism and D2 deformation of the Sausar Group (Chattopadhyay, Das, Hayasaka, & Sarkar, 2015).

In summary, petrological and geochronological data suggest an active continental margin setting and southward subduction of oceanic lithosphere at the southern margin of the CITZ at 1.6 Ga (Bhowmik et al., 2011). Further, from the foregoing description, the deformational history of the SSB can be related to Proterozoic thrust-nappe tectonics and the temporal change of polarity of subduction.

4.2.4 GRANULITE BELTS

4.2.4.1 Introduction
Two granulite belts occur on either side of the SSB and show different P–T–t paths of evolution and different age ranges. They are described as RKGB in the north and BBGB in the south. Distinct features of each belt are described below.

4.2.4.2 Ramakona-Katangi granulite belt
The RKGB occurs along a northerly dipping ductile shear zone at the northern margin of SSB (see Fig. 4.3). The RKGB extends over a strike length of ~240 km starting from Ramakona in the west through Khawasa and Katangi in the center and Bichia in the east. The RKGB is located to the south of GTSZ. A number of north-dipping thrust planes in the form of prominent ductile shear zones were described from RKGB (Roy & Hanuma Prasad, 2003). Syntectonic granite emplacement and down-dip stretching lineations characterize these shear zones. The RKGB is a distinct rock association consisting of mafic granulites, garnetiferous amphibolites, garnet–cordierite granulites, and porphyritic charnockite that occur as rafts and lenses within Tirodi gneisses (TBG). The granulite facies rocks are tectonically interleaved with Sausar supracrustals and younger granites. P–T estimates of 9–10 kbar and ~800°C indicate the existence of high-pressure granulites and upper amphibolite to granulite facies metamorphism with a clockwise metamorphic P–T path in the rocks of RKGB (Bhowmik, Wilde, Bhandari, Pal, & Pant, 2012). Combined petrology, mineral chemistry, and geobarometric studies reveal complete metamorphic history, indicating that the RKGB represents the reworked deep crustal section. Reworking is reflected by an isothermal decompression of 6–4 kbar at 750–800°C indicating the denudation of 13–17 km of crustal material, which was previously thickened through collision. These features were interpreted as an event

of early continental subduction (1.6 Ga) within the CITZ, followed by continent—continent collision, which is correlatable in age with Grenvillian orogeny. DSS profiles and gravity modeling results across the SSB reveal the presence of northerly dipping reflectors together with thicker and high density lower crust beneath the SSB. These features are comparable with the existence of steep, north-dipping ductile shear zones characterized by intense mylonitic fabrics in the region separating the SSB and the RKGB. These evidences and the presence of high-pressure granulites of the RKGB suggest a possible northerly subduction at around 1.0 Ga. However, further geological and geochronological studies are required to understand the geodynamics of the origin of the RKGB and the adjacent rock units in the region as well as the CITZ.

4.2.4.3 Balaghat-Bhandara granulite belt

The BBGB occurs at the southern margin of SSB and along the CISZ (see Fig. 4.3). The belt trends NE−SW and extends over a strike length of ∼190 km with variable width between 4 and 20 km. The BBGB is bound by the presence of northerly dipping shear zones on either side (Ramachandra & Roy, 2001). The entire lithological assemblage of the BBGB can further be divided into four distinct components: (1) a large part of the domain is occupied by a migmatitic felsic gneiss (locally with garnet) of tonalitic to granodioritic composition; (2) enclaves of garnet−cordierite gneiss, iron-formation granulite, quartzite, aluminous granulite, and felsic granulite occur within the felsic gneiss; (3) a mafic−ultramafic magmatic suite of metagabbro, metanoritic gabbro, metanorite, and metaorthopyroxenite occurs as concordant sheets in the felsic gneiss; and (4) mafic dykes of metagabbronorite and meta-olivine gabbro and amphibolites. They occur as detached lensoid bodies (0.5 × 2.0 km) within the intensely deformed migmatitic Amgoan gneisses. These rocks are enveloped in a strong ductile shear zone deformation that resulted in south verging isoclinal folds with a strong mylonitic fabric. A southerly tectonic transport at relatively high temperatures is indicated by the presence of sigma-type asymmetrical orthopyroxene porphyroclasts and the southerly vergence. The stretching lineations are well developed on the steeply dipping mylonite planes with a down-dipping nature. The lineation data, in conjunction with shear bands, suggest the top to the south. The shear sense indicators along the shear zone in the south also suggest upthrusting of BBGB over Amgoan gneisses or an extended part of TBG.

The major rock units in BBGB include two pyroxene granulites, charnockites, metaultramafites, BIF, cordierite granulite, and quartzites. The Bhandara part of the BBGB is represented by a large two-pyroxene granulite body elongated along ENE−WSW, extending along strike for about 5 km, with a maximum width of 1.5 km. It is locally associated with migmatitic gneisses and a thin band of quartzite. The two-pyroxene granulite is medium to coarse grained and shows uniform gabbroic composition with occasional relic ophitic and subophitic texture. The main foliation trend varies from NE−SW to ENE−WSW with steep southerly dips. Some local sinistral shear zones lying subparallel to the main foliation show protomylonitic character as evidenced by recovery and recrystallization of plagioclase and pyroxene grains, as well as bending of twin lamellae (Bhowmik & Roy, 2003). Thin gabbro−anorthositic gabbro layers of a few cm thick together with rare segregations of pyroxene up to 10 cm in size occur at places. Both the granite gneiss and low-grade meta-sediments of the SSB together at the northern margin of the BBGB were mylonitized showing development of protomylonitic fabrics. The granite gneiss at the southern margin of the BBGB also shows mylonitic and phyllonitic character, implying that both the northern and the southern margins of the BBGB are highly tectonized.

A full spectrum of rocks of the BBGB is best exposed in the Arjuni-Balaghat area and comprises a sequence of quartzite, BIF, metapelite, two-pyroxene granulite, charnockite and charnockitic gneiss, and migmatitic and granitic gneiss. Petrological studies suggest melt generation and P–T estimates of ~8 kbar and 700°C. Paleoarchean ancestry and mature crustal source were also recognized as the source for these gneisses. These are all concordant to the CISZ, which are mostly retrogressed. These granulites were metamorphosed under lower crustal, UHT conditions along a counterclockwise pressure–temperature (P–T) path at 1.6 Ga (Bhandari et al., 2011).

The age of peak metamorphism has been dated as 2672 Ma for charnockite, while the cooling event (represented by the development of coronal garnet) was around 1400 Ma from a mafic granulite of the BBGB (Abhijit Roy et al., 2006). The BBGB was interpreted as an allochthonous tectonic sheet between the low-grade (greenschist-facies) Sausar Group of rocks in the north and the cratonic domain of low- to medium-grade felsic gneisses of the Amgaon gneissic complex, south of the CISZ (Bhowmik & Roy, 2003). Geothermobarometric data and interpretation of reaction textures from the rocks of BBGB indicate the earlier UHT granulite–facies metamorphism at P–9 kbar, T–950°C, followed by a subsequent near-isobaric cooling event, which was terminated at 700°C. The electron microprobe dating of monazites yielded an age of 2040–2090 Ma for the UHT metamorphic event. The tectonothermal history of the granulites of the BBGB contrasts sharply with that from the northern granulite facies rocks of RKGB, and the latter is characterized by a clockwise P–T trajectory of possible Grenvillian age. Both the RKGB and the BBGB are separated by the relatively low-grade metamorphic SSB metamorphosed at c. 1000 Ma Grenvillian event (Pandey et al., 1998).

Different views were expressed for the origin and the tectonic setting of the BBGB. The BBGB is considered to be a part of oceanic crust and the belt represents a suture zone implying collisional tectonic processes (Yedekar et al., 1990). In contrast, based on the geochemical characteristics of mafic granulites, Ramachandra and Roy (2001) argued that the BBGB represents only intracontinental shear zone. They have also correlated BBGB with granulites of Gopalapatnam and Karimnagar belts located on either side of Godavari rift zone suggesting Archean ages of 2.6 Ga (Rajesham, Rao, & Murti, 1993). However, the relation between the BBGB and Karimnagar granulite belts remains unresolved due to lack of precise age data and detailed geological mapping.

It is significant to note that the recent studies revealed new constraints of metamorphism and geochronology. The amphibolites and granulites of BBGB display characteristic clockwise metamorphic P–T paths, implying a continent–continent collision setting. Using monazites from metamorphic overgrowths and metamorphic recrystallization domains from the felsic granulite of BBGB the metamorphism is dated at 1525–1450 Ma (Bhowmik, Sarbadhikari, Spiering, & Raith, 2005). Using geochronological and metamorphic constraints, they interpreted the metamorphic P–T conditions of 9 kbar, 950°C as the UHT rock assemblage from the BBGB. Such high temperatures are consistent with the predicted biotite dehydration–melting reactions in both the felsic and aluminous granulites. The results provided the first documentation of UHT metamorphism and Palaeo- and Mesoproterozoic metamorphic processes in the CITZ. They have also established the allochthonous nature of the BBGB and two distinct granulite–facies tectonothermal events at the southern periphery of the CITZ.

The tectonothermal history of the granulite complex of the BBGB contrasts sharply with that of RKGB. The latter is characterized by a clockwise P–T trajectory of possible Grenvillian age. The two granulite domains are separated by the greenschist–amphibolite facies Sausar Group, which was metamorphosed during the c. 1000 Ma Grenvillian event (Lippolt & Hautman, 1994). It must also be noted that the latter deformations in the rocks of BBGB are also recorded in the Sausar Group of rocks. These observations indicate the following possibilities: (1) the deformation in both SSB and BBGB is of Grenvillian age, (2) the two granulite domains of RKGB and BBGB were juxtaposed with the SSB during the same collisional event at 1.0 Ga along the CITZ, and (3) the BBGB granulites were exhumed to shallow crustal levels during the same period, and (4) the Grenvillian orogeny appears to be responsible for the formation of the Indian subcontinent through collision of the BKC and BC along the CITZ. On a larger scale, the CITZ appears to record nearly 1000 Myr of multistage crustal evolutionary history during the Proterozoic period (Bhowmik et al., 2011).

4.2.5 TECTONIC SYNTHESIS

The CITZ extends from the west coast up to Shillong plateau gneissic complex in the northeast through CGGC in the center. However, detailed studies were mostly focused in the central sector probably due to widespread exposures with fewer cover sequences and relatively easier accessibility. All the tectonic models were proposed mostly based on the combined field, petrological, monazite chemical and SHRIMP U−Pb zircon dating in association with geophysical data sets available from the central sector of the CITZ.

Different tectonic models were proposed for the evolution of the CITZ. A southward-dipping subduction was first proposed, according to which Bundelkhand continent was subducted beneath the Bastar craton around 1.5 Ga producing arc-related magmatism in Sakoli and Dongargarh basins south of CISZ (Yedekar et al., 1990). The BBGB was considered as obducted oceanic crust along the CISZ, which was also supported by gravity signatures (Mishra, Singh, Tiwari, Gupta, & Rao, 2000). In contrast, the concept of northward subduction has gained importance by later geological and geophysical investigations (Acharyya & Roy, 2000; Mall, Reddy, & Mooney, 2008; Mandal, Sen, & Vijaya Rao, 2013; Roy & Hanuma Prasad, 2003).

4.2.5.1 Geophysical signatures across the CISZ

In view of the geodynamic significance of the CISZ, which is crucial in understanding the tectonic evolution of the CITZ, several geophysical techniques were applied to understand the nature of the CISZ. DSS results along a NW−SE profile across the CISZ delineated contrasting seismic reflection pattern and the tectonic features (Mall et al., 2008). The tectonic model predicts the Moho offset, exhumation of granulite belts surrounded by Tirodi gneisses, just north of the CISZ, and the geometry of shear zones to be listric in nature (Fig. 4.12) On either side of the CISZ, distinct crustal layers were interpreted with their variations in dipping geometries, consistent with the geometry of shear zones and associated lithological assemblages. Relics of oceanic lithosphere were also inferred beneath the CISZ involving northward subduction in the tectonic model. These features are consistent with the interpretation that the lower crust consists of metamorphosed subducted oceanic lithosphere together with the mantle wedge and arc-magma components (Naganjaneyulu & Santosh, 2010). Gravity modeling results show that all the crustal layers of the Bastar craton consistently dip northward just south of the CISZ supporting northward subduction (Nageswara Rao, Kumar, Singh, Prabhakar Rao, & Mall, 2011). Gravity results also suggest the increase of density form south to north, implying that the deeper crust was exhumed in the form of thrust stack including the granulites in the form of tectonic slivers. The subduction polarity is

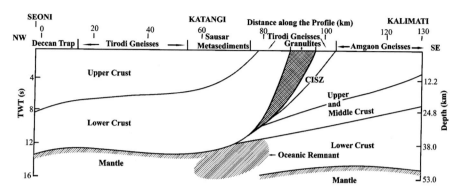

FIGURE 4.12

Interpreted seismic reflection along the Seoni-Katangi and Katangi-Kalimati transects of the N–S profile, showing the lithological assemblages, crustal layers, and the oceanic remnants associated with mantle around the CISZ.

Adapted from Mall, D.M., Reddy, P.R., & Mooney, W.D. (2008). Collision tectonics of the central Indian suture zone as inferred from a deep seismic sounding study. Tectonophysics, 460,116–123.

further supported by geological features such as: the northward dipping geometry of CISZ and other shear zones, Moho offset and overthrusting of SSB to south.

An advanced technique of common reflection surface stack method was applied to the multi-fold deep seismic reflection data acquired across the CISZ (Reddy et al., 1995), which led to the demarcation of well defined Moho and the existence of different crustal blocks (Mandal et al., 2013). The crustal blocks show distinct reflection fabrics on either side of the CISZ with a Moho offset by ~8 km, strongly supporting the northward subduction. Further, the features like oppositely dipping reflection fabrics, a bipolar gravity anomaly together with exposed geologic features across the CISZ suggest collisional tectonics and sinistral transpressional tectonic regime. The seismic sections obtained are comparable with those of well established many younger Phanerozoic collisional zones such as Alps and Pyrenees. The reasons for the seismic reflectivity could be the following: (1) sills of igneous parentage, (2) fine scale metamorphic foliation, (3) mylonite zones, (4) lithological layering, and (5) fluid traps. The seismic section was also interpreted in terms of multiple tectonic episodes; the earlier event during 1.6–1.5 Ga and the other at 1.1–1.0 Ga for the evolution of the CISZ.

The two granulite belts (RKGB and BBGB), fringing the SSB, had different crustal evolutionary histories at least up to 1100 Ma. The northern granulite (RKGB) belt is marked by a clockwise P–T–t metamorphic path. The peak metamorphism was associated with a strong compressional regime and high P and T conditions, representing the existence of a thicker crust (Abhijit Roy et al., 2006). Although the exact time of this event is not precisely known, but has been estimated to have occurred prior to 1100 Ma. The high-pressure granulites are believed to have resulted from the collision between the BKC in the north and Bastar craton in the south. It was followed by a steep isothermal decompression, which was related to the sudden collapse of the earlier thickened crust by strong extensional tectonics during either 1100 Ma or slightly earlier than 1100 Ma. The peak metamorphism in RKGB is constrained at 9–10 kbar and 750–850°C with a clock wise P–T, reflecting a major crustal thickening due to continental collision, followed by rapid thinning (Bhowmik et al., 2012).

The recent geochronological data obtained from robust techniques reveal that the UHT metamorphic event was at 1.6 Ga during the subduction−collision tectonics along the CITZ. Further, the duration of the progressive metamorphic event was estimated to be round 65−70 Ma, on the basis of SHRIMP zircon dating from the main metamorphic event in BBGB (Bhowmik, Alexanderwilde, Bhandari, & Sarbadhikari, 2014). Estimated geotherm derived from xenoliths in the CITZ suggests that the temperatures at the Moho could be 1050−1100°C (Dessai, Peinado, Gokarn, & Downes, 2010).

The rocks from TBG yielded Rb−Sr whole rock isochron age of 1525 ± 70 Ma with whole rock-mineral isochron of ∼860 Ma (Sarkar et al., 1986). The former marks the main phase of regional amphibolite facies metamorphism of Sausar Group and the later represents terminal thermal overprint. However, recent SHRIMP U−Pb zircon age data of 1618 ± 8 Ma from TBG was interpreted as magmatic crystallization of its protolith. Accordingly, the two tectonic units with contrasting isotopic compositions must have been sutured along the CISZ by an arc-continent collision between 1.57 and 1.54 Ga. Bhowmik et al. (2011) recently proposed an active continental margin setting for the southern margin of the CITZ at 1.6 Ga, owing to southward subduction of oceanic lithosphere beneath the Bastar craton, leading to the development of a magmatic arc (TBG) and a coeval back-arc basin (BBGB). Charnockite from RKGB yielded SHRIMP U−Pb zircon date as 938 ± 3 Ma indicating its magmatic crystallization and synmetamorphic emplacement of charnockite magma during the extensional collapse of thickened orogen (Bhowmik et al., 2012). The monazite from RKGB also recorded peak granulite facies metamorphism at 1.04 Ga, leading to the conclusion that the RKGB and the adjacent Sausar supracrustals must have witnessed same collisional orogen. Sausar peak metamorphism is also dated at ∼0.95 Ga from syntectonic granites (Chattopadhyay et al., 2015). Based on the available age data, it was concluded that the collisional process must have occurred along the CITZ between 1.04 and 0.94 Ga.

The limestones, BIFs, and the manganese-bearing horizons in association with quartzites and metacarbonates within the Mesoproterozoic SSB along the southern CITZ might represent a mixture of accreted oceanic and continental margin sediments (Naganjaneyulu & Santosh, 2010). The HP granulites and associated mafic and ultramafic sequences within the CITZ suggest subducted ocean floor as well as mantle wedge components. The protoliths of the orthogneisses, charnockites, and related rocks, which are widely distributed in this zone, correspond to TTG and adakitic magmas derived by slab melting during subduction. The K-rich granites mark postcollisional plutons emplaced along terrane sutures. These geological characteristics and lithological sequences provide robust evidence for the sutures within the CITZ to be zones of ocean closure where subduction−accretion−collision processes played a critical role. The identification of two spatially associated but temporally separate metamorphic domains and contrasting tectonothermal history in the southern margin of the CITZ are significant because they imply the presence of amalgamated crustal domains in the CITZ.

4.3 CITZ: EASTERN SECTOR

4.3.1 INTRODUCTION

The region lying between the Long. 83 and 88°E. and Lat. 22 and 25°N is considered here as the eastern sector of the CITZ, which partly overlaps the central sector (Fig. 4.13). The major

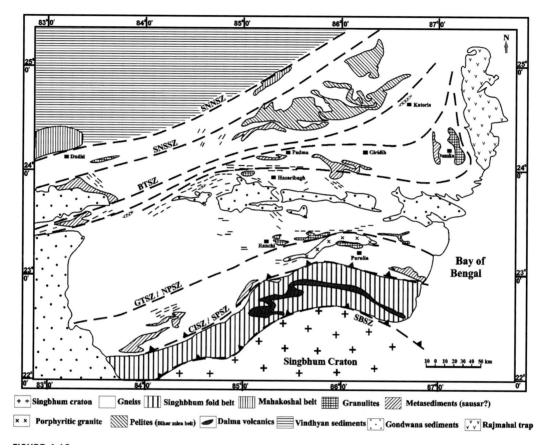

FIGURE 4.13

Geological and tectonic map of the eastern sector of the CITZ. *SNNSZ*, Son-Narmada north shear zone; *SNSSZ*, Son-Narmada south shear zone; *BTSZ*, Balrampur-Tattapani shear zone; *GTSZ*, Gavilgarh-Tan shear zone; *CISZ*, Central Indian shear zone; *NPSZ*, North Purulia shear zone; *SPSZ*, South Purulia Shear Zone; *SBSZ*, Singhbhum shear zone; *SFB*, Singhbhum Fold belt; *Sc*, Singhbhum craton.

Compiled from Acharyya, S.K. (2003). The nature of Mesoproterozoic Central Indian Tectonic Zone in central, eastern and northeastern India. Gondwana Research, 6, 197–214; Mazumdar, S.K. (1988). Crustal evolution of the Chhotanagpur gneissic complex and the mica belt of Bihar. Precambrian of the Eastern Indian shield Geological Survey of India Memoir, 8, 49–83.

geological units include: the vast spread of CGGC, the SFB, and the Sighbhum craton. The well-known Bihar mica belt occurs in the northern part, while the Damodar rift structure with coal bearing Gondwana sediments occurs in the central part of the region. The entire terrain is traversed by a number of near E–W—trending subparallel major shear zones. The shear zones in the southern part show a general convexity to north while the shear zones in the northern part show convexity to south, which are separated by the Damodar rift. Isolated granulite complexes with different sizes, shapes, and geometries are scattered all over the region. The details of each geological unit are given below.

4.3.2 CHOTANAGPUR GRANULITE-GNEISS COMPLEX

4.3.2.1 Introduction

The eastern sector of the CITZ is vastly occupied by Precambrian granulite-gneiss complex popularly known as CGGC due to its peak metamorphism. The CGGC extends over a large area (500×200 km) and is bound by the SPSZ in south and the NSL in the north. The N−S trending Rajmahal trap and Bengal basin mark the eastern boundary of the CGGC. Although both the central and eastern sectors of the CITZ are in continuation, the correlations of different geological and structural features is challenging in view of their capping with younger sediments.

The E−W trending CGGC mainly constitutes gneisses, migmatites, and isolated narrow belts of supracrustals and enclaves of boudin-type granulites and granites, which were metamorphosed to amphibolite facies to granulite facies grade representing deeper crustal section (Mahadevan, 2000). Several occurrences of gabbro−anorthosite complexes were reported from different parts of CGGC, associated with upper amphibolite to granulite facies rocks. Considering the characteristics such as: (1) the presence of khondalites, charnockites, pyroxene granulites; (2) extensive development of insitu migmatites; (3) the presence of augen gneisses containing k−feldspar and sillimanite; (4) pelitic assemblages like sillimanite, garnet, k−feldspar, staurolite etc.; and (5) the existence of calc−silicate rocks; the CGGC can well be defined as a high-grade metamorphic terrane that has attained granulite facies metamorphism (Mazumdar, 1988). The characteristics of CGGC are well correlatable with those of Limpopo metamorphic belt of South Africa. In view of the above, the terrain hereafter will be described as CGGC.

The important rock types include granitic gneisses, migmatites, and granites with enclaves of granulite facies rocks and a few basic intrusives. The stromatic migmatitic gneisses are widespread in the region. The granite gneiss complex is a composite mass, consisting mainly of granite gneiss, migmatites, and massive granite with enclaves of para- and orthometamorphics; dykes of dolerite; and innumerable veins of pegmatite, aplite, and quartz. The granite gneiss is further divisible and often mappable as biotite granite gneiss, porphyroblastic, and augen gneiss and charnockitic gneiss. The migmatites are heterogeneous mixtures of hornblende−biotite gneiss, sillimanite−biotite gneiss and K−rich leucocratic granite. The massive granite is also K−rich and clearly a later-phase intrusion into the granite gneiss and migmatites; in the north, they commonly form diapirs and are enclosed by quartzite and mica schist (Sarkar, 1982).

Detailed mapping in a small segment, around Bero-Saltora region in the east−central part of the eastern sector exhibits E−W trending fabrics typical of the CGGC (Fig. 4.14). Variably deformed gneisses and migmatites, metamorphosed to granulite facies conditions, consist of enclaves of mafic granulites and calc−silicates as discontinuous bands and lenses. The earliest recognizable fabric is the metamorphic gneissosity, which has been isoclinally folded. Subsequent deformation gave rise to noncylindrical folds often resembling sheath fold geometry. The fabrics show steep north-dipping mylonitic fabrics with pronounced strike-slip movement (Maji et al., 2008). The stretching lineations are moderately plunging to west. However, magmatic foliation is well preserved in the core of anorthositic pluton, while they are intensely sheared at the margins. The E-W−trending, north-dipping gneissic fabrics associated with Grenvillian metamorphism and ductile shearing with reverse sense of movements on the shear planes are all correlatable with the SSB and associated gneisses of the central sector of the CITZ.

The granulite enclaves and associated gneissic rocks show diversity in mineral assemblage with a spectrum of phase chemical characteristics. High Mg−Al granulite consists of

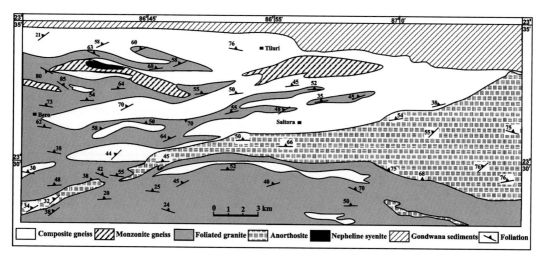

FIGURE 4.14

Detailed geological and structural map of Bero-Saltora area, representing a typical small segment of the CGGC.

After Maji, A. K., Goon, S., Bhattacharya, A., Mishra, B., Mahato, S., & Bernhardt, H. J. (2008). Proterozoic polyphasemetamorphism in the Chhotanagpur Gneissic Complex (India), and implication for trans-continentalGondwanaland correlation. Precambrian Research, 162, 385-402.

quartz—sillimanite—cordierite—orthopyroxene—garnet—rutile—plagioclase in different combinations. Often, rutile appears as a major mineral in some microdomains. Mafic granulites contain orthopyroxene, clinopyroxene, plagioclase, hornblende, ilmenite, biotite with or without garnet and quartz in different associations. The rocks, in general, show granoblastic fabric and are usually massive in appearance. However, a weak gneissic foliation is also locally observed, which is defined by alternate pyroxene-rich and plagioclase-rich layers. Migmatitic gneisses of stromatic variety consist of quartz, alkali feldspar, garnet, plagioclase, with variable amounts of biotite and minor amount of acicular sillimanite and tabular ilmenite. Large alkali feldspar grains include granular quartz and often show myrmekitic intergrowth at the peripheral regions.

One of the extensively worked and widely debated geological units in the eastern sector of the CITZ is the SFB or Singhbhum orogen, that occurs between the CGGC and the SC (Sarkar & Gupta, 2012). The SFB is bound by two important shear zones: the SPSZ or the Tamar-Porapahar shear zone to the north and the Sighbhum shear zone (SBSZ) to the south. Some researchers consider the SPSZ as the southern limit of the CITZ while some others consider the SBSZ as the southern boundary of the CITZ. However, it is considered here that the two shear zones represent the extended branches of the CISZ in the east representing a major suture zone in the form of SFB.

4.3.2.2 Shear zones

The eastern sector of the CITZ, like the central sector, shows similar tectonic frame work constituting a series of nearly east—west trending tectonic belts, separated by crustal scale shear zones (see Fig. 4.13), each of which has some distinctive geologic features (Mahadevan, 2002). The CGGC, the dominant lithological unit, constitutes a mosaic of structural blocks characterized by typical strike directions (Mazumdar, 1988). The structural gneiss domes (~150 km wide) are also common with in the CGGC, marked by the presence of near ENE−WSW trending boundary shear zones. The shear zones in the southern part show an arcuate form with convexity to north, while the shear zones in the northern part show the convexity to south.

The important shear zones in the eastern sector from north to south include: (1) the SNNSZ, (2) the SNSSZ, (3) the BTSZ, (4) the NPSZ, (5) the SSPZ, and (6) the SBSZ. The first two shear zones, SNNSZ and SNSSZ in the north, together described as NSL, continue in the eastern sector from the central sector with similar trends, geometry, and lithological characteristics. At the eastern extremity, they tend to swing toward the northeast. The geological map in the northern part of the CGGC around Chotanagpur and Dudhi (Fig. 4.15) shows the continuation of SNNSZ and SNSSZ and swinging toward the northeast.

The migmatitic gneisses are widespread in the region. Porphyritic granites and augen gneisses seem to be tectonic markers. The augen gneisses represent the transformed products of porphyritic granites with varying intensities of deformation. They are mostly related to the shear zones in the region. The porphyritic granite that occurs just north of Chotanagpur is confined to shear zones. The foliation trajectories constructed based on the available map of Mazumdar (1988), reveals the presence of broad structural domes and basins. For instance, a 30-km-wide domal structure can be visualized in the south western segment of the map and a 5-km structural basin could be inferred near Dudhi. Detailed mapping may bring out many more such domes and basins, which may be genetically connected to the development of mantle gneiss domes or mega-sheath fold structures. However, the inferences drawn are speculative until they are established in the field with more vigorous and detailed structural and petrological investigations. The northern domain constitutes a series of east−west trending subparallel shear zones dividing the domain into distinct terranes. The general geological characteristics of the CGGC are similar to other typical Precambrian high-grade terranes like the SGT and EGMB (Chatterjee, Mazumdar, Bhattacharya, & Saikia, 2007).

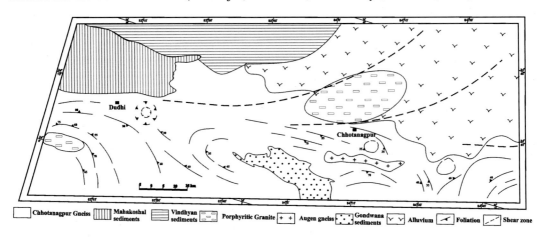

FIGURE 4.15

Geological and structural map of the northern part of the eastern sector of the CITZ. Foliation trajectories are constructed from the available structural measurements recognizing structural domes and basins (Mazumdar, 1988).

Mazumdar, S.K. (1988). Crustal evolution of the Chhotanagpur gneissic complex and the mica belt of Bihar.
Precambrian of the Eastern Indian shield Geological Survey of India Memoir, 8, *49−83.*

4.3.2.2.1 Balarampur-Tattapani shear zone

The BTSZ in the eastern sector branches into a number of shear zones and extend further east and north east (see Fig. 4.13). The northern most is well defined traversing through the southern margin of Bihar mica belt through Katoria. The middle one runs east–west up to Padma-Giridih and extends further in a N–NE direction. The southern most shear zone trends east–west via Hazaribagh and wraps around Dumka extending further northward fringing the western margin of Rajmahal traps. All these branches of shear zones that emerge from the BTSZ are restricted to the northern part of the CGGC. The shear zones are characterized by mylonites, phyllonites, and quartz veins. All the shear zones in the eastern extreme get deflected to near N–S, forming a major N-S–trending shear zone juxtaposed against the N-S–trending of the Rajmahal traps.

4.3.2.2.2 North Purulia shear zone

The NPSZ, apparently an extended part of the GTSZ, occurs to the north of SPSZ and trends ENE–WSW, just south of Ranchi. The NPSZ comprises augen gneisses and enclaves of charnock-ites and hypersthene bearing gneisses. Charnockitic rocks yielded upper intercept U–Pb zircon ages of 1624 ± 5 Ma and 1515 ± 5 Ma and lower intercept age of c. 1000 Ma. The former ages represent magmatic crystallization and metamorphic imprints while the later age reflects the age of emplacement of granites, probably suggesting the collisional event (Ray Barman, Bishui, Mukhopadhyay, & Ray, 1994). Sillimanite-hypersthene–bearing metapelite enclaves show P–T conditions of >10 kbar and $1200°C$ suggesting and substantiating a continental collision setting. Widespread granite emplacement all over the CGGC seems to be around 1.0 Ga. The Gumla-Purulia granulite belts occur in close vicinity of the NPSZ, which correlate well with BBGB of the central sector. The granulites are surrounded by the basement gneissic rocks comparable with those of TBG.

The NPSZ occurs to the south of Damodar rift structure and is believed to be an extension of the GTSZ from the central sector and extends through Jaipur and Purulia with varying width. The compiled regional scale map (Fig. 4.16) of Mazumdar (1988) shows the NPSZ distinctly with ENE–WSW trend with moderate dips to the north. The shear zone is marked by the development of pronounced porphyritic granite with high-grade enclaves of calc–granulites. Gradual transformation of porphyritic granites into augen gneisses suggests the intensity of deformation. The CGGC around the NPSZ show conformable relation with the regional east–west strike. The terrain consists dominantly of granite gneisses, migmatites, and granites (porphyritic and massive variety) with enclaves of metapelitic and metabasic granulites and intrusive basic rocks. Structural analyses from different parts of CGGC display complex fold interference patterns, which may represent sheath fold structures. The foliations in different parts of gneissic rocks, south of NPSZ, show different trends with dips mostly either to north or west. Foliation trajectories constructed from the available data exhibit broad east–west fold structures with gentle to moderate plunges to west with fold closures to east implying a major isoclinal fold structure. Several scattered and folded amphibolite bands are also mapped (Mazumdar, 1988).

4.3.2.2.3 South Purulia shear zone

The SPSZ occurs at the southern margin of the Meso- to Neoproterozoic CGGC and separates the Proterozoic SFB to the south (Fig. 4.16). It is also known as the Tomar-Porapahar shear zone and is believed to be the eastern extension of the CISZ. Mylonites, phyllonites, and cataclasites are the

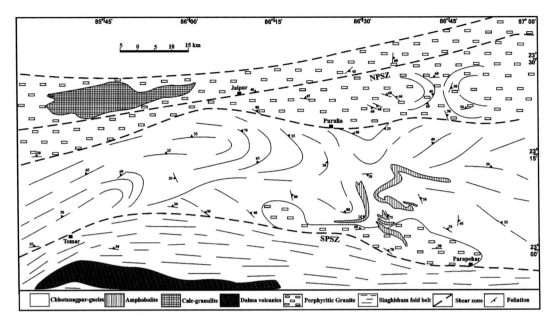

FIGURE 4.16

Geological and structural map of the southern part of the eastern sector of the CITZ. Foliation trajectories are constructed from the available structural measurements recognizing a broad isoclinal fold structure with northerly dipping axial plane (Recumbent?) *NPSZ*, North Purulia shear zone; *SPSZ*, South Purulia shear zone.

Source: Mazumdar, S.K. (1988). Crustal evolution of the Chhotanagpur gneissic complex and the mica belt of Bihar.
Precambrian of the Eastern Indian shield Geological Survey of India Memoir, 8, 49–83.

characteristic features of the SPSZ. Sedimentary associations, considered to be equivalents of Sausar supracrustals are exposed as discontinuous and isolated linear belts all along just north of the SPSZ. Migmatitic gneisses comprising both ortho- and para gneisses are the predominant rock association. The shear zone is dotted by several basic igneous bodies. The rocks to the north of the shear zone are mostly corundum–sillimanite–kyanite bearing schistose rocks, while phyllites occur dominantly to the south. Many bodies of porphyritic granites occur as a chain in the southern margin of the CGGC coinciding with the SPSZ implying the involvement of deeper tectonic processes.

4.3.2.3 Granulite belts

The high-grade granulites are mostly clustered around Purulia, Dumka, and Makrohar (see Fig. 4.13). Small bodies of anorthosite and relatively minor bodies of syenites, nepheline syenites, Rapakivi, and alkaline granitoids occur at several isolated places. A few isolated enclaves of granulites at the northern margin of CGGC show granulite facies rocks including migmatitic gneisses and charnockites. The high-grade rocks comprise concordant lenticular supracrustal enclaves of graphite-sillimanite–bearing pelites, calc–silicates and minor mafic–ultramafic and anorthositic rocks. Undeformed granite plutons are also mapped in the region.

The MGB occurs in the NW part, just south of the SNSSZ. The MGB occurs to the south of MSB in the form of enclaves surrounded by granitic rocks. The rocks in the enclaves were metamorphosed from amphibolite to granulite facies conditions (Pichai Muthu, 1990). The rock types include calc−silicates, marble, BIF, metapelites, and basic rocks and occur as mappable units in the form of small rafts along ENE−WSW direction. The presence of corundum-sillimanite−bearing meta-sediments indicates very high temperature-pressure conditions resembling those of the RKGB (Acharyya, 2003). They are intruded by gabbro−anorthosite and a granitic suite of rocks. The granulite facies metamorphism in MGB may be dated at 1.7 Ga. Similar kind of high-grade rock suite in the form of dismembered units was also recorded at the northern margin of the BSB. Both the high-grade units are in strike continuity along with lithological similarities, which may be considered as a part of the MGB.

The eastern margin of CGGC is marked by the presence of N−S trending ductile shear zone with possible sinistral kinematics truncating the E−W fabrics of CGGC and the high-grade metamorphic belts in the vicinity of Rajmahal traps (Chatterjee & Ghose, 2011). Petrographic, thermobarometric, and pseudosection analyses show that the granulite enclave suite of Purulia records a strong decompressive metamorphism from 11 to 5 kbar at 870−750°C possibly through a possible clockwise P−T trajectory. Texturally constrained monazite grains yielded an age bracket of c. 990−940 Ma from the high Mg−Al granulite, thus pinpointing the timing of decompression. The host migmatitic gneiss witnessed monazite growth at broadly the same time frame (c. 990−940 Ma), thereby implying the timing of anatexis by the same tectonothermal event affecting the granulites. The c. 990−940 Ma tectonothermal events can be correlated with the widespread mountain building activities in the Indo-East Antarctica sector during the assembly of Rodinia (Karmakar, Bose, Basu Sarbadhikari, & Das, 2011).

Granulite belts around Dumka occur in the north eastern part of the CGGC. Anorthosites are found as large crescent shaped outcrops near Dumka. The fabrics in the region get deflected from E−W to near N−S wrapping around Dumka merging with the N-S−trending shear zone abutting against the Rajmahal traps. The granulites formed in the region were formed at P−T conditions of 8−11 kbars and 650−850°C (Roy & Mukhopadhyay, 1992). It was also reported that the granulites are retrograded and migmatized.

4.3.2.4 Mica belt

A famous mica belt, popularly known as the Bihar mica belt, emplaced into a large meta-volcanic and sedimentary association, is a striking feature of the CGGC in the northern part (see Fig. 4.13). The mica belt extends in an ENE−WSW direction with two varieties of pegmatites: the older are barren and deformed with the host rocks, while the younger are mica-rich and undeformed. (Mahadevan & Maithani, 1967). Mica schists, migmatites, and granites of both porphyritic and nonporphyritic nature are important host rocks. Mica-bearing pegmatites are mostly clustered in the vicinity of nonporphyritic-type granites. Well-foliated, highly potassic granites vary from porphyritic monzogranite to nonporphyritic granite (Bhattacharya, 1988). These granites occur as domal, elliptical, and semicircular bodies, which may represent structural domes. The meta-sedimentary associations include amphibolites, quartzite, mica schist, and mica gneiss. The host rocks reached amphibolite facies conditions. All the host rocks including the basement gneisses and the meta-sedimentary associations witnessed a similar deformational history. The origin of the mica belt is considered to be related to 1000 Ma Grenvillian metamorphic event, also called a Satpura orogeny

(Mazumdar, 1988). The Bihar mica belts occur in the northernmost tectonic domain bound by shear zones. These shear zones swing to northeast and finally joining the N−S trending major shear zone at the western margin of Rajmahal traps, possibly with sinistral sense of kinematics.

4.3.2.5 Damodar rift

The E−W trending Damodar rift structure occurs in the central part of the eastern sector of the CITZ and divides the CGGC into two halves (see Fig. 4.13). It has a strike length of 375 km and a width of ∼85 km. The rift zone is represented by a series of basins hosting the Gondwana coal belt during the Carboniferous to early-Jurassic period. The rift consists of a thick pile of sediments along with mappable outcrops of granulite facies rocks that occur in the form of enclaves. Discontinuous and highly elongated narrow granulite belts occur on either side of the Damodar rift exhibiting contrasting metamorphic history (Acharyya, 2003). They occur in the vicinity of NPSZ in the south, while they are well preserved in the north eastern and northwestern parts, north of the rift structure.

The Gondwana basins occur in the form of half grabens, bound by large gravity faults at their southern margin, while a few are recorded in the northern margin (Sarkar, 1988). The faults show inward dips, developing into listric type at depth. They also occur subparallel to the tectonic grain of the basement gneisses. All the Gondwana basins are confined to the Damodar rift zone which is closely and genetically connected to the BTSZ, a major Precambrian ductile shear zone possibly with sinistral transtensional tectonics. It is possible that resurgent tectonics along a preexistng shear zone in the axial part of the graben structure must have given rise to the development of graben structure and associated Gondwana basins. This implies that it could be the result of reactivation tectonics of the BTSZ in the basement CGGC. The intrusive dolerites and lamprophyres are recorded with the Gondwana basins. The chemistry of lamprophyres indicates that they are mantle-derived and their K−Ar ages yielded at 120−105 Ma (Sarkar, Paul, Balasubrahmanyam, & Sengupta, 1980). The Damodar rift is in many ways similar to many other rift zones of the world such as the East African rift system. Isostatic gravity anomaly map shows that the rift structure is conspicuously marked by positive gravity anomaly (reaching 30 mgal and more), while the rest of the terrain of the CGGC is surrounded by negative anomaly contours conforming to the boundaries of the CGGC.

4.3.3 SINGHBHUM FOLD BELT

4.3.3.1 Introduction

The SFB, also described as north Sighbhum mobile belt or Singhbhum orogen, occurs at the southern margin of the CITZ in the eastern sector (see Fig. 4.13) (Gupta & Basu, 2000). The SFB occurs between the CGGC to the north and the SC to the south. It trends broadly east−west in an arcuate fashion with its convexity to north. The SFB extends for over 200 km along strike and is ∼60 km wide (Fig. 4.17). The southern boundary is marked by a well-known SBSZ, while the northern margin is marked by SPSZ (also described as Tomar-Porahar shear zone), which is believed to be an eastward extension of CISZ (Ramakrishnan & Vaidyanathan, 2008). The axial part of the SFB is occupied by an east−west folded Dalma ophiolite belt (DOB). The area between the Singbbhum shear zone and the DOB comprises a thick sequence of phyllites and mica schists with intercalated

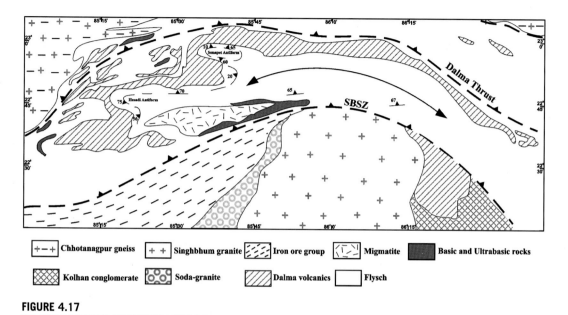

FIGURE 4.17

Geological and structural map of arcuate Singhbhum fold belt (SFB) with the Singhbhum craton to the south and the CGGC to the north. *SBSZ*, Singhbhum shear zone.

Simplified after Sarkar, A.N. (1982). Precambrian tectonic evolution of eastern India: A model of converging microplates. Tectonophysics, 86, 363–397.

bands and flat lenses of micaceous quartzites and gritty quartzites. They may represent an originally huge pile of shales with close intercalations of sandstone, clayey sandstone and micaceous sandstone. Excepting fine stratification planes, these rocks also preserve other sedimentary structures, e.g., cross-bedding (usually thin), convolute laminations, graded bedding, and current ripple laminations, which are often penecontemporaneously deformed. The rocks are intensely foliated with E—W trends and moderate to steep dips to north. Lineations show gentle plunges mostly to the east. Structures like load-casts, prolapsed bedding, sedimentary boudinage, pull-apart structures, slip bands, etc., indicate slumping during sedimentation. The lithological characters and sedimentary structures suggest an early synorogenic evolution of the sediments (Sarkar, 1982).

A thick sequence of meta-flysch occurs to the north of DOB, which extends up to the SPSZ. The meta-flysch contains phyllite and mica schist and thin and impersistent bands of quartzite, quartz-schist, carbon—phyllite, impure gray to black cherts and talc-silicate rocks. There are also numerous lensoid bodies of metamorphosed mafic and ultramafic rocks of varying size within the meta-flysch, particularly in the northwestern part. These may represent the original exotic blocks and slabs of ophiolite compositionally similar to the DOB (Sarkar, 1982). The ultramafic blocks are mostly altered to talc—schist. The margins of these masses as well as the enveloping phyllites and mica schists are also marked by intense deformation.

The domain between the Dalma volcanic belt and the SPSZ comprises pelitic and tuffaceous metasediments but differs from the rock type of the region south of Dalma belt in that they have abundant chert and black shale and larger mafic—ultramafic bodies. Impure lime stone and carbonatite, syenite, nepheline syenite, and granite plutons occur along the SPSZ. However, the rocks show an over all conformity with the structural architecture of SFB. Tungsten bearing granites occur near Kuilapal in the eastern part, which yielded a K—Ar age of 1163 Ma (Sarkar, Saha, & Miller, 1969) and a whole rock Rb—Sr age of 1638 ± 38 Ma (Sengupta et al., 1994). A fine-grained rhyolitic rock from the eastern part near Chandil also yielded Zircon Pb—Pb age of 1631 Ma and was considered as the emplacement age (Nelson, Bhattacharya, Misra, Dasgupta, & Altermann, 2007; Sarkar & Gupta, 2012).

Structural observations show that the SFB must have had an arcuate disposition originally with dominantly E—W folds with recumbent nature and subhorizontal axial planes in the central part, gradually veering to the northeast in the west and ESE to SE in the east. These fold structures are all pervasive throughout the SFB. They are, however, best developed in the DOB and the adjoining meta-flysch regions.

The SFB consists of a number of major subparallel thrusts and shear zones besides the marginal shear zones. All of them are longitudinal in nature with a few transverse faults. The longitudinal thrust zones also occur on both sides of the DOB, e.g., the "Dalma thrust" (Dunn & Dey, 1942), and also at the contact zones of meta-flysch and the molasse belt. All these thrust zones seem to have formed in the early stages of the orogenesis as they occur parallel to the broad trend of the SFB, which might have acted as the conduits for emplacement of a number of some early mafic and ultramafic rocks (Banerjee, 1975).

The broad structural pattern between the SBSZ and the DOB show E—W gently plunging antiforms and synforms with their axial planes dipping to north. The eastern sector of the SFB is dominated by N-S—trending folds and associated fabrics giving rise to a series of structural domes and basins (Sarkar & Saha, 1962). This is further supported by the recognition of steeply plunging sheath fold—like structures with arcuate hairpin curvature of the fold axis defining a culmination in the region. These features suggest a distinct possibility that they may be related to mega-sheath fold structures.

4.3.3.2 Singhbhum Shear Zone

The SBSZ, an arcuate belt at the northern contact of the SC, defines the southern margin of the SFB. The SBSZ represents a major geotectonic suture demarcating two distinct geological terranes (Sarkar & Gupta, 2012). It separates the Mesoarchean SC to the south from the Proterozoic SFB to the north. The SBSZ is well known in Indian geology as the "Copper Belt Thrust" (Dunn & Dey, 1942) or "Singhbhum thrust" or "SBSZ" mainly because of the occurrence of copper and other base metal sulfides and uranium minerals and evidence of strong thrusting in the zone. The thrusting has also affected the rocks of the molasse belt in the central sector and the northernmost part of the SC. The SBSZ is more than 25 km wide, comprising three major thrust slices in the western part, while it is a narrow zone of about one km in the central part and more than five km in the eastern part. In the central and eastern parts of the belt, the rocks are highly sheared with penetrative planes of differential movements, dipping steeply to north. The SBSZ is composite in character with arrays of thin layers of mylonites. The SBSZ, in general, comprises mainly phyllonites, chlorite—sericite schist and sheared granite gneiss and some lenticular bodies of sheared conglomerate and sandstone. The phyllonites represent the equivalents of shale and granites, while the chlorite—sericite schist is mostly sheared and altered product of basic and ultrabasic rocks. A full spectrum of mylonites and possible sheath fold structures are also recorded. The northern margin of the SBSZ is, however, marked by a string of kyanite-bearing quartzite bodies with occasional veins of massive posttectonic kyanite. The SBSZ is also dotted by many lensoid bodies of ultramafic rocks as well as with glucophane schists (?).

The SBSZ is characterized by well developed down-dip stretching lineations and folds of different geometries indicating the tectonic transport from north to south and the deformation was noncoaxial. S−C fabrics and sigmoidal curvature of foliations are conspicuous indicating the intensity and kinematic sense of shearing. Deformation along the zone is heterogeneous giving rise to variedly deformed rocks. The SBSZ also constitutes a typical tectonic melange comprising granitic mylonite, quartz-mica phyllonite, quartz-tourmaline rock with widely distributed deformed and fragmented volcanic and volcanoclastic rocks. The granitic rocks show a textural gradation from less deformed variety having coarse-to-medium−grained granitoid texture through augen-bearing protomylonite and mylonite to ultramylonite. The structural features in the field suggest successive sets of mylonitic foliation, folding of the earlier fabrics, and their truncation by the later ones resulting from the progressive thrust-type of shearing movement (Mukhopadhyay & Deb, 1995).

4.3.3.3 Dalma ophiolite belt

The DOB is a conspicuous geological and geomorphological unit in the SFB (see Fig. 4.17). It shows the tortuous pattern extending over a distance of about 200 km and forms a number of spectacular fold closures toward the east and west of Singhbhum, described as "Dalma Volcanics." The DOB varies in thickness from one to four km and is marked by the presence of thrust zones on either side. The rocks are intensely sheared at many places (Dunn & Dey, 1942). The western sector of the DOB exhibit spectacular regional folds with east-west−trending fold axes. The rocks dip steeply to north and are concordant with the adjacent sedimentary layers. At places, the DOB splits into smaller bands with conformable intercalations of several sedimentary rocks that include carbon−phyllite, bedded chert, quartzite, purple phyllite, and conglomerate with boulders and pebbles of ophiolite (Sarkar, 1982). The DOB comprises predominantly basaltic flows mixed with minor amounts of pyroclasts and tuffs, erupted under subareal conditions. Abundant darker fragments and lenses of mafic−ultramafic rocks consisting of peridotite, serpentinite, gabbro, and pyroxenite are also mapped. The ultramafics show cumulate textures. The meta-basalt is believed to have originated under submarine conditions in an early stage and occurs as a steeply dipping extensive sheet within the flysch sediments. Although the basalts are mainly extrusive, they present an apparent intrusive or subvolcanic relationship at places due to the sinuous disposition of the DOB. The volcanism in DOB seems to be violent at times as is evident from the profuse development of agglomerates and breccias in the central segment (Fig. 4.18). The sedimentary rocks occurring in the immediate vicinity are fragmental meta-graywacke, thinly bedded siltstone, carbon−phyllite, purple phyllite, and a few isolated lime stone patches. They are multiply folded with greenschist- to-amphibolite facies metamorphism with well-preserved mafic and ultramafic rocks containing serpentinites and horizons of pillow lava, interbanded with deep sea sediments. The depth of DOB was also estimated to be around 10 km, which are intercalated with mafic volcanic rocks and lensoid granite−granodiorite plutons (Rekha et al., 2011).

The post deformational structures in the Dalma volcanosedimentary pile are similar to those found else where in SFB. A thrust zone with well-developed fold structures was recognized in the vicinity of the Dalma volcanic belt around Sonepet. Profuse development of volcanic breccias and

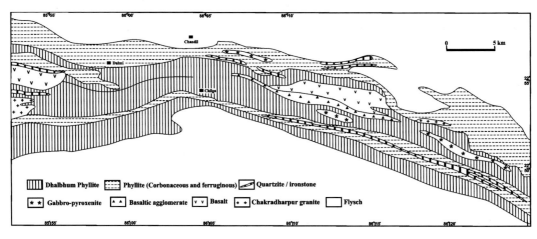

FIGURE 4.18

Detailed geological map of the central segment of the Dalma ophiolite belt showing the basaltic agglomerates.

After Gupta, A. & Basu, A. (2000). North Singhbhum Proterozoic mobile belt, eastern India—A review.

MS Krishnan Centenary Volume, Geological Survey of India Special Publication 55, *195—226.*

agglomerates were recorded along the belt. The rocks are metamorphosed to greenschist-facies conditions. The detailed structures around Sonepet (Fig. 4.19) show remarkable anticlinorium associated with intense fold structures with E—W fold axes and sinistral kinematics (Gupta & Basu, 2000). Detailed structural mapping around Sonepet and Hesadih—Ragod (Fig. 4.20) established the geometry of early fold structures and their variations along and across the fold belt. They display spectacular regional fold structures with east—west fold axes—all the structural features associated with DOB broadly suggest transpressional tectonic regime.

Based on the geochemical signatures, the Dalma volcanics were interpreted as the ocean floor basalts or arc tholeiites developed on the ocean floor and an admixture of ophiolites (Yellur, 1977). The chemistry of Dalma volcanics also shows close correspondence with komatiites and tholeites belonging to island arc type of the known Archean green stones of South Africa and Canada (Sarkar & Gupta, 2012). They show characteristic features of subduction associated with island arc type (Bose, Chakrabarty, & Saunders, 1989). The K—Ar age of volcanic rocks yielded 1547 ± 20 Ma (Sarkar et al., 1969) and the Sm—Nd date of 1.6 Ga was obtained for an ultramafic body with MORB signatures (Roy, 1998). The gabbro—pyroxenite rocks of DOB yielded Rb—Sr age of 1619 ± 38 Ma, which is considered as both emplacement as well as metamorphic age (Roy et al., 2002).

4.3.3.4 Tectonic models

The SFB, a curvilinear fold thrust belt of intense deformation, was considered to be a southerly overturned "geo-anticline" with a prominent south-verging thrust, followed by a narrow syncline to the north (Dunn & Dey, 1942). The SFB is restricted between SBSZ in the south and the SPSZ in the north with repeated reactivation tectonics over a prolonged period of time. The SFB shows typical structures of fold thrust belts of modern orogenic belts (Mukhopadhyay, 1984). The foliation

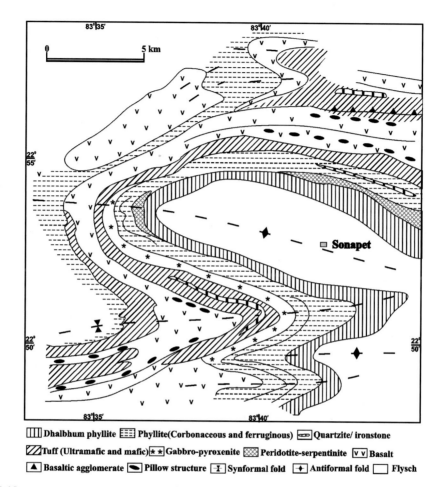

FIGURE 4.19

Detailed structural map of the Sonepet antiformal structure in Dalma ophiolite belt.

After Gupta, A. & Basu, A. (2000). North Singhbhum Proterozoic mobile belt, eastern India—A review.
MS Krishnan Centenary Volume, Geological Survey of India Special Publication 55, *195–226*.

fabrics all over the SFB resemble a typical "flower structure" as represented in implying oblique convergence and transpressional tectonic regime (Sarkar & Gupta, 2012). The Dalma volcano-sedimentary belt, bound by thrust shear zones on either side, was interpreted as back-arc setting (Bose & Chakrabarty, 1981), while Sarkar (1982) ascribed them to ophiolitic origin.

The intraplate subduction model was proposed involving northward subduction of the SC along the SBSZ, the southern boundary of the fold belt (Sarkar & Saha, 1977). Although they have considered that the Singhbhum thrust zone as the convergent plate junction, they did not consider the structural, metamorphic, and petrochemical data of Dalma ophiolite and the different sedimentary facies in the context of plate tectonics.

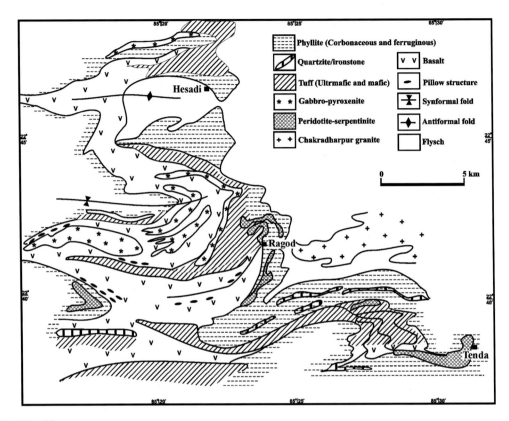

FIGURE 4.20

Detailed structural map of the Hesadi–Ragod sector of Dalma ophiolite belt.

After Gupta, A. & Basu, A. (2000). North Singhbhum Proterozoic mobile belt, eastern India—A review.
MS Krishnan Centenary Volume, Geological Survey of India Special Publication 55, 195–226.

The real plate tectonic model involving subduction and collisional processes (Fig. 4.21) was first proposed by Sarkar (1982) keeping the following points in view: (1) the existence of a linear and folded DOB, sandwiched between thick flysch wedge that contains exotic blocks and slabs of mafic and ultramafic rocks; (2) the involvement of thrusting for the emplacement of DOB; (3) the chemical signatures of basic rocks of DOB show abyssal tholeiite in composition; (4) the sediments, closely associated with the DOB, represent a deep-sea euxinic facies conditions and show an evidence of submarine gravity sliding; (5) the restriction of the SFB between two boundary thrust/shear zones; (6) display of the zonal pattern of the earlier metamorphism, showing uniform low-grade (green schist) in the ophiolite belt and distinctly higher grades (amphibolite and hornblende-hornfelse to granulite) on its either side; and (7) a bouguer gravity anomaly map reveal a series of elongated minima over the flysch belt south of the Dalma ophiolite (NGRI Map No. GPH-5, 1975), while the isostatic (Airy-Heiskanen) anomaly map (NGRI Map No. GPH-3, 1975) register positive anomalies over the CGGC and a steep fall from the DOB. In the above model, the

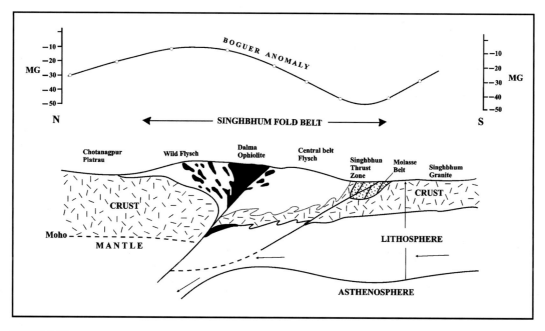

FIGURE 4.21

A schematic cross section through Singhbhum craton—Singhbhum fold belt—Chotanagpur region displaying a northward subduction model. Bouguer gravity anomaly profile computed from Map No. NGRI/GPH-5 of Natl. Geophys. Res. Inst., India, 1975.

After Sarkar, A.N (1988) Tectonic evolution of the Chotanagpur Plateau and the Gondwana basins in Eastern India: An interpretation based on supra-subduction geological processes. Gelogical Society of India, Memoir 8, 127–146.

SC and CGGC were referred to as two different microplates and their convergence took place along the SFB. Further, the geological features of the SFB are comparable with many geological features akin to well-known collision mountain belts of the world, and the plate tectonic model of converging microplates is significant for the evolution of SFB.

The SFB was described as an orogen bound by inward-dipping thrust planes and the DOB emplaced in the central core (Mahato, Goon, Bhattacharya, Mishra, & Bernhardt, 2008). The structural framework and thermal structure across the SFB is correlated with two thermal events: (1) collisional thickening of the orogen (>1.5 Ga), followed by convective removal of the lithospheric root resulting in postburial heating, and (2) syncollision and exhumation of the orogen at 1.3 Ga. The exhumation was caused by an outward-directed extrusion of the crustal slices along imbricate foreland-vergent thrusts dipping toward the center of the orogen. In the intracratonic extensional tectonic model, plume-driven ensialic basin formation and mafic magmatism followed by crustal shortening (Gupta & Basu, 2000; Mukhopadhyay, 1990) were envisaged. In these ensialic rifting models, the SFB represents the basin formed between the SC and the CGGC due to incipient rifting.

Laser ablation inductively coupled plasma mass spectrometry U–Pb zircon and U–Th–Pb monazite chemical ages of monazites from northern part of the SFB show Grenvillian ages similar to that of the southern parts of the CGGC. However, zircons from the central parts of the CGGC preserve older concordant ages (up to 2.6 Ga), while the 1.0−0.90 Ga age population is scarcely represented. The Grenvillian ages are lacking to the south of the DOB and the monazites from several samples of SFB yielded a tightly constrained age of 1.59−1.56 Ga, representing high−P metamorphic event, possibly due to northward subduction of the meta-flysch belt (Rekha et al., 2011). These two terranes must have been accreted to the CGGC during Grenvillian time resulting in vast emplacement of widespread granitoids in the CGGC.

4.3.4 **TECTONIC SYNTHESIS**

The eastern sector of the CITZ includes the widespread CGGC and the SFB. The SNNSZ represents the northern boundary while the SBSZ marks the southern boundary. The structural trends in the eastern sector lie subparallel to the trends observed in the central sector and are considered as continuous in both the sectors. Some researchers considered the SPSZ as the southern boundary of the CITZ and correlated with the CISZ (Ramakrishnan & Vaidyanathan, 2008). Some others regarded the SBSZ as the southern boundary of the CITZ (Jain et al., 1995; Rekha et al., 2011). Extension of the CISZ eastward is a tricky issue, which is crucial to have a comprehensive tectonic understanding of the eastern sector of the CITZ.

According to Rekha et al. (2011), the CISZ at its eastern extremity, splays into three E−W trending subparallel thrust zones. The southernmost is represented by south vergent SBSZ, separating the Mesoarchean SC in the south from the Proterozoic SFB in the north. The other shear zone is inferred to be located within the Meso to Neoproterozoic CGGC, while the intermediate thrust zone is marked by SPSZ. In the light of the afore said description, the following features in SFB are crucial in identifying the southern boundary of the CITZ: (1) The SBSZ represents the craton-mobile belt contact in the form of SBSZ hosting copper, molybdinum and uranium; (2) the rocks of the SFB as well as the CGGC were subjected to similar deformational history and metamorphism; (3) Recognition of Dalma volcanics as equivalents of ophiolitic assemblages; (4) Both the marginal shear zones SPSZ and SBSZ of the SFB show north ward dips and typical characteristic features of suture zones, similar to that of the CISZ; (5) Presence of intensely sheared volcano-sedimentary assemblages in both SFB and the CGGC; (6) Wide spread occurrence of mafic-ultramafic assemblages; and (7) Tectonic models involving the northward subduction of oceanic crust of SC and amalgamation of CGGC. All the characteristics described here are similar to that of the CISZ and hence the SFB in its totality can be considered as the eastward extension of the CISZ. The CISZ extends from the western margin of the Aravalli-Delhi orogen thrust contact, through the SBSZ marking the junction between the BKC and the DC, BC, and SC.

Sarkar (1988) proposed a model of convergent tectonics for the evolution of CGGC based on their characteristics in terms of structural architecture, lithological assemblages and the metamorphic history. According to him, the CGGC block represents an over riding plate and the SC beneath it during the 1600 Ma with the collision of the two continental blocks. Considering the occurrence of metasedimentary enclaves within the granitic rocks all over the region together with other characteristic features, the northward subduction of oceanic lithosphere of SC was invoked during early Proterozoic and considered subsequent long term geodynamic processes to be responsible for the present architecture of the CGGC (Fig. 4.22). The tectonic model involving the northward subduction of oceanic lithosphere comprising basalt, harzburgite, iherjolite and depleted pyrolite is

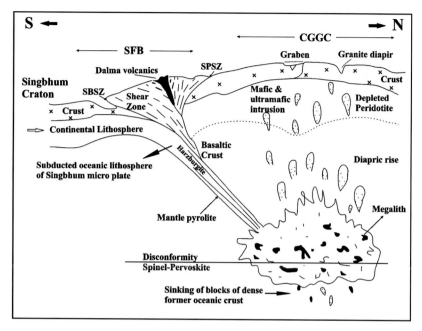

FIGURE 4.22

Tectnoic model proposed for the evolution of SFB in terms of continental collision tectonics involving the subduction of the Singhbhum microplate under the Chotanagpur plate.

After Sarkar, A.N (1988) Tectonic evolution of the Chotanagpur Plateau and the Gondwana basins in Eastern India: An interpretation based on supra-subduction geological processes. Geological Society of India, Memoir 8, 127-146.

significant in view of the limited data available at that time. This model is broadly consistent with the present concepts of tectonic evolution of the CITZ with several new data sets and the advancement in literature.

Geochemically, migmatitic gneisses of CGGC represent meta- to peraluminous and the granites show both I and S type. The CGGC, like the TBG of central sector of the CITZ, also show 1.7−1.6 Ga tectonothermal history, affected later by another major event (1.1−0.9 Ga) of high-grade Grenvillian metamorphism (Chatterjee, Crowley, & Ghose, 2008). U−Th−Pb chemical dating of xenotime in granitic gneisses indicate that Grenvillian metamorphism was pervasive through the entire northern part of CGGC. The absence of high-grade Grenvillian metamorphism in MSB and the presence of porphyritic granites in the eastern sector of the CITZ suggest that the belt remained unaffected by the terminal collision between the two blocks (Chatterjee, Bhattacharya, Duarah, & Mazumdar, 2011).

Rekha et al. (2011) proposed a two stage Proterozoic accretion model for the juxtaposition of CGGC, SFB and the SC based on the lithostructural and geochronological constraints (Fig. 4.23). The first amalgamation episode at c. 1.56 Ga is interpreted in terms of the northward subduction of the southern part of SFB (meta-flysch belt) (Mahato et al., 2008) against the microplate consisting of the DOB. The second episode of accretion is related to the c. 1.0 Ga amalgamation of the

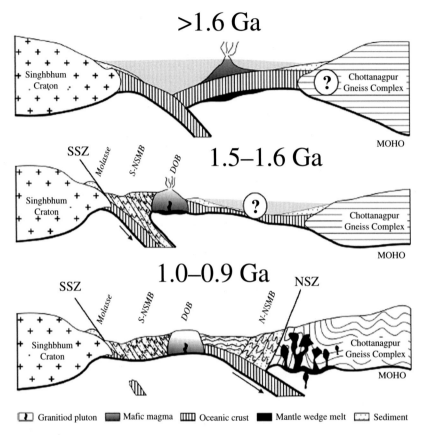

FIGURE 4.23

A schematic representation of possible Proterozoic tectonic evolution of the Singhbhum fold belt and the adjacent regions.

After Rekha, S., Upadhyay, D., Bhattacharya, A., Kooijman, E., Goon, S., Mahato, S., & Pant, N.C. (2011). Lithostructural and chronological constraints for tectonic restoration of Proterozoic accretion in the eastern Indian Precambrian shield. Precambrian Research, 187, 313–333.

CGGC with the northern part of the SFB as well as with the SC, giving rise to the emplacement of expansive syncollision Grenvillian age granitoids with in the CGGC. This event is linked to the convergence of the CGGC, SFB and the SC involving northward subduction of oceanic crust beneath the CGGC.

In summary, the tectonic setting, geological, structural and metamorphic characters together with geochemical characteristics and geochronological constraints of the SFB and the CGGC in the eastern sector of the CITZ, are closely comparable to those associated with many of the Phanerozoic collisional belts of the world (Dewey & Bird, 1970). The variable tectonic models are constrained by lack of adequate geochronological data base, full scale structural cross sections along with geophysical data set and the regional perspectives involving the entire length and breadth of the CITZ. Hence, there is a strong need for further detailed geological, geophysical studies and geochronological studies with an appropriate integration.

4.4 CITZ: NORTHEASTERN SECTOR

The CITZ extends further to north east into the Meghalaya-Shillong plateau exposing dominantly quartzo-feldspathic gneisses and high-grade migmatites, mostly covered by younger sequence. The gneissic rocks show similar characteristics of those of the central and eastern sectors of CITZ (Acharyya, 2003). The granite plutons in the region comprise xenoliths of metaigneous as well as metasedimentary rocks. The migmatitic gneisses yielded Rb–Sr whole rock isochron age of c. 1.7 Ga, while the granite plutons yielded ages ranging from 885 to 480 Ma (Ghosh et al., 1991). It is also emphasized that there is a strong imprint of Pan-African tectonics involving granite emplacement during c. 550 Ma, which has not been reported so far from the central and eastern sectors of the CITZ. The CITZ in this part does not show any evidence of 1000 Ma event (Chatterjee et al., 2011). However, Mesoproterozoic granulites of Shillong plateau were clustered at c. 1600 and 500 Ma and show similar characteristics with those temporally related rocks in the CITZ. In the light of the above, it is inferred that the Shillong plateau may have been continuous forming a part of the CITZ since Mesoproterozoic (Chatterjee et al., 2007).

4.5 TECTONIC SYNTHESIS

The structural architecture of the CITZ, compiled from the available literature together with the newly interpreted shear zone systems, shows a set of crustal-scale ENE-WSW trending shear zones throughout the entire stretch (see Fig. 4.1). The NSL constitutes two bounding shear zones namely, the SNNSZ and the SNSSZ. The other shear zones that follow from north to south include: the BTSZ that branches into three distinct shear zones in the eastern sector; the GTSZ that extends in the form of NPSZ in the eastern sector; the CISZ, which extends in the eastern sector as the SFB, which is bound by SPSZ in the north and the SBSZ in the south. All the shear zones divide the CITZ into distinct geological terranes with contrasting geologic histories. The western extensions seem to be extending westward up to the Arabian Sea, where it merges with NE–SW trending Aravalli-Delhi orogenic belt. It is speculated that both of them may coincide with the East African orogen.

Several contrasting tectonic models were proposed with wide variation on many critical issues like the polarity of subduction, timing of collision, and location of suture zone, etc., but the basic commonality was the recognition of an ancient plate tectonic regime involved in the tectonic evolution of the CITZ. Yedekar et al. (1990) were the first to propose a major Palaeoproterozoic collisional orogeny involving southward subduction along the CISZ, leading to the amalgamation of the BKC in the north and the South Indian cratons including the Dharwar, Bastar, and Singhbhum in the south. This model is supported by the following lithological, geochemical, and geophysical considerations: (1) the existence of rocks of contrasting modes of occurrence, mineral assemblages and metamorphic grades across the CISZ, with low-grade volcanogenic sequences lying to the south and the high-grade metasediments and granulites to its north; (2) the presence of calc–alkaline geochemical signatures in the Dongargarh granitoids presuming that they were related to southward subduction; (3) the interpretation of the granulites of the BBGB, occurring to the north of the CISZ as the exhumed oceanic crust during the collision orogeny; and (4) the results of seismic profile and gravity–modeling studies show a sharp variation in the crustal thickness and gravity anomaly across the CISZ (Mishra et al., 2000). Accordingly, the subduction was initiated at c. 2.4 Ga, and culminated with the continent–continent collision at c. 1.5 Ga. However, the new geological and geochronological data acquired later challenged the model. The origin of the some of the features like the evolution RKGB, MGB, BSB and the vast expanse of Mesoproterozoic (c. 1.8–1.0 Ga) granitic magmatism lying to the north of CISZ cannot be explained. Further, the structural vergence in the SSB is consistent with the model of thick-skinned, fold-and-thrust belt with hinterland toward the north. In view of the above and the failure of the southward subduction model in satisfying the present architecture of the CITZ, some other new models were invoked with the generation of new additional data.

Roy and Hanuma Prasad (2003) proposed an alternative plate tectonic model involving the northward subduction with emphasis on the collision-related RKGB, lithological aspects of various supracrustal belts and large-scale arc to continent–continent, collision-related granitic magmatism. The model presents a two-stage model for the accretion of the northern and the southern cratonic blocks (Fig. 4.24), following north-directed subduction of the oceanic crust of the BC below BKC, with Mahakoshal rift basin (c. 2 Ga) representing a back-arc rift (1). The subduction system was initiated at c. 2.2 Ga and was followed by the closure of the Mahakoshal basin at c. 1.8 Ga accompanied by calc–alkaline granitic magmatism channeled through reverse-slip ductile shear zones in a contractional tectonic regime (2). Calc–alkaline nature of the magmatism, contractional tectonic regime, and low-pressure metamorphism are akin to continental margin-arc setting. This magmatism continued up to c. 1.5 Ga. The BSB that occurs to the south of the MSB was interpreted as an intraarc belt, filled with sediments and bimodal volcanics (1 & 2). The back-arc Mahakosal basin closed at c. 1.5 Ga, as evident by the presence of syntectonic granitic rocks. This event was also accompanied by large scale mantle melting, resulting in copious hydrous ultramafic–mafic magmatism in the region.

The granitic magmatism in the region between Mahakoshal and BSBs contain slivers of medium-to-granulite–grade rocks. The collision ended at c. 1.5 Ga with the continent–continent collision (3). This event is also marked by the development of collision-related RKGB, reflecting the existence of the suture between BC and BKC. This is further supported by the presence of northerly dipping structural fabrics along the RKGB, implying the south-directed thrusting during the collision. This collision resulted in anomalous crustal thickening and attendant migmatization and lower crustal melting. During the subsequent rapid decompression 15-km-thick crust was removed and granulites were brought to middle crustal levels (Bhowmik et al., 1999). The amalgamated BC and BKC formed the basement for the Sausar supracrustal rocks of psammite–pelite–carbonate association, deposited in a continental shelf setting (4). During continued south verging thrusting, the Sausar basin might have closed at c. 1.1 Ga. In summary, according to above model, the imprints of c. 1.5 Ga events are widespread and can be seen throughout the width of CITZ. Accordingly, the c. 1.5 Ga event is represented by collisional orogeny and terminal suturing, which stitched the southern BC and northern BKC. The absence of Pan-African imprints in the other regions of CGGC was interpreted that the CGGC largely escaped this event (Maji et al., 2008).

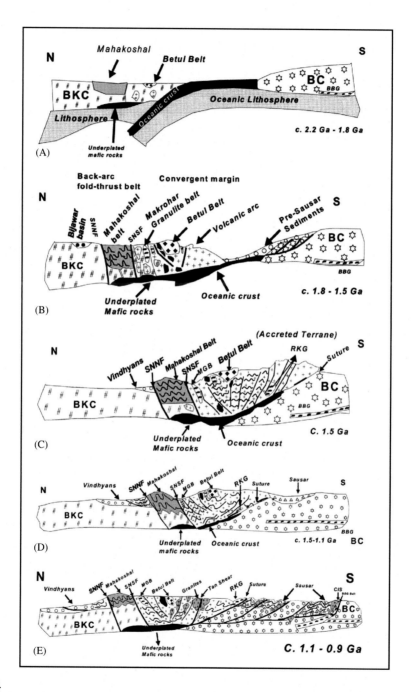

FIGURE 4.24

A tectonic model for the evolution of the CITZ involving northward subduction through different temporal stages.
After Roy, A. & Hanuma Prasad, M. (2003). Tectonothermal events in Central Indian Tectonic Zone (CITZ) and its implications in Rodinian crustal assembly. Journal of Asian Earth Science, 22, 115–129.

Consistent with a plate tectonic regime, Acharyya (2003) invoked another kind of tectonic model with some temporal variations in the polarity of subduction. He subdivided the entire CITZ broadly into two major longitudinal tectonic belts: northern and southern belts. The MSB represents the northern belt, while the SSB together with CGGC represents the southern belt. According to him, the earliest tectonic event in CITZ must have been the closure of Mahakoshal rift by southward subduction leading to the collision together with the emplacement of sheet like bodies of calk—alkaline granitoids at around 1.7 Ga. While the MSB represents an uplifted block by resurgent tectonics, the Vindhyan basin located to its north might form a foreland-type basin. The presence of rocks with high P—T conditions from the southern belt reflect the continental collision around 1.6—1.5 Ga. Based on the high-grade granulite facies metamorphism, exhumation, and shearing of basement gneisses such as TBG, the age of 1.4 Ga was considered to represent the age of continental collision. In the light of these observations, Acharyya (2003) argued that the concept of CISZ representing a Mesoproterozoic suture with the exhumed oceanic crust and southward subduction may not be valid. According to him, there must be a switch over of the polarity of subduction from southward to northward during around 1.0 Ga, and the rocks must have witnessed renewed nappe-thrust tectonics marking the tectonic transport from north to south.

The concept of southward subduction of oceanic lithosphere was also supported by Bhowmik et al. (2011) with an active continental margin setting for the southern margin of the CITZ at $1 \cdot 6$ Ga resulting in a magmatic arc. The southward subduction resulted in the development of TBG in the central and northern domains. Mantle melting and mafic magmatic accretion produced lower crustal UHT granulite facies metamorphism, attendant crustal anatexis, and the formation of the BBGB in the form of a coeval back-arc basin. These two tectonic units with contrasting isotopic compositions were sutured along the CISZ by an arc-continent collision at 1.6 Ga (Bhowmik et al., 2012), giving rise to the Proto-Greater Indian landmass.

After significant advancement in terms of geological and geophysical aspects related to understanding the evolution of the CITZ, Naganjaneyulu and Santosh (2010) proposed a double sided subduction history along the CITZ based on the identification of mafic—ultramafic layered intrusives as highly conductive bodies. This was further supported by the identification of a two-layered lithospheric mantle structure, a highly resistive nature on the top, and a low resistive nature at the bottom (Patro & Sarma, 2009). These results are consistent with the recognition of anomalous conductive features at mid-lower crustal depths, which might be due to partial melts and fluids (Dhananjayanaidu & Harinarayana, 2009; Gokarn, Rao, Gupta, Singh, & Yamashita, 2001). The results of the magnetotelluric profiles across the CITZ reveal low resistivity in the upper crust that probably correspond to the TTG and felsic accretionary complexes. Similar features were also recorded along the shear zones of NSL that represents accreted high-pressure ultramafic cumulates based on xenoliths data (Dessai et al., 2010). The best example of double- sided subduction history in modern times is the on-going subduction—accretion process in the western Pacific region. The process resulted in the development of paired collision type and Pacific type orogens giving rise to high grade metamorphic belts and the high-pressure and UHT assemblages.

Magmatic zircons from tuffs of Chattisgarh basin, located south of the CITZ, yielded SHRIMP U—Pb zircon ages 990 ± 23 Ma to 1020 ± 15 Ma (Patranabis-Deb, Bickford, Hill, Chaudhuri, & Basu, 2007). In contrast, magmatic zircons derived from tuffs of Vindhyan basin, north of the CITZ, yielded SHRIMP U—Pb ages of 1628 ± 8 Ma to 1631 ± 5 Ma (Ray, Martin, Veizer, & Bowring, 2002). Both the above magmatic events may be related to far field effects to c. 1.6 and

1.0 Ga tectonothermal events with in the CITZ. It is possible that similar magmatic and metamorphic events have been recorded all along the CITZ, suggesting that the orogen constitutes reworked early Mesoproterozoic crust (Chatterjee & Ghose, 2011; Karmakar et al., 2011). From the foregoing description, it can be concluded that the final amalgamation between the BKC and BC must have occurred between 1.06 and 0.94 Ga collisional orogeny along ~1500 km long orogen of CITZ, witnessing multistage crustal evolutionary history in the Proterozoic.

Two events of ocean closure involving the accretion of the Eastern Indian Precambrian terranes at 1.55 and 1.0 Ga along the CITZ is broadly similar to those advocated by Acharyya (2003) and Roy and Hanuma Prasad (2003). Apparently, terrane accretion processes along the CITZ were fairly uniform along the CITZ in spite of the large number of disparately evolved lithostratigraphic units involved within and neighboring the CITZ (Rekha et al., 2011). The lithological association of limestones, BIFs, and the manganese-bearing horizons in association with quartzites and metacarbonates within the Mesoproterozoic Sausar Group occurring along the southern domain of the CITZ probably represents a mixture of accreted oceanic and continental margin sediments (Santosh, 2012). Manganese-bearing chert horizons have been widely reported from the Jurassic accretionary complex in southwest Japan and are thought to constitute markers to identify ocean plate stratigraphy and their imbrication within an accretionary belt. Thus, the geological characteristics and lithological sequences within the CITZ are typical of ancient subduction—accretion—collision tectonic models.

LIST OF ABBREVIATIONS

ADOB	Aravalli-Delhi Orogenic Belt
BBGB	Bhandara-Balaghat Granulite Belt
BC	Bastar craton
BKC	Bundelkhand craton
BSB	Betul Supracrustal Belt
BTSZ	Balarampur-Tattapani shear zone
CGGC	Chotanagpur granulite-gneissic complex
Ch.B	Chattisgarh Basin
CISZ	Central Indian shear zone
CITZ	Central Indian tectonic zone
DOB	Dalma Ophiolite Belt
EGMB	Eastern Ghats Mobile Belt
GTSZ	Gavilgarh-Tan shear zone
MGB	Makrohar Granulite Belt
MSB	Mahakoshal Supracrustal Belt
NPSZ	North Purulia shear zone
NSL	Narmada-Son lineament
RKGB	Ramakona-Katangi granulite belt
SBSZ	Singhbhum shear zone
SC	Singhbhum craton
SFB	Singhbhum Fold Belt
SGT	Southern Granulite Terrane
SNNSZ	Son-Narmada north shear zone

SNSSZ	Son-Narmada south shear zone
SPSZ	South Purulia shear zone
SSB	Sausar Supracrustal Belt
TBG	Tirodi biotite gneiss

REFERENCES

Acharyya, S. K. (2003). The nature of Mesoproterozoic Central Indian Tectonic Zone in central, eastern and northeastern India. *Gondwana Research*, *6*, 197−214.

Acharyya, S. K., & Roy, A. (2000). Tectonothermal history of the Central Indian Tectonic Zone and reactivation of major faults and shear zones. *Journal of the Geological Society of India*, *55*, 239−256.

Banerjee, A. K. (1975). On the evolution of the Singhbhum nucleus, eastern India. *Quarterly Journal, Geological Mining and Metal-lurgical Society of India*, *47*, 51−60.

Bhandari, A., Pant, N. C., Bhowmik, S. K., & Goswami, S. (2011). 1.6 Ga ultrahigh temperature granulite metamorphism in the Central Indian Tectonic Zone: Insights from metamorphic reaction history, geothermobarometry and monazite chemical ages. *Geological Journal*, *45*, 1−19.

Bhattacharjee, D., Vikrant Jain, Chattopadhyay, A., Biswas, R. H., & Singhvi, A. K. (2016). Geomorphic evidences and chronology of multiple neotectonic events in a cratonic area: Results from the Gavilgarh Fault Zone, central India. *Tectonophysics*, *677−678*, 199−217.

Bhattacharya, B. P. (1988). *Sequence of Deformation, Metamorphism and Igneous Intrusions in Bihar Mica Belt* (pp. 113−126). Geological Society of India Memoir 8.

Bhowmik, S. K., Alexanderwilde, S., Bhandari, A., & Sarbadhikari, A. B. (2014). Zoned monazite and Zircon as monitors for the thermal history of granulite terrances: An example from the Central Indian Tectonic Zone. *Journal of Petrology*, *55*, 585−621.

Bhowmik, S. K., Bernhardt, H. J., & Dasgupta, S. (2010). Grenvillian age high-pressure upper amphibolite−granulite metamorphism in the Aravalli−Delhi Mobile Belt, North-western India: New evidence from monazite chemical age and its implication. *Precambrian Research*, *178*, 168−184.

Bhowmik, S. K., Pal, T., Roy, A., & Pant, N. C. (1999). Evidence of pre-Grenvillian high-pressure granulite metamorphism from the northern margin of the Sausar Mobile Belt, Central India. *Journal of the Geological Society of India*, *53*, 385−399.

Bhowmik, S. K., & Roy, A. (2003). Garnetiferous metabasites from the Sausar Mobile Belt: Petrology, P−T path and implications for the tectonothermal evolution of the Central Indian Tectonic Zone. *Journal of Petrology*, *44*, 387−420.

Bhowmik, S. K., Sarbadhikari, A. B., Spiering, B., & Raith, M. M. (2005). Mesoproterozoic reworking of Palaeoproterozoic ultrahigh-temperature granulites in the Central Indian Tectonic Zone and its implications. *Journal of Petrology*, *46*, 1085−1119.

Bhowmik, S. K., Wilde, S. A., & Bhandari, A. (2011). Zircon U−Pb/Lu−Hf and monazite chemical dating of the tirodi biotite gneiss: Implication for Latest Paleoproterozoic to Early Mesoproterozoic Orogenesis in the Central Indian Tectonic Zone. *Geological Journal*. Available from http://dx.doi.org/10.1002/gj.1299.

Bhowmik, S. K., Wilde, S. A., Bhandari, A., Pal, T., & Pant, N. C. (2012). Growth of the Greater Indian landmass and its assembly in Rodinia: Geochronological evidence from the Central Indian Tectonic Zone. *Gondwana Research*, 2012. Available from http://dx.doi.org/10.1016/j.gr.2011.09.008.

Bose, M. K., & Chakrabarty, M. K. (1981). Fossil marginal basin from Indian shield: A model for the evolution of Singhbhum Precambrian belt, eastern India. *Geolisch Rundschau*, *70*, 504−518.

Bose, M. K., Chakrabarty, M. K., & Saunders, A. D. (1989). Petrochemistry of the lavas from Proterozoic Dalma volcanic belt, Singhbhum, eastern India. *Geolisch Rundschau*, *78*, 633−648.

Chakraborty, K., & Roy, A. (2012). Mesoproterozoic differential metasomatism in subcontinental lithospheric mantle of Central Indian Tectonic zone: Evidence from major and trace element geochemistry of padhar mafic-ultramafic complex. *Journal of the Geological Society of India*, *80*, 628−640.

Chatterjee, N., Bhattacharya, A., Duarah, B. P., & Mazumdar, A. C. (2011). Late Cambrian reworking of Paleo-Mesoproterozoic granulites in Shillong-Meghalaya Gneissic Complex (Northeast India): Evidence from PT pseudosection analysis and monazite chronology and implications for East Gondwana assembly. *Journal of Geology*, *119*, 311−330. Available from http://dx.doi.org/10.1086/659259.

Chatterjee, N., Crowley, J. L., & Ghose, N. C. (2008). Geochronology of the 1.55 Ga Bengal anorthosite and Grenvillian metamorphism in the Chotanagpur Gneissic Complex, eastern India. *Precambrian Research*, *161*, 303−316.

Chatterjee, N., & Ghose, N. C. (2011). Extensive Early Neoproterozoic high-grade metamorphism in North Chotanagpur Gneissic Complex of the Central Indian Tectonic Zone. *Gondwana Research*, *20*, 362−379.

Chatterjee, N., Mazumdar, A. K., Bhattacharya, A., & Saikia, R. R. (2007). Mesoproterozoic granulites of the Shillong−Meghalaya Plateau: Evidence of westward continuation of the Prydz Bay Pan-African suture into Northeastern India. *Precambrian Research*, *152*, 1−26.

Chattopadhyay, A., Bandyopadhyay, B. K., & Khan, A. S. (2001). *Geology and Structure of the Sausar Fold Belt: A Retrospection and Some New Thoughts* (64, pp. 251−263). Geological Survey of India, Special Publication.

Chattopadhyay, A., Das, K., Hayasaka, Y., & Sarkar, A. (2015). Syn- and post-tectonic granite plutonism in the Sausar Fold Belt, central India: Age constraints and tectonic implications. *Journal of Asian Earth Sciences*, *107*, 110−121.

Chattopadhyay, A., Huin, A. K., & Khan, A. S. (2003). Structural framework of Deolapar area, central India and its implications for Proterozoic nappe tectonics. *Gondwana Research*, *6*, 107−117.

Chattopadhyay, A., & Khasdeo, L. (2011). Structural evolution of Gavilgarh-Tan Shear Zone, central India: A possible case of partitioned transpression during Mesoproterozoic oblique collision within Central Indian Tectonic Zone. *Precambrian Research*, *186*, 70−88.

Chattopadhyay, A., Khasdeo, L., Holdsworth, R. E., & Smith, S. A. F. (2008). Fault reactivation and pseudotachylite generation in the semi-brittle and brittle regimes: Examples from the Gavilgarh−Tan Shear Zone, central India. *Geological Magazine*, *145*, 766−777.

Chaudhuri, A., & Basu, A. (1990). *Nature of volcanism and related gold mineralization at Suda, Sidhi district, Madhya Pradesh. Precambrian of Central India* (pp. 549−562). Geological Survey of India, Special Publication.

Das, L. K., Naskar, D. C., Roy, K. K., Majumdar, R. K., Choudhury, K., & Srivastava, S. (2007). Crustal structure in Central India from gravity and magnetotelluric data. *Current science*, *192*, 200−208.

Dessai, A. G., Peinado, M., Gokarn, S. G., & Downes, D. (2010). Structure of the deep crust beneath the Central Indian Tectonic Zone: An integration of geophysical and xenolith data. *Gondwana Research*, *17*, 162−170.

Dewey, J. F., & Bird, J. M. (1970). Mountain belts and the new global tectonics. *Journal Geophysical Research*, *75*, 2625−2647.

Dhanajayanaidu, G., & Harinarayana, T. (2009). Deep electrical imaging of the Narmada-Tapti region, central India from magnetotellurics. *Tectonophysics*, *476*, 538−549.

Dhanajayanaidu, G., Veeraswamy, K., & Harinarayana, T. (2011). Electrical signatures of the Earth's crust in central India as inferred from magnetotelluric study. *Earth Planets Space*, *63*, 1175−1182.

Dunn, J. A., & Dey, A. K. (1942). *The geology and petrology of eastern Singhbhum and surrounding areas*. Geological Survey of India, Memoir 69, (2).

Ghosh, B., & Praveen, M. N. (2008). Indicator minerals as guides to base metal sulphide mineralisation in Betul Belt, Central India. *Journal of Earth System Science*, *117*, 521−536.

Ghosh, K. K., Raj, J., & Nandy, K. (1998). On the intrusive suite from Bilaspur, Madhya Pradesh. *Journal of the Geological Society of India*, *51*, 97−102.

Ghosh, S., Chakraborty, S., Paul, D. K., Sarkar, A., Bhalla, J. K., Bishui, P. K., & Gupta, S. N. (1991). Geochronology and geochemistry of granite plutons from East Khasi Hills, Meghalaya. *Journal of the Geological Society of India*, *137*, 331−342.

Gokarn, S. G., Rao, C. K., Gupta, G., Singh, B. P., & Yamashita, M. (2001). Deep crustal structure in central India using magnetotelluric studies. *Geophysical Journal International*, *144*, 685–694.

Golani, P. R., Bandyopadhyay, B. K., & Gupta, A. (2001). Gavilgarh-Tan Shear: Prominent ductile shear zone in Central India with multiple reactivation history. *Geological Survey of India Special Publication*, *64*, 265–272.

Gupta, A., & Basu, A. (2000). *North Singhbhum Proterozoic mobile belt, eastern India—A review. MS Krishnan Centenary Volume* (55, pp. 195–226). Geological Survey of India, Special Publication.

Jain, S. C., Nair, K. K. K., & Yedekar, D. B. (1995). *Geology of the Son-Narmada-Tapi lineament zone in the central India. Geoscientific studies of the Son Narmada-Tapi lineament zone* (10, pp. 1–154). Geological Survey of India, Special Publication.

Kaila, K. L. (1988). Mapping the thickness of Deccan Trap flows in India from DSS studies and inferences about a hidden Mesozoic basin in the Narmada-Tapti region. In K. V. Subba Rao (Ed.), *Deccan Flood Basalts* (pp. 91–116). Geological Society of India Memoir 10.

Kaila, K. L., Reddy, P. R., Dixit, M. M., & KoteswaraRao, P. (1985). Crustal structure across Narmada-Son lineament, central India from deep seismic soundings. *Journal of the Geological Society of India*, *26*, 465–480.

Karmakar, S., Bose, S., Basu Sarbadhikari, A., & Das, K. (2011). Evolution of granulite enclaves and associated gneisses from Purulia, Chhotanagpur Granite Gneiss Complex, India: Evidence for 990–940 Ma tectonothermal event(s) at the eastern India cratonic fringe zone. *Journal of Asian Earth Sciences*, *41*, 69–88.

Kumar, M. R., Singh, A., Kumar, N., & Sarkar, D. (2015). Passive seismological imaging of the Narmada paleorift, central India. *Precambrian Research*. Available from http://dx.doi.org/10.1016/j.precamres.2015.09.013.

Lippolt, H. J., & Hautmann, S. (1994). 40Ar/39Ar ages of Precambrian manganese ore minerals from Sweden, India and Morocco. *Mineral Deposita*, *18*, 195–215.

Mahadevan, T. M. (2000). *Precambrian lithospheric evolution and chemical and thermal regimes of the mantle* (55, pp. 23–34). Geological Survey of India, Special Publication.

Mahadevan, T. M. (2002). *Geology of Bihar and Jharkhand: Text Book Series*. Geological Society of India, 567p.

Mahadevan, T. M., & Maithani, J. B. P. (1967). *Geological and Petrology of Mica Pegmatites in parts of Bihar belt* (pp. 1–114). Geological Society of India, Memoir 93.

Mahato, S., Goon, S., Bhattacharya, A., Mishra, B., & Bernhardt, H. J. (2008). Thermo- tectonic evolution of the North Singhbhum Mobile Belt: A view from the western part of the belt. *Precambrian Research*, *162*, 102–107.

Maji, A. K., Goon, S., Bhattacharya, A., Mishra, B., Mahato, S., & Bernhardt, H. J. (2008). Proterozoic polyphase metamorphism in the Chhotanagpur Gneissic Complex (India), and implication for trans-continental Gondwanaland correlation. *Precambrian Research*, *162*, 385–402.

Mall, D. M., Reddy, P. R., & Mooney, W. D. (2008). Collision tectonics of the central Indian suture zone as inferred from a deep seismic sounding study. *Tectonophysics*, *460*, 116–123.

Mandal, B., Sen, M. K., & Vijaya Rao, V. (2013). Newseismic images of the Central Indian Suture and their tectonic implications. *Tectonics*, *32*, 908–921. Available from http://dx.doi.org/10.1002/tect.20055.

Mazumdar, S. K. (1988). *Crustal evolution of the Chhotanagpur gneissic complex and the mica belt of Bihar. Precambrian of the Eastern Indian shield* (pp. 49–83). Geological Survey of India Memoir 8.

Mishra, D. C. (1977). Possible extension of Narmada-Son lineament towards Murray Ridge and eastern syntaxial bend. *Earth Planetari Science Letter*, *36*, 301–308.

Mishra, D. C., Singh, B., Tiwari, V. M., Gupta, S. B., & Rao, M. B. S. V. (2000). Two cases of continental collision and related tectonics during the Proterozoic period in India: Insight from gravity modelling constrained by seismic and magnetotelluric studies. *Precambrian Research*, *99*, 149–169.

Mukhopadhyay, D. (1984). The Singhbhum shear zone and its place in the evolution of the Pre-cambrian mobile belt, north Singhbhum. *Indian Journal of Earth Science, CEISM*, , 205–212.

Mukhopadhyay, D. (1990). *Precambrian plate tectonics in the Eastern Indian Shield. Crustal Evolution and Metallogeny* (pp. 75–100). New Delhi: Oxford and IBH Publishing Co.

Mukhopadhyay, D., & Deb, G. (1995). Structural and Textural development in Singhbhum shear zone, eastern India. *Proceeding of the India Academy of Science, Earth and Planetary Science, 104*, 385−405.

Naganjaneyulu, K., & Santosh, M. (2010). The Central India Tectonic Zone: A geophysical perspective on continental amalgamation along a Mesoproterozoic suture. *Gondwana Research, 18*, 547−564.

Nageswara Rao, B., Kumar, N., Singh, A. P., Prabhakar Rao, M. R. K., & Mall, D. M. (2011). Crustal density structure across the Central Indian Shear Zone from gravity data. *Journal of Asian Earth Science, 42*, 341−353.

Narayanaswamy, S., Chakravarthy, S. C., Vemban, N. A., Shukla, K. D., Subramaniam, M. R., Venkatesh, V., & Nagarajaiah, R. A. (1963). *The geology and manganese ore deposits of the manganese belt in Madhya Pradesh and adjoining parts of Maharashtra. Part I: General introduction*. Bulletin Geological Survey of India A-22, (I).

Nelson, D.R., Bhattacharya, H.N., Misra, S., Dasgupta, N., & Altermann, W. (2007). New Shrimp U-pb dates from the Singhbhum Craton, Jharkhand-Orissa region, India. (Abst) International Conference on Precambrian sediments and Tectonics. 2nd GPSS meeting, IIT Bombay, p. 47.

Pandey, B.K., Krishna, V., & Chabria, T. (1998). An overview of Chhotanagpur gneiss granulite complex and adjoining sedimentary sequences, Eastern and Central India. International seminar on Precambrian crust in eastern and central India: UNESCO-IUGS-IGCP-368, 131−135.

Patranabis-Deb, S., Bickford, M. E., Hill, B., Chaudhuri, A. K., & Basu, A. (2007). SHRIMP ages of zircon in the uppermost tuff in Chattisgarh basin in central India require 500 Ma adjustments in Indian Proterozoic stratigraphy. *Journal of Geology, 115*, 407−415.

Patro, P. K., & Sarma, S. V. S. (2009). Lithospheric electrical imaging of the Deccan trap covered region of western India. *Journal Geophysical Research, 114*, B01102. Available from http://dx.doi.org/10.29/2007JG005572.

Pitchai Muthu, R. (1990). *The occurrence of gabbroic anorthosites in Makrohar area, Sidhi district, Madhya Pradesh* (28, pp. 320−331). Geological Survey of India, Special Publication.

Qureshy, M. N. (1982). Geophysical and Landsat lineament mapping—an approach illustrated from West Central and South India. *Photogrammetria, 37*, 161−184.

Rajesham, T., Rao, B. Y. J., & Murti, K. S. (1993). The Karimnagar granulite terrane - A new sapphirine-bearing granulite province, South India. *Journal of the Geological Society of India, 41*, 51−59.

Ramachandra, H. M., & Roy, A. (2001). Evolution of the Bhandara−Balaghat granulite belt along the southern margin of the Sausar Mobile Belt of central India. *Proceedings of the Indian Academy of Sciences, Earth and Planetary Sciences, 110*, 351−368.

Ramakrishnan, M., & Vaidyanadhan, R. (2008). *Geology of India* (volume1Geological Society of India, 556p.

Ray Barman, T., Bishui, P. K., Mukhopadhyay, K., & Ray, J. N. (1994). Rb−Sr geochronology of the high-grade rocks from Purulia, West Bengal and Jamua-Dumka sector, Bihar. *Indian Minerals, 48*, 45−60.

Ray, J. S., Martin, M. W., Veizer, J., & Bowring, S. A. (2002). U-Pb zircon dating and Sr isotope systematic of the Vindhyan Supergroup, India. *Geology, 30*, 131−134.

Reddy, P. R., Murthy, P. R. K., Rao, I. B. P., Prakash Khare, G. K., Rao, D. M., Mall, P. K., & Reddy, M. S. (1995). *Deep crustal seismic reflection pattern in central India-Preliminary interpretation* (pp. 537−544). Geological Society of India Memoir 3.

Rekha, S., Upadhyay, D., Bhattacharya, A., Kooijman, E., Goon, S., Mahato, S., & Pant, N. C. (2011). Lithostructural and chronological constraints for tectonic restoration of Proterozoic accretion in the eastern Indian Precambrian shield. *Precambrian Research, 187*, 313−333.

Roy, A., Kagami, H., Yoshida, M., Roy, A., Bandyopadhyay, B. K., Chattopadhyay, A., & Pal, T. (2006). Rb−Sr and Sm−Nd dating of different metamorphic events from the Sausar Mobile Belt, central India: Implications for Proterozoic crustal evolution. *Journal of Asian Earth Sciences, 26*, 61−76.

Roy, A. (1998). Tectonothermal history of the Proterozoic supracrustal belts and their implication on the crustal evolution in the Central Indian Shield. In: UNESCO-IUGS- IGCP-368 Seminar on Precambrian Crust in Eastern and Central India, Bhubaneswar, India (Abstract), 30−35.

Roy, A., & Bandyopadhyay, B. K. (1990). *Tectonic and structural pattern of the Mahakoshal belt of Central India: A discussion* (28, pp. 226−240). Geological Survey of India, Special Publication.

Roy, A., & Devarajan, M. K. (2000). *A reappraisal of the Stratigraphy and tectonics of the palaeo-proterozoic Mahakoshal supracrustal belt, Central India* (57, pp. 79–97). Geological Survey of India, Special Publication.

Roy, A., & Hanuma Prasad, M. (2001). *Precambrian of Central India: A possible Tectonic model* (64, pp. 177–197). Geological Survey of India, Special Publication.

Roy, A., & Hanuma Prasad, M. (2003). Tectonothermal events in Central Indian Tectonic Zone (CITZ) and its implications in Rodinian crustal assembly. *Journal of Asian Earth Science, 22,* 115–129.

Roy, A., Ramachandra, H. M., & Bandopadhyay, B. K. (2000). Supracrustal belts and their significance in the crustal evolution of Central India. *Geological Survey of India, Special Publication, 55,* 361–380.

Roy, A., Sarkar, A., Jeyakumar, S., Aggarwal, S. K., & Ebihara, M. (2002). Mid- Proterozoic plume-related thermal event in the Eastern Indian Craton: Evidence from trace elements, REE geochemistry and Sr-Nd isotope systematic of basic-ultrabasic intrusive from Dalma Volcanic Belt. *Gondwana Research, 5,* 133–146.

Roy, A., Sarkar, A., Jeyakumar, S., Aggarwal, S. K., Ebihara, M., & Satoh, H. (2004). Late Archean mantle metasomatism below eastern Indian craton: Evidence from trace elements, REE geochemistry and Sr-Nd-O isotope systematics of ultramafic dykes. *Proceedings of the Indian Academy of Sciences, Earth and Planetary Sciences, 113,* 649–665.

Roy, J. N., & Mukhopadhyay, K. (1992). Study of Chhotanagpur gneissic complex along north–south transects with special reference to the nature and evolution of the various enclaves in the complex. *Records, Geological Survey of India, 125,* 81–83.

Santosh, M. (2012). *India's Paleoproterozoic legacy* (365, pp. 263–288). Geological Society of London, *Special Publications.*

Sarkar, A., Paul, D. K., Balasubrahmanyam, M. N., & Sengupta, N. R. (1980). Lamprophyres from Indian Gondwanas –K-Ar ages and chemistry. *Journal of the Geological Society of India, 21,* 188–193.

Sarkar, A. N. (1982). Precambrian tectonic evolution of eastern India: A model of converging microplates. *Tectonophysics, 86,* 363–397.

Sarkar, A. N. (1988). *Tectonic evolution of the Chotanagpur Plateau and the Gondwana basins in Eastern India: An interpretation based on supra-subduction geological processes* (8, pp. 127–146). Geological Society of India, Memoir.

Sarkar, S. C., & Gupta, A. (2012). *Crustal Evolution and Metallogeny in India* (p. 840). Cambridge University Press.

Sarkar, S. N., & Saha, A. K. (1962). A revision of the Precambrian stratigraphy and tectonics of Singhbhum and adjacent regions. India. *Quarterly Journal, Geological Mining and Metal-lurgical Society of India, 34,* 97–136.

Sarkar, S. N., & Saha, A. K. (1977). The present status of the Precambrian stratigraphy, tectonics and geochronology of Singhbhum-Keonjhar-Mayurbhanj region, eastern India. *Indian Journal of Earth Science, (S. Roy volume),* , 37–65.

Sarkar, S. N., Saha, A. K., & Miller, J. A. (1969). Geochronology of the Precambrian rocks of Singhbhum and adjacent regions, eastern India. *Geological Magazine, 106,* 15–45.

Sarkar, S. N., Trivedi, J. R., & Gopalan, K. (1986). Rb-Sr whole-rock and mineral isochron age of the Tirodi Gneiss, Sausar Group, Bhandara district, Maharashtra. *Journal of the Geological Society of India, 27,* 30–37.

Sengupta, S., & Gangopadhyay, K. K. (1994). *Preliminary investigation for wolframite around Satnala-Arhala area and extending Kuilapal Granite body, west of Chhendapathar in parts of Bankura district, West Bengal* (pp. 114–117). Records, Geological Survey of India.

Subba Rao, M. V., Narayana, B. L., Divakara Rao, V., & Reddy, G. L. N. (1999). Petrogenesis of the protolith for the Tirodi gneiss by A-type granite magmatism: The geochemical evidence. *Current Science, 76,* 1258–1264.

Venkat Rao, K., Srirama, B. V., & Ramasastry, P. (1990). *A Geophysical appraisal of Mahakoshal Group of upper Narmada valley in Precambrian of central India* (28, pp. 99–117). Geological of the Survey of India, Special Publication.

West, W. D. (1936). Nappe structure in the Archean Rocks of the Nagpur District. *Transactions of National Institute of Sciences India, 1*, 93–102.

Yedekar, D. B., Jain, S. C., Nair, K. K. K., & Dutta, K. K. (1990). *The Central Indian collision suture* (28, pp. 1–37). Precambrian of Central India: Geological Survey of India, Special Publication.

Yellur, D. D. (1968). Carbonatite complexes as related to the structure of the Narmada valley. *Journal of the Geological Society of India, 9*, 118–123.

Yellur, D. D. (1977). Geochemical clues in the investigation of the tectonic environment of the Dalma greenstones, Bihar, India. *Chemical Geology, 20*, 345–363.

FURTHER READING

Chattopadhyay, A., & Ghosh, N. (2007). Polyphase deformation and garnet growth in politic schists of Sausar Group in Ramtek area, Maharashtra, India: A study of porphyroblast–matrix relationship. *Journal Earth System Sciences, 116*, 423–432.

Santosh, M., Maruyama, S., & Yamamoto, S. (2009). The making and breaking of supercontinents: Some speculations based on superplumes, superdownwelling and the role of tectosphere. *Gondwana Research, 15*, 324–341.

THE ARAVALLI-DELHI OROGENIC BELT

CHAPTER OUTLINE

5.1 INTRODUCTION

The Aravalli-Delhi Orogenic Belt (ADOB), a prominent physiographic unit, is located in the northwestern part of the Indian shield (Fig. 5.1). The ADOB occurs from the vicinity of Delhi in the north to Palampur in Gujarat in the south and extends for more than 700 km in NE−SW

Proterozoic Orogens of India. DOI: http://dx.doi.org/10.1016/B978-0-12-804441-4.00005-5

267

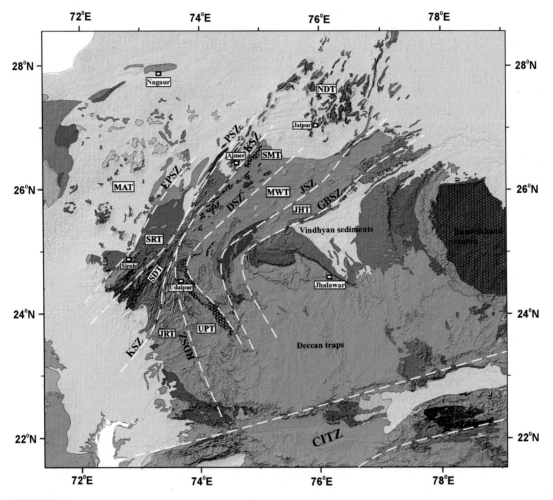

FIGURE 5.1

Geological map showing major rock types and tectonic features of Aravalli Delhi Orogenic Belt (ADOB) superposed over the digital elevation model. *MAT*, Marwar Terrane; *EPSZ*, Erinpura shear zone; *PSZ*, Phulad shear zone; *SDT*, South Delhi Terrane; *KSZ*, Kaliguman shear zone; *SMT*, Sandmata Terrane; *DSZ*, Delwara shear zone; *MWT*, Mangalwar Terrane; *JSZ*, Jahazpur shear zone; *JHT*, Jahazpur Hindoli Terrane; *GBSZ*, Great Boundary shear zone; *NDT*, North Delhi Terrane; *JRT*, Jharol Terrane; *RDSZ*, Rakhabdev shear zone; *UPT*, Udaipur Terrane; *CITZ*, Central Indian Tectonic Zone.

direction with a width varying from 120 km to 250 km. The ADOB trends NNE−SSW and is a collage of two variably metamorphosed volcano-sedimentary belts, which are described as the Paleoproterozoic Aravalli Fold Belt (AFB) and the Mesoproterozoic Delhi Fold Belt (DFB). The basement of these fold belts is considered to be the Banded Gneissic Complex (BGC), a heterogeneous Archean terrain. The term BGC was originally defined by Heron (1953) as a heterogeneous mixture of migmatites, infolded metasedimentary sequences, granites, pegmatites, aplites, and metabasic rocks and considered this as the oldest stratigraphic unit of ADOB. Subsequent studies over the years revealed that the BGC is extremely diverse in terms of rock associations, metamorphic grade, and geological antiquity. The BGC consists of amphibolite and granulite facies gneisses, migmatites, amphibolite bodies, and rneta-sedimentary enclaves. It is locally unconformably overlain by the weak to moderately metamorphosed Mesoproterozoic Aravalli and Neoproterozoic Delhi Supergroups. The BGC was further divided into two regions, BGC-I and BGC-II by Gupta (1934). The BGC, in the southern part of ADOB, comprising tonalite−trondjhemite−granodiorite (TTG) gneiss-migmatite−intrusive granitoid−amphibolite rock association with supracrustal components was termed as BGC-I (Gupta, 1934), which was later decribed as Mewar gneiss by Roy (1988). The amphibolite gneiss constitutes a major constituent of the Mewar Gneiss, showing low Zr, Nb, and Rb, and moderate Cr, Ni, V, and Sr, which have a tholeiitic parentage with probable oceanic affinity (Kataria, Chaudhari, & Althaus, 1988). Isolated patches of small bodies of altered ultramafic rocks occur in some parts of the Mewar Gneiss. High-alumina para-gneisses occur as comformable bodies with other lithologic units of the Mewar gneiss and gradually pass on to migmatites of various types. Fuchsite quartzite and marble also occur as isolated linear bodies along with a few minor occurrences of calc−silicate rocks and para-amphibolites. Patches of very high pressure granulites, presumably emplaced from deeper crustal levels, have also been reported from this belt. This cratonic block, which acted as a basement for the overlying Paleoproterozoic supracrustal successions of the Aravalli Supergroup, remained largely unaffected by the succeeding Proterozoic tectonothermal events. The tonalite−trondhjemite−granodiorite suites and granites of BGC-I have yielded ca. 3.3−2.5 Ga emplacement ages (Gopalan, Macdougall, Roy, & Murali, 1990; Roy & Kröner, 1996; Weidenbeck & Goswami, 1994).

In contrast, the BGC-II occurs in the northern domain and comprises a range of high-grade supracrustal rocks, and metamorphosed felsic (TTG, granite, and charno−enderbite suites) and basic igneous rocks. These rocks were previously considered to have a similar Archean origin (e.g., Roy, Kröner, Bhattacharya, & Rathore, 2005; Sharma, 1988). Limited U−Pb zircon provenance data (Buick, Allen, Pandit, Rubatto, & Hermann, 2006) suggests that some part of BGC-II metasediments were deposited in the Paleoproterozic. These observations, together with the lack of a clear unconformity between the BGC-II and the Aravalli Supergroup, have even led to the suggestion that the relationship between the two is transitional, being a function of increasing metamorphic grade from the latter to the former (Naha & Roy, 1983). There exists several uncertainities and questions regarding the nature and relationship of the BGC and the overlying Aravalli and Delhi supergroup of rocks. It is also unclear whether or not different proposed BGC domains can be correlated. It is interesting to note that the status of BGC as an older basement was also challenged. Crookshank (1948) and Naha and Halyburton (1974) were of the opinion that the BGC is not stratigraphically older but represents the migmatized equivalents of the younger supracrustals. However, evidences like the presence of local angular unconformities between the BGC and the supracrustals (Roy & Paliwal, 1981) and the occurrence of the basal Aravalli conglomerates at many places (Naha & Mohanty, 1988; Roy et al., 1988) suggest that the BGC represents the basement. Further, the occurrence of small pockets of white mica (commercially known as pyrophyllite) along the BGC−Aravalli contact, indicate the possible palaeosols developed along the peneplained surface of the basement (Roy, Somani, & Sharma, 1981).

The ADOB is flanked by the Marwar terrane (MAT) in the west and Mewar craton in the east. The Mewar craton forms a promontory structure on the western edge of the Bundelkhand Protocontinent and the ADOB wraps around it. The Mewar craton is dominantly composed of Mesoarchean tonalite−TTG gneisses and the region to the east is covered by Vindhyan sediments. The MAT is the least studied, and the earlier geological records were not accessible because of a hostile environment and the coverage either with sands of the Thar Desert or younger sediments. The region exhibits only Neoproterozoic-Phanerozoic geological history and there is no reliable evidence for rocks older than 1100 Ma.

The current status of our knowledge on the crustal evolution of the ADOB originated from numerous structural, petrological, geochemical, and geochronological investigations by scores of workers (Gupta, 1934; Heron, 1953; Roy & Jakhar, 2002). Excellent reviews have been published recently in the form of books that include: Ramakrishnan and Vaidyanathan (2008); Sharma (2012); Sarkar and Gupta (2012). A special publication by the Geological Survey of India provides excellent maps including geological, geophysical, and aeromagnetic anomaly maps with more emphasis on metallogeny and mineral exploration (Golani, 2015). The maps provide well-defined regional tectonic features, basement configuration, potential mineralized zones, and possible targets for the deep-seated mineral deposits. The maps also provide insights into the tectonic evolution of ADOB.

Understood from a modern perspective, the ADOB can be described as a mosaic of juxtaposed geological terranes within a reworked Archean basement complex based on the lithological assemblages, structural architecture, and the currently available literature. For the sake of brevity and clarity, the ADOB can be divided and described in terms of major terranes separated by crustal-scale shear zones (Fig. 5.2). The ADOB constitutes a series of arcuate and near N-S−trending sub-parallel major shear zones. The most prominent among them from east to west include: (1) the Great Boundary shear zone (GBSZ), (2) the Jahazpur shear zone (JSZ), (3) the Delwara shear zone (DSZ), (4) the Kaliguman shear zone (KSZ), (5) the Phulad shear zone (PSZ), (6) the Rakhabdev shear zone (RDSZ), and (7) the Banas shear zone (BSZ). These major ductile shear zones represent either Proterozoic sutures or ophiolite mélange zones (Sinha-Roy, Malhotra, & Guha, 1995), delineating many subterranes within the ADOB based on distinct lithological assemblages and structural styles. The ADOB constitutes the following distinct geological terranes from east to west: (1) the Jahazpur-Hindoli terrane (JHT), (2) the Mangalwar terrane (MWT), (3) the Sandmata terrane (SMT), (4) the Udaipur terrane (UPT), (5) the Jharol terrane (JRT), (6) the South Delhi terrane (SDT), (7) the North Delhi terrane (NDT), (8) the Sirohi terrane (SRT), and (9) the MAT.

5.2 SHEAR ZONES

5.2.1 GREAT BOUNDARY SHEAR ZONE

The most conspicuous tectonic boundary at the eastern margin of the ADOB is known as the GBSZ. It separates the Archean/Paleoproterozoic supracrustals of the Hindoli Group of ADOB and the Paleo-Mesoproterozoic cover sequence belonging to the lower and middle parts of the Vindhyan group of sediments. The GBSZ is well exposed in and around the Chittaurgarh region

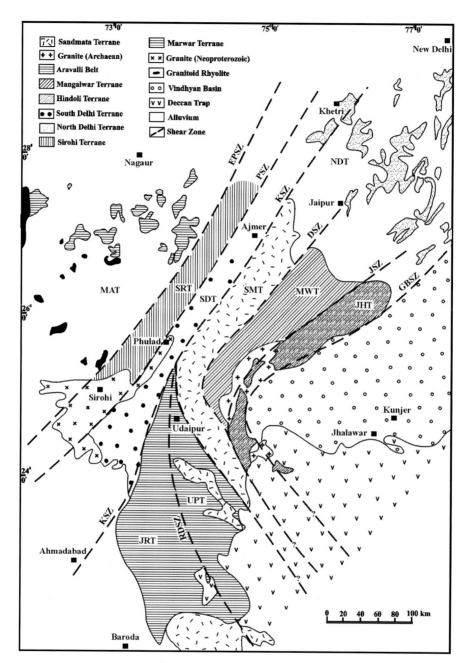

FIGURE 5.2

Simplified structural architecture of ADOB displaying important lithologies, major tectonostratigraphic terranes and crustal-scale shear zones. *MAT*, Marwar Terrane; *SRT*, Sirohi Terrane; *EPSZ*, Erinpura shear zone; *PSZ*, Phulad shear zone; *SDT*, South Delhi Terrane; *KSZ*, Kaliguman shear zone; *SMT*, Sandmata Terrane; *DSZ*, Delwara shear zone; *MWT*, Mangalwar Terrane; *JSZ*, Jahazpur shear zone; *JHT*, Jahazpur Hindoli Terrane; *GBSZ*, Great Boundary shear zone; *NDT*, North Delhi Terrane; *JRT*, Jharol Terrane; *RDSZ*, Rakhabdev shear zone; *UPT*, Udaipur Terrane.

Modified after Heron, A. M., (1953). The geology of Central Rajputana. Memoir of the Geological Survey of India, 79, *339;* Gupta, S. N., Arora, Y. K., Mathur, R. K., Iqbaluddin, Prasad, B., Sahani, T. N., & Sharma, S. B., (1980). Lithostratigraphy map of Aravalli region. *Hyderabad: Geological Survey of India publication.*

and the granitic basement is juxtaposed against the cover rocks. It strikes NNE—SSW and preserves a zone of intense deformation within the cover sequence of the Vindhyan Supergroup.

In the northeastern sector, the GBSZ comprises two subparallel faults, one of which cuts through the Neoproterozoic upper Vindhyan rocks and the other runs along the interface of the Hindoli supracrustals and upper Vindhyan rocks (Fig. 5.3). Characteristic morphotectonic features are common all along the GBSZ in the form of distinct fault scarps. Truncation of beds, silicification, brecciation, and iron oxidation of Vindhyan sediments are significant. Fermor (1930) was the first to propose that the GBSZ is a pre-Vindhyan structure, which has been reactivated as a reverse fault during and after the sedimentation of the Vindhyan System. The large-scale folds in the cover-sequence trend oblique to the trend of the GBSZ.

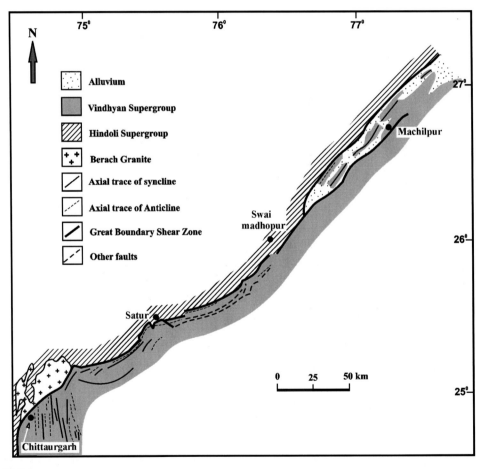

FIGURE 5.3

Geological and structural map of Great Boundary shear zone.

Modified after Srivastava, D. C., & Amit Sahay, (2003). Brittle tectonics and pore-fluid conditions in the evolution of the Great Boundary Fault around Chittaurgarh, Northwestern India. Journal of Structural Geology 25, *1713–1733.*

The Vindhyan cover rocks, exposed in the vicinity of the GBSZ are characteristically deformed into a series of open, N-S—trending and nonplunging folds. The development of mylonic foliation in Beach granites is prevalent in the vicinity of GBSZ. There are many meter scale ductile shear zones in a well-exposed Berach river section with lensoid or tabular zones of well developed mylonitic foliations and stretching lineations with wide scatter in orientations which colud be due to intensity of strain. Some of the fold structures show nonplane and noncylindrical geometry. Tight isoclinal folds with overturning to east occur along the GBSZ. The GBSZ is an imbricate zone comprising a series of steeply dipping reverse faults and intervening slices of sheared rocks in the Precambrian terrane of northwestern India (Sinha-Roy et al., 1995). It extends for more than 400 km up to Gangetic alluvium with a width of more than 1 km. The GBSZ is characterized by the presence of: (1) a thin band (~ 100 m) of phyllonites, (2) a thick zone (<2 km) of abundant fractures, faults and en échelon vein arrays, and (3) mesoscopic thrusts and thrust-related kink folds (Srivastava & Amit Sahay, 2003). Geophysical studies indicate that the GBSZ dips steeply to northwest, and cuts a thick section of the crust up to the Moho boundary (Reddy et al., 1995). Based on the above description, it can be surmised that the GBSZ is a northwesterly dipping major boundary shear zone with dextral transpressional movements affecting both the basement complex and cover sediments.

5.2.2 JAHAZPUR SHEAR ZONE

The JSZ, also described as thrust zone, separates the low grade metamorphic Hindoli-Jahazpur Terrane to the east and medium- to high-grade metamorphic MWT to the west. The JSZ trends NE—SW and dips steeply to northwest with dextral sense of movements. The major rock types in the JSZ include migmatites, conglomerate, and a variety of phyllites associated with different lithologies, quartzites, dolomites, and occasional Banded Iron Formation (BIF). (Fig. 5.4). They show tight-isoclinal folds. All lithologic units show a variation in thickness along strike which could be due to shearing. Evidence for synsedimentary sulphide mineralization is recorded along the JSZ. Several prominent band-like outcrops of metamorphosed shale—sand—carbonate association with black shales and ironstones occur along the belt and are well exposed near Jahazpur.

The western margin of the JSZ is marked by a thrust zone that has brought up an older sequence of migmatites of the Mangalwar Complex with enclaves of metamorphosed cover sequence. The conglomerate bed does not exist on the western margin of the belt probably due to thrusting. It is interesting to note the presence of small bands of BIF associated intermittently at the interface of orthoquartzite and dolomite. The BIF consists of thin rhythmic layers of chert and hematite, the latter extensively altered to limonite. The BIF unit locally shows high values of Zn (4000 ppm), Ni (1300 ppm), and Co (400 ppm) with the Mangalwar Complex.

This thrust zone is an imbricate structure with horse-like features developed at many places along the JSZ. The orthoquartzite unit has developed into a quartz mylonite at the thrust contact. These features, together with the metabasic components, present in the eastern part of the JSZ was interpreted as the development of the Jahazpur basin through ensialic rifting and rift failure (Sinha-Roy & Malhotra, 1989). The ramp and flat geometry of the Jahazpur thrust indicates that it has developed in a compressional environment and represents a major crustal boundary with great tectonic significance in the evolution of the ADOB.

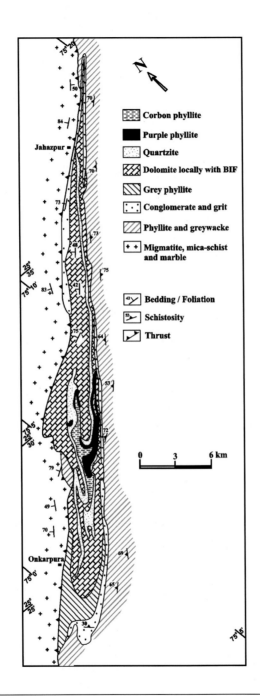

FIGURE 5.4

Geological map of Jahazpur shear zone.

Modified after Sinha-Roy, S., & Malhotra, G., (1989). Structural relations of proterozoic cover and its basement: An example from the J-ahazpur Belt, Rajasthan. Journal of the Geological Society of India, 34, 233–244.

5.2.3 **DELWARA SHEAR ZONE**

The DSZ is arcuate in nature and separates the SMT to the west and the MWT to the east (see Fig. 5.2). The sheared rocks from SMT are juxtaposed with the deformed felsic orthogneisses of the MWT at the DSZ. The orthogneisses are migmatitic in nature and are strongly mylonitized. Away from the shear zone similar orthogneisses of MWT are generally not migmatitic. These observations suggest that the DSZ must have played significant role for the fluid flow that allowed: (1) hydration and deformation in the SMT granulites, and (2) partial melting in the MWT orthogneisses (Roy & Rathore, 1999; Roy et al., 2005). The observed relationships between melting and shearing in the MWT are similar to those attributed to shear-zone–focused, fluid-present melting in orthogneisses elsewhere.

The DSZ hosts one of India's largest Pb–Zn deposits in Rampura-Agucha region. The host rocks of mineralization include: garnet–biotite–sillimanite gneiss with lenses of amphibolites, quartzo–feldspathic bands and calc–silicate rocks and migmatites with intrusions of pegmatites and aplites. The other important rock types are mica schist, graphite schist, calc–silicate–marble–association, banded ferruginous rocks, and pegmatites. Norite intrusions also occur at a few places (Fig. 5.5), and the metamorphism is of upper amphibolite facies. The rock formations trend NE–SW with medium to steep dips to the southeast. Regionally, the area is traversed by a series of subparallel shear zones that are characterized by complex folding including km-scale sheath folds (Roy & Jakhar, 2002). Mylonitic fabrics and stretching lineations are well developed and were related to thrust tectonics. The amphibolites are geochemically interpreted as the derivatives of plume-generated basalts (Deb & Sarkar, 1990). Amphibolites yielded single zircon age of ∼1450 Ma, while a Pb–isotope model age of galena from this belt is around 1800 Ma. However, Ar–Ar mineral ages from these rocks yielded between 800 and 900 Ma.

5.2.4 **KALIGUMAN SHEAR ZONE**

The KSZ occurs between the SMT in the east and the SDT in the west (see Fig. 5.2). The KSZ trends NE–SW and extends for over 700 km and lie subparallel to the PSZ. The KSZ in the south splits into two: one continues southwest separating the JRT and the SDT and the other continues due south in the form of the RDSZ that occurs between the JRT and the UPT. The KSZ strikes NE–SW with steep dips dominantly to SE. The deformational history of the KSZ reveals initial thrusting with subsequent dextral strike-slip shearing along NE–SW direction with low dip-slip component, thus suggesting dextral transpressional movements (Bhattacharya, Nagarajan, Shekhawat, & Joshi, 1995). Intense transpositional fabrics are common in the eastern part of the shear zone and the rocks are often seen grading into granulite facies conditions. The rocks along the KSZ are intensely tectonized and was described as paleo suture because of its association with highly deformed ultramafic rocks at a few places (Sen, 1981).

The KSZ is also characterized by a staurolite–kyanite zone between Ajmer and Nasirabad and all the structural elements are highly oriented parallel to the strike of the KSZ. The high grade granulite facies rocks also occur as discontinuous bodies over a strike length of 40 km around the Pilwa-Chinwali region, which may extend further north beneath the alluvium cover. The area is characterized by ridges of intensely deformed quartzites that overlie the migmatized paragneisses. The quartzite markers display NNE-SSW trending fold structures on map scale. The area is

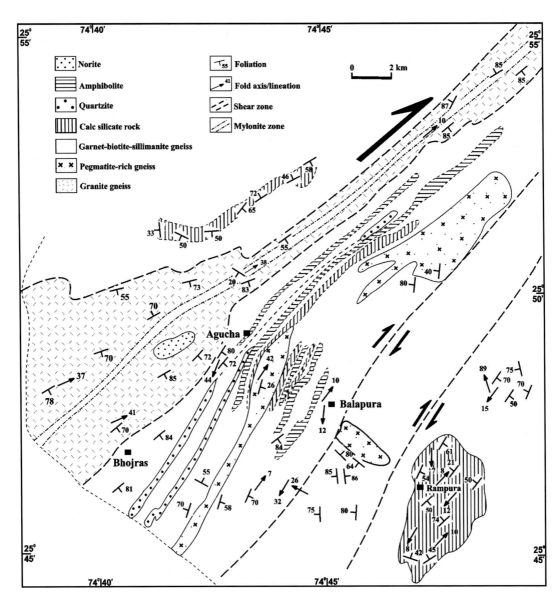

FIGURE 5.5

Geological and structural map of Rampura-Agucha region.

Modified after Roy, A. B., & Jakhar, S. R., (2002). Geology of Rajasthan (Northwestern India), Precambrian to Recent. *Scientific Publishers, p. 421.*

traversed by extensive occurrence of ductile shear zones. High-grade rocks are exposed between the quartzites and the migmatites. Subvertical bedding planes and well-developed stretching lineations are common features in quartzites. The high-grade rocks include charnockites, pelitic granulites, two-pyroxene granulites, leptynites, garnetiferous amphibolites, and augen gneisses. A P—T estimate reveals pressures of 5.5—9.5 kbars and temperatures of ~630°C for the metapelites. Zircon geochronology of these granulite bodies indicate three periods at ca 1435 Ma, 1128 Ma, and 1000 Ma (Fareeduddin & kroner, 1998), and the migmatites yielded an age of 2.8 Ga (Tobisch, Collerson, Bhattacharya, & Mukhopadhyay, 1994).

5.2.4.1 Shyamgarh antiformal structure

The KSZ is well defined in the field with a spectacular large scale antiformal structure associated with a prominent zone of mylonitic fabrics near Shyamgarh (Fig. 5.6). A typical asymmetrical structure associated with very steep axial surface with prominent lithological and structural discontinuity is well mapped from aerial photographs (Heron, 1953). The axis of Shyamgarh antiform plunges at 35 degrees toward the northeast. The core of the antiformal fold is occupied by semipelitic schist with pegmatite intrusions. The eastern limb of the fold is cut off by the NNE-SSW trending KSZ. The KSZ is well exposed here for a strike length of about 10 km and is ~400 m wide with easterly dipping intense mylonitic fabrics (Bhattacharya et al., 1995). The fabrics are characterized by ribbon quartz, feldspar augens with asymmetric pressure shadows. The western margin of the KSZ is marked by a chain of elongated quartzite hill ridges. Mesoscopic scale fold structures of early generation show coaxial geometry of fold hinges with that of the large antiformal structure implying its coeval nature with the shear zone. This structure was also described later as a zone of tectonic mélange. A dark-to-black—colored calcareous breccias (mélange?) with angular fragments of chert and siliceous marble of 3—10 mm occur for a length of 3 km. The asymmetry of augens and associated pressure shadows, abundant dextral folds, and other kinematic indicators suggest dextral sense of movements along the KSZ.

5.2.4.2 Kishangarh syenite body

The KSZ also hosts alkaline magmatism, which is well exposed and elongated in a region north of Kishangarh. An interesting suite of alkaline rocks with nepheline syenite are the dominant rocks around Kishangarh, east of Ajmer (Fig. 5.7). The western contact of the nepheline syenite body is marked by pre-Delhi supracrustals of quartztite, mica schist, marble, and amphibolites in the region. The syenite body extends for about 25 km long and 1—5 km wide and seems to be completely restricted to the KSZ (Fig. 5.8). The syenite body consists of feldspar minerals and variable amounts of mafic minerals. A narrow discontinuous zone of fenitized rocks represented by syenite and alkali diorite forms a capping over the structurally conformable nepheline syenite pluton. The gneissic foliation at the marginal parts of the syenite pluton gradually becomes diffused and turns into massive granitoid in the central parts. The intrusion is deformed into an elongate flattened domal structure. Complex folding and gneissic foliation at the margins suggest that the syenite body was synkinematically emplaced at ~1500 Ma. Aravalli metasediments and metabasites hosting Cu mineralization are also associated with nepheline syenites.

A chain of albitite intrusions in association with ultramfic rocks occur along a strike length of ~300 km, east of Khetri belt, which could be a possible extension of the KSZ in the northern part. These intrusions are well known for uraninite—fluorite—magnetite—chalcopyrite—molybdenite mineralization. In another part of the KSZ, a few kilometers east of Shyamgarh, near Masuda, well-developed transposition fabrics and the highly stretched prominent lithological bands of amphibolites and calc—silicate show NE—SW trends. The presence of early folds of noncylindrical geometry on different scales indicates that the rocks were subjected to ductile shearing.

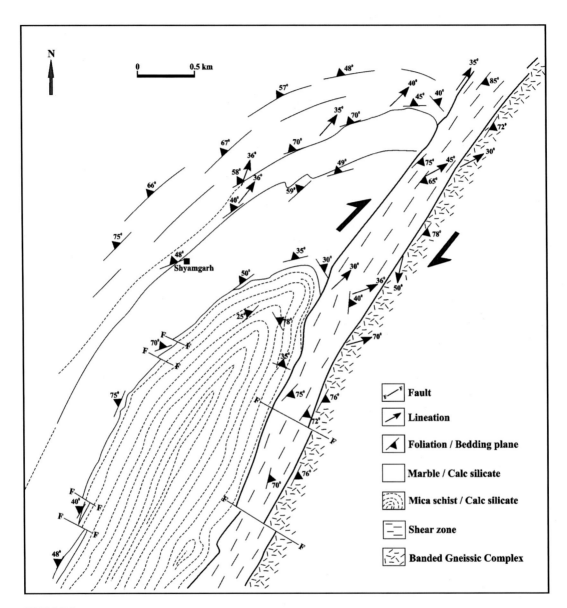

FIGURE 5.6

Geological map of the region around Shyamgarh antiformal structure.

Modified after Roy, A. B., & Jakhar, S. R., (2002). Geology of Rajasthan (Northwestern India), Precambrian to Recent. *Scientific Publishers, p. 421.*

The fold limbs were attenuated, exhibiting dextral sense of movements. Mylonitic fabrics with a width of ~1 km and the attenuation of fold limbs are well developed, substantiating dextral kinematics. The variations in the behavior of stretching lineations from subhorizontal

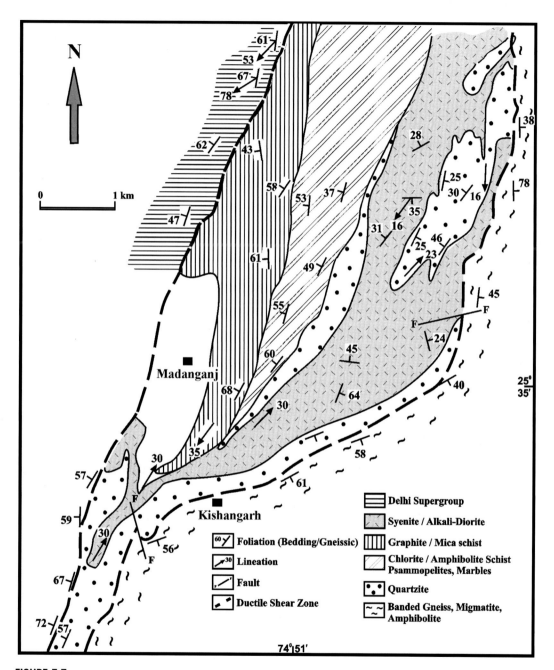

FIGURE 5.7

Geological map of alkaline rocks around Kishangarh.

Modified after Roy, A. B., & Dutt, K. (1995). Tectonic evolution of the Nephelene syenite and associated rocks Kishangarh,
District Ajmer, Rajastahan. Memoir of the Geological society of India, 31, 231–257.

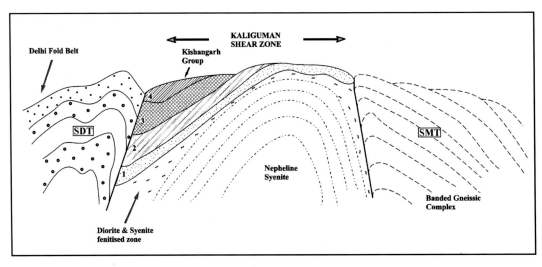

FIGURE 5.8

A schematic east—west cross section across the Kaliguman shear zone bound domal-shaped nepheline syenite body in Mandawaria Hill.

Modified after Roy, A. B., & Dutt, K. (1995). Tectonic evolution of the Nephelene syenite and associated rocks Kishangarh, District Ajmer, Rajastahan. Memoir of the Geological society of India, 31, *231–257.*

to subvertical suggest large-scale dextral transpressional tectonics in the region of Masuda and along the KSZ.

5.2.5 PHULAD SHEAR ZONE

The PSZ is one of the most prominent shear zones that occurs at the western margin of the ADOB. The PSZ also marks the western margin of the DFB (see Fig. 5.2) and is well exposed for more than 500 km at the contact between the SDT to the east and the SRT to the west. The PSZ comprises ~ 500 m wide rock assemblage including the basement gneisses, cover rocks and also the younger intrusions of mafic—ultramafic complexes and Erinpura granite intrusions (Golani, Reddy, & Bhattacharjee, 1998). Profuse development of mylonitic fabrics and well developed down-dip stretching lineations are common through out the PSZ. Refolded folds, curvilinear fold hinges, deformed sheath fold structures and lineations are significant along the shear zone. At many places of the PSZ, thick mafic—ultramafic complex including volcanic flows, gabbros, and ultramafic bodies are exposed. They are exposed discontinuously in a ~ 300 km long and 20-to-50-km—wide linear belt running from Barr-Sewaria region in the north to Palampur in the south. This sequence along the belt has been considered as remnant Mesoproterozoic oceanic crust (Sugden, Deb, & Windely, 1990). The PSZ has been dotted by numerous granite intrusions that yielded whole-rock Rb—Sr ages between 900—800 Ma (Choudhary, Gopalan, & Anjaneya Sastry, 1984).

The detailed mapping of the central part of the PSZ in the area between north of Phulad and south of Karwara reveal that the important rock types along the zone are amphibolites, quartz mica schist (altered felsic tuff), calc—silicate rock, quartzite, and biotite schist. Rock assemblage of granite gneiss, amphibolites, quartzite, and calc-arenite occurs as tectonic mélange in association with pre-Delhi supergroup to the west in the vicinity of the PSZ (Fig. 5.9). The PSZ is characterized by extensive development of mylonites with a width of ∼20 m, which can be traced along the strike for more than 30 km with a width of ∼500 m. The most conspicuous structural feature is the occurrence of refolded down-dip lineations and deformed sheath fold structures. The PSZ is characterized by progressive ductile shearing with a complex history of folding with the development of planar, nonplanar, and refolded sheath folds (Ghosh, Hazare, & Sengupta, 1999). The folds are very tight isoclines and are deformed into a variety of geometric patterns like "eyed" folds. Zircon geochronology of these granulites indicate that the PSZ is a deep level shear zone with complex evolutionary history (Fareeduddin & Kroner, 1998). Zircons yielded three periods of crystallization at 1435 Ma, 1000 Ma, and 1128 Ma.

The region between Phulad-Ranakpur-Basanthgarh represents an important sector of the PSZ, where Precambrian ophiolite complexes were reported (Gupta et al., 1980). Structurally, the lowest part of the complex comprises a discontinuous band of intensely deformed harzburgite followed by layered cumulus gabbroic rocks. These are overlain by hornblende schists, gabbros, sheeted dykes, and pillowed basalts, which are intruded by large bodies of diorite. Geochemical classification suggests that all noncumulus mafic rocks are subalkaline basalts, which are tholeiitic to calc-alkaline with boninite affinity (Khan, 2005). A forearc environment related to subduction tectonic regime was suggested. The complex petrological association of chromite-bearing serpentinite, gabbro, amphibolite, and chert represents the major component of Precambrian ophiolitic rocks intruded by granodiorites and granites along the PSZ. Metabasalts (amphibolites) around Phulad have average minor and trace element compositions characteristic of modern mid-ocean ridge basalts, whereas those around Ranakpur have characteristics of oceanic arc basalts (Volpe & Macdougall, 1990). The geochemical characteristics of Phulad ophiolites are related to suprasubduction zone tectonics.

An ophiolitic melange made up of pillowed metabasalt, pillow breccia, layered metagabbro, serpentinite, talc—chlorite schist, metagreywacke, mafic metatuff, and turbiditic metasemipelite is also documented in the Basantgarh area (Fig. 5.10). The rocks are strongly deformed with several thrust zones in between, defining the boundaries of the basal volcanic units and the metasedimentary rocks (Fig. 5.11). The plagiogranite occurs as tectonic slivers emplaced on top of the pillowed metabsalts. Basantgarh ophiolite mélange was reported to contain relict blueschist facies mineral assemblages in the matrix of metagreywacke (Sinha-Roy & Mohanty, 1988). The blueschist assemblages are likely to have been developed during the subduction process. The MAT represents the volcanic arc setting of the region and the high-pressure blueschist facies rocks were developed during convergence, subduction and collision at the western boundary of the SDT marking the PSZ as a distinct suture (Sinha-Roy & Mohanty, 1988). The intervening Delhi sediments between the PSZ and the KSZ were subjected to intense deformation resulting in the development of SDT. Serpentinite mélanges with relics of grospydite, and talc—chlorite—kyanite rocks indicate high pressure metamorphism to be expected in a subduction zone (Sychanthavong & Merh, 1984). Further south, amphibolites with relic pillows in the Sirohi area, geochemically resemble low K—tholeiites occurring nearest to trenches in modern island arcs (Bhattacharya & Mukherjee, 1984). The PSZ marks the site of paleosubduction and collision. The bivergent reflection fabric observed at PSZ can almost be traced to a depth of ∼40 km. The PSZ separates the highly folded

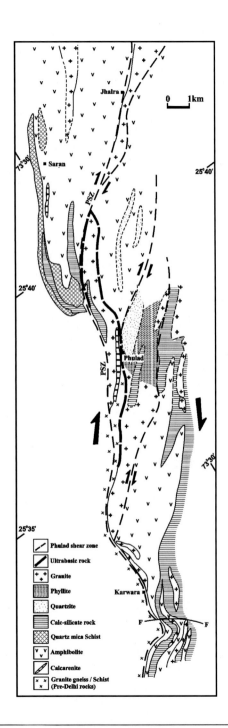

FIGURE 5.9

Geological and structural map of Phulad shear zone.

Modified after Golani, P. R., Reddy, A. B., & Bhattacharjee, J. (1998). The Phulad shear zone in central Rajasthan and its tectonostratigraphic implications. *Jodhpur: Scientific Publishers (India), pp. 272–278.*

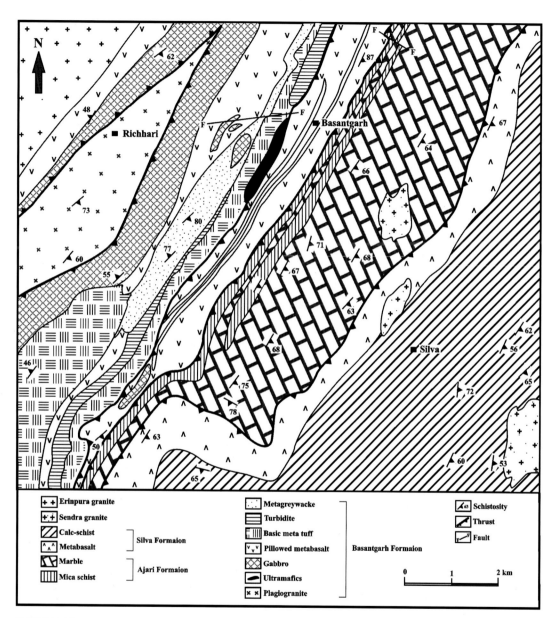

FIGURE 5.10

Geological and structural map of Basantgarh area showing ophiolite complex.

Modified after Sinha-Roy, S., & Mohanty, M. (1988). Blueschist facies metamorphism in the ophioliticmelange of the late Proterozoic Delhi fold belt, Rajasthan, India. Precambrian Research, 42, 97–105.

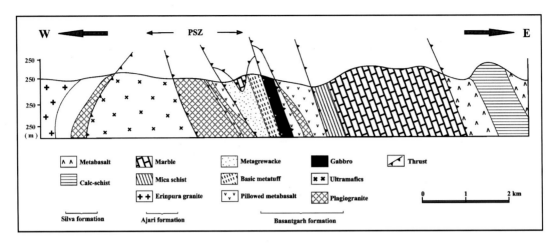

FIGURE 5.11

Geological cross section across the Basantgarh ophiolite complex.

Modified after Sinha-Roy, S., & Mohanty, M. (1988). Blueschist facies metamorphism in the ophioliticmelange of the late Proterozoic Delhi fold belt, Rajasthan, India. Precambrian Research, 42, 97–105.

and deformed SDT from the MAT. Geomorphologically these two terranes are also at different elevations, and the transition is at the PSZ (Vijaya Rao & Krishna, 2013).

5.2.6 BANAS SHEAR ZONE

The BSZ occurs in the basement rocks and is distinct from other shear zones in its orientation, which was described as domain boundary (Fig. 5.12). The shear zone occurs just to the south of popularly known "hammer head syncline" within the host BGC via Nathdwara-Khamnor. The BGC—Aravalli contact along the BSZ is interpreted to represent a migmatite front. On the basis of structural similarities between the BGC and the cover rocks, the migmatites were interpreted to have been derived from the Aravallis and Raialo sequence. However, other features, like mineral paragenesis and metamorphism of both the units, do not support the contention.

The BSZ trends broadly E—W with mild sigmoidal geometry at its ends and is marked by intense mylonitic fabrics defined by quartz-rich and carbonate-rich layering. The kinematic indicators like S—C fabrics, fish-like mineral elongations, bent lamellae of plagioclase grains, high-strain nature of rocks showing dimensional and crystallographic orientations and the development of ultramylonites suggest that the rocks of the BSZ were subjected to ductile shearing (Sinha-Roy, Mohanty, Malhotra, Sharma, & Joshi, 1993). The stretching lineations vary from gentle to moderate plunges. The regional map pattern and the associated lithological contacts suggest dextral transptressional tectonics dominantly with wrench component along the BSZ.

The BSZ is the zone of juxtaposition of the Bhilwara Supergroup in the north and the Aravalli Supergroup in the south. The major rock units in the south is semipelite of Debari Group containing quartz, biotite, chlorite, muscovite and minor amounts of carbonates at a few places suggesting upper greenschist facies conditions. To the north of the BSZ, popularly known "hammer head

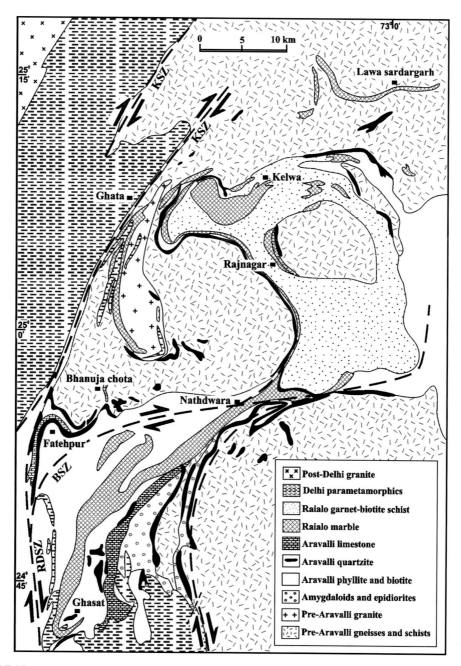

FIGURE 5.12

Geological and structural map of Banas shear zone displaying the popular structure of "hammer head syncline". RDSZ, Rakhabdev shear zone; KSZ, Kaliguman shear zone.

Modified after Heron, A. M., (1953). The geology of Central Rajputana. Memoir of the Geological Survey of India, 79, *339* syncline," which belong to BGC of MWT is observed. The rocks are polymetamorphic with estimated P−T conditions of 600°C and 5−6 kbar (amphibolite facies) conditons. The rock sequence and metamorphic characteristics vary across the BSZ indicating that the rocks on either side were tectonically juxtaposed. Regionally, the BSZ is connected to the major shear zones such as the KSZ, RDSZ in the southwest and the JSZ in northeast, both of which are considered as suture zones. The preexisting structures must have been rotated and reoriented and the BSZ may be related to thrust tectonic regime.

5.2.7 RAKHABDEV SHEAR ZONE

The RDSZ, a significant branch of the KSZ, occurs in the central part of the AFB and separates the JRT to the west and UPT to the east. The RDSZ appears to be an important tectonic divide, demarcating the shallow water shelf facies rocks to west and the deep water facies to east indicating highly contrasting depositional histories. The RDSZ is a strong zone of shearing extending in N−S direction for over a length of 100 km with a width of about 2−3 km (Fig. 5.13). This large outcrop of ultramafic rocks along with cherts and chromite-bearing serpentinites are the dominant rock types. The other major rock types include quartzite, phyllite, and mica schist. The ultramafic rocks are represented by talc−chlorite (antigorite) schist and serpentinite with variable amounts of actinolite−tremolite, talc−tremolite, asbestos, and dolomite. These rocks were described as dismembered ophiolite complex and the RDSZ is also described as ophiolite decorated suture zone (Deb & Sarkar, 1990). The BGC represents the surrounding rock type of the RDSZ and was inferred to be the site for asthenospheric upwelling and subsequent ocean crust formation. The chromite-bearing serpentinites along the RDSZ, delineating the shelf-rise boundary represent this ocean floor material, interthrust during partial obduction against the continental rise sediments. A thick apron of phyllites and turbidites were deposited on the continental rise, now represented by the Jharol belt and probably buried the spreading center.

5.3 TERRANES

The ADOB, which was visualized traditionally as a synclinorium by Heron (1953), has undergone major revisions in understanding its evolution. It is currently described as a Proterozoic orogen consisting of a collage of NE-SW trending terranes juxtaposed along major ductile shear zones (see Fig. 5.2). Broadly, the ADOB has been traditionally divided into three classes: the Aravalli Group in the south, Bhilwara Group in the central part, and Delhi Group of rocks in the western and northen parts (Gouda et al., 2015). On the basis of lithological assemblages and the tectonic attributes, the entire ADOB can be divided into the following tectonic domains or terranes. The important terranes of the ADOB from east to the west include: JHT; MWT; SMT; SDT, NDT, JRT, UPT, SRT, and MAT occur to the west of ADOB. All the terranes comprise thick sequences of Proterozoic metasedimentary and metaigneous rocks unconformably overlying the basement gneisses, popularly known as the BGC. The BGC comprises inextricable mixture of polyphase metaigneous and metasedimentary enclaves of varying dimensions (Gupta, 1934; Heron, 1953; Raja Rao, 1971). The basal unconformity between the BGC and the overlying supracrustal rocks has been extremely

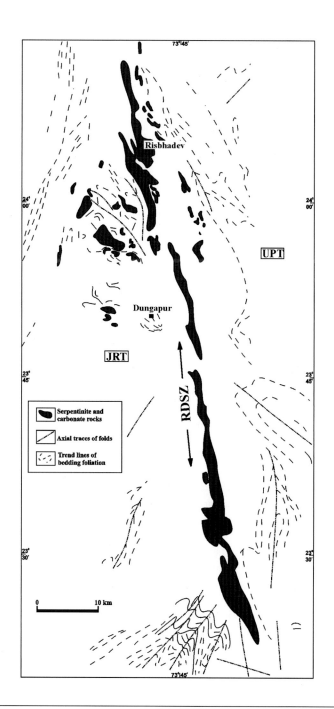

FIGURE 5.13

Structural map of a part of the Rakhabdev shear zone separararting the Jharol terrane and the Udaipur terrane.

Modified after Gupta, S. N., Arora, Y. K., Mathur, R. K., Iqbaluddin, Prasad, B., Sahani, T. N., & Sharma, S. B. (1980).

Lithostratigraphy map of Aravalli region. *Hyderabad: Geological Survey of India publication.*

tectonized and emplaced into various tectonic levels during Meso- and Neoproterozoic orogenesis. The details of each terrane are described below.

5.3.1 JAHAZPUR-HINDOLI TERRANE

The JHT is the easternmost terrane in ADOB. The terrane is sandwitched between the MWT to the west and the Vindhyan sediments of the Bundelkhand craton to the east (see Fig. 5.2). The JHT is marked by the presence of JSZ in the west and GBSZ in the east (Fig. 5.14). The terrane is characterized by low grade metamorphic rocks dominated by the presence of both felsic and mafic volcanics and volcanoclastic sediments interbedded with turbidite sequence of quartz−wacke and pelites and interlayered lava flows and tuff of basalt, basaltic andesite and dacite-rhyodacite composition (Bose & Sharma, 1992). The rocks are metamorphosed mainly in greenschist facies conditions grading locally into amphibolite facies. These rocks were interpreted by a few to indicate continental margin sequence coupled with a volcanic arc. Rhyodacite yielded zircon Concordia age of 1850 Ma. These rocks were also considered as the Gwalior facies rocks of Aravalli system by Heron (1953).

The JSZ is marked by ∼80 km long sulphide mineralization spread mainly along two dextral-subparallel belts of arenite−pelite—carbonate association with black shales and BIF (Fig. 5.15). The host rocks are dolomitic carbonate, chert, banded ferruginous chert and carbonaceous phyllite and are metamorphosed to greenschist facies conditions. Jahazpur granite, underlying the basal conglomerate, intrudes on the Hindoli Group and is traditionally compared with ∼2600 Ma old Berach granite, which represents a composite body consisting of older gray granite and younger pink alkali granite. The Berach granites, located in the southeastern part of the JHT, extend for a length of ∼140 km with a width of ∼25 km. They represent both granitoid as well as foliated gneissic types and were suggested to form the basement rocks upon which the Aravallis were deposited (Gupta, 1934). They vary in their texture from coarse porphyritic to highly foliated gneissic type and trend N−S to NE−SW. Some bands of ferromagnesian minerals like actinolite, chlorite, and epidote also occur along the foliation planes. Patches of dolerite and amphibolites were found with in the Berach granites. These granites were found to be transforming into unfoliated reddish or pink granite at several places. Rb−Sr age of these granites is given as ∼2445 Ma (Choudhary et al., 1984), which is also later confirmed by single zircon age.

The geochemical signatures of the rocks from JHT suggest the presence of an ocean in the region (Deb, Thorpe, Cumming, & Wagner, 1989), which separated the eastern Bundelkhand craton from the western Marwar craton. The intervening oceanic crust along with the sediments was subducted under the western Marwar (Sinha-Roy, 1988). Further convergence was accommodated by an imbricated Jahazpur thrust zone. The convective hydrothermal solutions, generated during the subduction process, carry conductive minerals (sulphide and graphite) to the surface through subduction zone representing as an example of large-scale metallogenic province associated with subduction zones.

There exists divided opinion about the correlation of the Hindoli-Jahazpur terrane. Some considered them as a part of the Aravallis, while others interpret them to be Archean and suggest them to be a part of the Bhilwara Supergroup along with the Sandmata-Mangalwar terrane. Some others regard it as an independent sequence of Paleoproterozoic age (Porwal, Carranza, & Hale, 2006). The issue remains unresolved. It is also interesting to note that the rocks of the

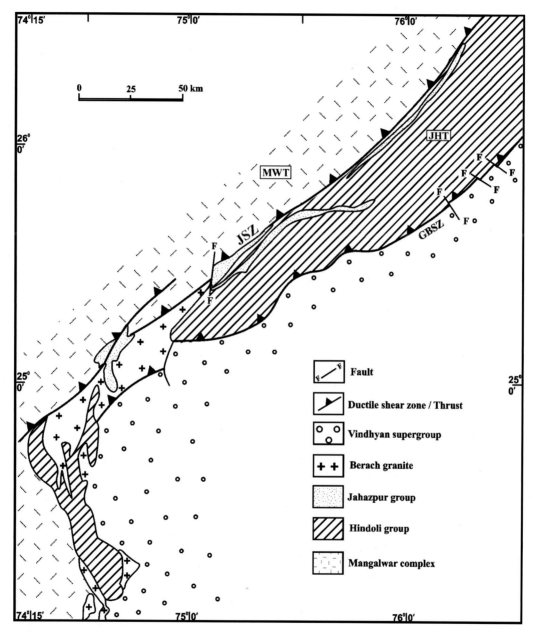

FIGURE 5.14

Geological map of the Jahazpur-Hindoli terrane. *MWT*, Mangalwar terrane; *JSZ*, Jahazpur shear zone; *JHT*, Jahazpur Hindoli Terrane; *GBSZ*, Great Boundary shear zone.

Modified after Sinha-Roy, S., Malhotra, G., & Mohanty, M. (1998). Geology of Rajasthan. Bangalore:
Geological Society of India, p. 278).

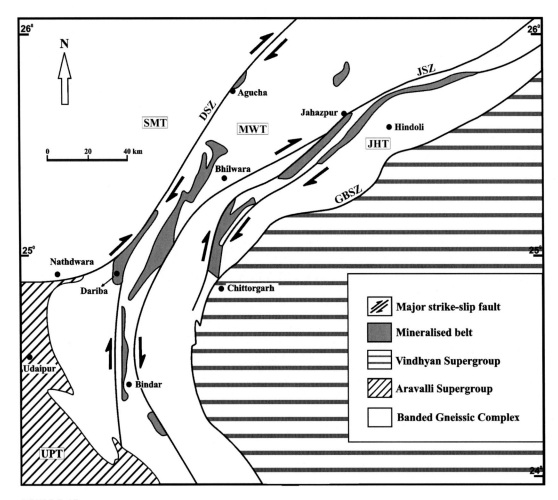

FIGURE 5.15

Map showing the mineralized belts of JHT and MWT.

Modified after Sinha-Roy, S., Malhotra, G., & Mohanty, M. (1998). Geology of Rajasthan. *Bangalore: Geological Society of India, p. 278.*

JHT are correlated with the Mahakoshal belt of the Central Indian Tectonic Zone (CITZ), suggesting their possible connection in the form of Satpura mobile belt beneath the cover of Deccan traps.

5.3.2 MANGALWAR TERRANE

The MWT is situated between the JSZ in the east and the DSZ in the west. It is also limited by the BSZ in the south. The MWT comprises a hetereogeneous assemblage of amphibolite facies supracrustal rocks (metapelites, metacarbonates, quartzites), meta-granitoids and amphibolites, which are considered as a greenstone sequence and was described as the Tanwar group of rocks. The MWT also consists of tonalitic and granodioritic gneisses and intrusive granites, known as the Ran igneous complex. The other important rockypes include sillimanite—mica schist, calc—silicate gneiss and marble, fuchsite quartzite, quartzite, amphibolites, ultramafics, and several norite and gabbro plutons (Guha & Bhattacharya, 1995). Amphibolites occur as thick bands as well as enclaves. Some of the amphibolite bands extend for over 7 km along the strike. The effusive nature of these rocks is evident from the presence of amygdules filled mainly by epidote and quartz and rarely by calcite. The metavolcanics show relict igneous textures and tholeitic to calk—alkaline chemical affinity. The Bouguer anomaly map shows significant high on the order of 20 mGal above the regional level and was interpreted to be the result of subsurface basic bodies or high density contrast (Varma, Mitra, & Mukhopadhyay, 1986). The basement for all these rocks is profusely tectonized, termed as Asan bimodal gneiss exposing a tectonic mélange (Roy & Jakhar, 2002). The gneiss may represent a metamorphic bimodal volcanic suite. The area is traversed by several ductile shear zones, mostly lying at lithological contacts, which were often filled with albite. There is no unconformity between the basement gneiss and the other supracrustals. In general, the rocks were subjected to amphibolite grade metamorphism. The garnet—sillimanite schist occupying the major part of the "hammer-head syncline" region was interpreted as metagreywacke (Sinha-Roy et al., 1993).

The association of dolomites, quartzites, ferruginous and carbonaceous cherts, graphitic shales, early basic volcanic flows, with characteristic geochemical signatures along with long, linear zones of base metal mineralization are significant in the MWT. In the context of comparison between the ADOB and the CITZ, the MWT is correlatable with Betul Group of rocks that comprises base metal mineralization.

A high-grade metapelite from the MWT, near Bhinai yielded concordant weighted mean zircon populations at 1837 ± 14 Ma, 1703 ± 5 Ma, and 938 ± 32 Ma, with an oldest concordant detrital zircon component dated at 2300 Ma (Buick et al., 2006). Based on the available data, it appears that the volumetrically dominant banded quartzofeldspathic gneisses of MWT are broadly time equivalent to the Sandmata Complex magmatic suite and must have been juxtaposed some time after the 1692 Ma emplacement age of the protolith. The timing of local anatexis in the banded gneisses cannot therefore be related to the 1720 Ma age of magmatism and granulite-facies metamorphism recorded by the granulite facies rocks of SMT (Buick et al., 2006). Based on textural criteria, metapelitic rocks within the amphibolite-grade rocks of MWT are polymetamorphic. The earliest metamorphism reached mid-amphibolite facies conditions, while the subsequent event reached upper amphibolite—facies conditions (6–8 kbar, 650°C–700°C) and was associated with anatexis of metapelites and the emplacement of extensive pegmatite suites (Sharma, 1988). Fareeduddin and Kroner (1998) obtained a single zircon Pb evaporation age of 1690 Ma from a trondhjemite body in the MWT near the Sandmata Hill. The total available data therefore suggests that felsic magmatism and, possibly local in situ anatexis in the MWT, occurred during the Paleoproterozoic, broadly contemporaneously with in situ anatexis and felsic magmatism in the SMT.

The Amet granite, inferred to have formed by syn-kinematic partial melting of the Mangalwar Complex, yielded an ion probe 207Pb/206Pb zircon age of 1641 ± 4 Ma (Wiedenbeck, Goswami, & Roy, 1996). But the zircon Electron Probe Micro Analyser (EPMA) age obtained from the Amet granite itself shows 1753 ± 30 Ma (Biju-Sekhar, Yokoyama, Pandit, Okudaira, Yoshida, & Santosh, 2003). However, the metamorphic zircon from two samples of MWT have yielded identical early Neoproterozoic weighted mean ages of 949 ± 11 Ma and 938 ± 32 Ma. Considering these ages, the juxtaposition of the SMT and MWT was inferred to have occurred at 950—940 Ma (Buick et al., 2006). This is in contrast with the previous notion that the exhumation of the granulites of SMT and localized melting of their country rocks occurred during the Paleoproterozoic.

5.3.3 SANDMATA TERRANE

The SMT occurs in the northwestern part of the AFB and to the west of MWT. The SMT is a long linear belt with a strike length of about 200 km with a width of ~ 40 km and is bounded by the KSZ to the west and DSZ to the east. The SMT is largely composed of high-grade granulite facies garnet—sillimanite pelitic gneisse, quartzofeldspathic gneiss with enderbite interbands, enderbite—charnockite massif, two-pyroxene granulite, leptynite, cordierite—garnet pelitic gneiss, pyroxenite, and norite bodies (Fig. 5.16). Small "klippe-like" bodies of garnet—sillimanite pelitic gneisses occur in the region with tectonic contact particularly in the vicinity of major ductile shear zone (KSZ) at the western margin (Guha & Bhattacharya, 1995). The garnet—cordierite pelitic gneisses occur as thin bands along the shear zones. The rocks have been metamorphozsed to high-grade facies conditions giving rise to penetrative gneissic foliation, which appears to be the earliest deformational fabric in the region. Estimated pressure—temperature conditions are 7—10 kb and 800—1000°C. During the subsequent deformation, the rocks were reworked by shearing and intense migmatization giving rise to extensive stromatic migmatites and seggregations of quartzofeldspathic mineral assemblages. The migmatitic gneisses often grade across and along strike to granite and granodiorite bodies suggesting their association in space and time. One such granite body yielded Rb—Sr whole rock age of 1900 Ma (Choudhary, Gopalan, & Anjaneya Sastry, 1984). There are many subparallel ductile shear zones in the region and some of them were interpreted as thrust zones because of the presence of imbrication and stacking associated with mesoscopic sheath folds.

The SMT is dominated by migmatitic gneisses (2.83 Ga, Tobisch et al., 1994) with sporadic enclaves of amphibolite and metapelite. The SMT is marked by shear zone bounded pockets of granulites known as the Sandmata Granulite Complex (Guha & Bhattacharya, 1995; Sharma, 1988). The high-grade granulites occur as shear-zone bounded pods and lenses of various dimensions within the amphibolite facies felsic gneisses (Roy et al., 2005). These can be traced on a regional scale from Sandmata in the south to Bhinai in north. Metapelitic to metapsammitic granulites dominate in the region of Sandmata Hill (Fig. 5.17) with subordinate calc—silicate rocks, garnet-two pyroxene metagabbros, charnockite and rare metanorite dykes (Sharma, 1988). The metapelitic granulites are stromatic migmatites, comprising thin aluminous layers rich in garnet, biotite, and sillimanite and decimeter to meter-scale, locally discordant garnet leucogranite layers. In contrast, Sandmata granulites occurring further to the north, near the town of Bhinai are dominated by a suite of magmatic charno-enderbites, which show extensive hydration and recrystallization resulting in strongly foliated and lineated amphibolite—facies rocks. Porphyroclasts of relict magmatic ortho- or clinopyroxene are common in them. The fabrics in these rocks are multiply coaxially folded into sheath-like geometries. Fold axes and mineral elongation lineations plunge variably to the WSW to SW (Roy & Rathore, 1999).

The Sandmata Complex is traversed by several shear zones and often grades into deformed felsic orthogneisses of the MWT. These orthogneisses are strongly sheared migmatites that are

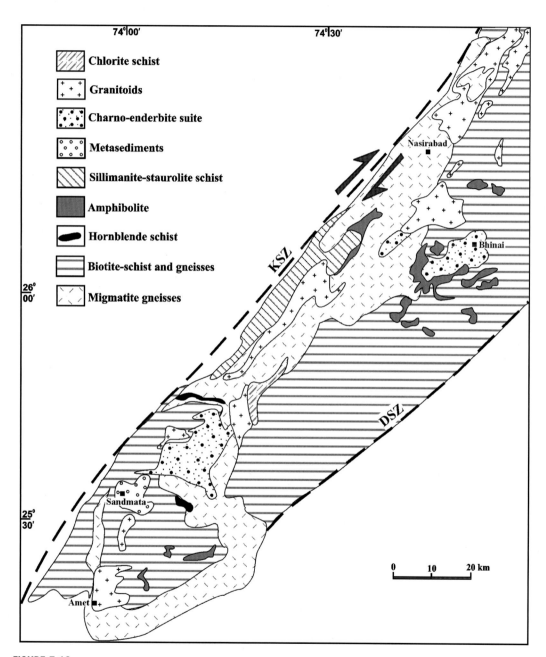

FIGURE 5.16

Geological map of Sandmata Terrane. *KSZ* − Kaliguman shear zone.

Modified after Heron, A. M., (1953). The geology of Central Rajputana. Memoir of the Geological Survey of India, 79, *339;*
Gupta, S. N., Arora, Y. K., Mathur, R. K., Iqbaluddin, Prasad, B., Sahai, T. N., & Sharma, S. B. (1997). The Precambrian geology of
the Aravalli region, southern Rajasthan and north-eastern Gujarat. Memoir of the Geological Survey of India, 123, *262.*

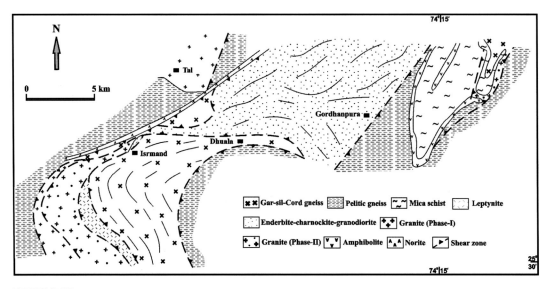

FIGURE 5.17

Geological and structural map of Sandmata granulite facies rocks around Sandmata Hills.

Modified after Gupta, S. N., Arora, Y. K., Mathur, R. K., Iqbaluddin, Prasad, B., Sahani, T. N., & Sharma, S. B. (1980). Lithostratigraphy map of Aravalli region. Hyderabad: Geological Survey of India publication.

variably mylonitized (Roy et al., 2005). However, the orthogneisses in MWT are generally not migmatitic, suggesting the presence of a major shear zone described as the DSZ, which was interpreted as a thrust by Sharma (1988). Although the rocks have suffered intense grain size reduction and recrystallization along this zone, stretching lineations are generally not well preserved due to annealing at relatively high temperatures.

The significant feature of granulite facies rocks of Sandmata hills is the development of strong foliation along the margins of massif granulites, apparently giving rise to large scale fold structures. Large-scale structures show appressed isoclinal folding, often assuming sheath fold geometries. Considering the field relationships in some sectors of the SMT the granulite bodies are interpreted as folded allochthonous sheets emplaced within the gneissic rocks (Roy & Rathore, 1999). Sinha-Roy, Malhotra, and Mohanty (1998) believed that the Sandmata granulites constitute entirely of high pressure granulite facies rocks and that the isograd separating the granulite facies from the amphibolites facies forms the eastern boundary of Sandmata granulites. All the lines of evidence suggest that the Sandmata granulites represent folded "klippe-like bodies" separated from the gneissic rocks by prominent thrust zones, implying that the rocks of SMT must have witnessed thrust-nappe tectonics. Further, the lithological assemblages, structural features and intense migmatization are closely comparable with those of the region in the axial parts of the Cauvery suture zone, where large scale structural domes and basins were interpreted, resulting from constrictive deformation in a larger transpressional tectonic regimes (see Chapter 2: The Southern Granulite Terrane in the present book).

The emplacement of these granulites and the granulite facies metamorphism are broadly synchronous at c. 1.7−1.8 Ga (Fareeduddin & Kröner, 1998). The above thermal event is believed to have reset the host rock of Archean migmatitic gneisses to younger ca. 1.8 Ga ages. The Amet granite, also extending into MWT, occurs at the south extreme of the SMT. The texture of the granite changes from augen gneiss to typical porphyritic granite with coarse grained, which could be indication of variation in the intensity of deformation. There are many other similar bodies around the granulites but dominantly at the southwestern margin of Sandmata granulites displaying the mylonitization of different degrees. The amphibolites that occur around Amet, are seen as alternate bands with biotite gneiss. Chemically, the amphibolites of SMT support the theory of ocean tholeite magma (Kataria et al., 1988).

Several opinions were expressed about the evolution of granulite facies rocks in SMT. They represent tectonically interleaved bodies with the supracrustal rocks that show amphibolite facies metamorphism and that the entire ensemble constitutes the basement of Aravalli−Delhi sequences (Sinha-Roy, Guha, & Bhattacharya, 1992). Sharma (1995) suggested a mechanism of thickening of the continental crust due to underplating of asthenosphere-derived basaltic partial melts. The granulites were subsequently exhumed to shallower levels through major shear zones.

5.3.3.1 Timing of magmatism and metamorphism in the Sandmata Complex

The timing and extent of polymetamorphism in the Mangalwar and Sandmata Complexes from the ADOB remains contentious, with Archean, Paleoproterozoic, and Neoproterozoic events having previously been postulated (Buick, Clark, Rubatto, Hermann, Pandit, & Hand, 2010). The Sandmata granulite−facies rocks of SMT witnessed two episodes of metamorphic equilibration. Coarsegrained, peak-metamorphic granulite facies granoblastic assemblages record P−T conditions of 7−10 kbar; 800−900°C (Dasgupta, Guha, Sengupta, Miura, & Ehl, 1997), whereas the fine-grained, recrystallized assemblages within the shear zones formed at medium pressure, upper amphibolite- to lower most granulite-facies conditions probably at 1720 Ma (Sarkar, Ray Burman, & Corfu, 1989). Paleoproterozoic metamorphism and magmatism in the Sandmata granulite complex is also broadly synchronous with the timing of felsic magmatism in the granitic basement in the NDT.

Zircons from Sandmata granulites yielded two discordant apparent Th/U and 207Pb−206Pb ages of 1075 Ma and 888 Ma, suggesting polymetamorphic nature. A second tectonic event was also believed to have occurred during the early Neoproterozoic/late Mesoproterozoic (Dasgupta et al., 1997). However, from different lines of evidence a Neoproterozoic timing for the exhumation of the Sandmata Complex seems to be consistent with: (1) 1000 Ma isotopic disturbance recorded in zircon from the Sandmata Complex enderbite dated by Sarkar et al. (1989), (2) the evidence for 990−970 Ma extrusive and pre- to syn-kinematic intrusive felsic magmatism in the adjacent SDT. It is tentatively suggested that metamorphism in the high strain zone (DSZ) separating the SMT and MWT occurred at 950−940 Ma, not during the Paleoproterozoic as previously proposed.

The tectonic framework of the SMT and the adjacent region suggest the overthrusting of the Sandmata granulites and underthrusting of the low to medium−grade rocks of the MWT during the evolution of the ADOB and that the plate tectonic processes were operative during the Proterozoic−Archean time, similar to that of present day Himalayas (Sinha-Roy, 1988). In a larger perspective, the metamorphism seems to have taken place in the BGC during both the Paleoproterozoic and early Neoproterozoic indicating possible links between the evolution of the

SMT and some parts of the CITZ (Roy & Prasad, 2003) as well as the northern part of the Eastern Ghats Belt (Mezger & Cosca, 1999) and the 990—900 Ma Rayner Province of East Antarctica with which the latter has been correlated in Rodinia reconstructions.

5.3.4 JHAROL TERRANE

The rocks of the AFB represent a cover sequence and occur as a wedge-shaped form, which tapers to the north and widens toward the south (Fig. 5.18). The Aravalli supracrustal rocks occur in the midst of BGC, which is considered as the basement. Unconformable contacts between these two rock units are either marked by conglomerates or structural discordance. The Aravallis can be divided into two domains: the eastern domain with platformal facies and the western domain with deep water facies. The two domains are separated by the N-S—trending RDSZ. The Aravalli Terrane (both UPT and JRT) was considered to be an oceanic basin at 2.5 Ga, when sedimentation took place with shallow water stromatolite-bearing facies in the east and deep water carbonate-pelite facies in the west. These facies-domains have been separated by the ophiolite-bearing RDSZ that defines a subduction zone along which the Aravalli basin must have been closed at 1.8 Ga (Sarkar et al., 1989; Verma & Greiling, 1995).

The Terrane, west of the RDSZ, is described here as the JRT. It represents a wide triangular region and occurs between SDT and UPT (Fig. 5.18). The JRT consists of shelf facies rocks of Aravalli Super Group and exposes two major groups: Jharol group and Lunavada group of rocks. The JRT is sandwithched between the KSZ at the western margin and the RDSZ at the eastern boundary. In the south, different rock units of Godhra granites and gneisses are exposed in JRT as the basement rocks along with cover rocks such as the Cretaceous beds and Deccan traps. There are slices of ultrabasic rocks through out the terrane, which show highly tectonized with sheared contacts suggesting their tectonic mode of emplacement (Roy & Jakhar, 2002). Highly sheared and folded quartz veins are ubiquitous. The Jharol group of rocks include dominantly argillaceous units interbedded with thin bands of quartzite, which is also described as turbidite sequence. Quartzites occur as high structural ridges with varying thickness from a meter to 50 m and are isoclinally folded displaying the duplication of horizons. In general, the rocks are metamorphosed to amphibolite facies conditions. There are series of subparallel shear zones, characterized by intense mylonitic fabrics associated with phyllonites. A spectacular folding of complex geometry was defined by a thin and competent quartz-garnet rich band. In general, the Jharol formations are totally devoid of mafic metavolcanics. Ultramafic rocks occur as contemporaneous metavolcanics.

5.3.4.1 Bagdunda antiform

An oval-shaped antiformal—domal structure occurs in Bagdunda area of JRT (Sharma, Chauhan, & Bhu, 1988) revealing a circular form of gneissic basement structure (Fig. 5.19). Amphibolites along with domal-shaped basement gneisses occur in the NNE-SSW—oriented antiformal structure. This structure is centrally located between the KSZ and RDSZ. The Bagdunda domal structure (3×9 km) occurs in the northern part of JRT, just west of RDSZ. A band of ultramafic rocks occurs at the eastern margin of domal structure. The quartzites in association with gneisses show dips away from the core. The domal structure is inferred to be interference pattern of fold structures. There is also large variation with curvilinear hinge varying between 15 and 25 degrees along

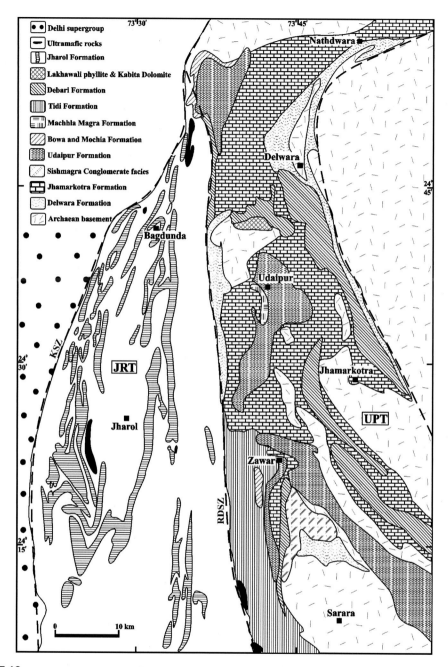

FIGURE 5.18

Regional geological map of Aravalli super group of rocks (JRT and UPT) around Udaipur, between Nathdwara and Sarara. KSZ, Kaliguman shear zone; RDSZ, Rakhabdev shear zone.

Modified after Heron, A. M., (1953). The geology of Central Rajputana. Memoir of the Geological Survey of India, 79, 339; Roy, A. B., Paliwal, B. S., Shekhawat, S. S., Nagori, D. K., Golani, P. R., & Bejarniya, B. R., 1988. Stratigraphy of the Aravalli Supergroup in the type area. Memoir of the Geological Society of India, 7, 121–138.

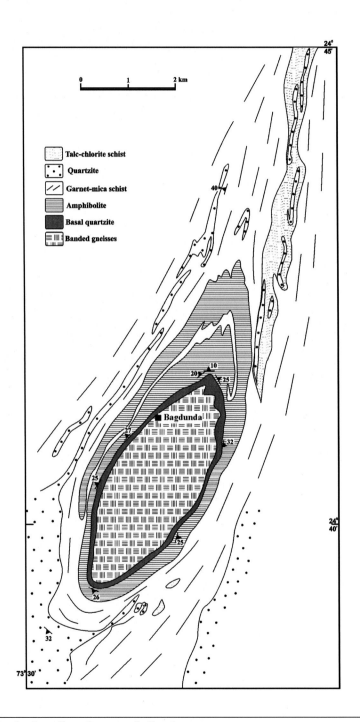

FIGURE 5.19

Strctural map of Bagdunda antiformal structure in Jharol terrane.

Modified after Sharma, B. L., Chauhan, N. K., & Bhu, H. (1988). Structural geometry and deformation history of the early Proterozoic Aravalli rocks from Bagdunda, district Udaipur, Rajastahan. Memoir of the Geological Society of India, 7, 169–192.

the elongation. The gneissic foliation in gneisses in the core of the dome are conformable with the bedding in quartzites, which are near horizontal suggesting flat ductile shearing during the decoupling of the cover sequence. The interference type of fold structures is common. Considering the field observations and the structural pattern, the Bagdunda domal structure can be compared with the Perundurai domal structure occurring in the axial part of the Cauvery suture zone, where the origin was inferred to be constrictive deformation in a larger transpressional tectonic regime (Chetty & Bhaskar Rao, 2006).

5.3.4.2 Lunavada region-structural geometry

The Lunavada region is an apparent and joint continuation of JRT and UPT forming the southern part of Aravalli Super Group without any tectonic break. The lithologies of both the UPT and the JRT continue into the region with a major part of the Lunavada region occurring to the west of RDSZ. The rocks of the Lunavada region shows intricate structural geometry of superposed folding exhibited by the thin bands of quartzite along with thick bands of phyllite (Fig. 5.20). The major rock types in the region are a sequence of metapelitic and quartzitic layers with bands of calc—silicates in the form of a series of overturned folds with NW-SE—trending axial traces. However, the interpreted geometry of the structure suggest the superimposition of NW-SE—trending open folds on a chain of NE-SW—trending antiformal domes and basins (Mamtani, Karant, Merh, & Greiling, 2000). The structural map of the Lunavada region show the presence of regional-scale superposed folds with extreme complexity of deformation. The dips of foliations are predominantly to north with gentle to moderate values and the mineral stretching lineations show gentle plunges toward the NNW. The early generation folds are coaxial, both having NE-SW—trending fold axes, giving rise to the formation of a Type-I11 interference fold pattern (Ramsay & Huber, 1987).

A number of shear zones traverse the JRT dividing the region into blocks of contrasting structural patterns. The shear zone that separates the Lunavada region from the Jharol formations was interpreted as a thrust that aided the intricately folded quartzite—phyllite sequence to ride over the sequence of Jharol formations. There are also many subparallel shear zones that bound the isolated bodies of serpentinite, which lie close to the RDSZ. In general, the Lunavada region appears to be bound by the shear zones on either side and the regional structure seem to be constituting domes and basins, which are likely to have been derived from transpressional tectonics. Another interesting observation is that the Bagdunda antiformal structure in the north and the domal structure in Lunavada region are aligned along a N-S—trending zone.

5.3.4.3 Godhra granite

With in the metasedimentary rocks of Lunavada Group, granites represent an important rock type that occur in a vast migmatitic terrain of granites and gneisses showing intrusive relationship. They are described as Godhra granites in the north and Champaner group in the south (Fig. 5.21). The granites were dated as 955 ± 20 Ma with the help of Rb-Sr whole-rock dating technique (Gopalan, Trivedi, Merh, Patel, & Patel, 1979). Lunavada group of rocks terminate in the southwestern part against the contact of the Godhra granite and gneiss. Further south of these granitic rocks, low-grade phyllites and quartzites of Champaner group are exposed around Champaner town. These granitic rocks are characterized by a simple deformation history compared to the Lunavada group of rocks. The gneisses around Godhra granite display banded appearance and comprise light-colored quartzofeldspathic alternating with dark biotite-rich layers. These banded gneisses trend

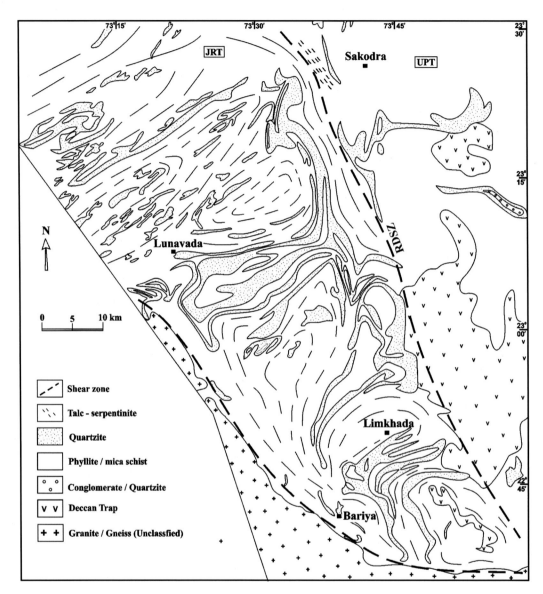

FIGURE 5.20

Structural map of Lunavada region in Jharol terrane. *JRT*, Jharol Terrane; *RDSZ*, Rakhabdev shear zone; *UPT*, Udaipur Terrane.

Modified after Gupta, B. C., & Mukherjee, P. N. (1938). The geology of Gujarat and South Rajputana. Records, Geolological Survey of India, 73, 163–208.

NE−SW to E−W with gentle northerly dips preserving the evidences of polyphase deformation (Mamtani, Karmakar, & Merh, 2002). Tight northwest plunging isoclinal recumbent/reclined folds are well preserved with their axial planes subparallel to the gneissic foliation in these rocks. The

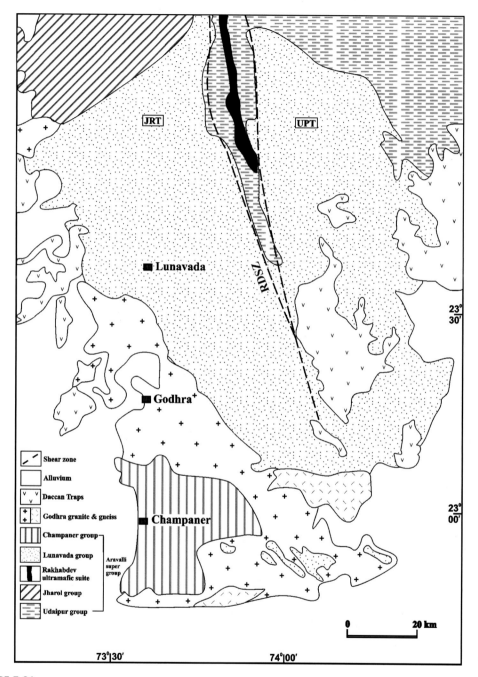

FIGURE 5.21

Geological map of Godhra granite in the southern part of the Jharol terrane. *JRT*, Jharol Terrane; *RDSZ*, Rakhabdev shear zone; *UPT*, Udaipur Terrane.

Modified after Mamtani, M. A., Karmakar, B., & Merh, S. S. (2002). Evidence of polyphase deformation in gneissic rocks around devgadh bariya: Implications for evolution of godhra granite in the Southern Aravalli Region (India).
Gondwana Research, 5, 401–408.

gneisses were interpreted to have been subjected to high strain and melting that led to the formation of the granitic rocks at deeper crustal levels and the granite were synkinematically emplaced along the foliation planes. The Champaner group of granites and gneisses show open to tight E-W−trending folds without the effect of any superposed folding (Merh, 1995) suggesting that the deformation may be related to secondary deformation in the Lunavada region.

5.3.5 UDAIPUR TERRANE

The UPT, the type area of Aravalli Super Group of rocks, occurs in the south eastern part of the ADOB, sandwitched between the RDSZ in the west and an extended branch of the KSZ in the east. It is also limited by BSZ in the north (see Fig. 5.18). The Aravalli Supergroup includes not only the "Aravalli system" but also the outliers of Delhi system, and the Raialo marbles and mica schists. The Aravalli Supergroup occurs in close association with the Archean basement represented by BGC (Mewar gneiss) and associated granitoids. The gneiss metasediment interface is marked by unconformity (Roy & Jakhar, 2002). There were different views on stratigraphic schemes with internal contradictions, essentially resulting from misidentification of field relationships between the metavolcano-sedimentary rocks and granite−gneiss complex. It also suffered from incorrect assumptions of intrusive relationship of granites and the surrounding gneisses. In general, the rocks of UPT represent large linear outcrops with NNW−SSE trends with steep dips. Detailed mapping in different sectors reveal steeply dipping high-strain belts separate areas of low strain with moderate dips.

The UPT comprises low grade metamorphic rocks preserving a host of depositional sedimentary features, which can be conveniently described in the form of four rock units: the basement gneisses, and the lower-, middle-, and upper-Aravalli group of rocks. The lower Aravalli group occurs in the form of two subparallel linear belts. The easternmost belt extends from southeast of Nathdwara in the north to Banswara in the southeast, which lies between the BGC in the east and Debari formation in the west. The important rock types in the belt are metabasalts with thin bands of dolomite−quartzite and veins of barite (Fig. 5.22). A thin veneer of metavolcanic rocks locally intervenes between the two. Discontinuous and lenticular pockets of white mica deposits are always observed at the contact zone between the basement and the overlying metasediments. These deposits were found to be metamorphosed clay pockets and probable paleosols based on their chemical characteristics (Roy & Paliwal, 1981). A number of lava flows were identified and characterized by the presence of vesicles. Individual flows contain thin (10−20 cm) discrete ironstone and carbonate bands. The chemistry of metavolcanics shows broad variation in composition from continental tholeite to oceanic to high Mg−basalt and komatiite (Ahmad & Rajamani, 1991).

The rocks in the eastern belt are overlain by carbonate rocks that are dominantly dolomitic, followed by orthoquartzite, ferruginous dolomite, manganese-bearing dolomite, and carbonaceous phyllites. In addition, a persistent horizon of stromatolitic rock−posphate also exists within the dolomitic sequence. These rocks are typically exposed at Jhamarkotra and hence the entire sequence was described as Jhamarkotra formation. A characteristic feature in these rocks is the presence of N-S−trending linear belt of carbonaceous phyllite with rock phosphate-bearing dolomite-quartzite, possibly coinciding with the site of the extended branch of the KSZ. There are also known occurrences of copper−uranium minerals in carbonaceous phyllite, while manganese minerals and stromatolitic phosphorite are known in dolomitic limestones.

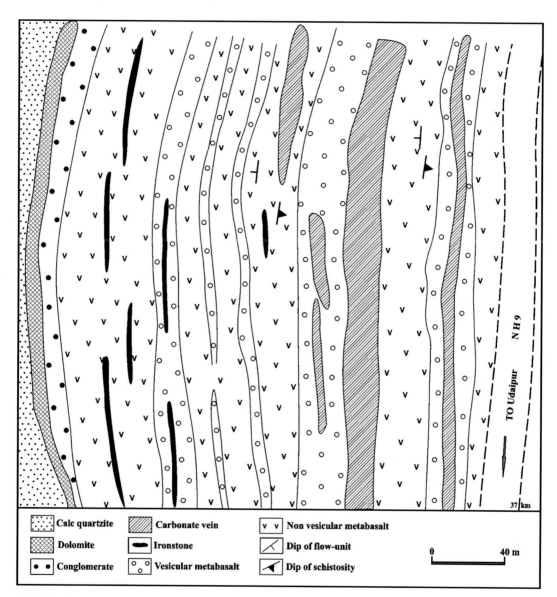

Calc quartzite   Carbonate vein   v v Non vesicular metabasalt

Dolomite   Ironstone   Dip of flow-unit

• • Conglomerate   o o Vesicular metabasalt   Dip of schistosity

0   40 m

FIGURE 5.22

Flow bands of volcanic flows in Delwara formations near Nagaria.

Modified after Sinha-Roy, S. (1988). Proterozoic Wilson cycles in Rajasthan. In: Roy, A.B. (Ed.), Precambrian of the Aravalli Mountain, Rajasthan, India. Memoir of the Geological Society of India 7, *95–108.*

The middle Aravalli group of rocks consisting of a thick sequence of greywacke overlies the lower group. Rhythmic layering is well preserved in graded thick beds of greywacke. Several isolated lenses of pebbly and boulder-bearing conglomerate with phyllitic matrix occur within greywacke. The pebbles belong to quartzite, granite, carbonate, and phyllite. Chemical characteristics indicate that the greywackes are continent-derived sediments (Sreenivas, 1999). Thinly bedded slate and phyllite with interbeds of dolomite and quartzite constitute the youngest of the middle Aravalli group. Although the rocks of the UPT belong to a low grade that reaches up to the upper amphibolite facies metamorphism, there are variations with local high grade rocks in some sectors that lie mostly along the ductile shear zones.

The major rock types in the Upper Aravalli group are polymictic conglomerates, quartzites, dolomites, and phyllites. The conglomerates contain pebbles and boulders of quartzite, granite, granite gneiss, marble, amphibolites, tourmaline rich rock, and mica schist. Conglomerates grade upward into arkosic quartzites which seem to have been highly deformed. Ultramafic rocks are also common in UPT particularly, south of Nathdwara.

Like any orogenic belt, the rocks of the UPT were subjected to polyphase deformation and gave rise to complex structural architecture. The isoclinal nature of first-generation folds show extremely drawn out hinges with synchronous development of schistosity. The second generation folds show variation in their trend often attaining noncylidrical geometry. The dominant structural grain in UPT is NNW−SSE with steep dips on either side. Detailed mapping in different sectors reveal that both high strain and low strain areas are well preserved. The dips are gentle to moderate in low-strain areas in contrast to high-strain areas. The earliest deformation resulted in the development of sinistral asymmetric folds implying simple shear (Naha & Mazumdar, 1971). Subsequent deformation resulted in the development of vertical axial planes but the folds show extreme variation in their plunge from near horizontal in the limb regions of early reclined folds to vertical at the hinge zones. Such complex outcrop pattern was described as fold interference structures generally of Type II of Ramsay (1967). The fold interference structures show varied forms such as hook-shaped, eyed folds, and irregular domes and basins of different scales. Majority of shear zones in UPT are characterized by mylonitic fabrics and occur at the contact zones between the basement and the cover sequences. Detailed structural mapping in different parts of the UPT reveals formation of complex outcrop patterns, which could be due to the superposition of different folding phases or the heterogeneous movements along ductile shear zones.

5.3.5.1 *Zawar mining region*

The Zawar region is located in the southern part of UPT and is known for large deposits of zinc−lead ores. The host rocks belong to Middle Aravalli group of rocks. The major rock types in the region include: carbonaceous phyllite with dolomite, greywacke, dolomite, quartzite, and phyllite and mica schist. The ore mineralization is strata bound and is confined to dolomite (Banerjee & Sarkar, 1998). The differential movements along the basement-cover sequence generated thick zones of ductile shearing and intense mylonitic fabrics with some relict low strain zones givingrise to complex outcrop patterns. The zawar region represents such a zone of typical complex outcrop pattern (Fig. 5.23). The early folds in the region are, in general, asymmetric and show a thinning of shorter limbs and detachment in some localities that led to the formation of typical "fish-hook−type" geometry and often attaining sheath fold geometry. A spectacular example is the presence of well-exposed sheath fold near Salumbar (Mukhopadhyay & Sengupta, 1979). Mega sheath fold structures were also reported from NW of Udaipur town. Small scale kinks and chevron folds are well developed in the Zawar area. It is possible that the complex outcrop pattern around Zawar region could be due to the presence of bounding shear zones.

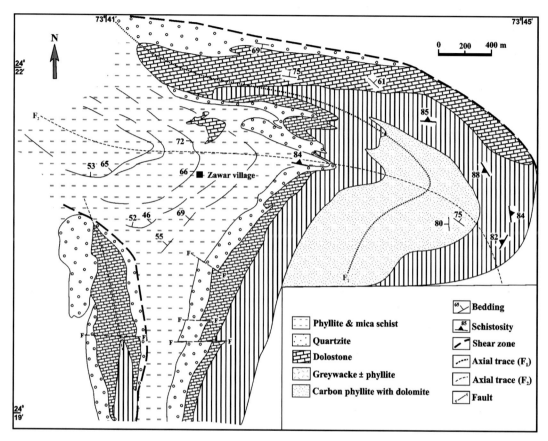

FIGURE 5.23

Detailed structural map showing complex fold structures in Zawar mining district in Udaipur terrane.

Modified after Roy, A. B. (1995). Geometry and evolution of superposed folding in Zawar lead-zinc mineralised belt, Rajasthan.
Proceedings of Indian Academy of Sciences (Earth & Planetary Sciences), 104, *349–371.*

5.3.5.2 *Untala granite*

The Untala granite, located east of Udaipur, is a typical K-feldspar—bearing granite grading into diorite. It is a medium-grained, pink in color, massive leucocratic granite without any deformation and is surrounded by migmatitic gneisses in the form of a domal structure. The gneisses constitute dominantly of granodiorite—quartz diorite sequences. The enclaves of tonalite—tonjhemite and adamellite to granodiorite occur both with in the granite as well as in the migmatitic gneisses. Small carbonatite intrusions are also mapped. The entire body of Untala granite is not homogeneous but includes patches of biotite gneisses varying in composition from adamallite to granodiorite and tonalite along its margins (Fig. 5.24). The granite is criss-crossed by pink granite veins and pegmatites. The gray-colored gneiss conatins megacrysts of plagioclase. The trondhjemite gneiss yielded zircon age of ~2900 Ma. From the field relationships it is possible that the Untala granite was syntectonically emplaced.

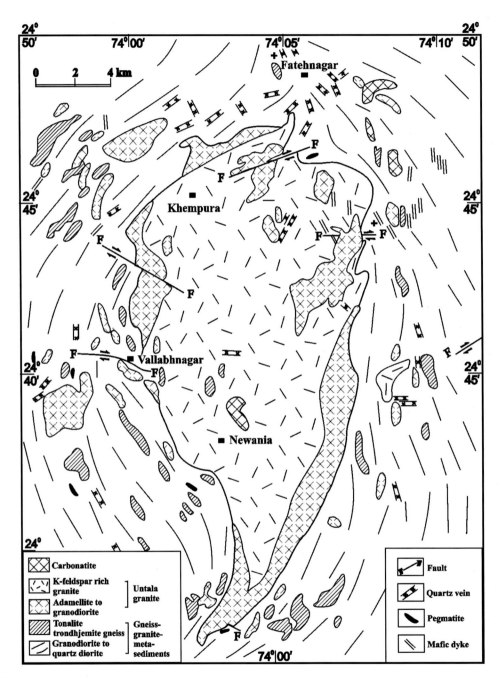

FIGURE 5.24

Map of Untala granite, Udaipur terrane.

Modified after Roy, A. B., & Jakhar, S. R., (2002). Geology of Rajasthan (Northwestern India), Precambrian to Recent.
Scientific Publishers, p. 421.

After detailed structural analyses of rocks in Udaipur region, two structural profiles covering the entire Aravalli fold system including the JRT and UPT were constructed by Sugden, Deb, and Windely (1990), as shown here (Fig. 5.25). Structurally, both the UPT and JRT including the Bhilwara formations record a similar tectonic history, intense E–W stretching bringing early (F1) folds in parallelism. The basement-cover contacts were interpreted as shear/thrust zone and the AFB must have been thrust bodily transported over the craton supporting the hypothesis of allochthonous nature of the Aravalli rocks relative to the basement complex. This is substantiated by the presence of the high-strain pyrophyllite–muscovite horizon along the contact zone, believed to have acted as a glide horizon (decollement). The hypothesis is further supported by the presence of subhorizontal fabric and E–W stretching lineations accommodating 10 km of strain with a change in stress pattern from simple shear to dominant pure shear. The bulk of the kinematic axes during D2 remained roughly parallel to the D1 axes, indicating that the two phases record an essentially coaxial stress history. The final deformational event was dominated by subhorizontal shear parallel to the trend of the orogen resulting in the development of a series of shallow,

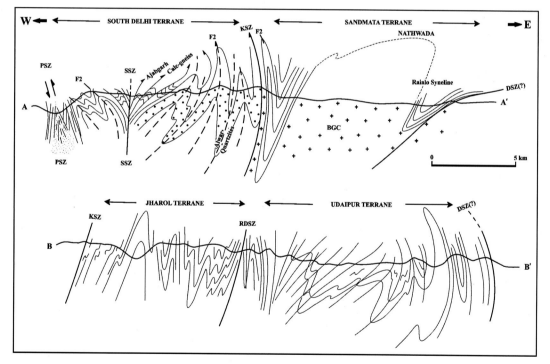

FIGURE 5.25

Map showing the structural profiles across the ADOB. *PSZ*, Phulad shear zone; *SDT*, South Delhi Terrane; *SSZ*, Sabarmati shear zone; *KSZ*, Kaliguman shear zone; *DSZ*, Delwara shear zone; *RDSZ*, Rakhabdev shear zone; *BGC*, Banded gneissic complex.

Modified after Sugden, T. J., Deb, M., & Windely, B. F. (1990). The tectonic setting of mineralization in Proterozoic Aravalli Delhi orogenic belt, NW India. vol. 8. Amsterdam: Developments in Precambrian Geology, Elsevier, pp. 367–390.

southeasterly dipping shears and NE-trending recumbent folds. This phase seems to be important in terms of collisional tectonics because it represents the time when colliding lithospheric plates had effectively "locked" and further convergence was accommodated by wrenching parallel to the plate boundaries (Sugden et al., 1990).

Regarding the evolution of the AFB, Roy (1990) followed traditional geosynclinal concepts comparable to the eugeosynclinal—miogeosynclinal couplet of an Alpine type. Many others initiated and proposed modern concepts of continental collision tectonics following Wilson cycle in the relam of plate tectonic models (e.g., Sen, 1981; Sychanthavong & Merh, 1984). However, all these models were heavily dependent on the classic work of Heron (1953), from where new thoughts and new concepts emerged.

5.3.6 SOUTH DELHI TERRANE

The SDT and the NDT were delineated from the DFB based on the Rb—Sr whole-rock data from synkinematically emplaced granites that yielded 1.65—1.45 Ga and ~0.85 Ga, respectively (Choudhary et al., 1984). Based on the lithofacies association and available age data, the DFB is diachronous and the evolutionary histories of the NDT and SDT are different (Sinha-Roy et al., 1995). However, there is also another opinion that the DFB represents a single belt of stratigraphically homologous and tectonically contiguous in nature. In the present context, for the sake of brevity and clarity, the DFB has been described as two units SDT and NDT. The SDT represents a highly elongated terrane in the south and the NDT is a widespread region in the north. Many details of the DFB as well as ADOB can be found in excellent reviews of Gupta (1934); Heron (1953); Gupta et al. (1997) and (Roy & Jakhar, 2002).

The SDT occurs as a linear fold belt over a strike-length of about 750 km; along the western edge of the AFB (including the SMT) and is confined between the KSZ in the east and PSZ in the west. The rock formations in SDT are dominated by arenaceous facies in the eastern sector and calcareous facies in the western sector (Heron, 1953). The major rock types in SDT are broadly described as the Gogunda and Kumbhalgarh groups of rocks along with magmatic phases of the Phulad ophiolite suite, syn-sedimentational acid flows and tuffs and synorogenic granites (Gupta et al., 1997). In general, the rocks are marked by amphibolite facies metamorphism and multiple stages of folding. However, the terrane shows sporadic occurrences of granulites, tectonic slices of ophiolite, blue schist, and basement migmatitic gneisses. Several conglomerate horizons are also reported within SDT.

The SDT is bound by the major boundary shear zones along with many other shear zones within the supracrustal belts, thus making the SDT divisible into many longitudinal tectonic zones (Fig. 5.26). Another important shear zone that occurs at the axial zone of SDT, which was recognized earlier as a thrust (Heron, 1953), is described here as the Sabarmati shear zone (SSZ). The SSZ is characterized by the presence of mylonites, ultramafic rocks and tectonic slivers of the basement gneiss. The SSZ broadly separates the pelitic and psammitic rocks of continental slope facies and platformal pelite—carbonate sequence of the Bhim Group in the eastern sector, and the basic and felsic volcanic with shallow water clastics forming Barotiya and Sendre Groups in the western sector (Sinha-Roy et al., 1998). In general, the eastern sector is devoid of volcanics in contrast to the western sector. Considering its separate geochronologic and lithotectonic identity, the western sector is described as the Ambaji-Sendra belt (Deb, 2001).

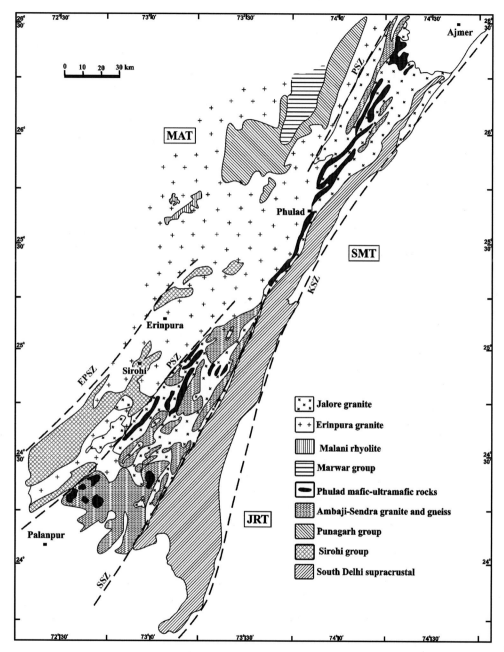

FIGURE 5.26

Gelogical map of the South Delhi Terrane. *EPSZ*, Erinpura Shear zone; *PSZ*, Phulad shear zone; *SSZ*, Sabarmati shear zone; *KSZ*, Kaliguman shear zone; *MAT*, Marwar terrane; *SMT*, Sandmata terrane; *JRT*, Jharol terrane.

Modified after Deb, M., Thorpe, R. I., Krstic, D., Corfu, F., & Davis, D. W. (2001). Zircon U–Pb and galena Pb isotope evidence for an approximate 1.0 Ga terrane constituting the western margin of the Aravalli–Delhi Orogenic Belt, northwestern India. Precambrian Research, 108, *195–213.*

The Barotiya Group of the western sector is further divided into three tectonic units: the western unit consists of a Barr conglomerate and associated mica schist, impure marble, calc schist, subarkose, and bimodal volcanic rocks. The central unit contains marble and metavolcanics and the eastern unit is represented by metavolcanic, subarkose, and impure marble. The Barotiya Group also comprises metachert, marble, pillow basalt, plagiogranite, gabbro and serpentinite, pyroxenite, interpreted as marginal sequence. The Sendre Group is represented by metachert, marble, meta rhyolite, and metabasalt, interpreted as volcanic arc sequence. A thick complex of mafic and ultramafic rocks closely associated with the Barotiya-Sendra sediments occur all through the length of SDT and have been collectively referred to as Phulad Ophiolite Suite (Gupta et al., 1980).

The contact between different sequences is defined by prominent ductile shear zones and thrusts. A structural cross section from Phulad to Devgarh across the SDT (Fig. 5.27), reveals the existence of the PSZ on the west and the KSZ in the east and the SSZ in the axial part separating the distinct lithological units (Sinha-Roy, 1988). A thick turbiditic sequence of calc–schist, marble, and semipelitic schist occur near Kamlighat, which is interpreted as an accretionary trench sequence. This sequence is tectonically overlain by conglomerate grit with pre-Delhi gneisses and minor carbonate sequence that were interpreted as sediments of arc-trench gap. These rocks are further followed by volcanic pile represented by metamorphosed rhyolite, andesite, and basalt

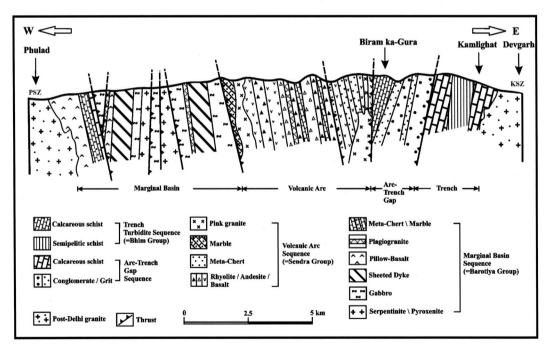

FIGURE 5.27

Structural cross section across the South Delhi Terrane.

Modified after Sinha-Roy, S. (1988). Proterozoic Wilson cycles in Rajasthan. In: Roy, A. B. (Ed.), Precambrian of the Aravalli Mountain, Rajasthan, India. *Memoir of the Geological Society of India 7, 95–108.*

with minor metachert and marble, which is intruded by K-rich granites. Further to the west, serpentinite, pyroxenite, layered gabbro, sheeted dyke complex, and pillowed basalt occur in the form of an imbricate zone, which has been described as the Phulad ophiolite complex.

The earliest folds in the SDT are generally isoclinal and recumbent in nature with NNE-trending axes, which were described as "gravity-induced" structures by Naha, Mukhopadhyay, Mohanty, Mitra, and Biswal (1984). These folds were described contemporaneous with thrusting and were coaxially refolded by progressive deformation. The later folds, although of the similar age, show variations in different regions. Vertical extrusion of high-grade calc—gneisses was accommodated by displacement by discrete upward-splaying shear zones during the subsequent deformation and the development of major discontinuities (e.g., the PSZ/Ranakpur shear zone) on the western margin of the SDT. The regional structure of the SDT is an antiformal structure with gentle plunge to NE. All the structural and tectonic features that constitute the SDT indicate that it is an imbricate thrust zone subjected to dextral transpressional tectonic regime (Gupta, Mukhopadhyay, Fareeduddin, & Reddy, 1991; Gupta et al. 1995). Considering the structural architecture, the SDT was also interpreted as "flower structure" (Sugden et al., 1990).

A gneiss domal structure was also mapped from Anasagar area with in the axial zone of the SDT (Mukhopadhyay, Chattopadhyay, & Bhattacharyya, 2010) and was considered as a thrust-related gneiss dome and not as a metamorphic core complex. The dislocation zone along the gneiss-supracrustal contact is invariably parallel to the foliation in the underlying gneiss as well as with the overlying bedding planes. It was interpreted that the dislocation zone had a ramp-and-flat geometry exposing flat structure at the present level of exposure with respect to the footwall. The truncations in the upper unit represent hanging wall ramps, and the parallel parts represent hanging wall flats. The ramps and flats on the fault surface were inferred to be located at depth. It is therefore likely that the Anasagar gneiss dome and its enveloping supracrustal rocks form a part of the pre-Delhi basement (?Aravalli Supergroup) caught up within the Delhi deformation. Available geochronological data indicate that the emplacement of the Anasagar gneiss dome predated the formation of volcanic rocks in SDT.

The regional metamorphism in SDT represents, in general, amphibolite facies conditions with variation largely from amphibolite to granulite grade metamorphism at a few localities in isolated domains through out (Sharma, 1988). The presence of granulites is mostly restricted to the longitudinal shear zones, possibly representing the deeper levels of magmatic arc. One such important location of granulites in the southern part of the SDT is the Balaram-Kui-Surpagla-Kengora granulites (BKSK granulites); the details are described below.

The BKSK granulites occur in the form of a lensoid body marked by several large-scale shear zones (Fig. 5.28). The Surpagla shear zone defines the eastern margin with low-grade rocks of the Ambaji basin while the Kui-Chitraseni shear zone marks the western margin. The other shear zones like Ghoda-, Jogdadi-Balaram shear zones occur within the BKSK granulite terrane (Singh, De Waele, Karmarkar, Sarkar, & Biswal, 2010). The shear zones vary in width from a few meters to hundreds of meters and extend for a few kilometers along the strike exhibiting variably mylonitized different rock units. The shear zones are characterized by ductile shearing with dominant thrust displacements implying the exhumation of granulites through thrusting.

The Ambaji basin in the region represents a large-scale southwesterly plunging F2 fold with both limbs showing southeasterly dip (see Fig. 5.28; Cross-sections A—B and C—D as inset). The BKSK granulites comprise pelitic and calcareous metasedimentary granulites; a gabbro—norite—basic

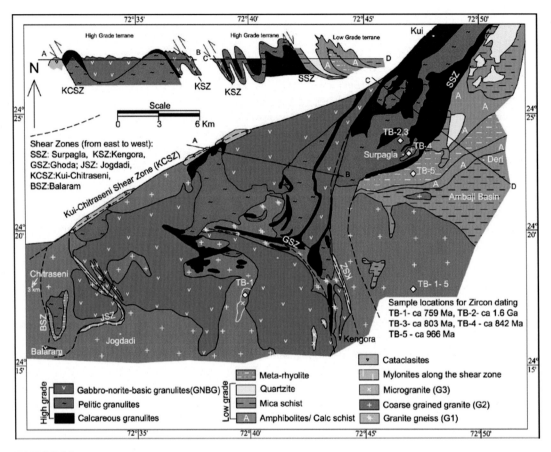

FIGURE 5.28

Geological and structural map of the Balaram-Kui-Surpagla-Kengora granulites, southern part of the South Delhi Terrane.

Adapted from Singh, Y. K., De Waele, B., Karmarkar, S., Sarkar, S., & Biswal, T. K. (2010). Tectonic setting of the Balaram–Kui–Surpagla–Kengora granulites of the South Delhi Terrane of the Aravalli Mobile Belt, NW India and its implication on correlation with the East African Orogen in the Gondwana assembly. Precambrian Research, 183, *669–688.*

granulite suite, and several phases of granitoid rocks termed the Ambaji granites. The pelitic granulites exhibit prominent migmatitic structures and calcareous granulites are interlayered with metarhyolites and metabasalts, representing rift-related synsedimentary lava flows. The metabasalts were metamorphosed to amphibolites with highly flattened pillow structures in some places. The gabbro–norite plutons exhibit magmatic layering with mineral segregation. The layers range in thickness from a few mm to a meter, and vary widely in composition from anorthosite, troctolite, to pyroxenite. The rocks show a calc-alkaline affinity of the gabbro–norite–basic granulite suite, which were interpreted as ophiolite suite emplaced in a magmatic arc setting during 860–750 Ma (Khan, Smith, Raza, & Huang, 2005). The peak temperature and pressure were estimated as $\geq 850°C$ and 5.5–6.8 kb.

The volcanogenic massive sulphide deposits and their host volcanic rocks within the western sector of the SDT yielded an age of 987 ± 6 Ma for rhyolite. A Sm–Nd whole-rock isochron age of 1012 ± 78 Ma was obtained for diorites associated with amphibolites at Ranakpur (Volpe & Macdougall, 1990). This suggests that the rocks of the western sector of SDT are much younger to the rocks of the eastern sector of the SDT as well as the Aravalli group of rocks. All of this evidence indicates that the region is a distinct metallogenic province of volcanic associated massive sulphide deposits in the Ambaji-Sendra terrane with an age of about 1.0 Ga with characteristics of island arc environment (Deb et al., 1989).

5.3.6.1 Tectonic synthesis

Tradionally, the geosynclinal concept was in vogue for the evolution of the SDT since the days of Heron (1953). Following the modern concepts of subduction and collision, many researchers proposed varied tectonic models. Sinha-Roy (1984) proposed and applied plate tectonic regime for the evolution of SDT involving rifting, formation of ocean basin, and closure analogous to modern orogenic belts. Sinha-Roy (1988) suggested two stages of westerly subduction during the collision between an arc sequence in the west and a trench sequence in the east. This resulted in the development of carbonate–turbidite trench sequence in the eastern sector and an island arc sequence in the western sector of the SDT. Further, Phulad ophiolite mélange zone at the westen margin of the SDT was developed as a consequence of obduction during dextral transpressional tectonic regime.

Contrary to westward subduction, easterly subduction of oceanic crust in a spreading intracratonic rift basin was proposed with the initial development of the Ambaji-Sendra belt as an arc-trench system (Sen, 1981). Accordingly, all the tectonic zones of the SDT were accreted against the continental mass to the east. The eastward subduction along the western side of the SDT was further supported by the terminal collision of the arc terrane with the Aravalli continental margin, involving intense progressive strain and oblique convergence (Deb & Sarkar, 1990). The SDT must have been subjected to a late Proterozoic accretion event that may have involved closure of an ocean basin, and the Pan-African orogeny of eastern Africa and the Arabian Peninsula (Sugden et al., 1990). The available geochronological data in SDT presents the range between ca. 1.7 and 0.8 Ga (Choudhary et al., 1984), and the ocean was closed through eastward subduction along the KSZ that occurs along the eastern contact of the SDT. Based on deformational history alone, a common evolutionary history was invoked for the rocks of both SDT and NDT and the variations in structural styles, fold forms were ascribed to the development shear zones and later plutonic intrusions.

The base metal mineralization in the SDT show certain similarities with those formed in an island arc environment. The narrow time-stratigraphic interval of the ore zones and alignment of the ore bodies in small, linear fault controlled basins appear to suggest that they were generated over steeply dipping subduction zones. The rich polymetallic sulfide lenses in the Ambaji-Deri zone in the southern part of the SDT constitute an important base metal concentration. The stratiform ore bodies are hosted by magnesian sediments overlying massive to schistose metavolcanic amphibolites. Geochemically, the amphibolites suggest an ocean floor setting of emplacement while showing an overall island arc affinity. Such signatures, together with the presence of an alkali syenite pluton at Deri support the back-arc regime for this ore zone within the Ajabgarh Group. The subduction zone along the western fringe of SDT presumably became shallower and deeper-penetrating with time to produce the tungsten-tin deposits in S-type magmas. In this context, it is worth noting that the Ambaji-Deri ores formed around 1100 Ma while the tungsten-tin occurrences associated with the Erinpura, Jalor or Malani igneous phases range between 850 and 735 Ma, respectively (Crawford, 1975).

Broad periods of felsic magmatism could be delineated from the ADOB based on the Rb−Sr geochronology of different rock types (Choudhary et al., 1984). They established that the alkali granite and granodiorite pluton within the Alwar Basin, northeast of Ajmer, yielded an age of 1700−1500 Ma, while the plutons along the SSZ gave rise to ages between 900−800 Ma. The dichotomy in these ages could reflect separate magmatic episodes in independent terranes now juxtaposed along the suture zone. The geochemical, geochronological, and isotope studies on Phulad ophiolites and related rocks from the SDT led to the inference that they represent the fragments of the Proterozoic island arc complex (Volpe & Macdougall, 1990). With a high velocity of 7.3 km/s for the lower crust and the Moho depth of around 50 km, the SDT represents seismic expression of a paleo-arc signature (Tewari, Dixit, Rao, Venkateswarlu, & Vijaya Rao, 1997). All these evidences suggest the existence of an island arc in the region of SDT.

In summary, the structural, metamorphic and geochronological studies of SDT indicate that the granulites of SDT are early to mid-Neoproterozoic in age (900−700 Ma). Although, a late-Neoproterozoic/early-Cambrian overprint is not very prominent, the brittle deformation and pseudo-tachylite formation in the granulite facies rocks of the SDT can be interpreted to be Neoproterozoic-Cambrian in age (Sarkar & Biswal, 2005). The Meso-Neoproterozoic SDT must have been deformed during the collision of the Bundelkhand craton and MAT, resulting in the amalgamation of the northwest Indian blocks at ∼1 Ga (Bhowmik, Bernhardt, & Dasgupta, 2010).

5.3.7 NORTH DELHI TERRANE

The NDT is exposed like a huge fan at the northern part of the ADOB with a long handle in the form of the SDT extending to south (Fig. 5.29). The NDT has a maximum width of about 200 km north of Jaipur (Roy & Jakhar, 2002). The NDT is bounded by the GBSZ in the east and the Singhana-Jasrapur shear zone, a possible extension of PSZ, in the west. To the east, the Archean-Paleoproterozoic rocks of the Aravalli mountain range abut against the GBSZ, which separates them from the Mesoproterozoic platformal sediments of the Vindhyan Supergroup. The NDT is considered as a cover sequence over a Paleoproteorzoic basement consisting of the Aravalli fold belt and A-type granites of 1850−1700 Ma (Biju-Sekhar, Yokoyama, Pandit, Okudaira, Yoshida, & Santosh, 2003).

Traditionally, the rock units of the NDT were divided into three units (Heron, 1953). The lower unit (Raialo Group) is dominantly composed of carbonates and mafic volcanics, the middle unit (Alwar Group) is predominantly arenaceous with mafic volcanics, whereas the upper unit (Ajabgarh Group) consists of calcareous and argillaceous sediments with interlayered mafic volcanics. Polyphase deformation in NDT resulted in complex outcrop patterns with abundant structural duplication. The rocks are folded into appressed, upright NNE-SSW−trending antiforms and complementary synforms with moderate plunges. These folds were subsequently refolded on NW−SE and ENE−WSW axial planes, resulting in axial culminations and depressions. (Singh, 1984).

Broadly, the NDT can be described constituting three major basins from west to east. They include: the Khetri subbasin, the Alwar subbasin and the Lalsot-Bayana subbasin. The metamorphic history of the easternmost Lalsot-Bayana subbasin is considerably different from the other two as it

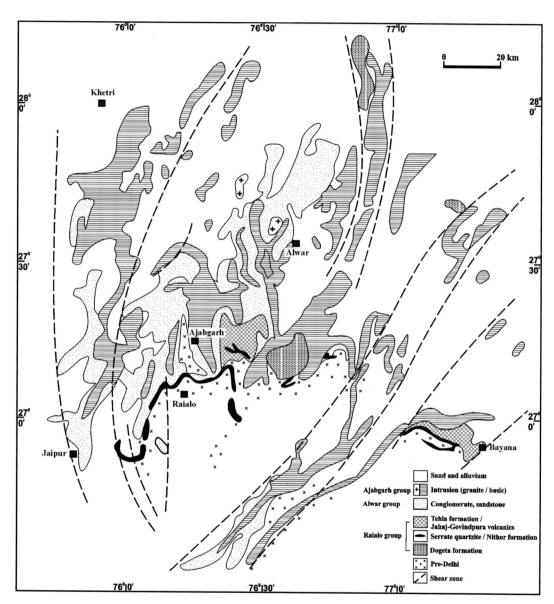

FIGURE 5.29

Geological map of the North Delhi Terrane.

Modified after Das Gupta, S. P. (1968). The structural history of the Khetri Copper Belt, Jhunjhunu and Sikar districts, Rajasthan.
Memoir of the Geological Survey of India, 98, *170.*

has been reported to be nearly unmetamorphosed, while the other two subbasins display medium grade metamorphic assemblages (Mehdi, Santosh, & Pant, 2016). The rocks in the eastern most basin are marked by the development of chemically distinct authigenic phengitic muscovite in metasiltstone. The rocks in Alwar and the Khetri subbasins were metamorphosed to amphibolite facies conditions and preserve evidence of ∼1000 Ma orogeny signifying the closure of the Delhi basin.

The Lalsot-Bayana subbasin constitutes all the three Groups of rocks namely Raialo, Alwar and Ajabgarh from bottom to top, which are separated by prominent unconformities. The Raialo Group is overlain by a sequence of basic volcanics and associated metasedimentary beds. The volcanics include tuff, aphanitic basalt, vesicular and amygdaloidal basalt, flow breccias, and inter beds of quartzite. The rocks show insignificant structural complexity compared to Jaipur-Alwar group of rocks (Singh, 1988) and very low-grade green schist facies metamorphism. The rocks of the region are characterized by the presence of broad open antiforms and synforms with upright geometry and moderate plunges to north east. Several shear zones occur subparallel to the axial planes.

The Alwar Group consists of a dominant arenaceous sequence with a large number of basic inter-calations. The former includes quartzite, sericite schist, quartzite (conglomerate) and quartzite with thin lenses of marble and conglomerate. The quartzites of the Alwar subbasin are mature (quartz are-nites) and commonly lack preservation of primary sedimentary structures. The rocks in the region show a complex and heterogeneous deformation pattern and metamorphism. Detailed structural mapping in the region, north of Jaipur, reveals that the sand-shale sequence of metasediments of shallow water depositional environment was intruded by basic rocks and granites. The early defor-mation resulted in a series of isoclinal folds of varying dimensions with NNE-SSW−trending axial planes, which were refolded coaxially giving rise to "hook-shaped" geometry similar to interference structures of Ramsay (1967). These are further overprinted and deformed generating near E−W non-plane, noncylindrical type of folds (Das, 1988). These folds appear to be large scale domes and basins and contributed to the current broad map pattern of the NDT. An interesting outcrop pattern was documented in a small sector showing the large scale interference fold structure around Khori−Bidara area, north of Jaipur (Fig. 5.30). The major axial directions vary in different seg-ments. Structural analysis shows southwesterly plunges with moderate values. It was interpreted that the initial folding was synformal structure with E-W−trending axial trend and the later deformation rotated the preexisting axial trend to NW−SE direction displaying flattened domal structure (Das, 1988) and a schematic three-dimensional view of the structure is presented (Fig. 5.31). A schematic E−W structural cross section from Ajitgarh to Barodia, north of Jaipur, depicts a regional antiform with moderate southerly plunge (Fig. 5.32). Higher peak metamorphic conditions (temperature around 645°C and pressure ∼7 kbar were also estimated (Kundu, Kazim, & Sharma, 2004). In the light of above, it needs to be further examined whether the antiformal structure represents a mantle gneiss dome structure.

The Ajabgarh assemblage comprises predominantly argillaceous in composition with imperis-tent carbonate horizons. The major rock units include quartzite, staurolite garnet schist, shale, amphibolite and metadolerite, quartzite with interbedded phyllites, carbon phyllites, tuffaceous, and marble. Many of the rock formations were brecciated along with the emplacement of lava flows. Impure marble with lenses of phosphorite, black flows, agglomerate tuff and intrusive granites are common. However, less-deformed, muscovite-bearing quartzite in the region is feldspathic, which has significant proportion of lithic fragments in matrix suggesting the derivation of clasts from adjacent granitic highlands (Singh, 1988). Based on the lithological assemblages and metamor-phism, the eastern most basin was interpreted to be a pre-Delhi intracratonic rift related volcanose-dimentary sequence.

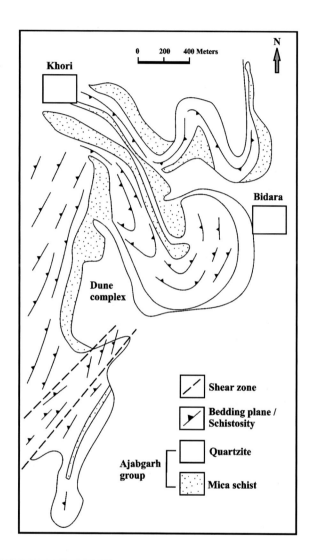

FIGURE 5.30

Detailed structural map of Khori-Bidara area, north of Jaipur.

Modified after Das, A. R. (1988). Geometry of the superposed deformation in the Delhi Supergroup of rocks, north of Jaipur, Rajasthan. Memoir of the Geological Society of India, 7, 247–266.

The Khetri subbasin occurs at the westernmost part of the NDT and is popularly known as Khetri Copper Belt (KCB) or Khetri fold belt (Fig. 5.33). The KCB constitutes the northernmost entity of the AFB and extends for about 80 km from Singhana in the NE to Raghunathgarh in the south. Large longitudinal folds with subhorizontal fold axes are common. The contact of most of the rock units are sheared. There are series of steeply inclined anticlines and synclines, some of which showing culminations and depressions. The folds plunge at a low to moderate angle to NE.

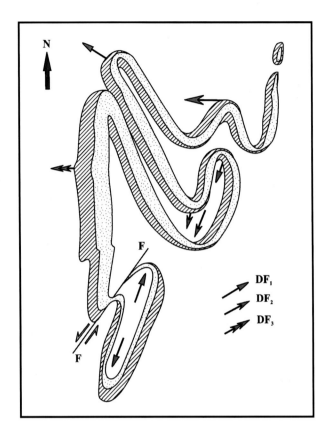

FIGURE 5.31

Schematic 3-D view of the fold pattern of Khori-Bidara area, north of Jaipur.

Modified after Das, A. R. (1988). Geometry of the superposed deformation in the Delhi Supergroup of rocks, north of Jaipur, Rajasthan. Memoir of the Geological Society of India, 7, 247–266.

Structural studies indicate that the metasediemntary rock assemblages in KCB must have been subjected to subhorizontal simple shear in a NW−SE direction on horizontal beds resulted in NE-trending isoclinals, recumbent or gently plunging reclined folds in the early deformation history (Naha, Mukhopadhyay, & Mohanty, 1988). Subsequent deformation led to the development of kinks and conjugate folds with near horizontal axial planes, and upright chevron folds with axial planes striking E−W caused by longitudinal shortening.

A major NNE-SSW−trending albitite zone (170 km × 15 km) occurs at the eastern part of KCB (Das Gupta, 1968). It represents an important tectonic/shear zone comprising quartzofelds-pathic rocks and albitic injections in association with widespread brecciation and alteration. The concept of albitite zone unfolds the understanding of genesis and control of copper-uranium mineralization. There are many such subparallel zones with similar characteristics, which were considered as a zone of regional scale metasomatism. The principal mineral of albitite rock is a fine-grained, red-colored aventurine variety of albite. They show complex undulose extinction

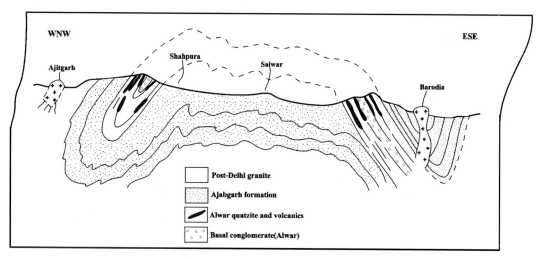

FIGURE 5.32

E—W structural cross section depicting the regional structure between Ajitgarh and Barodia region, north of Jaipur.

Modified after Das, A. R. (1988). Geometry of the superposed deformation in the Delhi Supergroup of rocks, north of Jaipur, Rajasthan. Memoir of the Geological Society of India, 7, 247–266.

and very rare twinning. Coarse-grained mafic cumulates (Pyroxenite) occurring as dismembered lenticular bodies are also common in the zone. The magnetite—ilmenite accumulations also occur as large segregated pockets within the pyroxenite bodies (Ray, 1990). The most notable feature of this area is that most of the granitoid rocks are albitized to varying extent; this was ascribed to a regional scale Na-metasomatic event. It is possible that the entire tectonic zone may represent a possible extension of the KSZ, which separates the SDT and the SMT. In addition, minor amount of felsic volcanic rocks were also reported. The intrusive rocks in the KCB are largely represented by mafic and granitoids rocks. The rocks obtained amphibolite facies metamorphism.

Several granite plutons intrude the KCB and their structural and petrochemical characteristics indicate both S-type and I-type nature. The rocks of the KCB can be further divided into (1) an older psammitic-dominated Alwar Group and (2) a younger pelitic-dominated Ajabgarh Group. However, some workers considered the rocks of KCB as pre-Delhi formations (Chakrabarti & Gupta, 1992).

Considering the distinct basement-cover relationships and lithological assemblages, the KCB is further subdivided into northern (NKCB) and southern Khetri Copper Belt (SKCB), separated by a NW—SE striking transverse Kantli Fault (Gupta & Bose, 2000). The KCB displays two episodes of prograde regional metamorphism (MI and M2), the andalusite—sillimanite facies in NKCB and a transition to kyanite—sillimanite facies in SKCB and an eastward increase in the grade of metamorphism was reported (Lal & Shukla, 1975). The region is characterized by a multiphase structural history with polyphase regional metamorphism of andalusite—sillimanite type in the northern and kyanite—sillimanite type in the southern domains (Lal & Ackermand, 1981) and sediment-hosted workable copper mineralization (Sarkar & Dasgupta, 1980). Besides, it also records

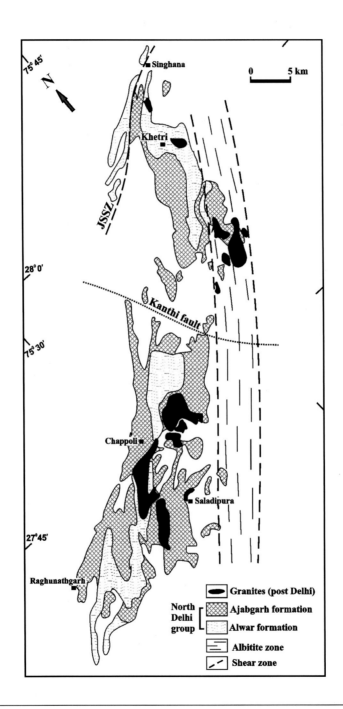

FIGURE 5.33

Geological and structural map of Khetri copper belt in the North Delhi Terrane.

Modified after Das Gupta, S. P. (1968). The structural history of the Khetri Copper Belt, Jhunjhunu and Sikar districts, Rajasthan.

Memoir of the Geological Survey of India, 98, 170.

an overprint by a Neoproterozoic thermal event with U–Th–Pb monazite ages of ~950–910 Ma (Kaur, Chaudhri, Biju-Sekhar, & Yokoyama, 2006). The inclusion monazite grain and the younger age of the matrix monazites provide a well-constrained age of 920–940 Ma, which is close to the age of the peak regional metamorphism in the area (Pant, Kundu, & Joshi, 2008).

The NNE–SSW to NE–SW striking metasedimentary rocks of the NKCB consists of feldspathic quartzite with magnetite, banded amphibole quartzite, garnetiferous chlorite schists, mica schists and quartzite, which are folded into a series of regional anticlines and synclines (Das Gupta, 1968). In contrast, Sarkar (2000) concluded that these rocks belong to a homoclinal sequence and are part of one limb of a regional fold. Recent work, however, classified the rocks in terms of an apparent Archean basement, overlain by Proterozoic cover sequences (Gupta, Guha, & Chattopadhyay, 1998). In this igneous–metamorphic complex, the igneous activity is primarily in the form of the late Paleoproterozoic (1.82–1.66 Ga) granitoid rocks (Kaur, Chaudhri, Raczek, Kröner, & Hofmann, 2009). The majority of these granitoids intrude a presumed Archean basement of metasedimentary (e.g., quartzites, metapelites, and calc–silicate rocks) and minor gneissic rocks. The basement-cover sequence in NKCB is marked by the development of sedimentary breccias, iron encrustation, and the presence of hematite–magnetite bands at the contact. The NKCB is represented by a basal unit of quartzite with sedimentary breccia and iron formation, followed by amphibolites and marbles in the middle level. The upper level assemblage constitutes andalusite–biotite schist, mica schist, orthoquartzite, metagreywacke with bands of carbon phyllite, dolomite, and quartz schist indicating deeper argillaceous marine sediments.

The basement cover in the SKCB is defined by a major shear zone known as the Chapoli fault (Fig. 5.34). The cover sequence is described as Shyamgarh group comprising pelitic schist, felsic volcanic, agglomerate, tuff, greywacke, orthoquartzite, BIF, conglomerate, calc–silicate rocks, and carbonaceous phyllites, suggesting deep water volcano sediments. North of Chapoli, the map shows nonconformable relationships between rock units associated with series of subparallel thrust/shear zones. The attenuation of series of isoclinal fold structures and the contact with the basement rocks is well preserved in the region. These are followed by shallow water sediments which is in contrast with that of the NKCB. Two distinct metamorphic facies rocks viz., andalusite–sillimanite facies to the north and kyanite–sillimanite facies occur to the south of the Chapoli fault.

A number of granite plutons occur both in SDT and NDT which show conformable foliations with the country rocks. They vary from tonalitic to syenitic in composition. The available age data shows a wide variation from 1800 Ma to 750 Ma. Further, petrochemical and srcutural data available on granitoids are suggestive of at least two generations of granitic emplacement in the region. Zircon and monazite ages suggest that the granitoid rocks range in age between ca. 1844 Ma and 1660 Ma, and were partly affected by an overprint of a Neoproterozoic thermal event at around 900 Ma (Kaur, Chaudhri, Raczek, Kroener, & Hofmann, 2007). Based on the structural data, Gupta et al. (1998) proposed that the Khetri granitoids were emplaced in an extensional tectonic regime. The upper part of subducting oceanic lithosphere, when interacting with the overlying mantle wedge, underwent dehydration and generated LILE-enriched fluids as indicated by the Ba-enrichment trend in the Jasrapur granitoids. The Jasrapur granite pluton (9×1 km) represents a sheet like body and is dominated by monzogranites with a subordinate amount of granodiorite and quartz monzodiorite. It has been characterized as peraluminous, I-type, and calc–alkaline granite with subordinate granodioritic component. A late phase of cross-cutting equigranular leucogranite dykelets and veins are also observed. The multi-stage development of the Jasrapur arc

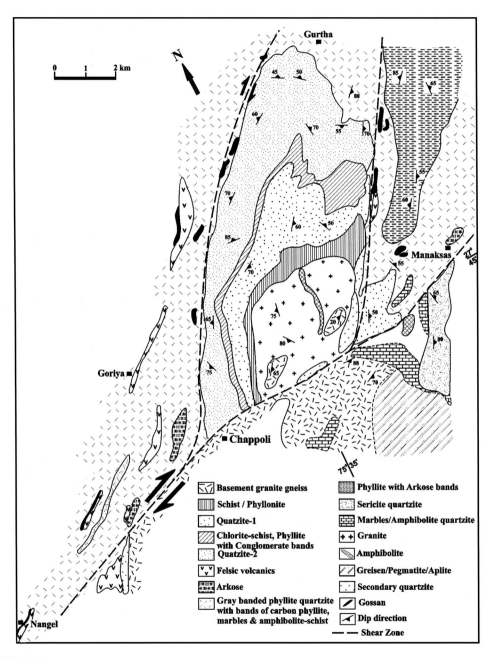

FIGURE 5.34

Geological map of South Khetri copper belt showing Chapoli fault and the basement-cover relationships.

Modified after Golani, P. R., Gathania, R. C., Grover, A. K., & Bhattacharjee, J. (1992). Felsic volcanics in south Khetri copper belt, Rajasthan and their metallogenic significance. Journal Geological Society of India, 40, *79–87.*

involving multisource components was suggested. Conventional geothermo−barometric estimates suggest high temperature ($\sim 800°C$) and pressure (≥ 600 MPa) conditions for the crystallization of the Jasrapur granitoids. A whole-rock Sm−Nd isotope dating yielded an isochron age of 1800 ± 59 Ma, in good agreement with a zircon Pb−Pb evaporation age of 1821.7 ± 0.4 Ma, which was interpreted as the age of emplacement (Kaur et al., 2009). This with other similar events documented farther south in the central ADOB indicate a 1850−1822 Ma event of continental arc magmatism in ADOB, which probably corresponds with the amalgamation of the Paleoproterozoic supercontinent Columbia.

Recent dating of medium grade regional metamorphism from NDT (Pant et al., 2008) indicates that the ~ 1000 Ma Delhi orogeny, well described from SDT (e.g., Bhowmik & Dasgupata, 2012; Deb et al., 2001) affected the entire Delhi fold belt including the NDT with the exception of the easternmost Lalsot-Bayana subbasin. This suggests that this subbasin may represent a pre-Delhi intracratonic rift basin and may not be a part of the DFB. However, the correlation of the KCB either with the SDT or the Aravalli formations still remain unresolved.

The division of DFB into SDT and NDT is also challenged inview of similar gravity and magnetic signatures all along the DFB (Amar Singh, Singh, & Golani, 2015). The qualitative gravity analysis also shows the Berach granite as the basement of Hindoli terrane. Despite several complexities and controversies, the structural and metamorphic history of rocks of the DFB (SDT and NDT) indicates a single stage orogenic evolution of the entire belt undergoing thermal reconstitutions at later periods (Roy & Jakhar, 2002). According to them, the nature of sedimentary package, volcanicity, and isotopic signatures of intrusive bodies and the metallogenic character attest to an intracratonic setting and rule out the Wilson cycle type of orogeny for the evolution of the DFB. Lack of detailed structural architecture around and along the shear zones pose difficulty in interpreting the geodynamic evolution of the ADOB despite the availability of detailed and classic geologic and structural maps since long (Heron, 1953).

5.3.8 SIROHI TERRANE

The SRT occurs to the west of SDT lying between the Erinpura shear zone (EPSZ) in the west and the PSZ in the east (Fig. 5.35). The SRT consists of low-grade metasedimentary rocks of metacarbonates and mica schists defining a tectonic contact with granites and share a common deformation history (Roy & Sharma, 1999). The SRT is characterized by steeply dipping shear zones and define a NE-SW−trending 20-km-long, eye-shaped structure with a maximum width of 6 km, which may represent a mega-sheath fold structure or a gneiss dome (Just, Schulz, de Wall, Jourdan, & Pandit, 2011). There is a prominent NE-SW−trending high strain zone in the northwestern part of the Mt. Abu that can be traced over a distance of c. 8 km with a maximum width (approximately 200 m) north of Delwara Temple, described as the Abu−EPSZ. The shear zone is characterized by mylonitic fabrics together with steep to subhorizontal stretching lineations indicating that the zone was subjected to dextral-ductile transpression. This is evident from a typical outcrop from a road section, east of Sirohi town where metasediments (metacarbonates, calc−silicates, metapelites) show steep dips and tight isoclinal folds with NE-SW−trending axes (de Wall, Pandit, Dotzler, & Just, 2012b).

The intrusions, generally characterized as S-type granites, are collectively termed as "Erinpura Granite" (Heron, 1953) often showing gneissic fabric. The Erinpura granite was believed to

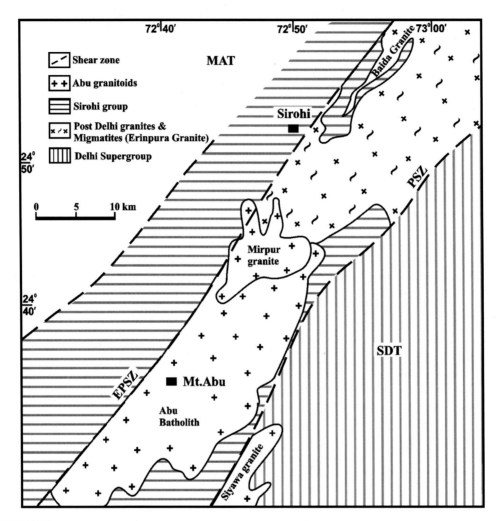

FIGURE 5.35

Geological map of Sirohi Terrane, west of South Delhi Terrane. *EPSZ*, Erinpura shear zone; *PSZ*, Phulad shear zone; *SDT*, South Delhi Terrane.

Modified after Just, J., Schulz, B., de Wall, H., Jourdan, F., & Pandit, M. K. (2011). Monazite CHIME/EPMA dating of Erinpura granitoid deformation: Implications for Neoproterozoic tectono-thermal evolution of NW India. Gondwana Research, 19, 402–412.

constitute two major batholoiths, the Abu-Erinpura and the Sewaria batholiths. There are wide variations in mineral assemblages, texture and the development of foliation between the two. Hence, the rocks in the south are described as "Abu-Erinpura batholith" and the gneiss-granite association in the north as Sewaria batholith. The Mt. Abu batholith, the most prominent geomorphic feature (elevation up to 1200 m above the ground level), occurs to the west of SDT. It is an elongated batholith with a prominent NE−SW trend parallel to the ADOB. The Abu-Erinpura batholith consists dominantly of both porphyritic granite–granodiorite in association with metasediments and amphibolites (Roy & Jakhar, 2002). The Erinpura granites are best exposed in the Sirohi-Jawai Dam region where they also show evidence of migmatization and shearing. Sediments of Sirohi Group were deposited over the Erinpura granite basement. Both the Erinpura granite and the Sirohi Group metasediments share a common deformation history and are overlain by the felsic volcanics and intruded by granites of the 770–750 Ma (Torsvik, Carter, Ashwal, Bhushan, Pandit, & Jamtyveit, 2001).

Erinpura granite displays enclaves of various sizes of massive gabbro, amphibolites, and metasediments. Prominent zones of ductile shearing occur at the contact of metasediements and the granites with well developed mylonitic fabrics and stretching lineations. The central part of the batholith comprises pink colored, coarse- to medium-grained, porphyritic and Schlieren-type granite described as "undeformed" by Gupta et al. (1997), despite the presence of poorly developed foliation. Granite gneisses are also reported to occur in the peripheral part of the batholith. The granitoids are intruded by 1–15 m wide, generally northwest-trending, steeply dipping, porphyritic rhyolitic dykes (K–feldspar–plagioclase–quartz–biotite), which are not deformed. The last magmatic phase in the regions is marked by subvertical (few cm to ~10 m wide) mafic dyke intrusions with chilled margins and sharp contacts with host rocks.

The granites yielded ages ranging between 873 to 820 Ma suggesting multiple episodes of emplacement (Choudhary et al.,1984; Purohit, Papineau, Kröner, Sharma, & Roy, 2012). However, the precise U–Pb zircon ages of Mt. Abu granite were obtained as 763 ± 3 Ma and 766 ± 4 Ma (Solanki, 2011; cf. de Wall et al., 2012b). The EPMA monazite dating on recrystallized and newly formed monazite constrained the timing of deformation and metamorphism of Erinpura granite at 775 ± 26 (Just et al., 2011) implying the formation of synkinematic monazite is younger than the magmatic monazite (863 ± 23 Ma; Just et al., 2011). Geochemically, the Mt. Abu magmatism shows a close similarity with that of the Malani Igneous suite (MIS). Recent geochemical, deformation, and geochronologic studies indicate that the Mt. Abu granitoids, traditionally considered as late kinematic in relation to Delhi orogeny, are part of the Malani magmatic episode (de Wall et al., 2012b).

The Siyawa granite batholith comprises a large rectangular outcrop of granite and gneissic granites (6×35 km). A thin body of conglomerate occurs at the eastern margin of Siyawa, characterized by intense shearing with well-developed down-dip stretching lineations. The western boundary is marked by persistent horizon of marble. The Siyawa granite also contain enclaves of quartz–mica schist, pelitic restite, migmatite, and conglomerate. The mineral association suggests that Sewaria granite occurs in a high-grade metamorphic environment. The age determined for the Siyawa granites ranges between 870–800 Ma (Choudhary et al., 1984). The chemistry of granites indicate that these rocks belong to S-type with tungsten mineralization.

A band of intensely deformed metasedimentary belt comprising marbles, phyllites, and quartzites occur in the western margin of the SRT. The U–Pb zircon age of 836 Ma was obtained for the emplacement of Siyawa granite (Deb et al., 2001). However, it has a different tectono-thermal history compared to the Erinpura granite from SRT inview of the obvious differences in the fabric development and mineral composition. The U–Pb zircon ages of 770–750 Ma for MIS magmatism (Gregory, Meert, Bingen, Pandit, & Torsvik, 2009) indicate an initial outpouring of minor basaltic and predominant rhyolitic lava flows, followed by granite emplacement in NE–SW direction at the western margin of the SDT and in the MAT, further west.

The NE-SW—trending shear zones in the central part of the Sirohi region deflect to an almost E—W orientation (80 degrees) in the southern part and coincide with the western thrust planes to form a set of shear zones, which are characterized by oblique-slip and strike-slip displacements. The metasediments and Erinpura granites in the region are cofolded and intensely sheared. The transition from dip-slip to strike-slip is preferentially observed in calc—silicate rocks in a road section to the south of Sirohi town. Contrasting deformational patterns are observed between the SRT and the adjacent terranes. Some parts of the Erinpura granites, such as the Siyawa granite and the granites exposed near Sirohi town also show evidence of migmatization, which is associated with anatexis and monazite crystallization at 779 ± 16 Ma. The end of the tectono-thermal event in SRT is constrained by a 736 ± 6 Ma Ar—Ar muscovite age data from near by ductile shear zone (Just et al., 2011). The age data indicates an overlap in timing between shearing and anatectic event. An interpreted cross section shows west-verging thrust stacking, tectonic inversion and diapiric rise of Balda granite (Fig. 5.36). The Balda granite might have been derived from low temperature melting of Sirohi sediments under hydrous flux and emplaced during the late stage of thrusting. Low-temperature deformation has overprinted the Balda granite and represents the terminal Cryogenian event in the Sirohi sector (Just et al., 2011). Deformation in the SRT gives an indication for high fluid activity, channelized along thrust planes. Some earlier studies described Pan-African resetting in Malani granites (Rathore, Venkatesan, & Srivastava, 1999) and granites of SDT (Crawford, 1975), which are substantiated by recent 509 ± 2 Ma, 514 ± 2 Ma (Ar—Ar) that overprint the Mt. Abu and type Erinpura granite (Ashwal et al., 2013) and argon loss at 550—490 Ma in mafic and felsic volcanics of the Sindreth Basin (Sen et al., 2013).

The SRT comprises a repetitive sequence of Erinpura granite—gneisses and Sirohi group of metasediments bounded by steep SE-dipping thrust planes. Deposition of the Sirohi sediments on an uplifted and denuded basement formed by Erinpura granite is most likely on account of the vast sedimentary coverage all over the western foreland of the southern SDT. The tectono-magmatic events in the SRT postdate the collision of ADOB with the Cryogenian closure of the Mozambique Ocean and Cryogenian terrane collision (Bhowmik et al., 2010). There is also increasing evidence for a later Pan-African imprint around 510 Ma recorded in the Mt. Abu and type Erinpura granite (Ashwal et al., 2013). In a broader perspective, the SRT constitutes South Delhi high-grade metamorphic terrane, the Sirohi anatectic terrane, and Sirohi fold and thrust terrane (de Wall et al., 2012b). These three major subterranes with distinct geological history share some common tectonic and magmatic imprints.

5.3.9 MARWAR TERRANE

The MAT represents the western most part of the NW-Indian shield and occurs to the west of ADOB. The terrane constitutes the vast expanse of Erinpura-Malani plutonic-volcanic acid magmatic suite that occurs to the west of SRT (Fig. 5.37). MIS comprises predominantly volcanic, represented by rhyolites, trachytes, dacites, and pyroclastic rocks, and local basaltic flows (Gupta et al., 1980). A number of isolated and widely spaced outcrops of dominantly felsic lava flows, tuffs (ignimbrites) and granitoids with small amounts of basic, alkaline and intermediate alkaline rocks are also exposed. The MAT, also described as the trans-Aravalli terrane, comprises a typical cratonic platformal sequence of sand—shale—carbonate with evaporite represented by

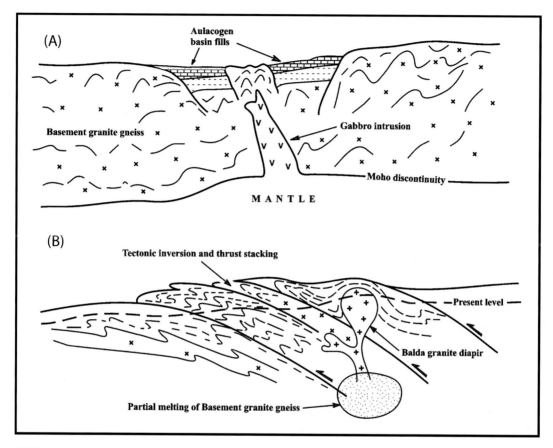

FIGURE 5.36

(A) a cartoon showing the evolution of Sirohi terrane as rift basins subsequent to crustal thinning and intrusion of gabbroic magma, and (B) the structural cross section of Sirohi terrane showing west-verging thrusts.

Modified after Roy, A. B., & Sharma, K. K. (1999). Geology of the region around Sirohi Town, western Rajasthan-a story of Neoproterozoic evolution of the Aravalli crust. *Jodhpur: Scientific Publishers (India), pp. 19–33.*

unmetamorphosed and undeformed sediments, overlying the MIS. Much of the terrane is covered with sands of the Thar Desert. In the east, these sediments overlie the steeply dipping mica schists, phyllites, and quartzites, which were correlated with Sirohi group of rocks. The MAT is bounded by a NE-SW—trending EPSZ in the east and the sediments show overturning of beds in its vicinity. The plutonic rocks represent within-plate A-type anorogenic magmatism with an initial extrusive phase of dominantly felsic volcanic, followed by intrusive phase of meta-aluminous to peralkaline granites (Siwana, Jalore and Tosham) and finally terminating with major mafic and minor felsic dykes swarm.

The magmatic activity of MIS commenced with an initial phase of basalt and felsic volcanic flows followed by a second phase of granitic plutons. The felsic and minor mafic dyke swarms

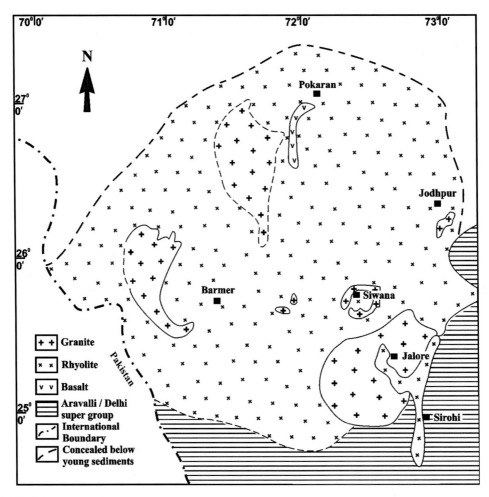

FIGURE 5.37

Regional geological map of Marawar Terrane.

Modified after Bhushan, S.K. (1995) Late Proterozoic continental Growth: Implications from geochemistry of acid magmatic events of West Indian Craton, Rajasthan. Mem. Geol. Soc. India, v. 34, pp. 339–355.

form the third and final phase of the igneous cycle. The Malani felsic volcanic rocks are unmetamorphosed, but slightly tilted and occasionally folded. Late-stage mafic dykes are all vertical to subvertical. The MIS unconformably overlies Paleo- to Mesoproterozoic metasediments, granites, and granodiorites (Pandit, Shekhawat, Ferreira, Sial, & Bohra, 1999) and unconformably underlies the Ediacaran-Cambrian Marwar Supergroup of red-bed and evaporatic sequences.

The MIS also includes high-level granitic plutons that have intruded the volcanic units of the same magma sources. Among them the Siwana ring complex is well known and comprises granites of peralakaline nature associated with basalt and gabbro (Fig. 5.38). However, the isotopic studies

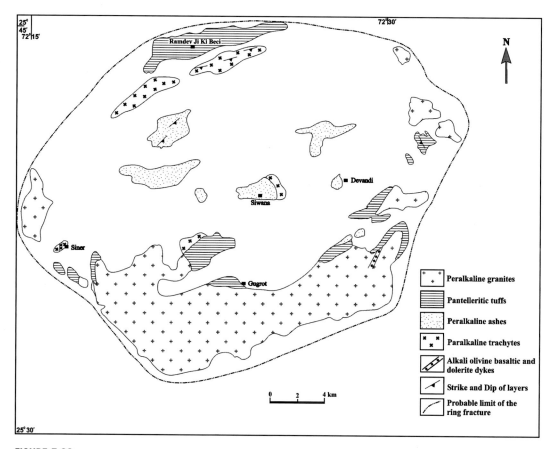

FIGURE 5.38

Geological map of Siwana ring structure, Marwar Terrane.

Modified after Rathore, S. S., Venkatesan, T. R., & Srivastava, R. K. (1999). Rb–Sr isotope dating of Neoproterozoic (Malani Group)
magmatism from Southwest Rajasthan, India: Evidence of younger Pan African thermal event by 40Ar–39Ar studies. Gondwana
Research, 2, 271–281.

indicate that all the rocks of Siwana complex were inferred to be the mantle- derived primary magma (Vallinayagam & Kochhar, 1998). Further, the Siwana granites and associated peralkaline trachytes and rhyolites are coeval and cogenetic.

There are also other granites at Jalore that belong to predominantly of a peraluminous type, but the minor association of peralkaline components was also reported (Kochhar & Dhar, 2000). The Malani volcanics around Gurapratap Singh and Diri include dacite, rhyodacite, and rhyolite with minor occurrences of basalt, andesite, and ultrapotassic rhyolite (Srivastava, 1988). All the felsic volcanics of the MAT with the exception of the ultrapotassic rhyolite are cogenetic and might have been formed by the fractional crystallization from a crustally derived magma. All these granites were considered to be anorogenic or A-type within plate granites (Kochhar, 1984). The MIS was possibly evolved in a postcollisional extensional environment. Widespread Pan-African thermal activity between 500−550 Ma was also recorded and the MAT represents a Neoproterozoic crustal fragment rather than the Archean, as earlier believed. Geochronological data from the Malani mafic dykes yielded a minimum 704 Ma age for the dykes which may represent the final pulse of MIS magmatism (Meert, Pandit, & Kamenov, 2013).

The MIS was variably interpreted as fissure type eruption (Bhushan, 1984), related to hot spot activity (Kochhar, 1984), or plume-related magmatism (Roy, 2001). Based on the characteristic ring structures and radial dykes, it is suggested that the Malani magmatism is evolved primarily by mantle plume activity (Kochhar, 1984). However, the plume concept was refuted in view of the absence of basic component in the dominantly felsic Malani magmatism. On the basis of major element chemistry, the parental magma for the Malani magmatism was suggested to be tholeiitic in nature (Pareek, 1981). The origin of felsic rocks is due to contamination by crustal mixing of tholeiitic magma with sialic crust. Further, it was interpreted that the emission was related to tectonic structures and attributed to post-Delhi orogenic activities. Some of these eruptions appearing as ring structures on the surface may deceptively indicate a deep mantle plume for their origin (Bhushan, 2000).

The deep crustal seismic images of the MAT reveal that the structures exhibit "thick-skinned" deformation and are related to the Mesoproterozoic collisional episode with the SDT (Vijaya Rao & Krishna, 2013). The architecture of the MAT exhibits ~90 km-wide imbricated crustal structure related to subduction−collision activity, and formation of a younger Moho and widespread juvenile volcano-plutonic magmatism covering the entire MAT due to postcollisional delamination process. This hypothesis supports the view that the MAT is a mobile belt constituting an assemblage of laterally extensive terranes/arcs, coeval with the global events that occurred during ~1100 Ma and 750 Ma. Several dipping reflectors were interpreted as thrust sheets possibly contiguous in the region west of the PSZ, which must have acted as the conduit for the magma transport.

5.4 NAGAUR-JHALAWAR TRANSECT

A multidisciplinary approach involving all geological and geophysical studies along a NW-SE−trending geotransect was attempted across the ADOB, covering the adjacent terranes. Geophysical data, in particular deep seismic reflection data, provide useful information in evaluating crustal architecture and mantle dynamics and have been recently employed in many studies in conjunction with the surface geological features to understand subduction−accretion−collision history in the Precambrian terranes. The transect starts from Nagaur in the northwest, cuts across all lithological assemblages and tectonic boundaries of the ADOB, and ends at Kunjer/Jhalawar (Fig. 5.39). The transect would here after be described as N−J transect. The transect runs nearly orthogonal to the regional tectonic trends of the ADOB. The important tectonic domains from Nagaur to Jhalawar include: the MAT, the SDT, the SMT, the MWT, the JHT, and the Vindhyan group of sediments of the Bundelkhand craton.

Geologically, the major rock types along the transect are: the Marwar group consisting dominantly of Malani igneous suite (MIS, Neoproterozoic); the Delhi group (middle to upper Proterozoic); the Aravalli-Bhilwara group (Archean-Proterozoic); and the Vindhyan sediments from northwest to south east (Fig. 5.40). The MAT consists of sand stone, silt stone and clay as a flat-

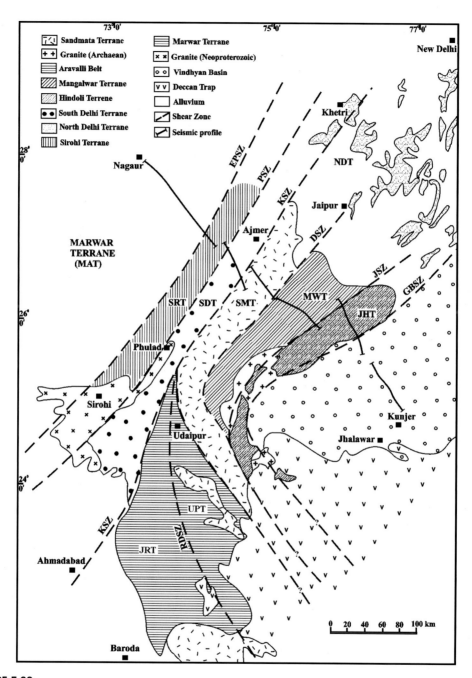

FIGURE 5.39

Geological and tectonic map of Aravalli Delhi Orogenic Belt showing the Nagaur– Jhalawar transect.
MAT, Marwar Terrane; *PSZ*, Phulad shear zone; *SDT*, South Delhi Terrane; *KSZ*, Kaliguman shear zone;
SMT, Sandmata Terrane; *DSZ*, Delwara shear zone; *MWT*, Mangalwar Terrane; *JSZ*, Jahazpur shear zone;
JHT, Jahazpur Hindoli Terrane; *GBSZ*, Great Boundary shear zone; *NDT*, North Delhi Terrane; *JRT*, Jharol
Terrane; *RDSZ*, Rakhabdev shear zone; *UPT*, Udaipur Terrane.

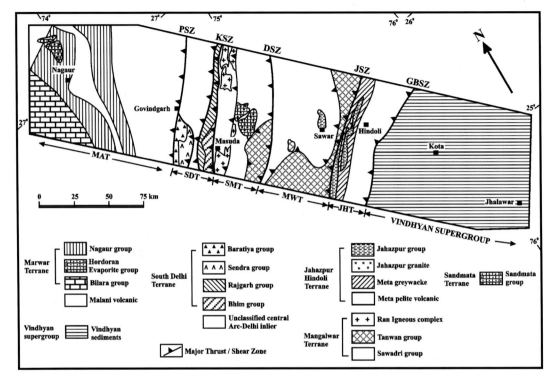

FIGURE 5.40

Geology along the Nagaur-Jhalawar transect. *MAT*, Marwar Terrane; *PSZ*, Phulad shear zone; *SDT*, South Delhi
Terrane; *KSZ*, Kaliguman shear zone; *SMT*, Sandmata Terrane; *DSZ*, Delwara shear zone; *MWT*, Mangalwar
Terrane; *JSZ*, Jahazpur shear zone; *JHT*, Jahazpur Hindoli Terrane; *GBSZ*, Great Boundary shear zone.

*Modified after Sinha-Roy, S., Malhotra, G., & Guha, D. B. (1995). A transect across Rajasthan Precambrian terrain in relation
to geology, tectonics and crustal evolution of south-central Rajasthan. In: Sinha-Roy and Gupta, K. R. (Eds.), Continental crust of
NW and Central India. Memoir of the Geological Society of India, 31, 63–89.*

lying undeformed cover sequence. The basement rocks here are the MIS constituting calcareous
facies (dolomite, lime stone, chert, and clay intercalations), evaporate facies and arenaceous facies.
At the eastern margin, lies the MIS and Erinpura granite in the form of Neoproterozoic magmatic
event. The igneous activity is represented by bimodal volcanic assemblage and intruded by granitic
emplacement followed by mafic and felsic dykes. Further east, the N—J transect cuts across one
of the prominent tectonic zones, the PSZ, decorated by Phulad-Basantgarh-Ratanpur ophiolite complexes, followed by SDT and its eastern margin defined by the KSZ.

The KSZ was believed to juxtapose the SDT in the form of a westerly dipping thrust, over
the SMT, the host for high grade metamorphic rocks in close association with the BGC. Further
east, the transect runs through MWT and JHT, which are separated by the major DSZ and the JSZ,
the characteristics of which were described earlier. The eastern-most tectonic element that occurs
along the transect is the GBSZ, followed by the Vindhyan group of sediments to the east.

The sediments contain undeformed and unmetamorphosed rock units of conglomerate, sand stone, shale and lime stone, which abut against the Hindoli group of rocks to the west. In summary, it can be mentioned that the N−J transect constitutes NE-SW−trending important tectono-stratigraphic terranes bound by major suture/shear zones with dextral transpressive characteristics.

5.4.1 STRUCTURAL CROSS SECTION

The NW-SE−trending geological and structural cross section along the N−J transect shows the different geologic terranes and associated shear zones (Fig. 5.41). The N−J transect runs through all the major geologic terranes and tectonic elements like PSZ, KSZ, DSZ, JSZ, and the GBSZ. All the shear zones were described as west-dipping thrusts penetrating through the basement continuing up to the Moho depths (Sinha-Roy et al., 1995). While the PSZ is characterized by ophiolite complexes, the KSZ is marked by granulite facies rocks. The JSZ in the eastern part of the transect shows the transportation of the amphibolites facies rocks of MWT on to greenschist facies rocks of the JHT.

5.4.2 GRAVITY AND MAGNETIC PROFILES

The geological and tectonic features are well reflected in gravity and magnetic profiles along the N−J transect with good correspondence. The Bouguer gravity anomaly map of the region (Reddy & Ramakrishna, 1988) along the N−J transect is, in general, characterized by a NE-SW−trending gravity high flanked by conspicuous lows on either side all along the strike of the 700-km long the Aravalli-DFB (Fig. 5.42). There is a sharp increase in gravity values from nearly −50 mGal to −10 mGal near Ajmer and the steep gravity gradient coincides with the location of the PSZ at the

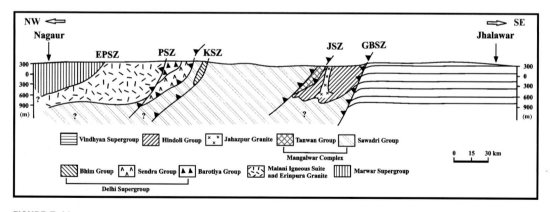

FIGURE 5.41

Geological and Structural cross section along the Nagaur-Jhalawar transect. *EPSZ*, Erinpura shear zone; *PSZ*, Phulad shear zone; *KSZ*, Kaliguman shear zone; *JSZ*, Jahazpur shear zone; *GBSZ*, Great Boundary shear zone.

Modified after Sinha-Roy, S., Malhotra, G., & Guha, D. B. (1995). A transect across Rajasthan Precambrian terrain in relation to geology, tectonics and crustal evolution of south-central Rajasthan. In: Sinha-Roy and Gupta, K. R. (Eds.), Continental crust of NW and Central India. Memoir of the Geological Society of India, 31, 63−89

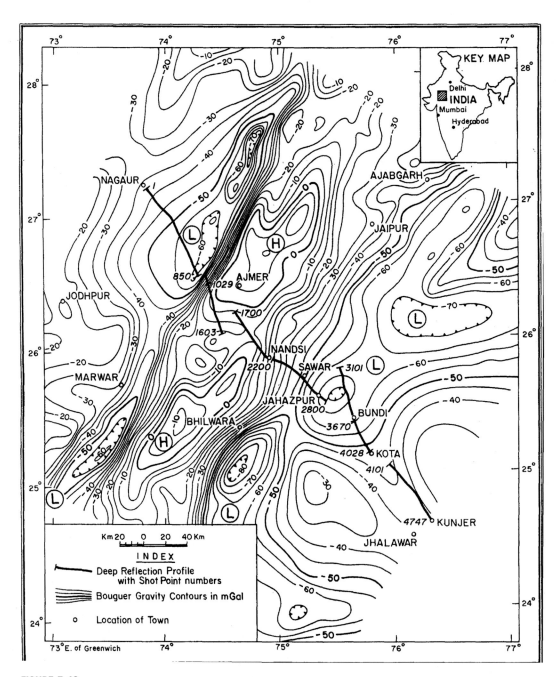

FIGURE 5.42

Regional gravity map around the Aravalli Delhi Orogenic Belt.

Adapted from Vijaya Rao, V., & Krishna, V.G. (2013). Evidence for the Neoproterozoic Phulad Suture Zone and Genesis of Malani magmatism in the NW India from deep seismic images: Implications for assembly and breakup of the Rodinia. Tectonophysics, 589, 172–185.

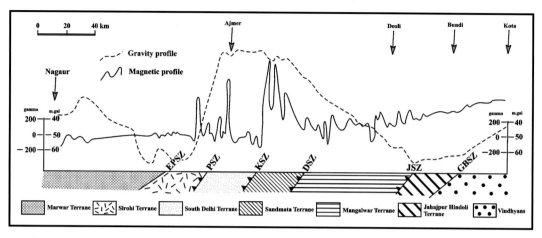

FIGURE 5.43

Gravity and Magnetic profiles along the Nagaur-Jhalawar transect. *EPSZ*, Erinpura shear zone; *PSZ*, Phulad shear zone; *KSZ*, Kaliguman shear zone; *DSZ*, Delwara shear zone; *JSZ*, Jahazpur shear zone; *GBSZ*, Great Boundary shear zone.

Modified after Sinha-Roy, S., Malhotra, G., & Guha, D. B. (1995). A transect across Rajasthan Precambrian terrain in relation to geology, tectonics and crustal evolution of south-central Rajasthan. In: Sinha-Roy and Gupta, K. R. (Eds.), Continental crust of NW and Central India. Memoir of the Geological Society of India, 31, 63–89

western boundary of the SDT. The gravity values, east of the KSZ, tend to decrease with a low gradient up to the JSZ. It is significant to notice the presence of lowest gravity values of −70 mGal at the JSZ from where the values start increasing toward the Vindhyan plateau. However, the GBSZ is not well reflected in gravity values. The modeling of the residual Bouguer anomaly along the transect suggests high-density/high-susceptibility rock formations in the upper crust at a depth of 15−16 km. The upward thrusting of the lower crustal rocks through different tectonic processes may be related to the evolution of AFB. Thickening of the lower crust under the Aravalli ranges may represent magmatism and underplating typical of mountain building processes.

Interestingly, the magnetic profile shows rather flat signature over the MAT in the west and over the Vindhyan sediments in the east with a number of significant spikes in between indicating the presence of crustal-scale shear zones (Fig. 5.43). It was also inferred that the shear zones like PSZ and JSZ extend well into the mantle. A low magnetic anomaly of significant amplitude and dimension in the aeromagnetic map shows a break in NE-SW−trending high magnetic linear feature together with a steep gravity gradient at the KSZ suggest its extension up to the Moho level (Gouda et al., 2015). The broadband magnetotelluric data reveals a 10-km thick NW-dipping conductor of 50 ohmmeter electrical resistivity at a depth of 3 km near the JSZ and the conductor extends to a depth of 25 km (Gokaran, Rao, & Singh, 1995).

5.4.3 DEEP SEISMIC REFLECTION STUDIES

Deep seismic reflection and limited-refraction/wide-angle reflection data were acquired along a 400-km-long Nagaur−Jhalawar transect. The major tectonic boundaries between the terranes are well reflected in the changes of reflectivity pattern (dip direction) all across the ADOB. Abrupt increase in middle and lower crustal refectivity is generally observed at the terrane boundaries. The steeply dipping reflections at various crustal depths cut across subhorizontal reflections and the Moho discontinuity is distinctly delineated all along the N−J transect with the exception of the regions below the SDT and SMT (Vijaya Rao, Rajendra Prasad, Reddy, & Tewari, 2000). The Moho was also inferred from 2-D gravity modeling at a depth of 48 km beneath SDT and SMT (Tewari et al., 1997).

The reflectivity pattern along the seismic profile from Nagaur to Jhalawar varies all along the N-J transect. The reflectivity pattern is moderate and the Moho is interpreted to be at a depth of 40 km below the MAT. The PSZ is characterized by divergent reflection fabrics. The regional Bouguer gravity anomaly map shows a broad gravity high of the order of 70 mGal over the SDT and a prominent low with a steep gradient across the PSZ. Such a feature is distinct all along the strike of the 700 km long PSZ. The steep gradient Bouguer gravity anomaly constitutes a positive−negative pair and is the characteristic feature of a suture zone (Vijaya Rao et al., 2000). The steep gradient at the suture zone may be due to the presence of a high-density body (2.82 g/cm^3) as delineated in the form of ophiolites exposed along the PSZ.

The SDT displays strong lower crustal reflections, while the boundaries are marked by steep dipping reflectors in the upper crust. The SMT and MWT are delineated by poor reflectivity in the upper crust while the middle crust contains domal-shaped reflectors with a near horizontal reflector. This feature was interpreted as the existence of double Moho at a depth of 45−50 km, which dips westward beneath the SMT and MWT. Further east of MWT, a highly reflective crustal- scale thrust zone (Jahazpur thrust or the JSZ) was inferred at the conatact between the MWT and JHT to a depth of 5 km in the form of a 25-km-thick pile of parallel dipping reflections with a lateral extension up to 80 km.

The bottom of the Jahazpur thrust is associated with a highly reflective and laterally extensive subhorizontal reflection band terminated at the Moho, located at a depth of 48 km. The Moho is thus interpreted as a younger feature relative to the Jahazpur thrust just above it (Reddy & Vijaya Rao, 2013). The Aravallis were believed to have witnessed westward subduction during the late Paaeoproterozoic, and the JSZ was interpreted to be a suture delineating the JHT and MWT (Vijaya Rao et al., 2000). The GBSZ was underlain by a highly reflective subhorizontal reflection fabric similar to that of Jahazpur thrust with a Moho depth of 42 km. It can be seen from the above that the Moho offset seems to be at least 6 km between the JSZ and the GBSZ. In general, the Moho depth is uniform in each of the terrane, but for the boundary zones. Such kind of Moho offsets were reported from the Arunta Province of Australia, the Svecofennian orogeny of the Baltic shield, the Pyrenees, and the Alps (BABEL Working Group, 1990; Goleby, Shaw, Wright, Kennett, & Lambeck, 1989).

The seismic reflection data was interpreted and modeled in terms of geological associations and plate tectonic controlled interactions in a broader framework of westward subduction. According to that model, the PSZ represents the boundary between the MAT and the ADOB and both the terranes were tectonically segmented at ~1.0 Ga. The interpreted structural cross section along the N−J transect and across the ADOB is presented in (Fig. 5.44). The following are the salient features of the model: (1) the crustal- scale shear zones delineating distinct tectonic domains are well reflected in seismic reflection data; (2) the shear zones are found to be listric with top to the SE. However, the imbricate thrust zone observed in the upper crustal level within the SMT shows

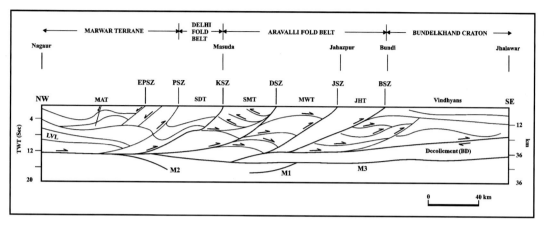

FIGURE 5.44

Structural cross section interpreted from the seismic reflection images of Nagaur-Jhalawar transect.
BD, Bundelkhand craton; *M1*, Moho beneath the BD extending into Sandmata terrane; *M2*, Moho beneath south Delhi Terrane, extending into Marwar Terrane; *M3*, present Moho; *MAT*, Marwar Terrane; *EPSZ*, Erinpura shear zone; *PSZ*, Phulad shear zone; *SDT*, South Delhi Terrane; *KSZ*, Kaliguman shear zone; *SMT*, Sandmata Terrane; *DSZ*, Delwara shear zone; *MWT*, Mangalwar Terrane; *JSZ*, Jahazpur shear zone; *JHT*, Jahazpur Hindoli Terrane; *GBSZ*, Great Boundary shear zone.

Modified after Sinha-Roy, S. (2008). Evolution of Precambrian terrains and crustal scale structures in Rajasthan craton, NW India: A kinematic model. IAGR Memoir, 10, 23–40.

top-to-NW, which can be interpreted as back thrust; (3) the domal disposition of the crustal layers between the DSZ and the JSZ was interpreted as SE-propogating antiformal duplex structure implying the faster and longer transport of successive and higher tectonic horses toward the JSZ; (4) the JSZ represents an important crustal detachment from the decollement ramp and defines the leading edge of the Bundekhand wedge indenting the Mewar craton along with the JSZ and GBSZ defining the roof and floor thusts respectively; (5) the PSZ truncates subhorizontal crustal layers at a depth of 20 km, probably marking the root zone of the SDT; (6) the KSZ reaches the mantle and exhibits a ramp structure at a depth of ~30 km from where the Sandmata complex was exhumed as a tectonic wedge. The SMT shows SE-verging structures in the mid-crustal level, while the upper crust is marked NW-verging thrusts; (7) the MAT is marked by steeply dipping normal fault system. The EPSZ truncates the upper crustal layer and a low velocity layer occurs at a depth of 20–30 km. The evolution of MAT along with westward subduction can be compared with a distensional and stretched hinter land of collision tectonics, similar to that of Tibet (Sinha-Roy, 2004); (8) the JSZ is characterized by oppositely dipping double Moho: Moho-M1 extending westward and the Moho-M2 extending eastward beneath MAT. Such deeper Mohos are suggestive of subduction and crustal thickening possibly related to Meso-Neoproterozoic collision tectonics and the current Moho (M3) may be related to later crustal processes; and (9) the segment east of the GBSZ shows gently disturbed and curved crustal layers at a depth of 12–20 km possibly representing an early duplex structure at the basement of Vindhyan sediments. However, the Vindhyan sediments are marked by shallow-dipping reflectors at a low angle between Jhalawar and GBSZ and shows a westerly dipping decollement zone of Bundelkhand craton at a depth of 20 km connecting to all major shear zones of the ADOB.

5.4.4 EASTWARD SUBDUCTION

In contrast to the westward subduction described above, eastward subduction of the oceanic lithosphere under the continental margin of Mewar craton with the trench being located at the western boundary of SDT was also proposed (Vijaya Rao & Krishna, 2013). They have re-analyzed and re-interpreted deep seismic reflection images along the Nagaur-Masuda segment and proposed eastward subduction. Their results show strong and continuous SE-dipping reflection bands extending from the upper to lower crust covering a distance of $\sim$80 km in the MAT. According to them, eastward subduction of oceanic lithosphere and subsequent collision of volcanic arc with MAT was responsible for the evolution of the SDT and the postcollisional delamination and orogenic collapse were responsible for the equilibrated younger Moho and evolution of Malani magmatism. They envisaged that the rocks of the SDT extend westward and form the basement for the Malani volcanics, supported by the complementary seismic refraction/wide-angle reflection data exhibiting significant differences in velocity structure and composition between the Marwar Terrain and the SDT. Eastward subduction of an oceanic plate with a mid-oceanic ridge was also favored by Sychanthavong and Desai (1977) and suppored by geological and geochemical data of SDT (Deb & Sarkar, 1990). This inference is in variation with earlier belief that the Archean BGC was the basement for the MIS (Bhushan, 2000), thereby implying the Archean age for the MAT. Further, some researchers opined that the MAT may be related to the Neoproterozoic Pan–African orogenic event (550 Ma) of East-African and Arabian-Nubian shield located further west (Volpe & Macdougall, 1990).

According to Vijaya Rao and Krishna (2013), the seismic reflection images across the late Mesoproterozoic SDT provide evidence for crustal-scale tectonic imbrication and collisional tectonism. The model also successfully resolves the ambiguity by correlating the Marwar Terrane with the Rodinia assembly rather than later Pan-African orogeny located further west. Evolution of the SDT and Malani magmatism are coeval with the Rodinia assembly and breakup. The South Delhi orogeny, located between the east- and the west-Gondwana fragments, plays an important role for reconstruction of the Gondwana.

In summary, the seismic images reveals possible subduction–collisional event in the form of SE-dipping reflection fabric throughout the Marwar terrane over which the 750 Ma MIS is emplaced. Considering the age of MIS the collision could be around 1000 Ma, suggesting that the MAT may represent a late-Mesoproterozoic crustal block. This event is coeval with the accretion of various other cratonic blocks of the world and formation of the Rodinia supercontinent during this period.

5.5 TECTONIC SYNTHESIS (ADOB)

The evolution of the ADOB was explained by widely divided tectonic models. The most significant tectonic models involving plate tectonic regimes include: (1) westward subduction and (2) eastward subduction. There is also yet another view and that there exist two orogens within the ADOB consisting of Paleoproterozoic—AFB and Mesoproterozoic DFB with out any signature of plate tectonic regime. However, a single stage orogenic belt evolution was proposed by some involving the evolution of basins by rifting and sedimentation in different cycles with thermal overprints and postorogenic shearing (Roy & Jakhar, 2002). The basic presumption in the non-plate tectonic model was that the ultramafic bodies do not represent ophiolites and it was the inversion of ensialic basins rather than plate tectonics. They also argued that the granulite facies rocks in ADOB were developed during the later stages of orogeny and exhumed from depth along shear zones. According to them, the Aravalli orogeny ceased at 1850 Ma and the Delhi orogeny developed later in terms of rifts, grabens, and aulacogens and that the rocks were metamorphosed at 1450 Ma.

Another tectonic model was proposed, following traditional lines, and suggested plume generated rifting, formation of intracratonic basins, deposition of shallow and deep water sediments, along with the emplacement of mafic and ultramafic lavas (Sharma, 2012). This was followed by further deepening of basins, crustal shortening and crustal thickening and subsequent westward subduction of the eastern block comprising Aravallis—Bundelkhand craton. He also described the anticlockwise P—T history for the Sandmata granulites, assumed to indicate crustal thickening by underplating and suggests ensialic orogenesis model for the evolution of the ADOB opposing the plate tectonic model.

Recent studies of geology, geophysics and geochronology led the researchers to understand the evolution of the ADOB by involving the plate tectonic processes. It is now almost established that the ADOB was subjected to subduction—accretion—collisional processes like any other orogenic belt all over the globe. However, within the plate ectonic scenarios, conflicting opinions exist about the polarity of subduction which would be detailed further.

Sen (1981) was among the first to invoke plate tectonic model for the evolution of ADOB invoking the initial development of rift basins and the development of Udaipur sea. He envisaged eastward subduction leading to the accretion of island arc and oceanic crust against the continent. In contrast, westward subduction was proposed and supported by studies on deformation, metamorphism, and magmatism in the region by Shychanthavong and Merh (1984). The model of westward subduction was further emphasized and supported by many with detailed geological studies and geophysical signatures (Sinha-Roy, 1995; Tewari et al., 1995). They also proposed the closure of the ocean and the final collision was at 1.5 Ga with the obduction of the oceanic crust along the RDSZ. A new marginal sea was further developed in the west and the mafic crust was subducted to west, ultimately leading to the collision and formation of DFB at 1.0 Ga (Fig. 5.45). It has generated Erinpura granitoids and forming a Cordilleran magmatic belt to the west of SDT at 0.9—0.8 Ga (Bhowmikt et al., 2010). According to Sinha-Roy (2008), the cold and flat lying Delhi lithospheric slab entrenched and got delaminated beneath the MAT resulting in the generation of Malani rhyolites at ca. 0.75 Ga, which could be the last Proterozoic lithospheric event in the region of ADOB. The westward subduction of SDT oceanic crust caused partial melting of the MAT crust, (A strong extensional tectonism was developed during the Neoproterozoic in the MAT and the MIS was developed in response to partial melting caused by intracontinental subduction of the Mewar terrane beneath the MAT) (Sinha-Roy, 2004).

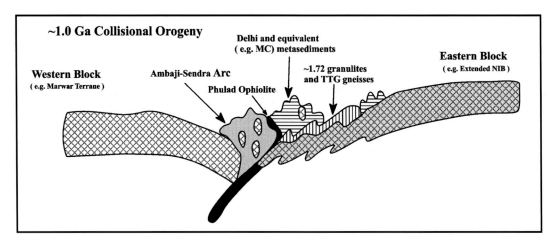

FIGURE 5.45

A diagram showing the collisional tectonics of the South Delhi Terrane at 1.0 Ga. With westward subduction.

Adapted from Bhowmik, S. K., Bernhardt, H.-J., & Dasgupta, S. (2010). Grenvillian age high-pressure upper amphibolite–granulite metamorphism in the Aravalli–Delhi mobile belt, northwestern India: New evidence from monazite chemical age and its implication.

Precambrian Research, *178, 168–184.*

A compression-dominated event associated with important magmatic pulse near 1.85 Ga was proposed by Kaur, Chaudhri, Raczek, Kröner, Hofmann, and Okrusch (2011) providing the evidence of subduction-accretion tectonics in the northern and central Aravalli mountain range. This event was followed by an extension dominated 1.72–1.70 Ga tectono-magmatic and tectono-thermal phase. In the northern domain, it is recorded by a number of rift related A-type granites (Biju-Sekhar et al., 2003), while in the central part, it is related to the development of coeval orthopyroxene-bearing granitoids, TTG gneisses and granulite facies metamorphism (e.g., Buick et al., 2006). However, it was also considered that the 1.72–1.70 Ga event marks the reworking of the Archean BGC crust along with the exhumation of granulites and the opening of the Delhi basin. The latest event of collision in the ADOB is of Grenvillian age at around 1.0–0.9 Ga. Lu–Hf isotope data of the detrital zircons additionally reveal that the oldest crust in the Aravalli mountain range or in its hinterland was formed from a depleted mantle source at ca. 3.7 Ga, and that new "juvenile" crust was formed at 3.25, 2.89, 2.67, 2.51 and 1.87–1.80 Ga (Kaur et al., 2011). The occurrence of 1.85–1.82 Ga subduction-related event and a widespread 1.72–170 Ga continental rift-related felsic magmatism in ADOB indicate the existence of an ∼1.85 Ga subduction–accretion–collisional orogen, similar to the one recently identified from the eastern periphery of the Indian plate (Dharma Rao, Santosh, & Wu, 2011), that underwent extensional tectonics at around 1.72–1.70 Ga manifested by A-type magmatic activity along with the exhumation of granulite-facies rocks. It was also speculated that 1.72–1.70 Ga was an important continental rift phase related to dispersal of the Columbia.

In summary, geological and geophysical studies established that the Plate tectonic processes were responsible for the evolution of Paleoproterozoic ADOB through an accretionary process of island arcs during subduction and/or by collision involving the Bundelkhand craton in the east and the MAT in the west. Individual terranes are recognized by differences in their seismic reflectivity characteristics and some of the crustal-scale shear zones may represent sutures that are characterized by various mineralized zones.

The timing of the tectonic juxtaposition of different terranes was suggested to be at c. 1 Ga (Buick et al., 2006). The lithologic association and their nature of distribution are in accordance with a Paleoproterozoic Pacific-type orogeny in the Aravalli region with a westward subduction of the Archean cratonic margin and development of a wide accretionary belt, imbricated ocean plate stratigraphy including ophiolites and the extrusion of a high-grade regional metamorphic belt at the orogenic core following the final collision (Santosh, 2010). The available geochronological data reinforces the suggestion that the tectonic history of the BGC-II is distinct from that of BGC-I; the former is dominated by the Paleoproterozoic and Neoproterozoic intrusive and metamorphic events, and the latter appears to be entirely Archean. This needs to be further resolved with systematic studies of geology and geochronology.

LIST OF ABBREVIATIONS

ADOB	Aravalli-Delhi Orogenic Belt
AFB	Aravalli Fold Belt
BGC	Banded gneissic complex
BKSK	Balara-Kui-Surpagla-Kengora
BSZ	Banas shear zone
CITZ	Central Indian tectonic zone
DFB	Delhi Fold Belt
DSZ	Delwara shear zone
EPSZ	Erinpura shear zone
GBSZ	Great Boundary shear zone
JHT	Jahazpur Hindoli Terrane
JRT	Jharol Terrane
JSZ	Jahazpur shear zone
KCB	Khetri Copper Belt
KSZ	Kaliguman shear zone
MAT	Marwar Terrane
MIS	Malani Igneous Suite
MWT	Mangalwar Terrane
NDT	North Delhi Terrane
NKCB	Northern Khetri Copper Belt
PSZ	Phulad shear zone
RDSZ	Rakhabdev shear zone
SDT	South Delhi Terrane
SKCB	Southern Khetri Copper Belt
SMT	Sandmata Terrane
SRT	Sirohi Terrane
SSZ	Sabarmati shear zone
UPT	Udaipur Terrane

REFERENCES

Ahmad, T., & Rajamani, V. (1991). Geochemistry and petrogenesis of the basal Aravalli volcanics near Nathdwara, Rajasthan, India. *Precambrian Research, 49*, 185–204.

Amar Singh, Singh, S. L., & Golani, P. R. (2015). Relevance of regional gravity features of Northwestern Indian Shield in localisation of base metal deposits of Rajasthan. *Geological Survey of India Special Publication, 101*, 247–265.

Ashwal, L. D., Solanki, A. M., Pandit, M. K., Corfu, F., Hendriks, B. W. H., Burke, K., & Torsvik, T. H. (2013). Geochronology and geochemistry of Neoproterozoic Mt. Abu granitoids, NW India: regional correlation and implications for Rodinia paleo-geography. *Precambrian Research, 236*, 265–281.

BABEL Working Group. (1990). Evidence for early Proterozoic plate tectonics from seismic refection profiles in the Baltic Shield. *Nature, 34*, 34–38.

Banerjee, S., & Sarkar, S. C. (1998). *Petrological and Geochemical background of the zinc-lead mineralisation at Zawar, Rajasthan-a study. The Indian Precambrian*. Jodhpur: Scientific Publishers (India), 312–326 pp.

Bhattacharya, A. K., Nagarajan, K., Shekhawat, L. S., & Joshi, D. W. (1995). Stratigraphy, structure and metamorphism of the Aravalli Fold Belt. *Records Geological Survey of India, 12*, 65–77.

Bhattacharya, P. K., & Mukherjee, A. D. (1984). Petrochemistry of metamorphosed pillow and the geochemical status of the amphibolites (Proterozoic) from the Sirohi district, Rajasthan India. *Geological Magazine, 121*, 465–473.

Bhowmik, S. K., Bernhardt, H.-J., & Dasgupta, S. (2010). Grenvillian age high-pressure upper amphibolite–granulite metamorphism in the Aravalli–Delhi mobile belt, northwestern India: New evidence from monazite chemical age and its implication. *Precambrian Research, 178*, 168–184.

Bhowmik, S. K., & Dasgupta, S. (2012). Tectonothermal evolution of the Banded Gneissic Complex in central Rajasthan, NW India: Present status and correlation. *Journal of Asian Earth Sciences, 49*, 339–348.

Bhushan, S. K. (1984). *Classification of Malani Igneous Suite. Symposium on three decades of developments in petrology, Mineralogy and Petrochemistry in India. Geological Survey of India. Special Publication, 12*, 199–205.

Bhushan, S. K. (2000). Malani rhyolite – a review. *Gondwana Research, 3*, 65–77.

Biju-Sekhar, S., Yokoyama, K., Pandit, M. K., Okudaira, T., Yoshida, M., & Santosh, M. (2003). Late Paleoproterozoic magmatismin Delhi Fold Belt, NWIndia and its implication: Evidence from EPMA chemical ages of zircons. *Journal of Asian Earth Sciences, 22*, 189–207.

Bose, U., & Sharma, A. K. (1992). The volcano-sedimentary association of the Precambrian Hindoli supracrustals in southern Rajasthan. *Journal of the Geological Society of India, 40*, 359–369.

Buick, I. S., Allen, C., Pandit, M. K., Rubatto, D., & Hermann, J. (2006). The Proterozoic magmatic and metamorphic history of the Banded Gneiss Complex, central Rajasthan, India: LA-ICP-MS U–Pb zircon constraints. *Precambrian Research, 151*, 119–142.

Buick, I. S., Clark, C., Rubatto, D., Hermann, J., Pandit, M., & Hand, M. (2010). Constraints on the Proterozoic evolution of the Aravalli-Delhi Orogenic belt (NW India) from monazite geochronology and mineral trace element geochemistry. *Lithos, 120*, 511–528.

Chakrabarti, B., & Gupta, G. P. (1992). Stratigraphy and structure of the North Delhi Basin. *Geological Survey of India Records, 124*, 5–9.

Chetty, T. R. K., & Bhaskar Rao, Y. J. (2006). The Cauvery Shear Zone, Southern Granulite Terrain, India: A crustal-scale flower structure. *Gondwana Research, 10*, 77–85.

Choudhary, A. K., Gopalan, K., & Anjaneya Sastry, C. (1984). Present status of geochronology of the Precambrian rocks of Rajasthan. *Tectonophysics, 105*, 131–140.

Crawford, A. R. (1975). Rb-Sr age determination for the Mount Abu granite and related rocks of Gujarat. *Journal of the Geological Society of India, 16*, 20–28.

Crookshank, H. (1948). Minerals of the Rajputana Pegmatites. *Transactions of Mining Geology and Metallurgical Institute of India*, *42*, 105−189.

Das, A. R. (1988). Geometry of the superposed deformation in the Delhi Supergroup of rocks, north of Jaipur, Rajasthan. *Memoir of the Geological Society of India*, *7*, 247−266.

Das Gupta, S. P. (1968). The structural history of the Khetri Copper Belt, Jhunjhunu and Sikar districts, Rajasthan. *Memoir of the Geological Survey of India*, *98*, 170.

Dasgupta, S., Guha, D., Sengupta, P., Miura, H., & Ehl, J. (1997). Pressure-temperature-fluid evolutionary history of the polymetamorphic Sandmata granulite complex, Northwestern India. *Precambrian Research*, *83*, 267−290.

de Wall, H., Pandit, M. K., Dotzler, R., & Just, J. (2012b). Cryogenian transpression and granite intrusion along the western margin of Rodinia (Mt. Abu Region): Magmatic fabric and geochemical inferences on Neoproterozoic geodynamics of the NW India block. *Tectonophysics*, *554−557*, 143−158.

Deb, M., & Sarkar, S. C. (1990). Proterozoic tectonic evolution and metallogenesis in the Aravalli−Delhi orogenic complex, NW India. *Precambrian Research*, *46*, 115−137.

Deb, M., Thorpe, R. I., Cumming, G. L., & Wagner, P. A. (1989). Age, source and stratigraphic implications of Pb isotope data for conformable, sediment-hosted, basemetal deposits in the Proterozoic Aravalli Delhi orogenic belt, NW India. *Precambrian Research*, *43*, 1−22.

Deb, M., Thorpe, R. I., Krstic, D., Corfu, F., & Davis, D. W. (2001). Zircon U−Pb and galena Pb isotope evidence for an approximate 1.0 Ga terrane constituting the western margin of the Aravalli−Delhi Orogenic Belt, northwestern India. *Precambrian Research*, *108*, 195−213.

Dharma Rao, C. V., Santosh, M., & Wu, Y.-B. (2011). Mesoproterozoic ophiolitic mélange from the SE periphery of the Indian plate: U−Pb zircon ages and tectonic implications. *Gondwana Research*, *19*, 384−401.

Fareeduddin, & Kröner, A. (1998). *Single zircon age constraints on the evolution of Rajasthan granulite. The Indian Precambrian* (pp. 547−556). India: Scientific Publishers.

Fermor, L. L. (1930). On the age of Aravalli Range. *Records, Geological Survey of India*, *62*(4), 391−409.

Ghosh, S. K., Hazare, S., & Sengupta, S. (1999). Planar, non-planar and refolded sheath folds in the Phulad Shear Zone, Rajasthan. *Journal of Structural Geology*, *21*, 1715−1729.

Gokaran, S. G., Rao, C. K., & Singh, B. P. (1995). Crustal structure in southeast Rajasthan using Magnetotelluric techniques. Continental Crust of NW and Central India. *Memoir of the Geological Society of India*, *31*, 373−381.

Golani, P. R. (2015). Recent development in metallogeny and mineral exploration in Rajasthan. *Geological Survey of India Special Publication*, *101*, 275.

Golani, P. R., Gathania, R. C., Grover, A. K., & Bhattacharjee, J. (1992). Felsic volcanics in south Khetri copper belt, Rajasthan and their metallogenic significance. *Journal Geological Society of India*, *40*, 79−87.

Golani, P. R., Reddy, A. B., & Bhattacharjee, J. (1998). *The Phulad shear zone in central Rajasthan and its tectonostratigraphic implications* (pp. 272−278). Jodhpur: Scientific Publishers (India).

Goleby, B. R., Shaw, R. D., Wright, C., Kennett, B. L. N., & Lambeck, K. (1989). Geophysical evidence for 'thick-skinned' crustal deformation in central Australia. *Nature*, *337*, 325−330.

Gopalan, K., Macdougall, J. D., Roy, A. B., & Murali, A. V. (1990). Sm−Nd evidence for 3.3 Ga old rocks in Rajasthan, north western India. *Precambrian Research*, *48*, 287−297.

Gopalan, K., Trivedi, J. R., Merh, S. S., Patel, P. P., & Patel, S. G. (1979). Rb-Sr age of Godhra and related granites, Gujarat (India). *Proceedings Indian Academy of Sciences (Earth Planet. Sci.)*, *88*, 7−17.

Gouda, H. C., Fareeduddin, Singh, R. K., Rajaram, H., Rajesh Kumar, Ramesh Kesavan, ... Rajendra Sharma (2015). *Aeromagnetic anomaly maps of the Aravalli Craton and their interpretation: some new insights on stratigraphy and metallogeny of region. Geological Survey of India. Special Publication*, *101*, 1−24.

Gregory, L. C., Meert, J. G., Bingen, B., Pandit, M. K., & Torsvik, T. H. (2009). Paleomagnetism and geochronology of the Malani igneous suite, northwest India: implications for the configuration of Rodinia and assembly of Gondwana. *Precambrian Research*, *170*, 13−26.

Guha, D. B., & Bhattacharya, A. K. (1995). Metamorphic evolution and high-grade reworking of the Sandmata Complex granulites. *Memoir of the Geological Society of India*, *31*, 163−198.

Gupta, B. C. (1934). The geology of central Mewar. *Memoir of the Geological Survey of India*, *65*, 107−168.

Gupta, B. C., & Mukherjee, P. N. (1938). The geology of Gujarat and South Rajputana. *Records, Geolological Survey of India*, *73*, 163−208.

Gupta, P., & Bose, U. (2000). An update on the geology of Delhi Supergroup in Rajasthan. *Geological Survey of India Specual Publication*, *55*(1), 287−306.

Gupta, P., Guha, D. B., & Chattopadhyay, B. (1998). Basement-cover relationship in the Khetri Copper Belt and the emplacement mechanism of the granite massifs, Rajasthan. *Journal of the Geological Society of India*, *52*, 417−432.

Gupta, P., Mukhopadhyay, K., Fareeduddin, & Reddy, M. S. (1991). Tectono-stratigraphic framework and volcanic geology of the south Delhi Fold Belt in central Rajasthan. *Journal of the Geological Society of India*, *37*, 431−441.

Gupta, S. N., Arora, Y. K., Mathur, R. K., Iqbaluddin, Prasad, B., Sahani, T. N., & Sharma, S. B. (1980). *Lithostratigraphy map of Aravalli region*. Hyderabad: Geological Survey of India publication.

Gupta, S. N., Arora, Y. K., Mathur, R. K., Iqbaluddin, Prasad, B., Sahani, T. N., & Sharma, S. B. (1995). *Geological map of the Precambrian of the Aravalli region, southern Rajasthan and northeastern Gujarat, India*. Geological Survey of India Publication.

Gupta, S. N., Arora, Y. K., Mathur, R. K., Iqbaluddin, Prasad, B., Sahai, T. N., & Sharma, S. B. (1997). The Precambrian geology of the Aravalli region, southern Rajasthan and north-eastern Gujarat. *Memoir of the Geological Survey of India*, *123*, 262.

Heron, A. M. (1953). The geology of Central Rajputana. *Memoir of the Geological Survey of India*, *79*, 339.

Just, J., Schulz, B., de Wall, H., Jourdan, F., & Pandit, M. K. (2011). Monazite CHIME/EPMA dating of Erinpura granitoid deformation: Implications for Neoproterozoic tectono-thermal evolution of NW India. *Gondwana Research*, *19*, 402−412.

Kataria, P., Chaudhari, M. W., & Althaus (1988). Petrochemistry of amphibolites from the Banded Gneissic Complex of Amet, Rajasthan, NW India. *Chemie de Erde*, *21*, 291−306.

Kaur, P., Chaudhri, N., Biju-Sekhar, S., & Yokoyama, K. (2006). Electron probe micro analyser chemical zircon ages of the Khetri granitoids, Rajasthan, India: Records of widespread late Palaeoproterozoic extension-related magmatism. *Current Science*, *90*, 65−73.

Kaur, P., Chaudhri, N., Raczek, I., Kroener, A., & Hofmann, A. W. (2007). Geochemistry, zircon ages and whole-rock Nd isotopic systematics for Palaeoproterozoic A-type granitoids in the northern part of the Delhi belt, Rajasthan, NW India: Implications for late Palaeoproterozoic crustal evolution of the Aravalli craton. *Geological Magazine*, *144*, 361−378.

Kaur, P., Chaudhri, N., Raczek, I., Kröner, A., & Hofmann, A. W. (2009). Record of 1.82 Ga Andean-type continental arc magmatism in NE Rajasthan, India: Insights from zircon and Sm−Nd ages, combined with Nd−Sr isotope geochemistry. *Gondwana Research*, *16*, 56−71.

Kaur, P., Chaudhri, N., Raczek, I., Kröner, A., Hofmann, A. W., & Okrusch, M. (2011). Zircon ages of late Palaeoproterozoic (ca. 1.72−1.70 Ga) extension-related granitoids in NE Rajasthan, India: Regional and tectonic significance. *Gondwana Research*, *19*, 1040−1053.

Khan, S. M., Smith, T. E., Raza, M., & Huang, J. (2005). Geology, Geochemistry and Tectonic Significance of mafic-ultramafic rocks of Mesoproterozoic Phulad Ophiolite Suite of South Delhi Fold Belt, NW Indian Shield. *Gondwana Research*, *8*, 553−566.

Kochhar, N. (1984). Malani Igneous Suite: Hotspot magmatism and cratonisation of the northern part of the Indian shield. *Journal of the Geological Society of India*, *25*, 155−161.

Kochhar, N., & Dhar, S. (2000). Comments on "Rb-Sr Isotope dating of Neoproterozoic (Malani Group) magmatism from Southwest Rajasthan, India: Evidence of younger Pan-African event by $^{40}Ar/Ar^{39}$ studies". *Gondwana Research*, *3*, 119.

Kundu, A., Kazim, K., & Sharma, S. (2004). Metamorphism in north Delhi Fold Belt- a case study from southwest Haryana and adjacent parts of Rajasthan. *Indian Minerals, 58*, 17–26.

Lal, R. K., & Ackermand, D. (1981). Phase petrology and polyphase andalusite-sillimanite type regional metamorphism in pelitic schist of the area around Akwali, Khetri Copper Belt, Rajasthan, India. *Neues Jahrbuch für Mineralogie-Abhandlungen, 141*, 161–185.

Lal, R. K., & Shukla, R. S. (1975). Low-pressure regional metamorphism in the northern portion of the Khetri Copper Belt, Rajasthan, India. *Jahrbuch für Mineralogie-Abhandlungen, 124*, 294–325.

Mamtani, M. A., Karant, R. V., Merh, S. S., & Greiling, R. O. (2000). Tectonic evolution of the southern Aravalli Mountain Belt and its environs–possible causes and time constraints. *Gondwana Research, 3*, 175–187.

Mamtani, M. A., Karmakar, B., & Merh, S. S. (2002). Evidence of polyphase deformation in gneissic rocks around devgadh bariya: Implications for evolution of godhra granite in the Southern Aravalli Region (India). *Gondwana Research, 5*, 401–408.

Meert, J. G., Pandit, M. K., & Kamenov, G. D. (2013). Further geochronological and paleomagnetic constraints on Malani (and pre-Malani) magmatism in NW India. *Tectonophysics, 608*, 1254–1267.

Mehdi, M., Santosh, K., & Pant, N. C. (2016). Low grade metamorphism in the Lalsot-Bayana Sub-basin of the North Delhi Fold Belt and its Tectonic Implication. *Journal of the Geological Society of India, 85*, 397–410.

Merh, S. S. (1995). *Geology of Gujarat* (p. 224). Bangalore: Geological Society of India.

Mezger, K., & Cosca, M. A. (1999). The thermal history of the Eastern Ghats Belt (India) as revealed by U–Pb and $^{40}Ar/^{39}Ar$ dating of metamorphic and magmatic minerals: implications for the SWEAT correlation. *Precambrian Research, 94*, 251–271.

Mukhopadhyay, D., Chattopadhyay, N., & Bhattacharyya, T. (2010). Structural evolution of a gneiss dome in the axial zone of the proterozoic South Delhi Fold Belt in Central Rajasthan. *Journal of the Geological Society of India, 75*, 18–31.

Mukhopadhyay, D., & Sengupta, S. (1979). "Eyed folds" in Precambrian marbles from southeastern Rajasthan, India. *Bulletin of the Geological Society of America, 90*, 397–404.

Naha, K., & Halyburton, R. V. (1974). Early Precambrian Stratigraphy of central and southern Rajasthan, India. *Precambrian Research, 1*, 55–73.

Naha, K., & Majumdar, A. (1971). Structure of the Rajnagar marble band and its bearing on the Precambrian stratigraphy of central Rajasthan, Western India. *Geologische Rundschau, 60*, 1550–1571.

Naha, K., & Mohanty, S. (1988). Response of basement and cover rocks to multiple deformations: A study from the Precambrian of Rajasthan, Western India. *Precambrian Research, 42*, 77–96.

Naha, K., Mukhopadhyay, D. K., & Mohanty, R. (1988). Structural evolution of the rocks of the Delhi Group around Khetri, northeastern Rajasthan. In A. B. Roy (Ed.), *Precambrian of the Aravalli Mountain, Rajasthan* (vol. 7, pp. 207–245). India: Memoir of the Geological Society of India.

Naha, K., Mukhopadhyay, D. K., Mohanty, R., Milra, S. K., & Biswal, T. K. (1984). Significance of contrast in the early stages of the structural history of the Delhi and pre-Delhi rock groups in the Proterozoic of Rajasthan, western India. *Tectonophysics, 105*, 193–206.

Naha, K., & Roy, A. B. (1983). The problem of the Precambrian basement in Rajasthan, Western India. *Precambrian Research, 19*, 217–223.

Pandit, M. K., Shekhawat, L. S., Ferreira, V. P., Sial, A. N., & Bohra, S. K. (1999). Trondhjemite and granodiorite assemblages from West of Barmer: Probable basement for Malani magmatism in Western India. *Journal of the Geological Society of India, 53*, 89–96.

Pant, N. C., Kundu, A., & Joshi, S. (2008). Age of Metamorphism of Delhi Supergroup rocks-electron microprobe age from Mahendragarh district, Haryana. *Journal of the Geological Society of India, 72*, 365–372.

Pareek, H. S. (1981). Petrochemistry and petrogenesis of the Malani igneous suite, India: summary. *Bulletin of the Geological Society of America*, *92*, 67−70.

Porwal, A., Carranza, E. J. M., & Hale, M. (2006). Tectonostratigraphy and base-metal mineralization controls, Aravalli province (western India): New interpretations from geophysical data analysis. *Reviews in Ore Geology*, *29*, 287−306.

Purohit, R., Papineau, D., Kröner, A., Sharma, K. K., & Roy, A. B. (2012). Carbon isotope geo-chemistry and geochronological constraints of the Neoproterozoic Sirohi Group from northwest India. *Precambrian Research*, *220−221*, 80−90.

Raja Rao, C. S., Poddar, B. C., Basu, K. K., & Dutta, A. K. (1971). Precambrian stratigraphy of Rajasthan-A review. Records. *Geological Survey of India*, *101*, 60−79.

Ramakrishnan, M., & Vaidyanadhan, R. (2008). *Geology of India* (volume 1, p. 556). Geological Society of India.

Ramsay, J. G. (1967). *Folding and fracturing of rocks* (p. 568). New York: McGraw-Hill.

Ramsay, J. G., & Huber, M. I. (1987). *The techniques of modern structural geology. Volume 2: Folds and Fractures* (pp. 309−700). London: Academic Press.

Rathore, S. S., Venkatesan, T. R., & Srivastava, R. K. (1999). Rb−Sr isotope dating of Neoproterozoic (Malani Group) magmatism from Southwest Rajasthan, India: Evidence of younger Pan African thermal event by 40Ar−39Ar studies. *Gondwana Research*, *2*, 271−281.

Ray, S. K. (1990). The albitite line of northern Rajasthan−a fossil intracontinental rift zone. *Journal of the Geological Society of India*, *36*, 413−423.

Reddi, A. G. B., & Ramakrishna, T. S. (1988). *Bouguer gravity atlas of northwestern (Rajasthan-Gujarat) shield of India*. Jaipur: Geological Survey of India.

Reddy, P. R., Rajendra Prasad, B. R., Vijaya Rao, V., Prakash Khare, Kesava Rao, G., Murthy, A. S. N., ... Sridher, V. (1995). *Deep seismic reflection profiling along Nandsi-Kunjer section of Nagaur-Jhalawar transect: Preliminary results* (vol. 31, pp. 353−372). Memoir of the Geological Society of India.

Reddy, P. R., & Vijaya Rao, V. (2013). Seismic images of the continental Moho of the Indian shield. *Tectonophysics*, *609*, 217−233.

Roy, A. B. (1990). *Evolution of the Precambrian crust of the Aravalli Mountain range*. Amsterdam: The Netherlands: Development in Precambrian Geology, Elsevier.

Roy, A. B. (1995). Geometry and evolution of superposed folding in Zawar lead-zinc mineralised belt, Rajasthan. . *Proceedings of Indian Academy of Sciences (Earth & Planetary Sciences)*, *104*, 349−371.

Roy, A. B. (1988). Stratigraphic and tectonic framework of the Aravalli mountain range. *Memoir of the Geological Society of India*, *7*, 3−32.

Roy, A. B. (2001). Neoproterozoic crustal evolution of northwestern Indian shield: Implications on break up and assembly of supercontinents. *Gondwana Research*, *4*, 289−306.

Roy, A. B., & Dutt, K. (1995). Tectonic evolution of the Nephelene syenite and associated rocks Kishangarh, District Ajmer, Rajastahan. *Memoir of the Geological society of India*, *31*, 231−257.

Roy, A. B., & Jakhar, S. R. (2002). *Geology of Rajasthan (Northwestern India), Precambrian to Recent* (p. 421). Scientific Publishers.

Roy, A. B., & Kröner, A. (1996). Single zircon evaporation ages constraining growth of the Aravalli craton, northwestern Indian shield. *Geological Magazine*, *133*, 333−342.

Roy, A. B., Kröner, A., Bhattacharya, P. K., & Rathore, S. (2005). Metamorphic evolution and zircon geochronology of early Proterozoic granulites in the Aravalli Mountains of northwestern India. *Geological Magazine*, *142*, 287−302.

Roy, A. B., & Sharma, K. K. (1999). *Geology of the region around Sirohi Town, western Rajasthan-a story of Neoproterozoic evolution of the Aravalli crust* (pp. 19−33). Jodhpur: Scientific Publishers (India).

Roy, A., & Prasad, M. H. (2003). Tectonothermal events in Central Indian Tectonic Zone (CITZ) and its implications in Rodinia crustal assembly. *Journal of Asian Earth Sciences*, *22*, 115−129.

Roy, A. B., & Paliwal, B. S. (1981). Evolution of lower Proterozoic epicontinental deposits: Stromatolite-bearing Aravalli rocks of Udaipur, Rajasthan, India. *Precambrian Research*, *14*, 49−74.

Roy, A. B., Paliwal, B. S., Shekhawat, S. S., Nagori, D. K., Golani, P. R., & Bejarniya, B. R. (1988). Stratigraphy of the Aravalli Supergroup in the type area. *Memoir of the Geological Society of India*, *7*, 121−138.

Roy, A. B., & Rathore, S. (1999). *Tectonic setting and model of evolution of granulites of the Sandmata Complex, the Aravalli Mountain, Rajasthan. International Symposium on Charnockite and Granulite Facies Rocks* (pp. 111−126). Geologists Association of Tamil Nadu.

Roy, A. B., Somani, M. K., & Sharma, N. K. (1981). Aravalli−Pre-Aravalli relationship − a study from Bhindar, south-central Rajasthan. *Indian Journal of Earth Science*, *8*, 119−130.

Santosh, M. (2010). Assembling North China Craton within the Columbia supercontinent: The role of double-sided subduction. *Precambrian Research*, *178*, 149−167.

Sarkar, G., Ray Burman, T., & Corfu, F. (1989). Timing of continental arc-type magmatism in NW India: Evidence from U-Pb zircon geochronology. *Journal of Geology*, *97*, 607−612.

Sarkar, S., & Biswal, T. K. (2005). Tectonic Significances of fissure veins associated with pseudotachylites of the Kui-Chitraseni Shear Zone, Aravalli Mountain, NW India. *Gondwana Research*, *8*, 277−282.

Sarkar, S. C. (2000). Crustal evolution and metallogeny in the Eastern Indian Craton. *Geological Survey of India. Special Publication*, *55*, 169−194.

Sarkar, S. C., & Dasgupta, S. (1980). Geologic setting, genesis and transformation of the sulphide deposits in the northern part of the Khetri copper belt, Rajasthan, India-an outline. *Mineralium Deposita*, *15*, 117−137.

Sarkar, S. C., & Gupta, A. (2012). *Crustal evolution and metallogeny in India*. Delhi: Cambridge University Press.

Sen, A., Pande, K., Sheth, S. C., Sharma, K. K., Sarkar, S., Dayal, A. M., & Mistry, H. (2013). An Ediacaran-Cambrian thermal imprint in Rajasthan, western India: Evidence from Ar−Ar geochronology of the Sindreth volcanics. *Journal of Earth System Sciences*, *122*, 1477−1493.

Sen, S. (1981). Proterozoic palaeotectonics in the evolution of Crust and location of metalliferous deposits, Rajasthan. *Quarterly Journal, Geological Mining and Metallurgy Society of India*, *53*, 162−185.

Sharma, B. L., Chauhan, N. K., & Bhu, H. (1988). *Structural geometry and deformation history of the early Proterozoic Aravalli rocks from Bagdunda, district Udaipur, Rajastahan*. Memoir of the Geological Society of India (vol. 7, pp. 169−192).

Sharma, R. S. (1988). Patterns of metamorphism in the Precambrian rocks of the Aravalli Mountain belt. *Memoir of the Geological Society of India* (vol. 7, pp. 33−75).

Sharma, R. S. (1995). An evolutionary model for Precambrian crust of Rajasthan: some petrological and geo-chronological considerations. *Memoir of the Geological Society of India* (vol. 34, pp. 91−115).

Sharma, R. S. (2012). *Cratons and Fold belts of India*. Springer publications.

Singh, S. P. (1984). Fluvial sedimentation of the Proterozoic Alwar Group in the Lalgarhgraben, Northwestern India. *Sedimentary Geology*, *39*, 95−119.

Singh, S. P. (1988). Sedimentation pattern of the Proterozoic Delhi Supergroup, NE Rajasthan, India and their tectonic implications. *Sedimentary Geology*, *58*, 79−94.

Singh, Y. K., De Waele, B., Karmarkar, S., Sarkar, S., & Biswal, T. K. (2010). Tectonic setting of the Balaram−Kui−Surpagla−Kengora granulites of the South Delhi Terrane of the Aravalli Mobile Belt, NW India and its implication on correlation with the East African Orogen in the Gondwana assembly. *Precambrian Research*, *183*, 669−688.

Sinha-Roy, S. (1984). Precambrian crustal interaction in Rajasthan, NW India. *Indian Journal of Earth Sciences*, *25*, 84−91.

Sinha-Roy, S. (1988). Proterozoic Wilson cycles in Rajasthan. In A. B. Roy (Ed.), *Precambrian of the Aravalli Mountain, Rajasthan, India* (7, pp. 95−108). Bangalore: Memoir of the Geological Society of India.

Sinha-Roy, S. (2004). Intersecting Proterozoic transpressional orogens, major crustal and suspect terranes in Rajasthan craton: A plate tectonic perspective. *Geological Survey of India Special Publication*, *84*, 207−226.

Sinha-Roy, S. (2008). Evolution of Precambrian terrains and crustal scale structures in Rajasthan craton, NW India: A kinematic model. *IAGR Memoir*, *10*, 23−40.

Sinha-Roy, S., Guha, D. B., & Bhattacharya, A. K. (1992). Polymetamorphic granulite facies pelitic gneisses of the Precambrian Sandmata Complex, Rajasthan. *Indian Minerals*, *46*(1), 1−12.

Sinha-Roy, S., & Malhotra, G. (1989). Structural relations of proterozoic cover and its basement: An example from the J- ahazpur Belt, Rajasthan. *Journal of the Geological Society of India*, *34*, 233−244.

Sinha-Roy, S., Malhotra, G., & Guha, D. B. (1995). A transect across Rajasthan Precambrian terrain in relation to geology, tectonics and crustal evolution of south-central Rajasthan. In Sinha-Roy, & K. R. Gupta (Eds.), *Continental crust of NW and Central India* (31, pp. 63−89). Bangalore: Memoir of the Geological Society of India.

Sinha-Roy, S., Malhotra, G., & Mohanty, M. (1998). *Geology of Rajasthan* (p. 278). Bangalore: Geological Society of India.

Sinha-Roy, S., Mohanty, M., Malhotra, G., Sharma, V. P., & Joshi, D. W. (1993). Conglomerate horizons in South-Central Rajasthan and their significance on Proterozoic stratigraphy and tectonics of the Aravalli and Delhi fold belts. *Journal of the Geological Society of India*, *41*, 331−350.

Sinha-Roy, S., & Mohanty, M. (1988). Blueschist facies metamorphism in the ophiolitic melange of the late Proterozoic Delhi fold belt, Rajasthan, India. *Precambrian Research*, *42*, 97−105.

Solanki, A., 2011. A petrographic, geochemical and geochronological investigation of deformed granitoids from SW Rajasthan: Neoproterozoic age of formation and evidence of Pan-African imprint. Masters Dissertation, University of Witwaters-rand, South Africa, 216 p.

Sreenivas, B. (1999). *Geochemistry of the Early Proterozoic metasedimentary rocks of the Aravalli Supergroup, Udaipur, Rajasthan, and its significance for understanding the secular changes across Archean-Proterozoic. Unpublished Ph.D. thesis* (p. 236). Hyderabad: Osmania University.

Srivastava, D. C., & Amit Sahay (2003). Brittle tectonics and pore-fluid conditions in the evolution of the Great Boundary Fault around Chittaurgarh, Northwestern India. *Journal of Structural Geology*, *25*, 1713−1733.

Srivastava, R. K. (1988). Magmatism in the Aravalli Mountain Range and its environs. *Memoir of the Geological Society of India*, *7*, 77−93.

Sugden, T. J., Deb, M., & Windely, B. F. (1990). *The tectonic setting of mineralization in Proterozoic Aravalli Delhi orogenic belt, NW India.* (vol. 8, pp. 367−390). Amsterdam: Developments in Precambrian Geology, Elsevier.

Sychanthavong, S. P. H., & Desai, S. D. (1977). Protoplate tectonics controlling the Precambrian deformations and metallogenic epochs of NW Peninsular India. *Mineral Science Engineering,*, *9*, 218−237.

Sychanthavong, S. P. H., & Merh, S. S. (1984). Proto-plate tectonics: The energetic model for the structural, metamorphic, and igneous evolution of the Precambrian rocks northwest peninsular India. *Geological Survey of India, Special Publication*, *12*, 419−457.

Tewari, H. C., Dixit, M. M., Rao, N. M., Venkateswarlu, N., & Vijaya Rao, V. (1997). Crustal thickening under the Paleo/Mesoproterozoic Delhi Fold Belt in NW India: Evidence from deep reflection profiling. *Geophysical Journal International*, *129*, 657−668.

Tewari, H. C., Vijaya Rao, V., Dixit, M. M., Rajendra Prasad, B., Madhava Rao, N., Venkateswarlu, N., ... Kaila, K. L. (1995). *Deep crustal reflection studies across the Delhi-Aravalli fold belt, results from the north-western part by controlled source seismic group. Memoir of the Geological Society of India* (vol. 31, pp. 383−402).

Tobisch, O. T., Collerson, K. D., Bhattacharya, T., & Mukhopadhyay, D. (1994). Structural relationship and Sm-Nd isotopic systematics of polymetamorphic granite gneisses and granitic rocks from central Rajasthan: Implications for the evolution of Aravalli craton. *Precambrian Research, 65*, 319–339.

Torsvik, T. H., Carter, L. M., Ashwal, L. D., Bhushan, S. K., Pandit, M. K., & Jamtveit, B. (2001). Rodinia refined or obscured: Paleomagnetism of the Malani igneous suite (NW India). *Precambrian Research, 108*, 319–333.

Vallinayagam, G., & Kochhar, N. (1998). *Geochemical characterisation and petrogenesis of A-type granites and the associated acid volcanic of Siwana Ring Complex, North Peninsular India* (pp. 460–481). Jodhpur: Scientific Publishers (India).

Varma, R. K., Mitra, S., & Mukhopadhyay, M. (1986). An analysis of gravity field over Aravallis and the surrounding region. *Bulletin Geophysical Research, 24*, 1–12.

Verma, P. K., & Greiling, R. O. (1995). Tectonic evolution of the Aravalli orogen (NW India): an inverted Proterozoic rift basin? *Geologische Rundschau, 84*, 683–696.

Vijaya Rao, V., & Krishna, V. G. (2013). Evidence for the Neoproterozoic Phulad Suture Zone and Genesis of Malani magmatism in the NW India from deep seismic images: Implications for assembly and breakup of the Rodinia. *Tectonophysics, 589*, 172–185.

Vijaya Rao, V., Rajendra Prasad, B., Reddy, P. R., & Tewari, H. C. (2000). Evolution of Proterozoic Aravalli Delhi Fold Belt in the northwestern Indian Shield from seismic studies. *Tectonophysics, 327*, 109–130.

Volpe, A. M., & Macdougall, J. D. (1990). Geochemistry and isotope characteristics of mafic (Phulad Ophiolite) and related rocks in the Delhi supergroup, Rajasthan, India: Implications for rifting in the Proterozoic. *Precambrian Research, 48*, 167–191.

Wiedenbeck, M., & Goswami, J. N. (1994). An ion-probe single zircon $^{207}Pb/^{206}Pb$ age from the Mewar Gneiss at Jhamarkotra, Rajasthan. *Geochimica et Cosmochimica Acta, 58*, 2135–2141.

Wiedenbeck, M., Goswami, J. N., & Roy, A. B. (1996). Stabilisation of the Aravalli craton of the northwestern India at 2.5 Ga: an ion-microprobe zircon study. *Chemical Geology, 129*, 325–340.

FURTHER READING

Ashwal, L. D., Demaiffe, D., & Torsvik, T. H. (2002). Petrogenesis of Neoproterozoic granitoids and related rocks from the Seychelles: evidence for the case of an Andean-type arc origin. *Journal of Petrology, 43*, 45–83.

Bhowmik, S. K., Basu-Sarbadhikari, A., Spiering, B., & Raith, M. M. (2005). Mesoproterozoic reworking of Paleoproterozoic ultrahigh-temperature granulites in the central Indian Tectonic Zone and its implications. *Journal of Petrology, 46*, 1085–1119.

de Wall, Pandit, M. K., & Chauhan, N. K. (2012a). Palaeosol occurrences along the Archean–Proterozoic contact in the Aravalli craton, NW India. *Precambrian Research, 216–219*, 120–131.

de Wall, H., Pandit, M. K., Sharma, K. K., Schöbel, S., & Jana, J. (2014). Deformation and granite intrusion in the Sirohi area, SW Rajasthan—Constraints on Cryogenian to Pan-African crustal dynamics of NW India. *Precambrian Research, 254*, 1–18.

Deb, M., & Chattopadhyay, A. (2004). *Westward extension of the Central Indian Tectonic zone into Aravalli–Delhi Orogenic Belt: A refinement of the earlier views. Uniformitarianism revisited: Comparison between ancient and modern orogens of India.* Geological Survey of India Special Publication (vol. 84, pp. 341–350).

Krause, O., Dobmeier, C., Raith, M. M., & Mezger, K. (2001). Age of emplacement of massif-stype anorthosites in the Eastern Ghats Belt, India: Constraints from U-Pb zircon dating and structural studies. *Precambrian Research, 109*, 25–38.

Naha, K., Choudhary, A. K., & Bhattacharyya, A. C. (1966). Superposed folding in the older Precambrian rocks around Sangat, central Rajasthan, India. *N. Jb. Geol. Palaont. Abh.*, *126*, 205−231.

Pandit, M. K., Carter, L. M., Ashwal, L. D., Tucker, R. D., Torsvik, T. H., Jamtveit, B., & Bhushan, S. K. (2003). Age, petrogenesis and significance of 1 Ga granitoids and related rocks from the Sendra area, Aravalli craton, NW India. *Journal of Asian Earth Sciences*, *22*, 363−381.

Pradhan, V. R., Meert, J. G., Pandit, M. K., Kamenov, G., Gregory, L. C., & Malone, S. J. (2010). India's changing place in global Proterozoic reconstructions: A review of geochronolgical constraints and paleomagmetic poles from the Dharwar, Bundelkhand and Marwar cratons. *Journal Geodynamics*, *50*, 224−242.

Rajendra Prasad, B., Tewari, H. C., Vijaya Rao, V., Dixit, M. M., & Reddy, P. R. (1998). Structure and tectonics of the Proterozoic Aravalli-Delhi fold belt in NW India from deep seismic reflection studies. *Tectonophysics*, *288*, 31−41.

Rathore, S. S., Venkatesan, T. R., & Srivastava, R. K. (1996). Rb-Sr and Ar-Ar systematics of Malani volcanic rocks of southwest Rajasthan: Evidence for a younger post crystalline thermal event. *Proceedings of the Indian Academy of Sciences (Earth & Planetary Sciences)*, *105*, 131−141.

Roy, A. B., Golani, P. R., & Bejarniya, B. R. (1985). The Ahar River Granite: Its stratigraphic and structural relation with the Proterozoic rock of south-eastern Rajasthan. *Journal of the Geological Society of India*, *26*, 315−325.

Sinha-Roy, S. (1989). Strike-slip fault and pull apart basins in Proterozoic fold belt development in Rajasthan. *Indian Minerals*, *43*(3&4), 226−240.

Torsvik, T. H., Amudsen, H., Hartz, E. H., Corfu, F., Kusznir, N., Gaina, C., … Jamtveit, B. (2013). A Precambrian microcontinent in the Indian Ocean. *Nature Geoscience*. Available from http://dx.doi.org/10.1035/NGEO01736.

Tucker, R. D., Ashwal, L. D., & Torsvik, T. H. (2001). U−Pb geochronology of Seychelles granitoids: A Neoproterozoic continental arc fragment. *Earth and Planetary Science Lettters*, *187*, 27−38.

Van Lente, B., Ashwal, L. D., Pandit, M. K., Bowring, S. A., & Torsvik, T. H. (2009). Neoproterozoic hydrothermally altered basaltic rocks from Rajasthan, northwest India: implications for late Precambrian tectonic evolution of the Aravalli Craton. *Precambrian Research*, *170*, 202−222.

THE GONDWANA CORRELATIONS

CHAPTER

6

CHAPTER OUTLINE

Proterozoic Orogens of India. **DOI: http://dx.doi.org/10.1016/B978-0-12-804441-4.00006-7**
© 2017 Elsevier Inc. All rights reserved.

6.1 INTRODUCTION

The Gondwana correlations were largely made by using the nature of continental margins, fauna and flora, lithological similarities, and metamorphic grades and other geological characteristics. During the last two decades, there has been a major shift and intense focus on the studies related to the origin, evolution and dispersal of the supercontinents through Earth's history, preserving the well-developed orogens marked by the presence of suture zones/shear zones. Several advancements in analytical techniques and concepts took place in different fields such as detrital zircon geochronology and Hf-isotopes, new data, and refinements in paleomagnetism, and new approaches in geophysical techniques including mantle tomography and numerical models that led to innovative proposals and global models on supercontinental reconstructions. New lines of thinking emerged with regard to the relationship between supercontinent history and solid Earth tectonics, metallogeny, surface environment and life. This chapter deals with the brief descriptions of the supercontinents with reference to their assembly- breakup history cycles and orogens of Gondwana supercontinent. Emphasis has been laid on the correlations and extensions of the shear zone network and suture zones associated with the Proterozoic orogens of India (POI) among different fragments of the Gondwana supercontinent. This is further substantiated with the help of the recent geochronological data that helped in understanding the assembly and breakup through different stages of Rodinia and Gondwana.

6.2 SUPERCONTINENTAL CYCLE

6.2.1 INTRODUCTION

The episodicity of "supercontinental cycle" in tectonic processes was recognized long before the advent of plate tectonics that provided the framework to account for the genesis and occurrence of supercontinents (Worsley, Nance, & Moody, 1982). Since the assembly of supercontinents requires the collision of continents, whereas a supercontinent breakup requires them to rift, the existence of a supercontinent cycle would be manifested (or "cycle would manifest") in the geologic record by episodic peaks in collisional orogenesis and rift-related mafic dike swarms. But, after the introduction of plate tectonics, recognition of the process of ocean closure by subduction provided an explanation for orogenesis and crustal growth (Dewey, 1969).

   Earth's history has been punctuated by the episodic assembly and breakup of supercontinents and has influenced the rock record more than any other geologic phenomena including Earth's tectonic, climatic, and biogeochemical evolution for billions of years. This hypothesis was considered as the most important advancement in Earth science since the advent of plate tectonics (Nance & Murphy, 2013). Of the five supercontinents proposed by Worsley et al. (1982), four supercontinents are now well recognized: the amalgamation of Pannotia (Gondwana), Rodinia, Columbia (or Nuna), and Kenorland. Based on the global synthesis of information on the formation of cratons and orogenic belts, the evolutionary history of different supercontinents could be traced (Rogers & Santosh, 2004). They include the oldest assemblies of "Ur" (c. 3.0 Ga cratons of Southern Africa and Western Australia), "Arctica" (c. 2.5 Ga cratons of Greenland, Fennoscandia, Laurentia, and Siberia) and "Atlantica" (c. 2.0 Ga cratons of Western Africa and South America), that remained coherent, which was described as "Columbia." The Columbia supercontinent was believed to contain nearly all of Earth's continental blocks at some time between 1.9 and 1.5 Ga. It was also proposed that the Columbia was formed when eastern India, Australia, and attached parts of Antarctica were sutured to western North America—the eastern margin of North America, the southern margin of Baltica/North China, and the western margin of the Amazon shield forming a continuous zone of continental building.

The supercontinent cycle indicates the assembly and breakup of Superia—Sclavia at ∼2.6 Ga, Columbia at ∼1.6−1.2 Ga, Rodinia at ∼1.1−0.8 Ga, and Pangea−Gondwana at 0.6−0.26 Ga (Nance, Murphy, & Santosh, 2014). While the Columbia supercontinent was constructed largely on the basis of aligning the 2.1−1.8 Ga orogenic belts, the Rodinia assembly is based, in part, on the presence of 1.1−0.9 Ga orogenic belts. In both the cases, the orogens are believed to have developed during the formation of the supercontinents. East Gondwana (India, Madagascar, Australia, Antarctica, and Sri Lanka) is depicted in Columbia, Rodinia, and Pangea as a united landmass (Fig. 6.1A−C). It was also suggested that supercontinents might undergo extroversion, in addition to introversion, where in the exterior margins of one supercontinent collide during the formation of the next younger supercontinent. For instance, extroversion is well reflected with the breakup of the archetypal Rodinia supercontinent and the subsequent formation of Gondwana (Hoffman, 1991). This extroversion model would be valid if the reconstruction model of Rodinia supercontinent is correct. However, in the context of Gondwana formation, no explanation is provided for why the individual elements that make up "East Gondwana" and "West Gondwana" maintain the same integral relationships for nearly 2 billion years until they were cleaved in the Mesozoic (Meert, 2014).

The assembly of supercontinents requires collision, accretion, and the activity of subduction systems along their margins, while their breakup requires rifting. Columbia started rifting in the Mesoproterozoic, and the final phase is assumed to be marked by extensive mafic dyke swarms, flood basalts, and layered intrusions at 1.3−1.2 Ga (Wang, Yang, Yang, & Xu, 2014). The Neoproterozoic and Phanerozoic eons are characterized by the initiation and eventual final breakup of the supercontinent known as Rodinia at about 900−800 Ma, a reassembly to form Pangea between 400 and 200 Ma and its subsequent breakup. The breakup of Pangea was heralded by the emplacement of continental flood basalts, layered intrusions and related magmatic Ni and PGE mineralization (Ernst, 2014). Continuing breakup and dispersal of continental fragments led to the inception of peripheral subduction systems, magmatic arcs, and metal deposits that are typically formed at convergent margins, such as porphyry and epithermal systems. Peripheral orogens form at continental margins adjacent to oceans like those formed along the Cordilleran and Andean side of the America. The concept of the supercontinent cycle has been further strengthened and established with the recent developments in understanding mantle tomography (e.g., Zhao, Yamamoto, & Yanada, 2013) and significant advances in precise geochronology (e.g., Condie, Belousova, Griffin, & Sircombe, 2009).

The hypothesis of the supercontinent cycle implies that Columbia would break up, continents would disperse, and then they would reamalgamate to form the next supercontinent, Rodinia. However, the breakup of Columbia is debated. Some authors infer the breakup based on numerous mafic dyke swarms and other intrusive rocks of various ages, all across several continents (e.g., Hou, Santosh, Qian, Lister, & Li, 2008). Later studies revealed that the birth of Gondwana involved the closure of a series of oceans between the converging crustal blocks during the mid-to-late Neoproterozoic, with Pacific-type subduction along a number of convergent margins progressively yielding to collisional orogeny accompanied by magmatic, metasomatic, and metamorphic processes characteristic of subduction−accretion−collision settings (Collins & Pisarevsky, 2005).

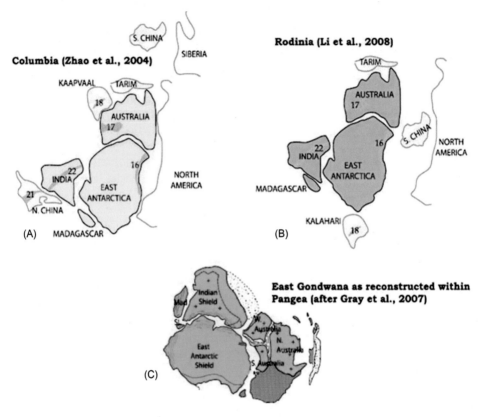

FIGURE 6.1

Configurations of different forms of supercontinents with east Gondwana (India, Madagascar, Australia, Antarctica, Sri Lanka): (A) the "Columbia" with a more or less traditional East Gondwana fit between India-Madagascar-East Antarctica and Australia (Zhao et al., 2004). Sri Lanka was not included in the Columbia configuration; (B) the "Rodinia" configuration showing a near identical fit between the East Gondwana elements in Columbia (Li et al., 2008), and (C) East Gondwana as it existed during the time of Pangea after Gray et al. (2008). The gray-shaded orogens (2.1–1.8 Ga) include: Transantarctic Orogen (16); Capricorn Orogen (17); Limpopo Belt (18); Trans-North China Orogen (21); Central Indian Tectonic Zone (22).

After Meert, J.G. (2014). Strange attractors, spiritual interlopers and lonely wanderers: The search for pre-Pangean supercontinents.

Geoscience Frontiers, 5, *155–166.*

6.2.2 ROLE OF PLUMES

It has been widely recognized in recent years that plumes play a major role in breaking apart supercontinents. Plume-driven uplift of continental crust, subsequent rifting, and subduction related nutrient systems in the late Neoproterozoic has also been speculated to have contributed to the birth of modern life on our planet (Santosh, Maruyama, Yusuke Sawaki, & Meert, 2014). Deep mantle tomography and numerical modeling have characterized the mantle plumes as thermal and chemical upwellings originating either from the core-mantle boundary, and/or from the 410 to 660 km discontinuity at the mantle transition zone. A model of supercontinent self-destruction in the vicinity of subduction associated with supercontinent assembly triggers superplume events from the 660 km mantle discontinuity leading to the fragmentation of the above continental mass. Such a supercontinent-triggered superplume mechanism is compatible with the Mesozoic fragmentation of Pangea-Gondwana. It was also concluded that the breakup of supercontinents, such as Columbia, Rodinia, and Gondwana resulted from the impact of mantle plumes, which led to the formation of giant rift systems in the separation of continental blocks (Şengör & Natal, 2001). There is also a likely relationship between superplumes, supercontinent breakup, and mass extinction. Upwelling plumes that break supercontinents apart generate large igneous provinces that may, in turn, affect the climate by producing large-scale volcanism and plume-induced "winters" with catastrophic effects on the atmosphere and life.

The assembly and breakup of the supercontinent Rodinia was ancestral to the long-lived supercontinent of Gondwanaland during the late Mesoproterozoic and Neoproterozoic period. The occurrence of possible global superplume events and the rapid true polar wander event(s), repeated low-latitude glaciations, the explosion of multicellular life, and the emergence of a plate dynamic and climatic system seem to be genetically related (Moores, 2002). These conceptual ideas enabled geoscientists to think beyond their specific disciplines, bringing together tectonicists, structural geologists, geochemists, stratigraphers, petrologists, geophysicists, climatologists, and paleontologists alike to work together towards a better understanding of Earth's system science at the dawn of the Phanerozoic.

6.2.3 GONDWANA SUPERCONTINENT

The Gondwana supercontinent was the largest unit of continental crust on Earth for more than 200 million years. The superterrane core of Gondwana included the modern continents of South America, Africa, and most of Antarctica and Australia, as well as Madagascar and the Indian subcontinent, which forms 64% of total landmass today and 19% of the total Earth surface (Torsvik & Cocks, 2011). The name "Gondwana" was originally coined by H.B. Medlicott (1869) of the Indian Geological Survey for a sedimentary sequence of nonmarine rocks in India, which hosted the distinctive Late Paleozoic Glossopteris flora. The various floral provinces had flourished on continents, which had always stayed in their present positions, but their connection by land bridges in the Late Paleozoic subsequently drowned beneath the oceans through isostatic readjustments. However, the concept of Gondwana was transformed and enlarged by Wegener (1912), who postulated for the first time that the major components of Gondwana were united as a single superterrane during the Late Paleozoic, as characterized both by the Glossopteris flora and also by the presence of glacial deposits normally formed in a polar region. Later, the different sectors of Gondwana had subsequently travelled apart across the oceans. However, there was much opposition to Wegener's ideas on continental drift and the concept of a previously united Gondwana, largely due to the lack of a plausible mechanism for the necessary continental movements. Despite the opposition, some of the authors, for instance, Du Toit (1937), continued to promote continental drift in his valued book. The skepticism and the debate continued to exist among most of the geological community until the advent of the plate tectonic theory in the 1960s.

The breakup of both Rodinia and Pangaea supercontinents started with broad mantle upwellings (or superplumes) beneath them, resulting in widespread bimodal magmatism (including plume magmatism) and continental rifting. This breakup mechanism provides us rare opportunities to gain insights into the 4-D geodynamic system marked by close interactions between the thermal dynamics of Earth's outer core, mantle convections, and lithospheric plate tectonics. It is fairly well established that the orogenic events related to the formation of the Gondwana supercontinent were described as the global "Pan-African" episode (Clifford, 1968). Subsequent studies suggest that the West Gondwana (i.e., the cratons of South America and Africa) was amalgamated between 650 and 600 Ma, and that the East Gondwana was assembled in at least two stages, between 750 and 620 Ma (East African Orogen (EAO)) and at 570–500 Ma (Kuunga Orogen) (Meert, 2001). The Kuunga Orogeny was originally defined on the basis of geochronological data and interpreted to be related to the collision between Australia/Antarctica and an already combined India/East Africa.

The crustal architecture of Indian continent constitutes several discrete crustal blocks and intervening shear zone systems, possibly representing collisional sutures manifest in the form of well-developed Proterozoic orogens that assume special significance in the form of a critical window when viewed in the geologic reconstruction of Gondwanaland (Fig. 6.2). The focal theme addressed in this chapter is to correlate and compare these POI and their constituents with similar features in other fragments of Gondwana. There is a general consensus that the late Neoproterozoic

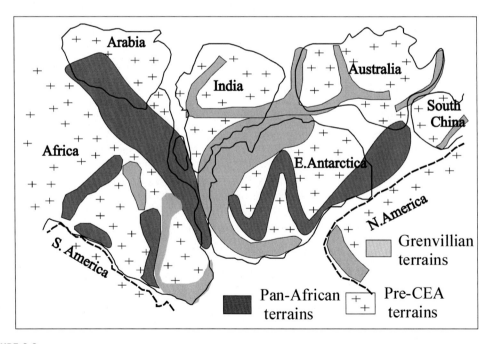

FIGURE 6.2

Pan-African and Circum-East Antarctica (Grenvillian) terrains in east Gondwana during c. 1000–500 Ma.

Modified after Yoshida, M., Windley, B.F., & Dasgupta, S. (2003). Preface-Proterozoic East Gondwana: Supercontinent Assembly and Breakup. Geological Society of London, *Special Publication, 206 (Yoshida, Windley, & Dasgupta, 2003).*

metamorphism and deformation reflects the Gondwana assembly. Therefore, there should be sutures that mark the boundaries of the various blocks that assembled during this period. However, disagreement exists over the location of such inferred sutures in the POI with similar orogens of Madagascar, Mozambique, and Dronning Maud Land of East Antarctica (e.g., Shackleton, 1996). It is challenging to identify sutures in such terranes from surface geology since granulite-facies metamorphism, partial melting, and pervasive ductile deformation mask many of the features used to identify sutures at shallower crustal levels, and the suture zones are also more likely to correspond to areas of poor outcrop than to regions of good exposure.

The junction between East and West Gondwana is marked by the Arabian-Nubian Shield (ANS) located at the northern end of the EAO (Fig. 6.3). The older components of the ANS include Archean and Paleoproterozoic continental crust, and the Neoproterozoic (c. 870−670 Ma) continental-marginal and juvenile intra oceanic magmatic-arc terranes that accumulated in an oceanic environment referred to as the Mozambique Ocean (Johnson & Wodehaimont, 2003). Gondwana assembly was initiated with subduction around 870 Ma, with an initial arc−arc convergence and terrane suturing at c. 780 Ma and the beginning of ocean-basin closure. Terrane amalgamation continued until c. 600 Ma, resulting in the juxtaposition of East and West Gondwana across the deformed rocks of the shield, and the final assembly of Gondwana was achieved by c. 550 Ma following overlapping periods of basin formation, rifting, compression, strike-slip faulting, and the creation of gneiss domes in association with extension and/or thrusting.

East Gondwana comprises the continents of Australia, India, and East Antarctica, which came together with the constituent continental blocks that now make up Africa and South America to form the supercontinent Gondwana (or Gondwanaland, Du Toit, 1937) during Neoproterozoic. East Gondwana was traditionally thought to have formed as a major part of Rodinia during the Mesoproterozoic Grenvillian-Circum-East Antarctic Orogeny (Yoshida, 1995) and survived until the Middle-Late Mesozoic when Pangaea broke up. An ice-covered Antarctica was the key component of this long-lived subsupercontinent. West Gondwana, on the other hand, got assembled during the Neoproterozoic and collided with preexisting East Gondwana at this time. However, the accumulation of Pan-African zircon ages mostly from Antarctica together with many of reliable paleomagnetic data from various parts of the globe, resulted in a reevaluation of the above classical model with the emergence of a radical new model. According to this new model, East Gondwana did not exist during the Neoproterozoic—along with West Gondwana, but got together during the Pan-African Orogeny synchronously with the whole of Gondwanaland (Meert & Powell, 2001). Both the models still command strong support and further data are required to constrain their future viability, although the new model is becoming increasingly popular. Further, a bi-modality in geochronological data from the east Gondwana was also recognized suggesting two orogenies for its assembly that include an earlier EAO ($\sim$750 and 620 Ma) and a later Kuunga orogeny ($\sim$570−530 Ma). The Kuunga orogeny was interpreted to mark the collision of Australia and Antarctica with the rest of Gondwana and was subsequently correlated with a broad belt of orogenesis from the Damara Orogen in the west to the Pinjarra Orogen in the east, with a southern spur to Dronning Maud Land (Meert, 2003). This paleogeographic model involves a Neoproterozoic continent consisting of Sri Lanka, Madagascar and India (SLAMIN) colliding with a combined Congo/Kalahari continent at $\sim$750−620 Ma, followed by Australia/East Antarctica colliding with the bulk of Gondwana at $\sim$570−530 Ma. However, there is also a variation in the SLAMIN model that India did not collide with west Gondwana until late Neoproterozoic-Early Cambrian times

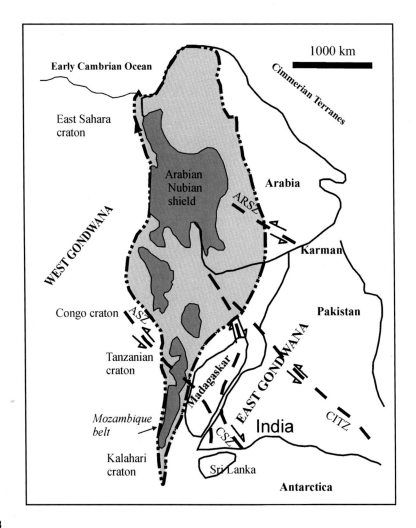

FIGURE 6.3

A sketch map of the East African Orogen showing the location of the Arabian-Nubian Shield (a collage of
Neoproterozoic terranes) relative to the Mozambique belt and adjacent cratonic margins caught up as the suture
between East and West Gondwana.

*Modified after Stern, R.J. (1994). Continental collision in the Neoproterozoic East African orogen: Implications for the consolidation
of Gondwana land. Annual Reviews of Earth and Planetary Sciences, 22, 319–351.*

(Boger & Miller, 2004). According to them, the EAO evolved as an accretionary orogen and was
partially superimposed by a 590−560 Ma orogen created by the collision of a combined India-
Madagascar and part of Antarctica with eastern Africa. They described this orogen as the
Mozambique suture and suggested that Australia-Antarctica (with an enlarged Antarctic component
including the Ruker Terrane) collided with India along the Kuunga suture at 535−520 Ma, leading
to the development of the Eastern Ghats Mobile belt (EGMB) along the east coast of India.

The issue of position of the Indian craton in Rodinia is a point of debate and discussion. Some of the paleomagnetic studies suggest that India was never a part of Rodinia. Based on paleo-magnetic and geochronological data a new model was proposed, in which the EGMB and the Rayner Complex of Antarctica were not a part of Rodinia, but collided with Gondwana at around 680−610 Ma. The amalgamation of the Rodinia supercontinent at around 1300−900 Ma involved worldwide orogenic events and that India became a part of Rodinia around 990−900 Ma involving the collision of the EGMB with the Rayner complex of East Antarctica (Li et al., 2008). The other proposal was that Rodinia was formed as a result of many collisional events of relatively long-lived (1100−700 Ma) orogens of Grenvillian age including the EGMB, East Antarctica and the Late Mesoproterozoic Albany Fraser belt of Australia. This model is consistent with the SWEAT (SW US-East Antarctic) hypothesis (Moores, 1991), where in the Grenvillian Belt of Laurentia continues around Antarctica and into India and Australia.

6.3 GONDWANA OROGENS

6.3.1 INTRODUCTION

The Gondwana forming orogens represent one of the most extensive orogenic systems comprising high-grade metamorphic rocks, representing the largest interconnected mountain belt of the last billion years (Fig. 6.4). These belts include: EAO and Mozambique Belt lying between the Congo craton and ANS, Azania and India; Kuunga Orogen separating western Antarctica, India and the Kalahari-Lurio-Vijayan craton. However, the Zambezi-Damaran Orogen between the Kalahari and Congo cratons, the Pinjarra Orogen between Australia-Mawson and India; and the Brasiliano Orogen between the São-Francisco-Rio de la Plata and Amazon cratons are considered separately.

The Gondwana orogenic systems currently offer us invaluable windows in unraveling the history of different supercontinents and the associated deep crustal processes. The structural evolution of orogens are generally thought to involve first, crustal thickening, and then, their collapse due to gravitational instability facilitating the reactivation of thrust structures into normal shear zones and faults. Therefore, the crustal-scale shear zones in Proterozoic orogens have a long-lived history of reactivation and are often suggested to originate as orogenic sutures (Petersson, Scherstén, Andersson, Whitehouse, & Baranoski, 2015). The continued reuse of preexisting crustal weaknesses through orogenic cycles is clearly a significant process of Earth's evolution.

6.3.2 GONDWANA CORRELATIONS

The global scale orogens and their processes with continuous acquisition of geological and geophysical data sets act as invaluable tools and provide insights into the reconstruction models of ancient supercontinents. For instance, the Grenvillian orogeny (c. 1200−1000 Ma) sutured most of the continental fragments of the Mesoproterozoic supercontinent Rodinia, where the orogeny manifested itself by a series of large-scale subduction and collisional processes.

The final assembly of Gondwana seems to have occurred as a simple process involving the closure of the Mozambique Ocean bringing together East Gondwana (Australia, Antarctica, India, Madagascar, and ANS) and West Gondwana (Africa and South America). The final suturing was

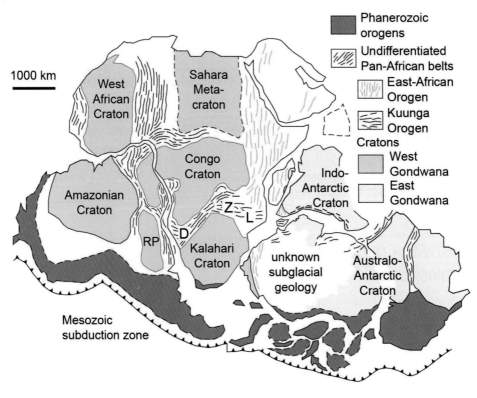

FIGURE 6.4

Map showing cratons (> 0.9 Ga) and Pan-African orogens (0.7–0.5 Ga) of Gondwana supercontinent. The East African orogen appears to be truncated by the Kuunga orogen comprising the Damara (D), Zamberi (Z), and Lurio (L) belts. RP is Rio de la Plata craton.

After Fitzsimons, I.C.W. (2016). Pan African granulites of Madagaskar and Southern India: Gondwana assembly and parallels with modern Tibet. Journal of Mineralogical and Petrological Sciences, 111, *73–88 (Fitzsimons, 2016).*

considered to be at around 600 Ma along the EAO and Mozambique Belt (Stern, 1994). However, in contrast, some recent studies suggested that East and West Gondwana may not have existed as independent and coherent continental masses in their own right and that the final amalgamation of Gondwana was involved in a more complex accretion of blocks along a variety of orogenic belts largely between 570 and 520 Ma (e.g., Meert, 2003).

India and east Antarctica were amalgamated through regionally extensive orogens of the EGMB and Rayner complex at around c. 1000–900 Ma ago according to a recent configuration of Rodinia (Phillips, Bunge, & Schaber, 2009). The Grenvillian orogenic event from Prydz Bay areas was also extended through EGMB-central Indian Tectonic Zone (CITZ)-Aravalli Delhi Orogenic Belt (ADOB). The north east extension of the CITZ through the NW segment of Shillong-Meghalaya (SM) plateau associated with the presence of Mesoproterozoic granulites shows similarly and temporally related rocks implying that both the regions must have been in contiguity together since

Mesoproterozoic. In the reconstructed model of Neoproterozoic-Cambrian assembly of the Rodinia supercontinent, the SM plateau was presumably the leading edge during an oblique collision of India with Australia-Antarctica. Further, the Pan-African imprints associated with Prydz bay region of east Antarctica may be extended northward into India through SM plateau (Chatterjee, Crowley, Mukherjee, & Das, 2008).

The orogens, by their nature, destroy preexisting geologic information through metamorphism, erosion, and subduction and thus reconstructing their paleogeographic histories is a challenging task. However, the POI play a critical role in understanding the reconstruction models of supercontinents such as Columbia, Rodinia, and Gondwana with a complementary record of rifting and passive margin development, and with quantitative kinematic data. The correlation and linkage of the Proterozoic orogens and associated shear zones with similar orogens of other adjacent Gondwana fragments is described below.

6.4 THE SOUTHERN GRANULITE TERRANE

6.4.1 INTRODUCTION

The Southern Granulite Terrane (SGT) forms a critical part of EAO and is located at the junction of EAO and the Kuunga orogeny. The rare presence of Grenvillian activity, the presence of imprints of a Paleoproterozoic tectono-metamorphic events, and the existence of crustal-scale shear zone features place the SGT together with Sri Lanka, most of Madagascar, and the Dronning Maud Land of East Antarctica within the EAO and the east Gondwana (e.g., Fitzsimons, 2003). The linkage of the continents through the piercing shear zones is significant in their correlations which would enable the better understanding of the reconstruction models of supercontinents.

6.4.2 CAUVERY-MERCARA-KUMTA-BETSIMISARAKA SUTURE ZONE

The Cauvery suture/shear zone (CSZ), the interface between the Archean Dharwar craton, and the Proterozoic Madurai granulite block, constitutes terranes of different ages and tectonothermal histories (Chetty, Yellappa, & Santosh, 2016). The connectivity of the CSZ with the Betsimisaraka suture zone (BSSZ) from eastern Madagascar can be traced for over 1000 km long reflecting the close juxtaposition of India-Madagascar during the Gondwana formation (Fig. 6.5) and is described as Rodinian suture zone (RSZ) (Ishwar-kumar et al., 2015). The RSZ can be extended further into east Africa and further north into the ANS as a part of EAO. The BSSZ occurs between the western India and Madagascar and is marked by the existence of the Archean-Neoproterozoic Antananarivo Block (2700−2500 Ma gneisses and 824−550 Ma granitoids) to the west and the Archean Antongil-Masora block (3300−2490 Ma gneisses and granitoids) to the east. The BSSZ is characterized by cataclastic/mylonitic, banded, and augen gneisses and was established as a suture zone and was considered to have formed during the amalgamation of East and West Gondwana during Neoproterozoic (Collins & Windley, 2002).

The BSSZ and the CSZ are connected by two important shear zones from the western India in the form of Kumta and Mercara suture zones. The continuation and correlation of these suture zones within the RSZ is validated by structural, geological, and geochronological data, along with

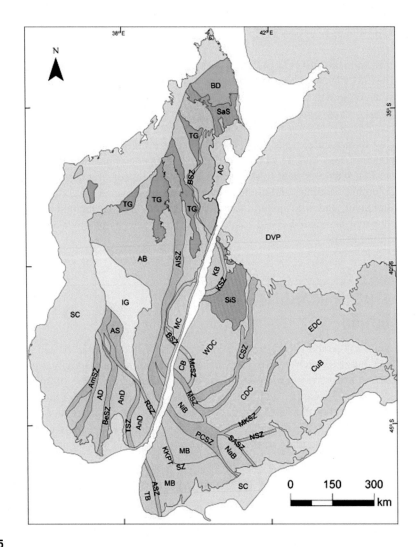

FIGURE 6.5

Reconstruction of India and Madagascar around 120 Ma and correlation of different shear zones between the western margin of Indian continent and Madagascar at around 120 Ma. Madagascar—*AC*, Antongil Craton; *MC*, Masora Craton; *AB*, Antananarivo block; *BSZ*, Betsimisaraka Suture Zone; *AISZ*, Angavo-Ifanadriana Shear Zone; *BD*, Bemarivo domain; *SaS*, Sahantaha Shelf; *TG*, Tsaratanana Sheet; *IG*, ItremoGroup; *RSZ*, Ranotsara Shear Zone; *AS*, Amboropotsy Sheet; *TSZ*, Tranomaro Shear Zone; *BeSZ,* Betroka Shear Zone; *AmSZ*, Amphanihy Shear Zone; *AnD*, Anosyan Domain; *SC*, sedimentary cover. India—*WDC*, Western Dharwar Craton; *EDC*, Eastern Dharwar Craton; *CDC*, Central Dharwar Craton; *DVP*, Deccan Volcanic Province; *CSZ*, Chitradurga Shear Zone; *CuB*, Cuddapah Basin; *KSZ*, Kumta Suture Zone; *SiS*, Sirsi Shelf; *McSZ*, Mercara Suture Zone; *KB*, Karwar Block; *CB*, Coorg Block; *MSZ*, Moyar Shear Zone; *NB*, Nilgiri Block; *SASZ*, Salem-Attur Shear Zone; *MeSZ*, Mettur Shear Zone; *NSZ*, Nallamalai Shear Zone; *PCSZ*, Palghat-Cauvery Shear Zone; *MB*, Madurai Block; *KKPTSZ*, Karur-Kambam-Painavu-Trichur Shear Zone; *ASZ*, Achankovil Shear Zone; *TB*, Trivandrum Block; *SC*, sedimentary cover.

After Ishwar-Kumar, C., Sajeev, K., Windley, B.F., Kusky, T.M., Fengc, P., Ratheesh-Kumar, R.T., ... Itaya, T. (2015). Evolution of high-pressure mafic granulites and pelitic gneisses from NE Madagascar: Tectonic implications. Tectonophysics, 662, 219–242.

mineral chemistry and thermodynamic modeling. Recent studies indicate that the RSZ had developed during the late-Neoproterozoic by the collision of the Antananarivo-Karwar-Coorg and Dharwar blocks. The Mesoarchean Coorg block, situated along the Mercara suture zone, was also recognized as an exotic microcontinent that amalgamated with the Dharwar Craton after c. 2500 Ma (Santosh et al., 2015).

The RSZ shows significant variations in terms of lithologies, structural styles, metamorphic histories and geochronological data. For instance, the CSZ is marked by Neoarchean granulites intruded by granitoids and dismembered ophiolites of ~780 Ma. The BSSZ records a wide range of detrital zircon ages (2950−1740 Ma) and also ages as young as c. 600 Ma. Similarly, the Kumta and Mercara suture zones also yielded detrital zircon ages ranging from 3420 to 1350 Ma. The structural, lithological and geochronological results from these suture zones suggest that closure of the Mesoproterozoic ocean occurred along the Kumta and Mercara suture zones. The Kumta suture, located at the western margin of peninsular India, separates c. 3200 Ma tonalite−trondhjemite−granodiorite (TTG) and amphibolite of the Karwar block in the west and quartzo−feldspathic gneisses(c. 2571 Ma) of the Dharwar Craton in the east. The curvilinear Kumta suture is ~15 km-wide and varies in strike from NW−SE to N−S to NE−SW with a dip of at 30−65 degrees to the west. The Kumta suture contains highly sheared and deformed amphibolite- to greenschist-facies schistose rocks that include quartz−phengite schist, chlorite schist, fuchsite schist, garnet−biotite schist, and marble. The Mercara Suture Zone is also curvilinear with a width of about 20 km, and is associated with highly sheared and deformed high-grade gneisses varying in strike from E−W to NW−SE to N−S (progressing southwards), with westward dips (Chetty, Mohanty, & Yellappa, 2012). The Mercara suture contains amphibolite- to granulite-facies garnet−kyanite−sillimanite gneiss, mylonitic quartzo−feldspathic gneiss, garnet−biotite−kyanite−gedrite−cordierite gneiss, garnet−biotite−hornblende gneiss, and calc-silicate granulite, which possibly represent the high-grade equivalents of the schistose rocks in the Kumta suture. The Mercara Shear Zone is further defined as a terrane boundary, and possible Mesoarchean (c. 3.0 Ga) suture that marks the collisional event between the Western Dharwar Craton and the Coorg Block. With dextral kinematics, the suture zone extends southwards and coalesce with the Moyar Shear Zone, which represents the northern margin of the CSZ. The accretion of these continental fragments might have coincided with the birth of the oldest supercontinent "Ur" (Amaldev et al., 2016).

The BSSZ with younger deformation in the east is interpreted as the site of the India-Azania collision and a provenance boundary (Malagasy Orogeny) between East African−derived terranes and Indian terranes and may be coeval with the Kuunga Orogeny the other side of India (Boger & Miller, 2004). During the final stage of amalgamation of Gondwana, an important high-strain zone of Angavo shear zone (1000 × 40 km), occurring to the west of BSSZ in Madagascar, was also formed during 561 and 532 Ma coevally with the final closure of the Mozambique Ocean between Dharwar Craton and the Madurai block in south India. The Angavo shear zone was found to be a preexisting rheological discontinuity formed by the Antogil−Dharwar suture and was correlated with the Moyar shear zone, the northern margin of the CSZ (Tucker, Roig, & Delor, 2011). The common manifestation in the contiguity of the RSZ from CSZ to BSSZ and associated structural fabrics imply subduction−collision processes in an oblique convergence during the assembly of Gondwana. The RSZ forms the eastern boundary of the EAO and can be termed as the eastern Gondwana suture.

The magmatism at 820–740 Ma arc-related gabbros and granitoids of Azania, the microcontinent preserved in EAO, was related to subduction of the Mozambique ocean along the site of the BSSZ (Collins & Windley, 2002). During this process, Azania was rifted off from East Africa because of roll-back of the Mozambique Ocean slab from a subduction zone. Interestingly, the subduction of the Mozambique Ocean also occurred during the same period along the west Indian margin leading to extrusion and emplacement of the Malani Igneous Suite in NW India that occurred between ∼770 and 730 Ma (Torsvik et al., 2001). The shear zones of the SGT in the eastern margin also extend to east and south through Sri Lanka and east Antarctica and witnessed Kuunga orogeny (570–500 Ma), which would be described later while dealing with the EGMB in the next section.

6.4.3 ACHANKOVIL SHEAR ZONE AND RANOTSARA SHEAR ZONE

The Achankovil shear zone (AKSZ), another significant crustal-scale shear zone in the SGT, holds a key position in juxtaposing the member terranes in East Gondwana reconstructions. The AKSZ has been extended and correlated with the sinistral Ranotsara shear zone of south Madagascar (Fig. 6.6). However, the correlation with Ranotsara shear zone was previously discarded because both dextral and sinistral movements were inferred along the AKSZ. The later structural studies revealed that the latest deformation along AKSZ is a sinistral transpression there by strengthening its correlation with the sinistral Ranotsara shear zone of Madagascar (Rajesh & Chetty, 2006). The initial dextral deformation along the AKSZ was correlated with the first event of orogeny (EAO) and the subsequent sinistral deformation with the second event (Kuunga orogeny). This inference is supported by the age data of c. 550 Ma obtained for undeformed granites and charnockites from the AKSZ. Further, the crustal-scale structures represented by the en échelon pattern of lineaments in the AKSZ can satisfactorily be explained as the features representing the latest sinistral transpressional tectonics, which could be of a Neoproterozoic age. However, more geochronological data is required to prove this hypothesis.

6.4.4 MADAGASCAR-SGT-SRI LANKA-ANTARCTICA

The EAO defines a ∼1000 km-wide belt of 650–500 Ma orogenesis that can be traced within Gondwana from northeast Africa through Madagascar-SGT-Sri Lanka upto the Donning Maud Land region of East Antarctica (Boger et al., 2015). The orogen involves multiple phases of collision and accretion and collectively defines one of the largest and most continuous orogenic belts within Gondwana (Fig. 6.7A). The EAO, extending for a few thousands of kilometers, must have formed during the waning stages of the Proterozoic when Gondwana was nearing the final stages of its formation, developing large collisional orogenic systems that produced varied structural and metamorphic imprints. It is in this context, another model of reconstruction invoked that the pre-Gondwana Indian plate probably extended into central and southern Madagascar and a single strand of the Mozambique Ocean possibly existed between pre-Gondwana India and the arc terranes of the ANS. The collision and docking of the pre-Gondwana terranes between the Indian plate and the arc terranes of the ANS is marked by the complete closure of the Mozambique Ocean giving rise to a late Ediacaran suture exposed in southwest Madagascar. This is supported by the similar ages of rocks reported along the western margin of the ANS and along the entire length of the EAO.

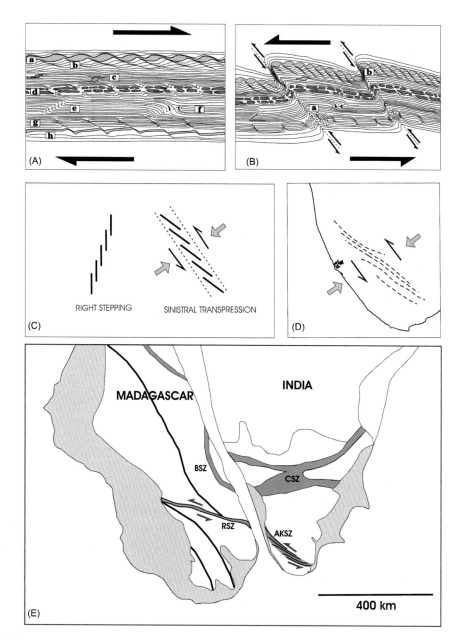

FIGURE 6.6

A schematic model showing the sequential development of dextral and sinistral structural features on mesoscopic scale. (A) showing the evidences for initial dextral non-coaxial deformation viz: a) backward rotating pinch and swells; b) S-C' fabrics; c) asymmetric porphyroclasts; d) asymmetric composite boudins; e) synthetic flanking folds; f) antithetic flanking folds; g) asymmetric foliation boudins and h) defection of foliations along ecc. (B) showing the development of a) trapezoidal boudins and b) sinistral shear bands by the reversal of shear sense in a subsequent deformation.(C) lineament patterns in transpressional regimes. Right overstepping of en echelon faults indicating sinistral transpression. (D) right over-stepping lineaments in AKSZ indicating sinistral transpression. (E) Strengthened correlation between sinistral Ranotsara Shear Zone (RSZ) of Madagascar with AKSZ of SGT, which suggests sinistral kinematics as the latest event of deformation. *BSSZ*, Betsimisaraka Suture Zone; *CSZ*, Cauvery Shear Zone System (after Rajesh, unpublished thesis).

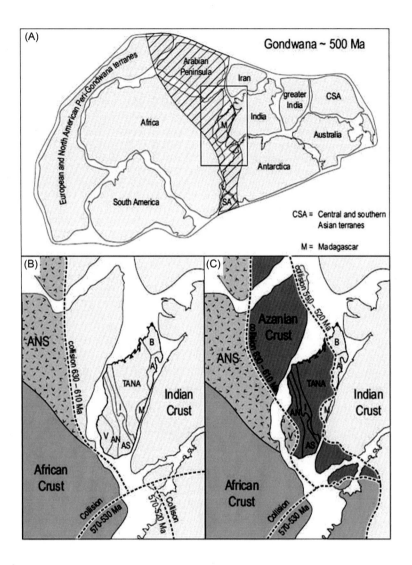

FIGURE 6.7

(A) Reconstruction of Gondwana illustrating the extent of the East African-Antarctic Orogen and the central location of Madagascar (M) within the orogenic belt, (B) showing single suture model assuming that Madagascar lay in the foreland of the pre-Gondwana Indian plate, and (C) Azania model assuming that central Madagascar existed as a separate microcontinent (Azania) prior to its multistage incorporation into Gondwana. Thick dashed lines mark the inferred traces of the sutures. Tectonic domains of Madagascar: *A*, Antongil; *AN*, Androyen; *AS*, Anosyen; *B*, Bemarivo; *I*, Ikalamavony; *M*, Masora; *TANA*, Antananarivo; *V*, Voribory; *ANS*, Arabian-Nubian Shield.

After Boger, S.D., Hirdes, W., Ferreira, C.A.M., Jenett, T., Dallwig, R., & Fanning, C.M. (2015). The 580–520 Ma Gondwana suture of Madagascar and its continuation into Antarctica and Africa. Gondwana Research, 28, 1048–1060.

In Madagascar, this suture lies along the Beraketa high-strain zone, the boundary between the Androyen and Anosyen domains (Fig. 6.7B). The event of 630−600 Ma is found on the western African ANS side of this structure, while it is absent on the Indian side in the east. Both sides of the suture, however, record subsequent 580−520 Ma orogenesis (Fig. 6.7C).

Another model of Paleogeographic reconstruction suggests that the Archean part of the Madurai Block may represent the southern extension of Azania, translated to east along the Malagasy Ranotsara shear zone in view of their similar geologic history of the SGT and Malagasy-Azania. However, the existence of Azania microcontinent was challenged recently and an alternative "Greater Dharwar" continent has been proposed, which incorporates all of eastern and most of central Madagascar on the periphery of Neoproterozoic India (Tucker, Roig, Moine, Delor, & Peters, 2014).

6.4.5 SGT AND NORTH CHINA CRATON

The eastern block of the North China Craton (NCC) and the southern block of the Indian shield share many similarities and represent unique cratonic blocks in the world that experienced a major Archean crust-forming event between 2.6 and 2.5 Ga (Kröner, Cui, Wang, Wang, & Nemchin, 1998). Both the blocks are characterized by the near contemporaneity of late Archean granitoid intrusive and metamorphic events, with the peak of metamorphism reached shortly (<50 Ma) after the widespread intrusion of granitoid suites. Moreover, dome and basin structures in the cratons and the anticlockwise P−T paths characterize the metamorphic evolution of the late-Archean granulites from both the blocks. The juxtaposition of the NCC with the Indian shield in the Paleoproterozoic was also proposed in the reconstruction model of the Columbia supercontinent configuration (Zhao, Cawood, Wilde, & Sun, 2002). A possible fit was contemplated for the NCC and the Indian shield, where the north margin of the eastern block of NCC is placed adjacent to the western margin of the south Indian block (Fig. 6.8). This kind of fit suggests the existence of two discrete continental blocks that developed independently during the Archean and Paleoproterozoic comprising the eastern block of the NCC with the south Indian block and the western block of NCC with the north Indian block. Both the blocks joined together along the Trans-North China Orogen (TNCO) and CITZ possibly during a Paleoproterozoic collisional event. This is further supported by similar structural styles, lithotectonic assemblages, and tectonothermal histories of TNCO and CITZ. However, a recent proposition states that the North China khondalite belt may have had its continuation with the Kerala Khondalite Belt of the Trivandrum granulite block rather than with the CITZ (Kröner et al., 2015) and varies from the previous hypothesis, indicating more probability of the SGT to have strong linkage with the Columbia supercontinent hypothesis. In the light of these divergent opinions, further studies are required and more focus should be done for detailed field observations combined with geochronological and other laboratory work for meaningful correlations.

In summary, it is well known that there are inconsistencies and disagreements about the existence and correlation of different shear/suture zones of different fragments of Gondwana. Therefore, understanding of tectonic evolution of East Gondwana requires further studies like the nature of the orogens and associated shear/suture zones and estimating their ages form a critical factor in Gondwana correlations.

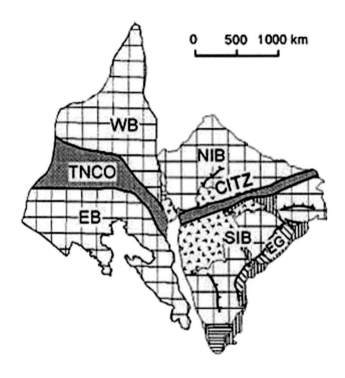

FIGURE 6.8

Reconstruction model showing the juxtaposition of CITZ-NCC; *EB*, Eastern Block; *WB*, Western Block; *TNCO*, Trans-North China Orogen; *NIB*, North India Block; *SIB*, South India Block; *CITZ*, Central India Tectonic Zone; *EGGB*, Eastern Ghats Granulite Belt.

After Zhao, G., Cawood, P.A., Wilde, S.A., & Sun, M. (2002). Review of global 2. 1–1. 8 Ga collisional orogens and accreted cratons: A pre-Rodinia supercontinent? Earth Science Reviews, 59, *125–162.*

6.5 THE EASTERN GHATS MOBILE BELT

6.5.1 INTRODUCTION

The EGMB, the prominent orogen occurring at the east coast of India, gained importance since long for its role in the reconstruction models of east Gondwana and its correlation with Antarctica and Australia. The close relationship between India-Antarctica in Gondwanaland was first suggested by Du Toit (1937) with the present east coast of India opposite and near that part of the coast of East Antarctica (Fig. 6.9A). It was later modified with minor changes by King (1950) with a slight counterclockwise rotation of Antarctica with Australia relative to India and Africa, placing Madagascar directly opposite the west coast of Peninsular India (Fig. 6.9B). Another reconstruction based upon the best geometrical fit of the continents at the 500-fathom line results in an assembly much closer to that of Du Toit (Fig. 6.9C), which is consistent with the paleomagnetic data (McElhinny & Embleton, 1974). Detailed examination of Indo-Antarctic relationships shows the

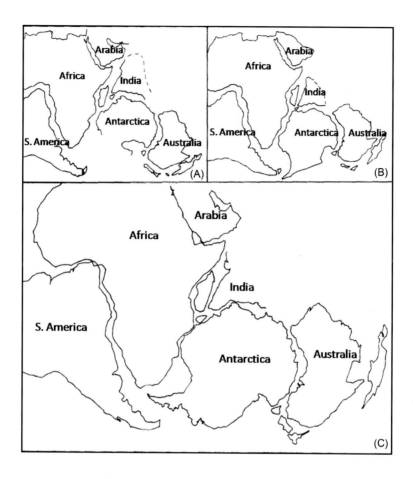

FIGURE 6.9

Earliest models of assembly of Gondwanaland: (A) Du Toit (1937); (B) B. King (1950); (C) Smith and Hallam (1970).

reassembly with excellently juxtaposed 1000-meter bathymetric contours with the bulge of Enderby Land fitting closely opposite the reentrant of the Indian coast at 15°N (Fig. 6.10). The geometrical fit of the EGMB of India and Enderby Land of east Antarctica in Gondwanaland, is supported by the presence of Precambrian high-grade metamorphic rocks with some partial recrystallization at both the margins suggesting the existence of a major belt of high-grade metamorphic rocks and possibly the original deposition and/or emplacement of the rocks. This implies that the two major crustal units of Indo-Antarctic part were associated together prior to their separation. Crawford and Campbell (1973) have suggested that shearing led to major horizontal displacement between old crustal blocks in Australia-Antarctica in Ordovician time, which was perhaps a consequence of very large-scale movement of the Gondwana supercontinent as a complex unit in relation to the dipole axis of the earth. Such movement equally explains well the effects seen in the

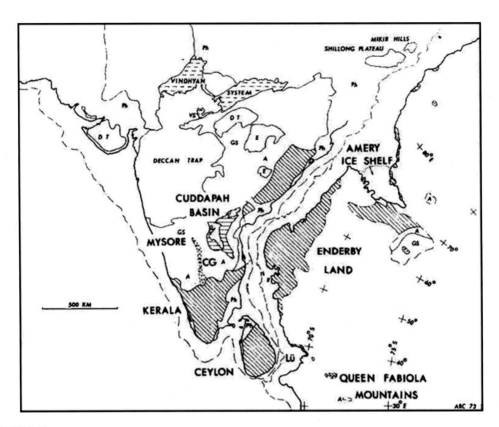

FIGURE 6.10

Reassembly of Peninsular India and part of East Antarctica by the juxtaposition of 1000 m bathymetric contours.

After Crawford, A.R. (1974). Indo--Antarctica, Gondwanaland, and the distortion of a granulite belt. Tectonophysics, 22, 141–157.

Indo-Antarctic part of Gondwanaland and may have occurred repeatedly during its long history (Crawford, 1974).

The Gondwana reconstruction model was further supported by the age determination of granulite-facies metamorphism and charnockite plutonism from the Rajahmundry-Visakhapatnam-Koraput-Terrane (RVKT) of the EGMB. U−Pb and Th−Pb ages obtained from perrierite and zircons of sapphirine-bearing granulite was found to be concordant at 1 Ga (Grew & Manton, 1986). This event was interpreted to be an extension of the EGMB of India into the late Proterozoic terrain of East Antarctica. The distribution of known exposures of Proterozoic and Archean rocks in India and Antarctica strongly support the Gondwanaland reassembly as originally proposed by Du Toit (1937).

The later geological and geophysical studies substantiated that the EGMB of India has strong linkage among the continents of East Antarctica and south western Australia. It was also well established that the EGMB forms a significant part of larger global orogenic belt system and was postulated to have developed as a result of the collision at 1000 Ma between eastern India and a part of

East Antarctica during the amalgamation of Rodinia (Hoffman, 1991). The Indo-Antarctic reconstruction model of the east Gondwana supercontinent was further strengthened essentially based on correlatable/extendable potential geologic piercing points across the conjugate continental margins. They include: major Neoproterozoic ductile shear zones Mahanadi rift (MR) and Lambert rift (LR), magnetic anomaly trends, and the Permian—Triassic ages and trends (Chetty, Vijay, Narayana, & Giridhar, 2003; Lisker, 2004).

6.5.2 SHEAR ZONE NETWORK OF INDIA-ANTARCTICA-MADAGASCAR

The prominent shear zones that connect the continuous network include the Nagavali-Vamsadhara Shear Zones (EGMB-central), Napier-Rayner Boundary Fault (NRBF) (Enderby Land, Antarctica), Cauvery Shear Zone (SGT), and BSSZ (eastern Madagascar) (Fig. 6.11). These shear zones are marked by Neoproterozoic (Pan-African) events of metamorphism and/or rejuvenation of preexisting Paleo-Mesoproterozoic crust. These are characterized by the preexisting thrust structures,

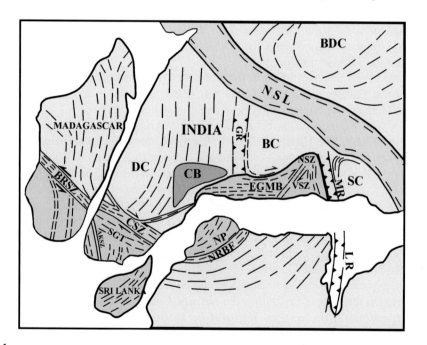

FIGURE 6.11

Reconstruction of East Gondwana supercontinent and the continuous chain of shear zone network: *BRSZ*, Bongolava-Ranotsara Shear Zone; *CSZ*, Cauvery Shear Zone; *NRBF*, Napier-Rayner-Boundary Fault; *NSZ-VSZ*, Nagavali-Vamsadhara Shear Zones; *BDC*, Bundhelkhand Craton; *BC*, Bastar Craton; *SC*, Singhbhum Craton; *DC*, Dharwar Craton; *CB*, Cuddapah Basin; *NSL*, Narmada-Son Lineament; *GR*, Godavari Rift; *MR*, Mahanadi Rift; *LR*, Lambert Rift; *EGMB*, Eastern Ghats Mobile Belt; *SGT*, Southern Granulite Terrane.

After Chetty, T.R.K., Vijay, P., Narayana, N.L., & Giridhar, G.V. (2003). Structure of the Nagavali shear zone, Eastern Ghats Mobile Belt, India: Correlation in the East Gondwana reconstruction. Gondwana Research, 6, 215—229.

emplacement of granite and syenite plutons, and compressive convergent tectonics. A few important observations of each shear zone are described below.

The extension and the continuity of major shear zones through the EGMB, SGT and the adjacent continents like Antarctica and Madagascar show a semicircular network around the Dharwar Craton. Further, the Cauvery shear zone (CSZ) and its extension into East Antarctica has been correlated with the NRBF of east Antarctica (Harris & Beeson, 1993).

The Nagavali-Vamsadhara shear zones (NVSZ) strike NNW−SSE and broadly divide the EGMB (central) into two parts. While the eastern part is correlatable with the Rayner Complex and the western part with the Napier Complex of Enderby Land of east Antarctica (Chetty, 1995a). The NVSZ is also described as a major tectonic boundary between the two crustal domains of distinct isotopic crustal history of Nd-model ages (Rickers, Metzer, & Raith, 2001). The NVSZ separates the Rajahmundry-Visakhapatnam-Koraput Terrane (RVKT) and the Berhampur-Bhaubaneswar-Phulbani Terrane (BBPT) of the EGMB. The BBPT shows consistently Neoproterozoic thermal imprints from Monazite age signatures, while the signatures of c. 1000, 800, and 550 Ma events are recorded from the NVSZ (Shaw et al., 1997). Significantly, the initiation of a clockwise P−T path at c. 990 Ma culminating in isothermal decompression at c. 780 Ma recorded in the Chilka Lake granulites distinguishes the BBPT from RVKT of the EGMB. The granitic magmatism along the NVSZ yielded 980−955 Ma while the Pan-African thermal overprint has also been recorded.

The NVSZ represents piercing points in the structural framework of Gondwana supercontinent that extends into the Enderby Land to join with the NRBF. The NRBF divides the Enderby Land into two distinct terranes: the Napier Complex with Nd-model ages between 3.2 and 4.2 Ga and the Rayner Complex with Nd-model ages between 1.4 and 2.4 Ga. Several similarities exist between the NVSZ and the NRBF. The NRBF is also a well defined shear zone predominantly comprising migmatitic gneisses and granitoids of ~960 Ma and a reworking event of 500−550 Ma (Black, Sheraton, & James, 1986). The NVSZ and NRBF were interpreted to continue and extend further into southern India to coincide with the Cauvery Suture/Shear Zone of the SGT.

In contrast to the earlier hypothesis of the established EGMB-Rayner Complex connection during the Rodinia assembly, the BBPT of the EGMB (Domain 3 of Rickers et al., 2001) was recently correlated with the exclusive Prydz Bay region of the Rayner Complex of East Antarctica since its inception (Bose, Das, Torimoto, Arima, & Dunkley, 2016). This conclusion negates the correlation of the NRBF with the Godavari rift structure as suggested by some earlier researchers and supports the hypothesis of the continuation of the NVSZ-NRBF. Further, the region of Rayner Complex occurring at the west of Prydz Bay, the Mawson Coast of Kemp Land also recorded high-temperature metamorphism in the time frame of c. 1150−920 Ma along clockwise P−T path (Halpin, Daczko, Milan, & Clarke, 2012). These evidences suggest that the BBPT seems to have evolved along with the Prydz Bay region of East Antarctica during c. 990−900 Ma time frame and came into contact with EGMB during the final assembly of Rodinia in the India-Antarctica sector.

The similarities of field, textural, and geochemical data between the BBPT of the EGMB and Prydz Bay are well preserved. The recent Pb isotopic data on feldspar strengthens the distinct similarity of Rayner Complex with that of BBPT (Flowerdew et al., 2013) while the Pb isotopic signatures from RVKT are ambiguous and less conclusive in their correlation. All this evidence strongly suggests that the BBPT of EGMB must have been a definite part of Rodinia. This may also show that the EGMB (along with BBPT) was located far from the main Pan-African tectonic belt.

The continuity of NVSZ-NRBF-CSZ and the BSSZ of Madagascar wraps around the Dharwar Craton in a semicircular shape with prominently outward dipping shear zones from the Dharwar Craton suggesting the polarity of subduction to be away from the Dharwar Craton at least during the Neoproterozoic (Chetty & Santosh, 2013). The associated accretion and collision history suggest a possible east—west convergence between Dharwar and Tanzanian cratons leading to the shortening of hotter and softened lithosphere beneath (Martelat, Lardeaux, Nicollet, & Rakotondrazafy, 2000), controlling the strain partitioning at the scale of supercontinents (Rodinia and Gondwana). However, the differential kinematics along shear zones and the precise timings need to be addressed in future studies. It is possible that different domains of the EGMB might have followed distinct evolutionary patterns during the early history, but share a common deformational history during the end Neoproterozoic, associated with the assembly of Gondwana. The common manifestation in the continuity of shear zones and associated structural fabrics imply subduction—collision process in an oblique convergence associated with the assembly of Gondwana.

6.5.3 RIFT STRUCTURES

The MR, India, and the LR, East Antarctica, represent two crustal-scale rift structures that were considered to represent a single Intra-Gondwana rift (see Fig. 6.11). Both the rift structures show strikingly similar characteristics such as crustal thickness, sedimentation pattern, structural architecture in the adjacent basement, kinematic, and paleocurrent indications, stratigraphic and Palynological features and contemporaneous 500 Ma dyke magmatism (Lisker & Fachmann, 2001). The MR, a part of ~600 km long Son-Mahanadi valley, trends WNW-ESE and lies orthogonal to the east coast of India. The MR is associated with two Gondwana basins located on either side of the Northern Boundary shear zone of the EGMB. There are also other small, isolated, and elongated Mesozoic upper Gondwana basins that occur in the vicinity of Mahanadi shear zone (MSZ). The LR, a typical half-graben structure in a thin continental crust of ~25 km, is associated with meridional fault systems at the margins (Fedorov, Ravich, & Hofmann, 1982). Similar to MR, the LR is also surrounded by Neoproterozoic granulites, which have undergone a multistage tectono-metamorphic history and the final metamorphic event during Neoproterozoic time (500 Ma, Boger, Carson, Wilson, & Fanning, 2000). In both the regions of MR and LR, the basement shear zones and their association with the Permian—Triassic coal bearing sedimentary sequence are closely correlatable. The cluster ages (820–1000 Ma) derived from detrital zircons from the northern parts of the EGMB and the Rayner complex of East Antarctica suggest that the MR and LR are a part of conjugate pair occurring on the two continents (Veevers, 2009).

After the cessation of transpressional tectonics during the Pan-African orogeny, the basement around MR remained relatively stable until the late Carboniferous. The rifting has initiated taking advantage of the preexisting tectonic framework reactivating the major shear zones. Deposition of Gondwana sediments started simultaneously during late Carboniferous and continued through the Permian into the Early Triassic (Veevers & Tewari, 1995). A new set of brittle fractures also developed at a later stage of reactivation associated with rifting. This kind of superposed brittle faulting and fracturing over the preexisting ductile shear zones were interpreted from SRTM data (Chetty, 2010). A new set of E—W trending large-scale fracture pattern parallel to the MSZ are mapped in the Phulabani-Daspalla domain, hosting the isolated small Gondwana basins. The development of upper Gondwana basins such as the Athgarh basin and a new set of fractures along the MSZ suggests a renewed rifting activity since the early Cretaceous ($\sim$140 Ma). The Athgarh basin was later intruded by a 117 Ma mafic dyke near Bhubaneswar.

Considering the standard reconstruction models, the position, orientation, and timing, the Mahandi rift strongly corresponds to the LR of east Antarctica during the pre-Gondwana breakup. Both of them are characterized by asymmetric half-graben structure, homoclinal basin tilts and Permo—Triassic growth faulting formed in the anisotropic basement. Detailed ocean floor topography through satellite altimetry also supported this model (Sandwell & Smith, 1997). The East Antarctic LR and the Indian MR are considered to represent segments of an intra-Gondwanan rift structure that was active at least since the Paleozoic. Fission track analyses of apatites collected across the shoulders of both the rift structures were used to compare their low temperature history, and to estimate the paleo-geothermal gradients before the onset of the last denudation/rifting stage during the Late Jurassic (Lisker, 2004). The paleo-geothermal gradients of both the juxtaposed Gondwana margins similarly increase from the basement towards the respective rift shoulder from 15.20 to 25.30°C/km. This trend of increasing paleo-geothermal gradients, together with a denudation episode commencing in the Early Cretaceous and coeval igneous activity, indicates a common rifting stage accompanying the breakup of Gondwana in the India-Antarctica sector.

6.5.4 EGMB-SOUTH AND ANTARCTICA

Several studies have attempted to correlate and compare the widespread mid-Mesoproterozoic granulite facies rocks of the EGMB-south with the similar metamorphic conditions in the rocks of Napier complex of east Antarctica in reconstructing the models of Gondwana. The reworking of granulite facies rocks of the region was correlated and a possible linkage between both the domains was suggested (Dasgupta & Sengupta, 2003). Another possible aspect in such comparisons is with regard to the source rocks for the sedimentary protoliths of EGMB-south. In contrast to the earlier belief that the source was from the adjacent Dharwar Craton, it was inferred that Napier Complex was the original source (Henderson, Collins, Payne, Forbes, & Saha, 2014). The interpretation was further extended that the EGMB-south represents a part of an exotic terrane, which was transferred to proto-India in the late Paleoproterozoic as part of a linear accretionary orogenic belt that may also have included south-west Baltica and southeastern Laurentia. Based on the isotopic, geological, and geochemical similarities, the proposed exotic terrane may represent an extension of the Napier Complex, Antarctica, and possibly connected to Proterozoic Australia (North Australian Craton and Gawler Craton). This hypothesis is supported by the similarities in U—Pb and Hf data between the protoliths of EGMB-south and the metasedimentary protoliths from the North Australian Craton, implying that the source could be the Antarctic for both the regions. The above-mentioned features indicate that the EGMB-south may represent an exotic terrane accreted to the Indian craton at c. 1.6 Ga, as a fragment of the Napier Complex of Antarctica and simultaneously connected to parts of Proterozoic Australia.

The recognition of c. 1760—1600 Ma events in EGMB-south can be related to the accretionary history of Columbia that formed at c. 2100—1800 Ma prior to Rodinia. Columbia had a long-lived history of subduction related growth via accretion at important continental margins for nearly 500 Myr and a correlation of major accretionary belts of c. 1760—1700 Ma orogenesis around the preexisting cratons involving Laurentia, Antarctica, South Africa, and Australia including the eastern margin of India was proposed (Zhao, Sun, Wilde, & Li, 2004). This proposal is supported by the recently obtained U−Pb zircon data from the EGMB-south, which preserves a record of a prolonged accretionary process that started in the arc-continent collision and culminated in the continent−continent collision (Pacific-type) during c. 1850−1600 Ma (Vijaya Kumar, Ernst, Leelanandam, Wooden, & Groves, 2009). All the tectonothermal events in the region including the UHT metamorphism can be considered as part of this accretionary process. During this model of accretionary orogenesis, the EGMB might be behaving like an advancing orogen at that time due to the closure of ocean basins. Thermal perturbation on a regional scale might have been induced by asthenospheric upwelling following slab break-off at relatively shallow depth and consequent production of basaltic magma emplaced in the lower crust. The thermal pulse recorded at c. 1600 Ma in the EGMB-south is similar to the reworking of the NRBF in East Antarctica and may indicate a connection involving the EGMB-south and Napier-Rayner complex at this time (Bose, Dunkley, Dasgupta, Das, & Arim, 2011).

The model of eastern India forming a part of the linear accretionary system encompassing south-eastern Laurentia and Baltica accounts for the magmatism and metamorphism recorded in the EGMB-south (Ongole Domain) between 1.72 and 1.6 Ga leading up to continent−continent collision at c. 1.6 Ga. Accretionary orogenesis might have been possibly terminated by the collision that developed between EGMB-central and Rayner Complex at c. 1 Ga (Mezger & Cosca, 1999). The model endorses that the EGMB−south represents a fraction of an exotic terrane that was transferred to proto-India in the late Paleoproterozoic (1.68−1.6 Ga).

The Grenvillian Provinces including the EGMB skirting east Antarctica were reworked, truncated and offset by Pan-African mobile belts, there by implying that east Gondwana underwent significant reorganization during the Cambrian, unlike in many models where it was assumed to have stabilized during the late Mesoproterozoic (Fitzsimons, 2000). Therefore, it is abundantly evident from the available geochronological data that the EGMB must have been affected by the Neoproterozoic-Cambrian events indicating that the EGMB was juxtaposed against east Antarctica during Gondwana assembly. The final breakup of Columbia at about 1.3−1.2 Ga was immediately followed by the assembly of the supercontinent Rodinia along the globally distributed Grenvillian orogens at 1.0 Ga (Dalziel, 1997). In most of the Rodinian reconstructions, the eastern margin of India is juxtaposed against East Antarctica with the Rayner complex being correlated with the EGMB (e.g., Yoshida, Funaki, & Vitanage, 1992). The formation of the EGMB−Rayner Complex and the other Grenvillian orogens girdling the Antarctic-South Australian Craton (Mawson continent) during the late Mesoproterozoic and the early Neoproterozoic (1300−900 Ma) was in response to the collision of Mawson continent with the marginal cratons such as Southern Africa, India, and Western Australia during Rodinia assembly. The opening of the Mesoproterozoic rift between India and east Antarctica subsequent to the breakup of Columbia may have formed a large oceanic basin between the two where in the sedimentary sequences of the EGMB were deposited (Zhao et al., 2004).

Veevers (2009) summarized and illustrated the assembly and breakup history of Antarctica and India (Fig. 6.12) in the following stages: (1) 1.7−1.5 Ga: Archean cratons (Dharwar, Bastar, and Napier) became sutured by fringing orogens (Karimnagar-Bhopalpatnam, CITZ, EGMB, Kemp Land), followed by narrow divergence of the Napier Craton from the Indian composite, (2) 1.3−0.9 Ga: initial 1.3 Ga convergence of India and Antarctica proceeding to final 1.0−0.9 Ga collision during the assembly of Rodinia. The main ocean closure was probably some where in

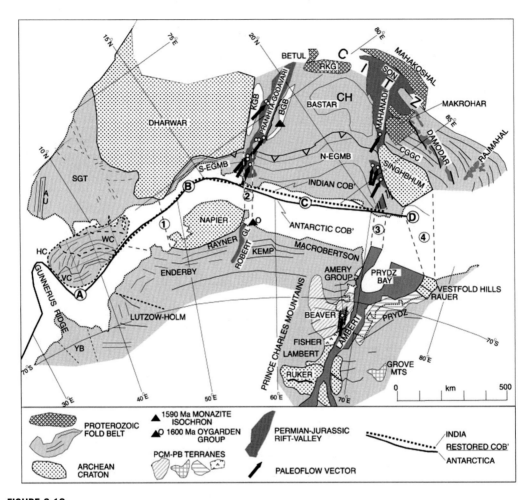

FIGURE 6.12

Summary of the assembly and breakup history of India-Antarctica through different stages.

After Veevers, J.J. (2009). Palinspastic (pre-rift and -drift) fit of India and conjugate Antarctica and geological connections across the suture. Gondwana Research, 16, *90–108.*

ice-covered central East Antarctica within inferred blocks of 1.2−0.8 Ga mafic granitoids. The EGMB−Rayner complex was metamorphosed to granulite-facies as it was thrusted away from interior East Antarctica over the Archean Napier, Dharwar, and Bastar cratons, (3) At 771 Ma, India had moved 40 degrees of latitude away from Antarctica (Meert & Lieberman, 2008), and (4) by 0.55−0.50 Ga, India and Antarctica got reunited along the same suture but without ocean closure, possibly by transpressional movement along a transform fault. During this final assembly of Gondwana, the orogen was further deformed through thrusting and strike-slip shearing of the

Rayner Complex against the Archean Napier Complex and the EGMB against the eastern margin of Archean Dharwar/Bastar cratons, (5) 0.3−0.2 Ga Permian−Triassic: the conjugate margins were crossed by a system of rift valleys, which accumulated the Gondwana succession of basal glacial sediments followed by coal measures and red beds, (6) 0.16−0.14 Ga Jurassic-Cretaceous: after preliminary rifting, continental breakup along the EGMB−Rayner complex, (7) to 0 Ma: the newly formed margins accumulated onlapping drift successions.

6.6 THE CENTRAL INDIAN TECTONIC ZONE

6.6.1 INTRODUCTION

The CITZ divides the Archean to Paleoproterozoic basement of the Indian shield into two major cratonic blocks, namely the South and North Indian Blocks. The South Indian Block includes the Dharwar, Bastar, and Singhbhum cratons, consisting predominantly of early-to-late Archean TTG gneisses and supracrustal rocks (namely greenstone sequences) with minor Paleoproterozoic rift-related formations. The North Indian Block represents a single crustal unit, known as the Aravalli-Bundelkhand Province and comprises Archean basement gneisses and Paleoproterozoic volcanic and sedimentary rocks metamorphosed from amphibolite to granulite facies (Mazumder, Bose, & Sarkar, 2000). The CITZ is marked by a number of parallel ductile shear zones with near east−west trend and subvertical to steep northerly dips (Jain, Yedekar, & Nair, 1991). The major lithologies are TTG gneisses, granitic intrusions, and supracrustal rocks including quartzites, mica schists, paragneisses, BIFs, calc-silicates, marbles, and basic to acid volcanics, ranging in age from the late Archean to the Proterozoic and metamorphosed from greenschist to granulite facies (Acharyya & Roy, 2000). A few high-pressure mafic granulites also occur as enclaves or boudins within TTG gneisses and preserve mineral assemblages and P−T evolution reflecting initial crustal thickening followed by near-isothermal decompressional exhumation and final cooling. This is interpreted to have resulted from collision between the South and North Indian Blocks to form the CITZ during a poorly defined period between 2.1 and 1.7 Ga (Bhowmik, Pal, Roy, & Pant, 1999).

The CITZ can be correlated not only with the well-studied Late Paleoproterozoic to Early Mesoproterozoic granulites from other segments of the Indian shield, but also from adjoining Gondwana continents (Bhowmik, 2014). The important features that are comparable in the adjacent fragments of Gondwana are: (1) the presence of Late Paleoproterozoic to Early Mesoproterozoic orogenesis along the vast stretches of the former Gondwanan fragments, (2) the common occurrence of several short-lived thermal pulses (20−30 Myr) in episodic orogenies, locally reaching UHT metamorphic conditions and spanning over 60−100 Myr, (3) the common presence of the granulite domains with high apparent thermal gradients at peak metamorphism, and associated counterclockwise metamorphic P-T paths, and (4) the duration of the granulite-facies metamorphism and subsequent cratonization between c. 1·72 and 1·54 Ga in many of the granulite belts. All these features suggest that the two supercontinent assembly events in the Paleoproterozoic (Columbia) and in the late Mesoproterozoic and early Neoproterozoic

(Rodinia), involve the development of a late Paleoproterozoic to early Mesoproterozoic landmass of the Indian Shield.

The comparison of the CITZ with the other Proterozoic orogens of Peninsular India reveals that the margins of the Archean cratons of Indian shield (Bundelkhand, Bastar, and Singhbhum and Dharwar cratons) were subjected to deformation, metamorphism, and magmatism, culminating in orogenesis during the period 1.7—1.5 Ga. It is significant to note the common association of 1.6 Ga high-T granulite facies metamorphism, commonly with a counterclockwise metamorphic P-T path in many of the orogens such as the CITZ, SM Plateau gneissic Complex, and EGMB-south. There is another feature of common orogenesis involving the occurrence of felsic plutonism, including charnockites all through the Proterozoic orogens of ADOB, CITZ, EGMB, and the SGT. Such a common association of Late Paleoproterozoic to Early Mesoproterozoic tectonothermal and tectono-magmatic events at an interorogen-scale can best be explained within the realm of accretionary orogenesis. In the light of the aforementioned information, the final amalgamation of the different cratonic blocks of the Greater Indian landmass and its integration with the Rodinia supercontinent must have taken place through continent—continent collisional processes during the Grenvillian orogeny at ∼1.0 Ga (Bhowmik, Bernhardt, & Dasgupta, 2010). However, it is significant to note the absence of these features in SGT, although its extension and continuation with the EGMB is well established.

The eastern segment of CITZ extends in NE direction and continues through the SM plateau and is presumed to abut against Indo-Mynmar mobile belt. In the reconstructions model of Gondwana supercontinent, it was postulated that the CITZ and EGMB merge together in the northeastern part of the Indian shield and continue to get juxtaposed with Albany-Fraser mobile belt and the Stirling Range formation of western Australia (Powell, McElhinny, Meert, & Park, 1993). Dextral transpression and northward thrusting, accompanied by granulite facies metamorphism at 1.2 Ga is a characteristic feature of the Albany belt, while dextral transpression with the ages of 590—540 Ma (Harris & Beeson, 1993) is a significant feature of the Sterling formation, located just north of Albany belt. The extensions of the CITZ, jointly with the northeast extensions of the EGMB in northeast India (Assam, Meghalaya and the northern Eastern Ghats Orogen) is associated with Neoproterozoic deformation, metamorphism and magmatism. Neoproterozoic granitoids intrude poorly dated schists in the basement exposed in far northeast India. These rocks are only dated by Rb—Sr methods, which yielded ages between 900 and 450 Ma. In the west, the CITZ extends beyond the Indian continent and coalesces with the ADOB to continue further probably to join the EAO (see Fig. 6.3).

The Mesoproterozoic rift zone related to CITZ was extended up to northeastern India and to Terre Adelie in Antarctica suggesting the existence of a single supercontinent known as Columbia (Rogers & Santosh, 2004). Based on isotopic studies in the CITZ, two important tectonothermal imprints were identified, which are coeval with major global tectonic events of Mesoproterozoic rifting and the Grenvillian orogeny. The collision between the two cratonic blocks must have occurred during the Grenvillian orogeny, one of the most extensive orogenic belts that ever existed (Mezger & Cosca, 1999). The imprints of the Grenvillian orogeny have already been reported in the EGMB of India and Rayner complex of Antarctica.

6.6.2 CITZ AND NORTH CHINA CRATON

The initial idea that the Eastern Block of the NCC and the South Indian Block of the Indian Shield are two unique cratonic blocks in the world that experienced a major Archean crust-forming event between 2.6 and 2.5 Ga was proposed by Kröner et al. (1998). Both blocks are characterized by the near contemporaneity of late-Archean granitoid intrusive and metamorphic events, with the peak of metamorphism reaching shortly (<50 Ma) after the widespread intrusion of granitoid suites. The idea was also supported by the presence of dome-and-basin structures and anticlockwise P–T paths characterizing the metamorphic evolution of the late Archean granulites from both the blocks. These similarities led to the postulation that both NCC and the Indian shield may have constituted a part of a single major active plate margin along which the juvenile crust was accreted onto an older landmass. This hypothesis also explains similar Paleoproterozoic formations between the two cratonic blocks. For instance, Paleoproterozoic (2.5–1.8 Ga) sedimentary-volcanic successions in both the blocks comprise lower clastic-rich, middle volcanic-rich and upper clastic + carbonate sequences, represented by the Liaohe Group in the Eastern Block and the adjoining Singhbhum, Dhanjori, and Kolhan Groups in the South Indian Block (Naqvi & Rogers, 1987). A possible continental fit for the NCC and the Indian Shield, where the northern margin of the eastern block is placed adjacent to the western margin of the South Indian block, with the TNCO and the western block of the NCC represent the continuations respectively of the CITZ (see Fig. 6.8). It was strongly argued that the CITZ was also earlier considered to be part of a Paleo-Mesoproterozoic system of collisional orogens leading to the supercontinent of Columbia (Zhao et al., 2002).

The CITZ and the TNCO share similar structural styles, lithotectonic assemblages, and tectonothermal histories. Both the orogens consist of low-grade volcanic–sedimentary (greenstone) sequences and granitoids in the center flanked by high-grade granulite–gneiss terrains on both the margins. Structurally, both the orogens have undergone large-scale thrusting and ductile shearing. While east-west–trending ductile shear zones and south-verging thrusts are prominent in the CITZ, a series of east-verging thrusts and north-south–trending ductile shear zones are dominant in the TNCO. Further, high-pressure mafic granulites (retrograded eclogites) were also discovered in both the orogens containing peak high-pressure and postpeak decompression and cooling mineral assemblages, defining near-isothermal decompressional, clockwise P–T paths. The metamorphic history suggests that both the orogens witnessed a similar tectonothermal process involving initial crustal thickening, subsequent rapid exhumation, and cooling. Summarizing all these, Zhao et al. (2002) proposed that the TNCO and the CITZ evolved together and suggested that the eastern and western blocks of the NCC were connected to the Southern Indian block and North Indian block, respectively, until they coalesced along the CITZ.

Recently, an alternate paleogeographic model between NCC and the CITZ (Fig. 6.13) has been invoked subsequent to the discovery of Paleo-, to Mesoproterozoic ophiolites from the Indian shield and the sapphirine-bearing Paleoproterozoic ultrahigh-temperature granulites from the NCC (Santosh, 2012). The model suggests that the close relationship between the eastern block in the NCC and South India block, as envisaged earlier, did not continue long, and the two blocks were separated possibly by the beginning of Mesoproterozoic with a different tectonic history since then. The southeast margin of the Indian plate underwent major convergent tectonics in the late Mesoproterozoic-Neoproterozoic with ocean closure and continental amalgamation associated with

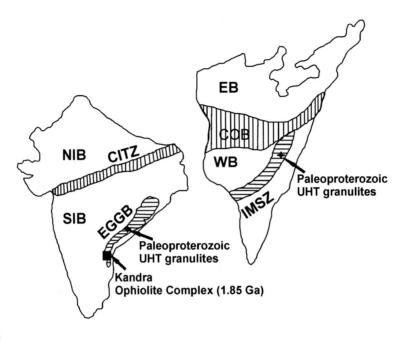

FIGURE 6.13

Possible configurations for the reconstruction of the North China Craton and the Indian Shield. *EB*, Eastern Block; *WB*, Western Block; *COB*, Central Orogenic Belt; IMSZ, Inner Mongolia Suture Zone; NIB, North India Block; SIB, South India Block; CITZ, Central India Tectonic Zone; EGGB, Eastern Ghats Granulite Belt.

Modified after Santosh, M. (2012). India's Paleoproterozoic legacy. Geological Society of London, Special Publications, 365, 263–288.

the Rodinia supercontinent assembly. According to this model, the southeast margin of Indian shield was in continuation of the western Block while the Central Indian block was in continuation with the eastern block of NCC making the CITZ and the Trans-China orogen a single continuous unit along which different blocks were collided and amalgamated.

6.7 THE ARAVALLI-DELHI OROGENIC BELT

6.7.1 INTRODUCTION

The Aravalli-Delhi Orogenic Belt (ADOB) occurs in the northwestern part of India and was inferred to join the Sausar supracrustal belt at the southern margin of the CITZ apparently in the form of major folding (Krishnan, 1961). Later studies also confirmed the continuation of the NE-SW–trending ADOB toward the east to join the CITZ with the presumption that both the orogens belong to Mesoproterozoic (Naqvi & Rogers, 1987). However, the recent age data shows that the South Delhi Terrane (SDT), the western part of ADOB is Neoproterozoic in age and that the Kaliguman shear zone (KSZ) demarcates the boundary between SDT and the

Mesoproterozoic-Aravalli terranes. The KSZ marks a suture zone along which the South Delhi basin closed through subduction. In contrast to the earlier belief of a major fold formed in the combined geometry of ADOB−CITZ, the structural fabrics of SDT (including the Phulad shear zone, PSZ) along with Sirohi Terrane (Neoproterozoic terranes) extend in southwest direction and possibly coalesce with the western extensions of the CITZ. Both ADOB and CITZ extend beyond the Indian continent before their final merger with similar terranes of the ANS, which represent an important part of the EAO. It is also conjectured that the northern extensions of the BSSZ in the northern part of Madagascar also join the ANS through the EAO. The ADOB in the NW India is generally shown to occupy the western margin of the supercontinent Rodinia in most of the paleogeographic reconstructions (Li et al., 2008).

The contrasting geophysical anomalies and geological dissimilarities are distinct across the PSZ and extend over a distance of about 1000 km along the strike in NE−SW direction probably upto the Himalayas in the north. In brief, the field relationships, mineralogy and geochemical data of the PSZ suggest that it represents the site of main subduction zone of the oceanic plate and the rock assemblage was comparable to those of a modern ophiolite, analogous to the Neoproterozic Central Asian fold belt ophiolites of Mongolia.

6.7.2 ADOB-ANS-EAO

The SDT in association with the PSZ marks a trace of the closure of proto-Mozambique Ocean within Gondwana. The ocean closure occurred when the Marwar craton (adjacent to the SDT), arc fragments (Bemarivo Belt in Madagascar and the Seychelles), and the components of ANS collided with the Aravalli-Bundelkhand Protocontinent at c. 850−750 Ma (Singh, De Waele, Karmakar, Sarkar, & Biswal, 2010). The SDT obtained granulite facies metamorphism through subduction during Neoproterozoic times. The available geochronological data shows that the granulite facies rocks, occurring at the southern margin of the SDT, seem to be much younger than the granulites of the Sandmata Terrane of the ADOB, the Sausar granulites of the CITZ, and those of the EGMB. However, the granulites of Balaram-Kui-Surpagla-Kengora (BKSK) region seem to be similar in terms of age, subduction zone tectonic setting, and exhumation through thrusting with those of the SGT. Further, the BKSK granulites are also comparable with the various granulite belts of the EAO with similarity in ages as well as structural styles, supporting the continuation of the SDT finally merging with EAO.

The evolution of SDT together with the Malani volcanic magmatism is closely related to the assembly and breakup of the Neoproterozoic Rodinia supercontinent at ~1100 Ma and ~750 Ma respectively. Thus, the SDT orogenic event has global significance, being part of the global network of Proterozoic orogenic events forming the late Mesoproterozoic Rodinia supercontinent representing the Grenvillian orogenic event. It is also well established that the Grenvillian orogenic event is evident from the CITZ, as well as from the EGMB, which are correlated with the Pinjara orogen of Australia, and the Circum-East Antarctic mobile belt of Antarctica. Together, these three orogenic belts represent one of the important phases of the Indian continental assembly during the late Mesoproterozoic.

In another reconstruction model, the Mt. Abu-Malani Igneous suite (MIS) occurring to the west of SDT, represents the northeastern continuation of the Neoproterozoic (800−700 Ma) magmatic belt traced from northern Madagascar, the Seychelles into NW part of India. Based on major, trace element geochemistry and U–Pb, Ar–Ar geochronology, the granitoids of MIS (at the western margin of the ADOB), are correlated with those of Seychelles and northern Madagascar and interpreted them to represent the western margin of Rodinia (Ashwal et al., 2013). This magmatic belt, located along the western margin of supercontinent Rodinia, was believed to have formed during the process of eastward subduction of the Mozambique Ocean. Transpressional forces induced during closure of the Mozambique Ocean must have been responsible for shaping the structural architecture of the Mt. Abu-Sirohi region. This kind of Paleoreconstruction of NW India-MIS-Seychelles-Madagascar at ∼750 Ma seems to be more probable (Torsvik et al., 2001). The findings of Just, Schulz, de Wall, Jourdan, and Pandit (2011) in Sirohi Terrane substantiate a Cryogenian (770–750 Ma) tectonic imprint in the southwestern sector of the ADOB, independent of the Delhi Orogeny (1–0.8 Ga). The above-described features strengthen the continuation hypothesis of late Neoproterozoic belts of central and north Madagascar into northwest India. It is also possible to draw analogies of tectonics, magmatism, and sedimentation between Neoproterozoic Marwar terrane and Cenozoic Tibet (Sinha-Roy, 1988).

Out of the three well known Precambrian crust-formation episodes at 2.7, 1.9, and 1.2 Ga, the 1.9 Ga event witnessed a significant growth of continental crust, mainly in the form of continental arc systems during the initial stage of supercontinent formation (Condie et al., 2009). The 1850–1822 Ma event of continental arc affinity recognized in the ADOB broadly coincides with many collisional orogens showing Andean-type arc affinities of the supercontinent Columbia (Kaur, Chaudhri, Racsek, Kroner, & Hofmann, 2009). Incidentally, the tectonic setting and the ages (1.85–1.82 Ga) of granitoids around the ADOB correspond well with the timing of accretion in Columbia.

The Proterozoic crustal evolution of ADOB, considered to be due to island arcs and characterized by high velocity and thick crust (50 km), seems to be comparable with the Baltic and Canadian Shields and is in contrast with the thickness of ∼35 km in Phanerozoic orogens (Nelson, 1992). The presence of dipping reflectors and the island arc signature in the ADOB suggest that originally a thick crust was formed and was not significantly altered after Proterozoic collision. Evolutionary periods of the Aravalli fold belt (1800 Ma) and Delhi fold belt (1100 Ma) correlate well with the global orogenic activity corresponding to the supercontinental cycles (Hoffman, 1988).

From the synthesis of the POI, it can be seen that the Indian continent was not a coherent block until the Neoproterozoic and possibly various crustal blocks finally amalgamated during the late Neoproterozoic-Cambrian (Pan-African) period. The suturing of the Africa-ANS with the western margin of Indian shield and their association with different Proterozoic orogens such as ADOB, CITZ, and the SGT signifies a remarkable event most probably reflecting the history of the collision intervening oceanic arcs between East and West Gondwana supercontinents. It is widely believed that the Pan-African tectonothermal event was wide spread throughout the Gondwana continents and resulted in large-scale crustal accretion and remobilization throughout the Gondwana prior to its fragmentation and plate tectonic movement (Kröner, 1981).

Regarding the position of India in the Paleo-, Mesoproterozoic supercontinent of Columbia, many models were proposed: (1) Placing Madagascar adjacent to western margin of India, (2) Connecting Madagascar with the SW margin of East Antarctica, (3) establishing a Paleoproterozoic terrane in south-central Madagascar considering that 2.2–1.8 Ga detrital zircons were derived from the combined terranes of South Madagascar-India-Sri Lanka, which were together at ∼1.8 Ga. The detrital zircon data of the supracrustal metasedimentary rocks from the northern parts of Madagascar show maximum depositional ages of around 1.8 Ga (De Waele et al., 2011). The ADOB was also considered as one of the potential source regions for this detritus. Madagascar detrital zircons revealed ages in the range of c. 3.27−1.72 Ga with notable peaks at 1.86 and 2.5 Ga matching closely with those of ADOB. This correlation favors the suggestion that the northern parts of Madagascar and northwest India formed a coherent crustal entity, and experienced a common tectonothermal history during the late Paleoproterozoic in the Columbia amalgam.

6.7.3 MALANI MAGMATISM AND ITS CORRELATIONS

There are multitude of tectonic settings proposed for Malani volcanism. The first stage of volcanism was associated with basaltic and felsic flows and was attributed to a hot spot source or lithospheric thinning and melting at the base of the crust, with an extended (over 100 million years) history (Rathore, Venkatesan, & Srivastava, 1999). However, this hypothesis is inconsistent with paleogeographic reconstructions that show India as an isolated fragment at ∼770 Ma. Geological evidence suggests that parts of Madagascar, East Antarctica, and Sri Lanka were part of this continental assembly (Collins & Pisarevsky, 2005).

The period of the Malani magmatic activity (750 Ma) coincides with the breakup of the Neoproterozoic Rodinia supercontinent. During the breakup of a supercontinent, rifting generally takes place at the preexisting suture zones, as they are rheologically weak and more susceptible to rifting. Interestingly, the location of rifting manifested in the form of Malani volcanic eruption in the Marwar Terrane is observed nearer and parallel to the PSZ. The paleoposition of India in relation to the ancient supercontinent Rodinia and proto-East Gondwana is in contrast with the hypothesis of a rift setting for the MIS and is more indicative of an Andean-type arc environment resulting from the subduction of the eastern Mozambique Ocean. The deep seismic images indicate that the tectonic fabrics of the SDT to be consistent with a long-lived "Andean-type" arc formed above an eastward-dipping subduction zone along the western margin of India during ∼1100 Ma (Vijaya Rao & Krishna, 2013).

The location of India within Gondwana is critical for evaluating the various tectonic models related both to the assembly of Gondwana and Rodinia. Some researchers believe that coherent and stable Gondwana existed from about 1.1 Ga until the Mesozoic breakup of Gondwana. However, the paleomagnetic data from MIS of India, Mundine dykes of Australia and Takamaka dykes of Seychelles place India and Seychelles at much higher latitudes, nearly 25 degrees latitude separation from Australia (Gregory, Meert, Bingen, Pandit, & Torsvik, 2009). It indicates that East Gondwana was not amalgamated at 750 Ma and was assembled later at the time of formation of Gondwana supercontinent during 550 Ma with the evolution of Kuunga orogeny. Thus, the MIS of NW India evidently plays an important role in the paleogeographic reconstruction of India between dispersal of the late Mesoproterozoic supercontinent Rodinia and Neoproterozoic assembly of Gondwana.

It is widely believed that the Pan-African tectonothermal event was widespread throughout the Gondwana continents and resulted in large-scale crustal accretion and remobilization throughout the Gondwanaland prior to its fragmentation and plate tectonic movement (Kröner, 1981). The widespread thermal signature between 500 and 550 Ma in the ADOB and the adjacent regions may be related to some large-scale magmatic event which is yet to be properly recorded. The Pan-African related events have earlier been reported from several places, particularly from the POI.

Although the direct linkage of shear zones in northern Madagascar with shear zones in ADOB cannot be established, the continuation of the Cryogenian suture between the two must have provided pathways for hot fluids to cause locally restricted Pan-African thermal resetting in northwest India.

In view of the multitude of tectonic settings proposed for Malani volcanisms, the paleoposition of India in relation to the ancient supercontinent Rodinia and proto-East Gondwana is not consistent with the hypothesis of a rift setting for the origin of MIS. Further, the MIS is more indicative of an Andean-type arc environment resulting from the subduction of the eastern Mozambique Ocean. From the above descriptions, it is challenging to compare all the orogens of India, particularly with the limited available field and age data. The correlations still remain open for discussion and further investigations. In order to resolve these contentious correlations, improved understanding of all the POI in terms of regional structural framework combined with multidisciplinary studies is essential.

6.8 SUMMARY

The POI, representing a critical part of Gondwana orogenic systems, delineate and juxtapose different Archean cratons of India (Dharwar, Bastar, Sighbhum, and Bundelkhand). These orogens, being in the central part of the Gondwana supercontinent, play a crucial role in the enhanced understanding of the reconstruction models of supercontinents such as Columbia, Rodinia, and Gondwana. The POI have attracted global attention in recent years for the reconstruction models related to timing and tectonics of the breakup and amalgamation of constituent fragments of Rodinia and Gondwana.

The structural and metamorphic characteristics of POI, in general, are highly varied and they exhibit a wide range of styles, magnitudes, and peak temperatures. There is no unique set of characteristics that defines these orogens, and their first-order structural framework and development appears to be comparable to those of orogens formed in the Phanerozoic. The episodic character of orogenies since late-Archean times has led to speculations that Phanerozoic-style plate tectonics can be applied to the Proterozoic, and that continental land masses have periodically assembled and dispersed since the Paleoproterozoic as a result of plate convergence and separation (Windley, 1995). These speculations have provided a major stimulus to the reconstruction of ancient supercontinents, including Meso-, to Neoproterozoic Rodinia, Neoproterozoic Pannotia/Gondwana, and Paleozoic Pangea (Dalziel, 1997) that led to the emergence of global-scale collisional orogenies (e.g., Mesoproterozoic Grenvillian and Phanerozoic Pan-African). It is in this context, that the correlation of POI with those of adjacent Gondwana continents would represent piercing points in establishing former linkages between separated continents.

The Indian lithospheric plate constitutes a plethora of geological events covering the entire history of our planet. The Proterozoic orogens of proto-Indian continent form an integral part of the Paleoproterozoic supercontinent Columbia. Rifting and separation of crustal Blocks, opening of oceans and their final closure are preserved in the travelogue of oceanic plate from mid oceanic ridge to trench in the form of "Ocean Plate Stratigraphy" in Proterozoic belts including the association of dismembered ophiolites, pelagic sediments and continental margin sequences which were

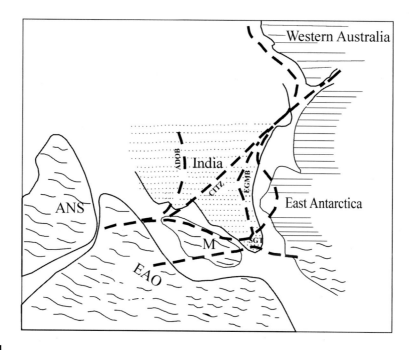

FIGURE 6.14

A sketch map showing the Proterozoic orogens of India and the constituent shear zones that extend into the adjacent dispersed Gondwana fragments exhibiting the pivotal role of India in the reconstruction models of Gondwana supercontitnent: *ANS,* Arabia Nubian Shield; *EAO,* East African Orogen; *M,* Madagaskar; *SGT,* Southern Granulite Terrane; *EGMB,* Eastern Ghats Mobile Belt; *CITZ,* Central Indian Tectonic Zone; *ADOB,* Aravalli-Delhi Orogenic Belt.

imbricated into POI (Santosh, 2012). The orogens include the SGT, EGMB, CITZ, and the ADOB in the form of accretionary belts.

The available geological and geophysical information from the Paleoproterozoic orogens of India, described in previous chapters, provide important clues on the broad architecture of these orogens and the subduction polarity. The U−Pb zircon geochronology of ophiolitic rocks from the SGT indicate both Neoarchean and Neoproterozic while the EGMB shows ages ranging from 1.85 to 1.33 Ga, all suggesting a prolonged Wilson cycle of subduction−accretion process before the final collisional event and extrusion of high-grade granulite facies rocks. While the Paleoproterozoic convergent orogens of ADOB and the EGMB of the Indian Peninsula were characterized by westward subduction of oceanic plates, ocean closure along the CITZ probably involved a double-sided subduction (Santosh, 2012). The Wilson cycle traces a continuum from Paleoproterozoic through Mesoproterozoic to Neoproterozoic in some of these orogens, with prolonged subduction−accretion history similar to the ongoing convergent margin processes in the western Pacific region.

The Indian continent (including the Dharwar, Bastar, Singhbhum and Bundelkhand cratons), the Antongil Block of east Madagascar, and the Napier and Rayner Provinces of East Antarctica are considered to have amalgamated before the middle Neoproterozoic. This amalgamation largely took place along the CITZ and the EGMB and the major collision occurred at ~1500 Ma although significant crustal shortening was also reported at ~1100 Ma. Similarly, the EGMB-south recorded crustal thickening and associated deformation and metamorphism in the late Paleoproterozoic/early Mesoproterozoic (1650−1550 Ma) that was likely to be the result of a collision with the Napier Complex of East Antarctica, which also recorded an ~1.6 Ga tectonothermal event. Late Neoproterozoic/Paleozoic deformation in the EGMB−Rayner Complex is mostly concentrated along major shear zones. The northern boundary of this complex is marked by the Neoproterozoic Lambert-Mahanadi Terrane and interpreted as a Neoproterozoic-Cambrian suture (Boger & Wilson, 2001). The magmatism along the eastern margin of India-Western Australia/Mawson is marked by the intrusive activity of anorthosites and granitoids with particular reference to the Chilka Lake Domain and the SM Plateau with characteristic Late Neoproterozoic deformation.

The POI exhibit a variety of both temporal and spatial scales and a variety of thermal conditions, similar to the Phanerozoic record. Reconstruction of the paleogeography of orogenic belts and cratons will not be complete without Paleomagnetism, which is currently not precise enough to resolve the comprehensive details among adjacent cratons, and certainly does not constrain the individual blocks or terranes that may build an orogen. As such, the spatial origin of accreted crustal terranes and colliding plates is therefore quite uncertain.

The POI are characterized by subduction−accretion and collision. Structurally, the orogens are generally composed of imbricated crustal terranes or nappes translated along low-, to moderately dipping shear zones, and include the flow of middle or lower crust within nappe-or crustal-scale channels through meling. The crustal-scale shear zone structures associated with the POI can be extended to other adjacent continents, now-dispersed segments of the Gondwana landmass (Fig. 6.14). Their linkage will provide a fresh basis for reconstructing the processes of supercontinent assemblage and breakup. It is well established that Mesoproterozoic rifting at the cratonic margins and subsequent crustal evolution along the Proterozoic orogens in the Indian shield can be correlated to plate tectonic processes linked to the assembly and breakup of supercontinents of Columbia, Rodinia, and Gondwana. However, more geochronological constraints together with comprehensive field observations are needed to strengthen the correlations and connections within the Gondwana and other supercontinents from the past.

LIST OF ABBREVIATIONS

AB	Antananarivo block
AC	Antongil craton
ADOB	Aravalli-Delhi Orogenic Belt
AISZ	Angavo-Ifanadriana shear zone
AKSZ	Achankovil shear zone
AmSZ	Amphanihy shear zone
AnD	Anosyan Domain
ANS	Arabian Nubian Shield
AS	Amboropotsy Sheet
BBPT	Berhampur-Bhubaneswar-Phulbani Terrane

BC	Bastar craton
BD	Bemarivo domain
BDC	Bundhelkhand craton
BeSZ	Betroka shear zone
BKSK	Balaram-Kui-Surpagla-Kengora
BRSZ	Bongolava-Ranotsara shear zone
BSSZ	Betsimisaraka suture zone
CB	Coorg Block
CDC	Central Dharwar craton
CITZ	Central Indian Tectonic Zone
CSZ	Cauvery Shear/suture zone
CuB	Cuddapah Basin
DC	Dharwar craton
DVP	Deccan Volcanic Province
EAO	East African Orogen
EB	Eastern Block
EDC	Eastern Dharwar craton
EGMB	Eastern Ghats Mobile Belt
GR	Godavari Rift
IG	ItremoGroup
KB	Karwar Block
KKPTSZ	Karur-Kambam-Painavu-Trichur shear zone
KSZ	Kaliguman shear zone
LR	Lambert Rift
MB	Madurai Block
MC	Masora craton
McSZ	Mercara suture zone
MeSZ	Mettur shear zone
MIS	Malani Igneous suite
MR	Mahanadi Rift
MSZ	Mahanadi shear zone
NB	Nilgiri Block
NBSZ	Northern Boundary shear zone
NCC	North China Craton
NIB	North India Block
NRBF	Napier Rayner Boundary Fault
NSL	Narmada-Son Lineament
NSZ	Nagavali shear zone
NVSZ	Nagavali-Vamsadhara shear zone
PCSZ	Palghat-Cauvery shear zone
POI	Proterozoic orogens of India
PSZ	Phulad shear zone
RSZ	Rodinian suture zone
RVKT	Rajahmundry-Visakhapatnam-Koraput-Terrane
SaS	Sahantaha shelf
SASZ	Salem-Attur shear zone
SC	Singhbhum craton
SDT	South Delhi Terrane

SGT	Southern Granulite Terrane
SIB	South India Block
SiS	Sirsi shelf
TB	Trivandrum Block
TNCO	Trans North China orogen
TSZ	Tranomaro shear zone
VSZ	Vamsadhara shear zone
WB	Western Block
WDC	Western Dharwar craton

REFERENCES

Acharyya, S. K., & Roy, A. (2000). Tectonothermal history of the Central Indian Tectonic Zone and reactivation of major faults/shear zones. *Journal of Geological Society of India*, *55*, 239−256.

Amaldev, T., Santosh, M., Li Tang, Baiju, K. R., Tsunogae, T., & Satyanarayanan, M. (2016). Mesoarchean convergent margin processes and crustal evolution: Petrologic, geochemical and zircon U−Pb and Lu−Hf data from the Mercara Suture Zone, southern India. *Gondwana Research*, *37*, 182−204.

Ashwal, L. D., Solanki, A. M., Pandit, M. K., Corfu, F., Hendriks, B. W. H., Burke, K., & Torsvik, T. H. (2013). Geochronology and geochemistry of Neoproterozoic Mt. Abu Granitoids NW India: Regional correlation and implications for Rodinia paleo-geography. *Precambrian Research*, *236*, 265−281.

Bhowmik, S. K., Alexanderwilde, S., Bhandari, A., & Sarbadhikari, A. B. (2014). Zoned monazite and Zircon as monitors for the thermal history of granulite terrances: An example from the Central Indian Tectonic Zone. *Journal of Petrology*, *55*, 585−621.

Bhowmik, S. K., Bernhardt, H. J., & Dasgupta, S. (2010). Grenvillian age high-pressure upper amphibolite−granulite metamorphism in the Aravalli−Delhi Mobile Belt, North-western India: New evidence from monazite chemical age and its implication. *Precambrian Research*, *178*, 168−184.

Bhowmik, S. K., Pal, T., Roy, A., & Pant, N. C. (1999). Evidence for Pre-Grenvillian high-pressure granulite metamorphism from the northern margin of the Sausar mobile belt in central India. *Journal of Geological Society of India*, *53*, 385−399.

Black, L. P., Sheraton, J. W., & James, P. R. (1986). Late Archean granites of the Napier complex, Enderby Land, Antarctica: A comparison of Rb-Sr, Sm-Nd and U-Pb isotopic systematic in a complex terrain. *Precambrian Research*, *32*, 343−368.

Boger, S. D., Carson, C. J., Wilson, C. J. L., & Fanning, C. M. (2000). Neoproterozoic deformation in the Radok Lake region of the northern Prince Charles Mountains, East Antarctica; evidence for a single protracted orogenic event. *Precambrian Research*, *104*, 1−24.

Boger, S. D., Wilson, C. J. L., & Fanning, C. M. (2001). Early Paleozoic tectonism within the East Antarctic craton: the final suture between east and west Gondwana? *Geology*, *29*, 463−466.

Boger, S. D., Hirdes, W., Ferreira, C. A. M., Jenett, T., Dallwig, R., & Fanning, C. M. (2015). The 580−520 Ma Gondwana suture of Madagascar and its continuation into Antarctica and Africa. *Gondwana Research*, *28*, 1048−1060.

Boger, S. D., & Miller, J. M. (2004). Terminal suturing of Gondwana and the onset of the Ross−Delamerian Orogeny: The cause and effect of an Early Cambrian reconfiguration of plate motions. *Earth and Planetary Science Letters*, *219*, 35−48.

Bose, S., Das, K., Torimoto, J., Arima, M., & Dunkley, D. J. (2016). Evolution of the Chilka Lake granulite complex, northern Eastern Ghats Belt, India: First evidence of ∼780 Made compression of the deep crust and its implication on the India−Antarctica correlation. *Lithos*. Available from http://dx.doi.org/10.1016/j.lithos.2016.01.017.

Bose, S., Dunkley, D. J., Dasgupta, S., Das, K., & Arim, M. (2011). India-Antarctica-Australia-Laurentia connection in the Paleoproterozoic−Mesoproterozoic revisited: Evidence from new zircon U-Pb and monazite chemical age data from the Eastern Ghats Belt, India. *Geological Society of America Bulletin, 123*(9/10), 2031−2049. Available from http://dx.doi.org/10.1130/B30336.1.

Chatterjee, N., Crowley, J. L., Mukherjee, A. B., & Das, S. (2008). Geochronology of the 983-Ma Chilka Lake anorthosite, India: Implications for Pre-Gondwana tectonics. *The Journal of Geology, 116*, 105−118.

Chetty, T. R. K. (1995a). A correlation of Proterozoic shear zones between Eastern Ghats, India and Enderby Land, East AntarcticaIn M. Santosh, & M. Yoshida (Eds.), *India and Antarctica during the Precambrian. Memoir* (vol. 34, pp. 205−220). Geological Society of India.

Chetty, T. R. K. (2010). Structural architecture of the northern composite terrane, the Eastern Ghats Mobile Belt, India: Implications for Gondwana tectonics. *Gondwana Research, 18*, 565−582.

Chetty, T. R. K., Mohanty, D. P., & Yellappa, T. (2012). Mapping of Shear Zones in the Western Ghats, Southwestern Part of Dharwar Craton. *Journal Geological Society of India, 79*, 151−154.

Chetty, T. R. K., & Santosh, M. (2013). Proterozoic orogens in southern Peninsular India: Contiguities and complexities. *Journal of Asian Earth Sciences, 78*, 39−53.

Chetty, T. R. K., Vijay, P., Narayana, N. L., & Giridhar, G. V. (2003). Structure of the Nagavali shear zone, Eastern Ghats Mobile Belt, India: Correlation in the East Gondwana reconstruction. *Gondwana Research, 6*, 215−229.

Chetty, T. R. K., Yellappa, T., & Santosh, M. (2016). Crustal architecture and Tectonic evolution of the Cauvery Suture Zone, Southern India. *Journal of Asian Earth Sciences, Journal of Asian Earth Sciences, 130*, 166−191.

Clifford, T. N. (1968). Radiometric dating and the pre-Silurian geology of Africa. In E. I. Hamilton, & R. M. Farquhar (Eds.), *Radiometric dating for geologists* (pp. 299−416). London: Interscience.

Collins, A. S., & Pisarevsky, S. A. (2005). Amalgamating eastern Gondwana: The evolution of the circum-Indian orogens. *Earth Science Reviews, 71*, 229−270.

Collins, A. S., & Windley, B. F. (2002). The tectonic evolution of central and northern Madagascar and its place in the final assembly of Gondwana. *Journal of Geology, 110*, 325−339.

Condie, K. C., Belousova, E., Griffin, W. L., & Sircombe, K. N. (2009). Granitoid events in space and time: Constraints from igneous and detrital zircon age spectra. *Gondwana Research, 15*, 228−242.

Crawford, A. R. (1974). Indo−Antarctica, Gondwanaland, and the distortion or a granulite belt. *Tectonophysics, 22*, 141−157.

Dalziel, I. W. D. (1997). *Overview: Neoproterozoic−Paleozoic geography and tectonics: Review, hypotheses and environmental speculations* (vol. 109, pp. 16−42). Geological Society of America Bulletin.

Crawford, A. R., & Campbell, K. S. W. (1973). Large-scale horizontal displacement within Australo-Antarctica in the Ordovician. *Nature (Phys. Sci.), 241*, 11−14.

Dasgupta, S., & Sengupta, P. (2003). Indo-Antarctic correlation: A perspective from the Eastern Ghats granulite belt, India. In M. Yoshida, B. F. Windley, & S. Dasgupta (Eds.), *Proterozoic East Gondwana: Supercontinent assembly and breakup* (vol. 206, pp. 131−143). Geological Society of London. Special Publication.

De Waele, B., Thomas, R. J., Macey, P. H., Horstwood, M. S. A., Tucker, R. D., Pitfield, P. E. J., … Bejoma, M. (2011). Provenance and tectonic significance of the Palaeoproterozoic metasedimentary successions of central and northern Madagascar. *Precambrian Research, 189*, 18−42.

Dewey, J. F. (1969). Evolution of the Appalachian/Caledonide orogen. *Nature, 222*, 124−129.

Du Toit, A. L. (1937). *Our wandering continents*. Edinburgh: Oliver and Boyd, 366 pages.

Ernst, R. E. (2014). *Large igneous provinces*. Cambridge University Press.

Fedorov, L. V., Ravich, M. G., & Hofmann, J. (1982). Geologic comparison of southeastern Peninsular India and Sri Lanka with a part of East Antarctica (Enderby Land, MacRobertson Land, and Princess Elizabeth Land). In C. Craddock (Ed.), *Antarctic Geoscience* (pp. 73−78). Madison: University of Wisconsin Press.

Fitzsimons, I. C. W. (2000). Grenville-age basement provinces in East Antarctica: Evidence for three separate collisional orogens. *Geology*, *28*, 879–882.

Fitzsimons, I. C. W. (2003). Proterozoic basement provinces of southern and southwestern Australia, and their correlation with AntarcticaIn M. Yoshida, B. F. Windley, & S. Dasgupta (Eds.), *Proterozoic East Gondwana: Supercontinent assembly and breakup* (vol. 206, pp. 93–130). Geological Society of London, Special Publication.

Fitzsimons, I. C. W. (2016). Pan African granulites of Madagaskar and Southern India: Gondwana assembly and parallels with modern Tibet. *Journal of Mineralogical and Petrological Sciences*, *111*, 73–88.

Flowerdew, M. J., Tyrrell, S., Boger, S. D., Fitzsimons, I. C. W., Harley, S. L., Mikhalsky, E. V., & Vaughan, A. P. M. (2013). Pb isotopic domains from the Indian Ocean sector of Antarctica: Implications for past Antarctica–India connections. In S. L. Harley, I. C. W. Fitzsimons, & Y. Zhao (Eds.), *Antarctica and supercontinent evolution* (vol. 383, pp. 105–124). Geological Society London, Special Publication.

Gray, D. R., Foster, D. A., Meert, J. G., Goscombe, B. D., Armstrong, R., Truow, R. A. J., & Passchier, C. W. (2008). *A Damaran perspective on the assembly of Southwestern Gondwana* (294, pp. 257–278). Geological Society of London. Special Publication.

Gregory, L. C., Meert, J. G., Bingen, B., Pandit, M. K., & Torsvik, T. H. (2009). Paleomagnetism and geochronology of the Malani Igneous Suite, Northwest India: Implications for the configuration of Rodinia and the assembly of Gondwana. *Precambrian Research*, *170*, 13–26.

Grew, E. S., & Manton, W. I. (1986). A new correlation of saphirine granulites in the Indo- Antarctic metamorphism terrane: Late Proterozoic dates from the Eastern Ghats Province of India. *Precambrian Research*, *33*, 123–137.

Halpin, J. A., Daczko, N. R., Milan, L. A., & Clarke, G. L. (2012). Decoding near-concordant U–Pb zircon ages spanning several hundred million years: Recrystallisation, metamictisation or diffusion? *Contributions to Mineralogy and Petrology*, *163*, 67–85.

Harris, L. B., & Beeson, J. (1993). Gondwanaland significance of Lower Palaeozoic deformation in central India and SW western Australia. *Journal Geological Society, London*, *150*, 811–814.

Henderson, B., Collins, A. S., Payne, J., Forbes, C., & Saha, D. (2014). Geologically constraining India in Columbia: The age, isotopic provenance and geochemistry of the protoliths of the Ongole Domain, Southern Eastern Ghats, India. *Gondwana Research*, *26*, 888–906.

Hoffman, P. F. (1988). United plates of America, the birth of a craton: Early Proterozoic assembly and growth of Laurentia. *Annual Review of Earth and Planetary Sciences*, *16*, 543–603.

Hoffman, P. F. (1991). Did the breakout of Laurentia turn Gondwanaland inside-out? *Science*, *252*, 1409–1412.

Hou, G., Santosh, M., Qian, X., Lister, G. S., & Li, J. (2008). Configuration of the Late Paleoproterozoic supercontinent Columbia: Insights from radiating mafic dyke swarms. *Gondwana Research*, *14* 395–509.

Ishwar-Kumar, C., Sajeev, K., Windley, B. F., Kusky, T. M., Fengc, P., Ratheesh-Kumar, R. T., ... Itaya, T. (2015). Evolution of high-pressure mafic granulites and pelitic gneisses from NE Madagascar: Tectonic implications. *Tectonophysics*, *662*, 219–242.

Jain, S. C., Yedekar, D. B., & Nair, K. K. K. (1991). Central Indian Shear Zone: A major Precambrian crustal boundary. *Journal of Geological Society of India*, *37*, 521–548.

Johnson, P. R., & Woldehaimanot, B. (2003). Development of the Arabian–Nubian Shield: Perspectives on accretion and deformaIion in the northern East African Orogen and the assembly of Gondwana. In M. Yoshida, B. F. Windley, & S. Dasgupta (Eds.), *Proterozoic East Gondwana: Supercontinent assembly and breakup* (vol. 206, pp. 289–325). London: Geological Society, Special Publications.

Just, J., Schulz, B., de Wall, H., Jourdan, F., & Pandit, M. K. (2011). Monazite CHIME/EPMA dating of Erinpura granitoid deformation: Implications for Neoproterozoic tectono-thermal evolution of NW India. *Gondwana Research*, *19*, 402–412.

Kaur, P., Chaudhri, N., Racsek, I., Kroner, A., & Hofmann, A. W. (2009). Record of 1.82 Ga Andean-type continental arc magmatism in NE Rajasthan, India: Insights from zircon and Sm—Nd ages, combined with Nd—Sr isotope geochemistry. *Gondwana Research*, *16*, 56—71.

King, L. C. (1950). Speculations upon the outline and mode of disruption of Gondwanaland. *Geological Magazine*, *87*, 353—359.

Krishnan, M. S. (1961). *The structure and tectonic history of India*, Memoir, Geological Survey Of India, 81, 131p.

Kröner, A. (1981). *Precambrian plate tectonics*. Amsterdam; New York: Elsevier Scientific Pub. Co.

Kröner, A., Cui, W. Y., Wang, S. Q., Wang, C. Q., & Nemchin, A. A. (1998). Single zircon ages from high-grade rocks of the Jianping Complex, Liaoning Province, NE China. *Journal of Asian Earth Sciences*, *16*, 519—532.

Kröner, A., Santosh, M., Hegner, E., Shaji, E., Geng, H., Wong, J., ... Nanda-Kumar, V. (2015). Palaeoproterozoic ancestry of Pan-African high-grade granitoids in southernmost India: Implications for Gondwana reconstructions. *Gondwana Research*, *27*, 1—37.

Li, Z. X., Bogdanova, S. V., Collins, A. S., Davidson, A., De Waele, B., Ernst, R. E., ... Vernikovsky, V. (2008). Assembly, configuration, and break-up history of Rodinia: A synthesis. *Precambrian Research*, *160*, 179—210.

Lisker, F. (2004). The evolution of the geothermal gradient from Lambert Graben and Mahanadi Basin—a contribution to the Indo-Antarctic rift debate. *Gondwana Research*, *7*, 363—373.

Lisker, F., & Fachmann, S. (2001). The Phanerozoic history of the Mahanadi region, India. *Journal Geophysical Research, B: Solid Earth*, *106*, 22027—22050.

Martelat, J.-E., Lardeaux, J., Nicollet, C., & Rakotondrazafy, R. (2000). Strain pattern and late Precambrian deformation history of southern Madagascar. *Precambrian Research*, *102*, 1—20.

Mazumder, R., Bose, P. K., & Sarkar, S. (2000). A commentary on the tectono-sedimentary record of their pre-2.0 Ga continental growth of India vis-à-vis Pre-Gondwana Afro-Indian supercontinent. *Journal of African Earth Sciences*, *30*, 201—217.

McElhinny, M. W., & Embleton, B. J. J. (1974). Australian palaeomagnetism and the Phanerozoic plate tectonics of eastern Gondwanaland. *Tectonophysics*, *22*, 1—29.

Medlicott, H. B. (1869). *Geological sketch of the Shillong plateau in NE Bengal* (pp. 151—207). Memoirs of the Geological Survey of India, 7(1), 151—207.

Meert, J. G. (2001). Growing Gondwana and rethinking Rodinia: A paleomagnetic perspective. *Gondwana Research*, *4*, 279—288.

Meert, J. G. (2003). A synopsis of events related to the assembly of eastern Gondwana. *Tectonophysics*, *362*, 1—40.

Meert, J. G. (2014). Strange attractors, spiritual interlopers and lonely wanderers: The search for pre-Pangean supercontinents. *Geoscience Frontiers*, *5*, 155—166.

Meert, J. G., & Lieberman, B. S. (2008). The Neoproterozoic assembly of Gondwana and its relationship to the Ediacaran-Cambrian radiation. *Gondwana Research*, *14*, 5—21.

Meert, J. G., & Powell, C., McA. (2001). Introduction to the special volume on the assembly and breakup of Rodinia. *Precambrian Research*, *110*, 1—8.

Mezger, K., & Cosca, M. A. (1999). The thermal history of the Eastern Ghats (India) as revealed by U—Pb and 40Ar/39Ar dating of the metamorphic and magmatic minerals: Implications for the SWEAT correlation. *Precambrian Research*, *94*, 251—271.

Moores, E. M. (1991). Southwest U.S.—East Antarctic (SWEAT) connection: A hypothesis. *Geology*, *19*, 425—428.

Moores, E. M. (2002). Pre-1 Ga (pre-Rodinian) ophiolites: Their tectonic and environmental: Implications. *Geological Society of America Bulletin*, *114*, 80—95.

Nance, R. D., & Murphy, J. B. (2013). Origins of the supercontinent cycle. *Geoscience Frontiers*, *4*, 439—448.

Nance, R. D., Murphy, J. B., & Santosh, M. (2014). The supercontinent cycle: A retrospective essay. *Gondwana Research, 25*, 4−29.

Naqvi, S. M., & Rogers, J. J. W. (1987). *Precambrian Geology of India* (pp. 1−233). New York: Oxford University Press.

Nelson, K. D. (1992). Are crustal thickness variations in old mountain belts like the Appalachians a consequence of lithospheric delamination? *Geology, 22*, 617−620.

Petersson, A., Scherstén, A., Andersson, J., Whitehouse, M. J., & Baranoski, M. T. (2015). Zircon U-Pb, Hf and O isotope constraints on growth versus reworking of continental crust in the subsurface Grenville orogen, Ohio, USA. *Precambrian Research*. Available from http://dx.doi.org/10.1016/j.precamres.2015.02.016.

Phillips, B. R., Bunge, H.-P., & Schaber, K. (2009). True polar wander in mantle convection models with multiple, mobile continents. *Gondwana Research, 15*, 288−296.

Powell, C., McElhinny, M. W., Meert, J. G., & Park, J. K. (1993). Paleomagnetic constraints on timing of the Neoproterozoic breakup of Rodinia and the Cambrian formation of Gondwana. *Geology, 21*, 889−892.

Rajesh, K., & Chetty, T. R. K. (2006). Structure and tectonics of the Achankovil Shear Zone, southern India. *Gondwana Research, 10*, 86−98.

Rathore, S. S., Venkatesan, T. R., & Srivastava, R. K. (1999). Rb−Sr isotope dating of Neopro-terozoic (Malani Group) Magmatism from Southwest Rajasthan, India: Evidence of younger pan-African thermal event by40Ar−39Ar studies. *Gondwana Research, 2*, 271−281.

Rickers, K., Metzer, K., & Raith, M. M. (2001). Evolution of the continental crust in the Proterozoic Eastern Ghats Belt, India and new constraints for Rodinia reconstruction: Implications for Sm−Nd, Rb−Sr and Pb−Pb isotopes. *Precambrian Research, 112*, 183−210.

Rogers, J. J. W., & Santosh, M. (2004). *Continents and supercontinents* (p. 289). New York: Oxford University Press.

Sandwell, D. T., & Smith, W. H. F. (1997). Marine gravity anomaly from GEOSAT and ERS 1 satellite altimetry. *Journal of Geophysical Research, 102*, 10,039−10,054.

Santosh, M. (2012). *India's Paleoproterozoic legacy* (vol. 365, pp. 263−288). Geological Society of London, *Special Publications*.

Santosh, M., Maruyama, S., Yusuke Sawaki, & Meert, J. G. (2014). The Cambrian Explosion: Plume-driven birth of the second ecosystem on Earth. *Gondwana Research, 25*, 945−965.

Santosh, M., Yang, Q. Y., Shaji, E., Tsunagae, T., Ram Mohan, M., & Satyanarayanan, M. (2015). An exotic Mesoarchean microcontinent: The Coorg block, southern India. *Gondwana Research, 27*, 165−195.

Sengoör, A. M. C., & Natal'in, B. A. (0030). Rifts of the worldIn R. E. Ernst, & K. L. Bucham (Eds.), *Mantle Plumes: Their identification through time* (vol. 352, pp. 389−482). Geological Society of America, (Special Paper).

Shackleton, R. M. (1996). The final collision between East and West Gondwana: Where is it? *Journal of African Earth Sciences, 23*, 271−287.

Shaw, R. K., Arima, M., Kagami, H., Fanning, C. M., Shairashi, K., & Motoyashi, Y. (1997). Proterozoic events in the Eastern Ghats granulite belt, India: Evidence from Rb-Sr, Sm-Nd systematics and SHRIMP dating. *Journal of Geology, 105*, 645−658.

Singh, Y. K., De Waele, B., Karmakar, S., Sarkar, S., & Biswal, T. K. (2010). Tectonic setting of the Balaram-Kui-Surpagla-Kengora granulites of the South Delhi Terrane of the Aravalli Mobile Belt NW India and its implication on correlation with the East African Orogen in the Gondwana assembly. *Precambrian Research, 183*, 669−688.

Sinha-Roy, S. (1988). Proterozoic Wilson Cycles in Rajasthan. In A. B. Roy (Ed.), *Precambrian of Aravalli Mountain, Rajasthan, India* (vol. 7, pp. 95−107). Memoir, Geological Society of India.

Smith, A. G., & Hallam, A. (1970). The fit of the southern continents. *Nature, 225*, 139−144.

Stern, R. J. (1994). Continental collision in the Neoproterozic East African orogen: Implications for the consolidation of Gondwana land. *Annual Reviews of Earth and Planetary Sciences, 22,* 319–351.

Torsvik, T. H., Carter, L. M., Ashwal, L. D., Bhushan, S. K., Pandit, M. K., & Jamtveit, B. (2001). Rodinia refined or obscured: Palaeomagnetism of the Malani igneous suite (NW) India. *Precambrian Research, 108,* 319–333.

Torsvik, T. H., & Cocks, L. R. M. (2011). *The Palaeozoic geography of central Gondwana* (vol. 357, pp. 137–166). London: Geological Society, Special Publication.

Tucker, R. D., Roig, J. Y., & Delor, C. (2011). Neoproterozoic extension in the Greater Dharwar craton: A reevaluation of the "Betsimisaraka Suture" in Madagascar. *Canadian Journal of Earth Sciences, 48,* 389–417.

Tucker, R. D., Roig, J. Y., Moine, B., Delor, C., & Peters, S. G. (2014). A geological synthesis of the Precambrian shield in Madagascar. *Journal African Earth Sciences, 94,* 9–30.

Veevers, J. J. (2009). Palinspastic (pre-rift and -drift) fit of India and conjugate Antarctica and geological connections across the suture. *Gondwana Research, 16,* 90–108.

Veevers, J. J., & Tewari, R. C. (1995). *Gondwana Master Basin of Peninsular India between Tethys and the interior of the Gondwanaland Province of Pangea.* Geological Society of America Memoir 187, 72 p.

Vijaya Kumar, K., Ernst, W. G., Leelanandam, C., Wooden, J. L., & Groves, M. J. (2009). *Assembly and fragmentation of the Proterozoic supercontinents Columbia and Rodinia: New U-Pb age data from the Eastern Ghats Belt, India* (vol. 7, p. 115). Geological Society of America Annual Meeting Abstracts with Programs 41.

Vijaya Rao, V., & Krishna, V. G. (2013). Evidence for the Neoproterozoic Phulad Suture Zone and Genesis of Malani magmatism in the NW India from deep seismic images: Implications for assembly and breakup of the Rodinia. *Tectonophysics, 589,* 172–185.

Wang, Q.-H., Yang, H., Yang, D.-B., & Xu, W.-L. (2014). Mid-Mesoproterozoic ($\sim$1.32 Ga) diabase swarms from the western Liaoning region in the northern marging of the North China Craton: Baddeleyite Pb-Pb geochronology, geochemistry and implications for the final breakup of the Columbia supercontinent. *Precambrian Research, 254,* 114–128.

Wegener, A. (1912). Die Herausbildung der Grossformen der Erdrinde (Kontinente und Ozeane), auf geophysikalischer Grundlage. *Petermanns Geographische Mitteilungen, 58,* 185–195, 253-256, 305-309.

Windley, B. F. (1995). *The evolving continents* (third ed). New York: Wiley.

Worsley, T. R., Nance, R. D., & Moody, J. B. (1982). Plate tectonic episodicity: A deterministic model for periodic "Pangeas". *Eos, Transactions of the American Geophysical Union, 65*(45), 1104.

Yoshida, M. (1995). *Assembly of East Gondwanaland during Mesoproterozoic and its rejuvenation during the Pan-African period* (vol. 34, pp. 25–45). Memoir, Geological Society of India.

Yoshida, M., Funaki, M., & Vitanage, P. W. (1992). Proterozoic to Mesozoic east Gondwanaland: The juxtaposition of India, Sri Lanka and Antarctica. *Tectonics, 11,* 381–391.

Yoshida, M., Windley, B. F., & Dasgupta, S. (2003). *Preface-Proterozoic East Gondwana: Supercontinent Assembly and Breakup* (p. 206). Geological Society of London, Special Publication.

Zhao, G., Cawood, P. A., Wilde, S. A., & Sun, M. (2002). Review of global 2. 1–1. 8 Ga collisional orogens and accreted cratons: A pre-Rodinia supercontinent? *Earth Science Reviews, 59,* 125–162.

Zhao, G., Sun, M., Wilde, S. A., & Li, S. (2004). A Paleo-Mesoproterozoic supercontinent: Assembly, growth and breakup. *Earth Science Reviews, 67,* 91–123.

Zhao, D., Yamamoto, Y., & Yanada, T. (2013). Global mantle heterogeneity and its influence on teleseismic regional tomography. *Gondwana Research, 23,* 595–616.

FURTHER READING

Cawood, P. A., & Buchan, C. (2007). Linking accretionary orogenesis with supercontinent assembly. *Earth-Science Reviews, 82*, 217–256.

Cawood, P. A., Kröner, A., Collins, W. J., Kusky, T. M., Mooney, W. D., & Windley, B. F. (2009). Accretionary orogens through Earth historyIn P. A. Cawood, & A. Kröner (Eds.), *Earth accretionary systems in space and time* (vol. 318, pp. 1–36). Geological Society of London, Special Publication.

Chetty, T. R. K. (1995b). *Strike-slip tectonics and the evolution of Gondwana basins of eastern India. Proceedings of 9th International Gondwana Symposium* (pp. 713–721). Geological Survey of India.

Chetty, T. R. K. (1996). Proterozoic shear zones in southern granulite terrain, India. In M. Santosh, & M. Yoshida (Eds.), *The Archean and Proterozoic terrains in Southern India within East Gondwana* (pp. 77–89). Gondwana Research Group Memoir 3.

Chetty, T. R. K. (2001). The Eastern Ghats Mobile Belt, India: A collage of juxtaposed terranes (?). *Gondwana Research, 4*, 319–328.

Chetty, T. R. K., & Bhaskar Rao, Y. J. (2006). The Cauvery Shear Zone, Southern Granulite Terrain, India: A crustal scale flower structure. *Gondwana Research, 10*, 77–85.

Collins, A. S., Clark, C., & Plavsa, D. (2014). Peninsular India in Gondwana: The tectonothermal evolution of the Southern Granulite Terrain and its Gondwanan counterparts. *Gondwana Research, 25*, 190–203.

Collins, A. S., Santosh, M., Braun, I., & Clark, C. (2007). Age and sedimentary provenance of the Southern Granulites, South India: U–Th–Pb SHRIMP secondary ion mass spectrometry. *Precambrian Research, 155*, 125–138.

Crawford, A. R. (1975). Rb–Sr age determination for the Mount Abu Granite and related rocks of Gujarat. *Journal of Geological Society of India, 16*, 20–28.

Golynsky, A.V. (2007). Magnetic anomalies in East Antarctica: A window on major tectonic provinces and their boundaries. In: Cooper, A.K., Raymond, C.R., et al. (Eds.), Proceedings of the 10th ISAES, USGS Open-file Report 2007–1047, Short Research Paper 006. 4 pages.

Hawkesworth, C. J., Dhuime, B., Pietranik, A. B., Cawood, P. A., Kemp, A. I. S., & Storey, C. D. (2010). The generation and evolution of the continental crust. *Journal of the Geological Society of London, 167*, 229–248.

Ishwar-Kumar, C., Windley, B. F., Horie, K., Kato, T., Hokada, T., Itaya, T., ... Sajeev, K. (2013). A Rodinian suture in western India: New insights on India– Madagascar correlations. *Precambrian Research., 236*, 227–251.

Kröner, A., Windley, B. F., Jaeckel, P., Brewer, T. S., & Razakamanana, T. (1999). New Zircon ages and regional significance for the evolution of the Pan-African orogen in Madagascar. *Journal of the Geological Society, London, 156*, 1125–1135.

Li, Z. X., Li, X. H., Kinny, P. D., & Wang, J. (1999). The breakup of Rodinia: Did it start with a mantle plume beneath South China? *Earth and Planetary Science Letters, 173*, 171–181.

Meert, J. G. (2012). What's in a name? The Columbia (Paleopangaea/Nuna) supercontinent. *Gondwana Research, 21*, 987–993.

Meert, J. G., & Van Der voo, R. (1997). The assembly of Gondwana 800–550 Ma. *Journal of Geodynamics, 23*, 223–235.

Murphy, J. B., & Nance, R. D. (1991). Supercontinent model for the contrasting character of Late Proterozoic orogenic belts. *Geology, 19*, 469–472.

Pisarevsky, S. A., Wingate, M. T. D., Powell, C., McA., Johnson, S., & Evans, D. A. D. (2002). Models of Rodinia assembly and fragmentationIn M. Yoshida, B. F. Windley, & S. Dasgupta (Eds.), *Proterozoic East Gondwana: Supercontinent assembly and breakup* (206, pp. 289–325). London: Geological Society, Special Publications.

Powell, C., McA., & Pisarevsky, S. (2002). Late Neoproterozoic assembly of East Gondwana. *Geology, 30,* 3−6.

Rajesh, K. (2005). Structure and tectonics of the Achankovil Shear Zone, southern Granulite Terrane, India (Unpublished Ph.D thesis).

Ratheesh-Kumar, R. T., Ishwar-Kumar, C., Windley, B. F., Razakamanana, T., Nair, R., & Sajeev, K. (2015). India-Madagascar paleo-fit based on the flexural isostasy of their rifted margins. *Gondwana Research, 28,* 581−600.

Runcorn, S. K. (1962). Convection currents in the Earth's mantle. *Nature, 195,* 1248−1249.

Schubert, G., Turcotte, D. L., & Olson, P. (2001). *Mantle convection in the earth and planets.* Cambridge: Cambridge University Press, 940 p.

Simmat, R., & Raith, M. M. (2008). U−Th−Pb monazite geochronometry of the Eastern Ghats Belt, India: Timing and spatial disposition of poly-metamorphism. *Precambrian Research, 162,* 16−39.

Index

Note: Page numbers followed by "*f*" refer to figures.

Printed in the United States
By Bookmasters